MW01517723

• Adopt-A-Book • A project of the LSAG and Georgian Alumni Association •

A-Book • A project of the LSAG and Georgian Alumni Association •

This book has been adopted by:

ROYAL   BANK

OF   CANADA

TORONTO

In honour of:

Analytical and Computational Methods in Electromagnetics

DISCLAIMER OF WARRANTY

The technical descriptions, procedures, and computer programs in this book have been developed with the greatest of care and they have been useful to the author in a broad range of applications; however, they are provided as is, without warranty of any kind. Artech House, Inc. and the author and editors of the book titled *Analytical and Computational Methods in Electromagnetics* make no warranties, expressed or implied, that the equations, programs, and procedures in this book or its associated software are free of error, or are consistent with any particular standard of merchantability, or will meet your requirements for any particular application. They should not be relied upon for solving a problem whose incorrect solution could result in injury to a person or loss of property. Any use of the programs or procedures in such a manner is at the user's own risk. The editors, author, and publisher disclaim all liability for direct, incidental, or consequent damages resulting from use of the programs or procedures in this book or the associated software.

For a listing of recent titles in the *Artech House Electromagnetic Analysis Series*, turn to the back of this book.

GEORGIAN COLLEGE LIBRARY 2501

250101(00)
$113.35

Analytical and Computational Methods in Electromagnetics

Ramesh Garg

Library Commons
Georgian College
One Georgian Drive
Barrie, ON
L4M 3X9

ARTECH HOUSE

BOSTON | LONDON
artechhouse.com

Library of Congress Cataloging-in-Publication Data
A catalog record for this book is available from the U.S. Library of Congress.

British Library Cataloguing in Publication Data
A catalogue record for this book is available from the British Library.

ISBN-13: 978-1-59693-385-9

Cover design by Igor Valdman

© 2008 ARTECH HOUSE, INC.
685 Canton Street
Norwood, MA 02062

All rights reserved. Printed and bound in the United States of America. No part of this book may be reproduced or utilized in any form or by any means, electronic or mechanical, including photocopying, recording, or by any information storage and retrieval system, without permission in writing from the publisher.

All terms mentioned in this book that are known to be trademarks or service marks have been appropriately capitalized. Artech House cannot attest to the accuracy of this information. Use of a term in this book should not be regarded as affecting the validity of any trademark or service mark.

10 9 8 7 6 5 4 3 2 1

QC661 .G37 2008
text 0134111710650
Garg, Ramesh,

Analytical and
  computational methods in
      c2008.

                        2009 06 19

QC661 .G37 2008
disc 0134111710668
Garg, Ramesh,

Analytical and
  computational methods in
      c2008.

                        2009 06 19

To my family Madhu, Geetika, and Prashant, and in the memory of my teacher and friend, the late Professor K. C. Gupta

Contents

CHAPTER 5

Fourier Transform Method   153

CHAPTER 6

Introduction to Computational Methods   199

CHAPTER 7

Method of Finite Differences   233

CHAPTER 8

Finite-Difference Time-Domain Analysis   281

CHAPTER 9

Variational Methods   355

CHAPTER 10

Finite Element Method   393

CHAPTER 11

Method of Moments   445

APPENDIX A

APPENDIX B

Preface

The interest in computational methods in electromagnetics is on the increase because of the availability of cheaper computational power, useful commercial software, and increasing demand on products. Most of the commercial software is based on the method of moments (MoM), finite-difference time-domain (FDTD), or finite element method (FEM) because of their versatility. The MoM and FEM are based on weighted residual method. MoM, in addition, requires considerable preprocessing in the form of Green's function, and therefore, a sound analytical background is needed. The FDTD method requires familiarity with finite-difference method. It is with this perspective and to make the book stand alone that the topics in this book have been selected.

The approach followed in the book is similar to that in a classroom. The selection of the material and the sequence of presentation has been influenced by the consideration that the interest of the students should be generated and maintained. As a result, the mathematical background necessary to understand the topics is also included. It happens in a topic like electromagnetics that physics and mathematics are coupled together. In presenting the book these two aspects are first treated individually so that purely mathematical aspects are not clouded by physical considerations. For this purpose, we have considered a number of examples which are completely mathematical in nature. Once the mathematical description of the method has been understood, problems in electromagnetics are attempted. Examples and solved problems are given in order to improve understanding. Most of the subject matter has been kept at the introductory level to help create interest in the subject. For this, and to clarify concepts without getting into complex mathematical details, there is emphasis on one-dimensional problems in Cartesian coordinates, and scalar wave equations. The one-dimensional problems, however, cannot include arbitrary shape. The two-dimensional problems are a trade-off between the complexity of the problem and its utility. The three-dimensional problems are mainly discussed in relation to the MoM.

In order to cover various applications in the electromagnetics area, examples from wave guidance, radiation, and microwave circuits are considered. The examples in the scattering area are a very few. We have tried to present a unified approach to the problems in electromagnetics in the form of solution of Sturm-Liouville differential equation, and the variational nature of the MoM and FEM. A book like this with limited scope cannot stand alone. Therefore, a number of references are included at the end of each chapter to help the reader bridge the gap.

One of the aims of the book is to create interest in computational methods. It is necessary, therefore, that the students should have access to the source codes so that they can generate a wealth of data and extract information from it. With this in mind, a number of source codes are included. As a next step, the students may be asked to write their own codes. It is thrilling for the students when they find that the results generated by their code compare with the expected values. Most of the problems included in the book range from simple electrostatic problems to the waveguide eigenvalue analysis for the cutoff frequency of waveguide modes. This requires the solution of scalar wave equation and is not difficult to solve analytically. The computed value can be compared with the exact analytical solution.

It is the observation of the author that the students do not like vector calculus. Visualization of field and current distribution is important in electromagnetics and difficult because of their vector nature. These problems are addressed through the solution of scalar wave equation and Poisson equation in the Cartesian coordinate system with which the students are comfortable. The essence of computational methods such as convergence, discretization error, numerical dispersion, effect of dielectric inhomogeneity, and arbitrary shape can be brought out in this limited scope. Visualization is improved through the plot of end results using MATLAB graphics color capability. While adopting this philosophy, it is kept in mind that analytical capability of the students should be improved as a part of the overall effort. This is achieved by a number of measures:

1. The students first solve the problems almost analytically using the computational methods. For this the size of the problem is deliberately kept small so that the resulting matrix size does not exceed three. The discretization assumed is therefore crude and symmetry considerations are used to reduce the size.

2. Once the students carry out this exercise in detail, they learn the various steps of implementation of the method, including implementation of the boundary conditions. This solution is expected to be inaccurate.

3. Based on this experience, the source code is developed with finer discretization of the same problem. Through this exercise they learn the effect of discretization on the accuracy of the solution and convergence.

4. The field or potential distribution is plotted to help in visualization.

5. The treatment of computational methods is presented as a work-book through details of the problems or examples discussed.

6. Instead of first presenting the generalized analysis of a particular topic from which the special cases follow, we straight away start with the analysis of a particular problem and then generalization follows. This way the attention of the students gets focused.

It is assumed that the reader has gone through undergraduate level courses in electromagnetics, microwave engineering, and numerical methods. Also, introductory level knowledge of MATLAB is assumed.

The book is organized into 11 chapters with two appendices at the end. Out of these, six chapters are devoted to analytical methods including a chapter on introduction to electromagnetics. The remaining five chapters are on computational methods: Chapter 6 is devoted to the common aspects of computational methods, and other four chapters cover FDM, FDTD, FEM, and MoM. There are a number of source codes in MATLAB. The codes are put on CD-ROM for the convenience of users.

Chapter 1 summarizes basic concepts in electrostatics and wave propagation, and is used as common background material for later chapters. It also introduces topics such as charge and current singularities on conductors. Diverse approaches available for the analysis of boundary value problems are discussed, bringing out the importance and limitations of theoretical analysis. Solution in the form of discretization of unknown functions using subdomain basis functions and the associated discretization of device geometry is suggested. Common analytical and computational methods are overviewed. Chapter 2 summarizes the most common method of separation of variables, defines orthogonality, and leads to orthogonal function expansion of arbitrary functions. Strum-Liouville differential equation represents all the differential equations of electromagnetics. Its solution provides the common analytical solution approach. The second half of the chapter is devoted to the eigenfunction based framework of function spaces. The approach is used to make the abstract subject of function spaces interesting. The concepts like completeness of a set, convergence of series, eigenfunctions of the operator, vector representation of function, and matrix representation of an operator are illustrated through the solution of one-dimensional differential equation. Limitations of the eigenfunction based function space are brought out and the generic solution in terms of subdomain basis functions is suggested. The chapter concludes with the Dirac delta function and its various representations, and prepares the reader for the next chapter on Green's function.

Chapter 3 on Green's function is essential to the development of MoM. The "direct construction" and eigenfunction expansion methods for constructing the Green's function are illustrated through several examples. Green's functions for the unbounded region are listed at the end of the chapter. Chapter 4 on contour integration and conformal mapping serves two purposes. The contour integration emphasizes the importance of analytical approach to determine singular integrals, which are essential to many analytical and computational methods. The conformal mapping analysis of planar transmission lines is discussed in detail and leads to closed-form analytical solutions.

The Fourier transform method, described in Chapter 5, is an analytical method useful for solving unbounded and semi-bounded regions, and layered dielectric geometry problems. Some examples of Fourier transform method include: excitation of unbounded transmission line, line excitation of grounded trough, probe excitation of modes in a rectangular waveguide, electric line source above a ground plane, radiation from apertures, Green's function for microstrip line, and free space Green's function.

Chapter 6 covers the essentials of computational methods. It includes the major steps in the analysis and the solution properties. The description of an unknown function in terms of subdomain basis functions is the essence of computational methods. The various types of subdomain basis functions are described. The discretization of operator is discussed. The basic characteristics of numerical solution such as convergence and accuracy are illustrated with numerical integration as an example. The phase error as a function of discretization size in FDM, FDTD, and FEM is compared, and the importance of higher order basis functions to reduce this error is brought out. The stability of the numerical solution is discussed and the origin of spurious solution is touched upon.

Chapter 7 on FDM introduces a simple and versatile differential equation solver. It also develops the basic background needed for FDTD. However, FDM is an inefficient method and results in large sized matrix. Efficient matrix solution technique in the form of iterative methods and successive over-relaxation are discussed. The effect of discretization on accuracy of the solution is addressed through numerical dispersion determination. The FDM is illustrated by its application to problems like inhomogeneously filled parallel plate capacitor, microstrip line, rectangular and ridge waveguide, and so forth.

The FDTD method of Chapter 8 is a very versatile and futuristic computational method. FDTD analysis simulates propagation of signal in time and space. Various concepts such as spatial step and numerical dispersion, time step and stability of the solution, excitation of grid, and absorbing boundary conditions are discussed. The one-dimensional FDTD analysis is applied to study basic phenomenon like dispersion, stability, reflection at a dielectric-air interface, determination of propagation constant in lossy medium, and the design of material absorbers. Analysis for two- and three-dimensional problems follows the pattern of one-dimensional analysis. References to the advances in FDTD are provided.

The variational method, described in Chapter 9, forms the foundation for FEM and MoM methods. The basic concepts of variational method such as stationarity, extremum, and functionals are discussed. The functionals are listed for the important differential equations of electromagnetics. The use of Ritz method to determine the stationary value of functionals is illustrated through a number of examples, which include capacitance of strip line, cutoff frequency of waveguides, resonant frequencies of cavities, and variational analysis of microstrip line. The method of weighted residuals is discussed in relation to point matching and Galerkin's solutions.

Chapter 10 is devoted to FEM, which is a simple, versatile, and very popular method. The formulation of FEM based on stationary functionals is discussed. Also, node based elements are employed for simplicity. After discussing the major steps in FEM, the method is illustrated by its application to the problems in inhomogeneously filled parallel plate capacitor, and cutoff frequency of waveguide modes. Both triangular and rectangular elements are employed for the analysis. Some of the hybrid FEM techniques are introduced.

The last chapter covers MoM. This method is partly analytical and partly computational. It is very efficient for the analysis of open region problems such as antennas, scatterers, and planar lines. The MoM is illustrated by solving a number of problems like, capacitance of strip line, charge distribution on a metal wire, current distribution and input impedance of a half-wave dipole, and RCS of a cylindrical scatterer. The various computational methods discussed in the book are compared qualitatively within the scope of the subject. The hybrid computational methods are referenced to bridge the gap between the scope of the book and the recent advances.

Appendix A discusses various methods for the solution of matrix equation in general, and Appendix B discusses some of the methods for the evaluation of singular integrals.

This book is based on my experience of teaching a large number of first-year graduate students at the Indian Institute of Technology, Kharagpur, India. Most of the contents of Chapters 6 through 11 were covered in one semester course work on computational methods. There is a good amount of extra material in these chapters from which the instructor can choose. The students attending this subject have contributed immensely to the development of the course material. Some of the students have contributed their source codes, which have been refined with usage by others. I am thankful to all the students and these students in particular. The subject of computational electromagnetics would be less interesting without the software. Critical comments by a number of reviewers have helped the author in the development of the book. Any suggestions towards the improvement of the book will be gratefully accepted. I would like to acknowledge the support of Rebecca Allendorf, Darrell Judd, and the staff at Artech House.

CHAPTER 1

Basic Principles of Electromagnetic Theory

Electromagnetic theory forms the foundation of high frequency (radio frequency and higher) electrical engineering. It is used to explain many wave phenomena like propagation, reflection, refraction, diffraction, and scattering. The high frequency circuits can be characterized more accurately when analyzed in terms of wave phenomena. As the clock rate of personal computers increases, the coupling between various subsystems of the computer also becomes important. The study of electromagnetic interference and compatibility is important to avert many disasters. In this chapter we shall review the fundamental aspects of electromagnetic theory. Common analytical and computational methods will be classified and overviewed, bringing out their advantages and limitations.

1.1 Maxwell's Equations

The physics of electromagnetic wave propagation is described mathematically by the Maxwell's equations. These equations can be written in differential as well as integral forms. For time-varying electromagnetic fields, the Maxwell's equations are as follows.

Differential Form

$$\nabla \times \mathbf{E} = -\frac{\partial \mathbf{B}}{\partial t}, \qquad \text{(Faraday's Law)} \tag{1.1a}$$

$$\nabla \times \mathbf{H} = \mathbf{J}_s + \frac{\partial \mathbf{D}}{\partial t}, \qquad \text{(Ampere's Law)} \tag{1.1b}$$

$$\nabla \cdot \mathbf{D} = \rho, \qquad \text{(Gauss's Law, electric)} \tag{1.1c}$$

$$\nabla \cdot \mathbf{B} = 0. \qquad \text{(Gauss's Law, magnetic)} \tag{1.1d}$$

where:

$\mathbf{E}(\mathbf{r}, t)$ is electric field intensity in volt/meter, (V/m);

$\mathbf{H}(\mathbf{r}, t)$ is magnetic field intensity in ampere/meter, (A/m);

$\mathbf{D}(\mathbf{r}, t)$ is electric flux density in coulomb/meter^2, (C/m^2);

$\mathbf{B}(\mathbf{r}, t)$ is magnetic flux density in weber/meter^2, (Wb/m^2);

$\mathbf{J_s}(\mathbf{r}, t)$ is the impressed electric surface current density in ampere/meter^2, (A/m^2);

$\rho(\mathbf{r}, t)$ is the impressed electric charge density in coulomb/m^3, (C/m^3).

The current density $\mathbf{J_s}$ and the charge density ρ for the time-varying fields are related to each other as

$$\nabla \cdot \mathbf{J_s} = -\frac{\partial \rho}{\partial t} \qquad (1.1e)$$

This equation is called the *equation of continuity* and it describes the principle of conservation of charge. The second term on the right side of (1.1b) is called the displacement current density and is defined as [1]

$$\mathbf{J}_d = \frac{\partial \mathbf{D}}{\partial t} \qquad (A/m^2) \qquad (1.2)$$

The differential form of Maxwell's equations holds true at each and every point in space.

Integral Form
The integral form of Maxwell's equations can be derived from their differential forms by employing Stoke's theorem and Green's divergence theorem. The integral form is given by

$$\oint \mathbf{E} \cdot d\ell = -\frac{\partial}{\partial t} \iint \mathbf{B} \cdot d\mathbf{S} \qquad (1.3a)$$

$$\oint \mathbf{H} \cdot d\ell = \iint \mathbf{J_s} \cdot d\mathbf{s} + \frac{\partial}{\partial t} \iint \mathbf{D} \cdot d\mathbf{s} \qquad (1.3b)$$

$$\oiint \mathbf{B} \cdot d\mathbf{s} = 0 \qquad (1.3c)$$

$$\oiint \mathbf{D} \cdot d\mathbf{s} = \iiint \rho \, dv \qquad (1.3d)$$

The Maxwell's equations are general and hold for fields with arbitrary time dependence in any medium and at any location. They reduce to the simpler form for special cases like static case, sinusoidal time varying (or time-harmonic) fields, and in source-free media.

Static or Quasi-Static Fields.   Under this assumption, we let $\partial/\partial t \equiv 0$, and the differential form of Maxwell's equations reduces to

$$\nabla \times \mathbf{E} = 0 \tag{1.4a}$$

$$\nabla \times \mathbf{H} = \mathbf{J_s} \tag{1.4b}$$

$$\nabla \cdot \mathbf{D} = \rho \tag{1.4c}$$

$$\nabla \cdot \mathbf{B} = 0 \tag{1.4d}$$

$$\nabla \cdot \mathbf{J_s} = 0 \tag{1.4e}$$

It may be noted that the static quantities are functions of space only and not time. In the static limit, electric and magnetic fields are independent of each other. Although the equations of (1.4) are valid for direct current (dc), they can also be used for quasi-static situations when the largest dimension of the electromagnetic device is about one-tenth of the operating wavelength. We shall use the word "device" in a general sense representing any transmission line section, microwave circuit, antenna, scatterer, and so on.

Time-Harmonic Fields.   For time-harmonic fields, the time variation is of the type $e^{j\omega t}$. Therefore, we can replace $\partial/\partial t$ by $j\omega t$, and the differential form of Maxwell's equations reduces to

$$\nabla \times \mathbf{E} = -j\omega \mathbf{B} \tag{1.5a}$$

$$\nabla \times \mathbf{H} = \mathbf{J_s} + j\omega \mathbf{D} \tag{1.5b}$$

$$\nabla \cdot \mathbf{D} = \rho \tag{1.5c}$$

$$\nabla \cdot \mathbf{B} = 0 \tag{1.5d}$$

$$\nabla \cdot \mathbf{J_s} = -j\omega\rho \tag{1.5e}$$

The time-harmonic Maxwell's equations are employed for the steady state behavior. The static case is the limiting case of time harmonic fields as $\omega \to 0$. The field quantities in (1.5) are functions of position only, and the phasor representation of fields has been employed. The relation between the phasor representation and its corresponding instantaneous version, with reference to $\cos(\omega t)$, is given by

$$\mathbf{E}(x, y, z, t) = \mathrm{Re}[\mathbf{E}(x, y, z)\,e^{j\omega t}] \tag{1.6}$$

Source-Free Case.   The fields in the region away from the sources may be obtained by substituting $\mathbf{J_s} = 0 = \rho$ in (1.5).

1.2   Constitutive Relations

The electric field $\mathbf{E}$ and the electric flux density $\mathbf{D}$ are not independent of each other, and are related through the material properties of the medium. Similarly, the magnetic field **H** and the magnetic flux density **B** are related. The relationship between the field strength and the corresponding flux density is called the constitutive relation. For a simple (linear, homogeneous, and isotropic) lossless medium, the constitutive relations, for the time-harmonic fields, are

$$\mathbf{D} = \epsilon(\omega)\mathbf{E} = \epsilon_0\,\epsilon_r(\omega)\mathbf{E} \tag{1.7}$$

$$\mathbf{B} = \mu(\omega)\mathbf{H} = \mu_0\,\mu_r(\omega)\mathbf{H} \tag{1.8}$$

where $\epsilon_0 = 8.854 \times 10^{-12}$ F/m, and $\mu_0 = 4\pi \times 10^{-7}$ H/m are the permittivity and permeability, respectively, of vacuum. In practice, free space is normally assigned the material properties of vacuum. ϵ and μ are the permittivity and permeability, respectively, of the medium. Similarly, ϵ_r and μ_r are the relative permittivity or relative dielectric constant and relative permeability, respectively of the medium. We shall mostly consider linear, isotropic medium.

The constitutive relations are exact for free space and can be used approximately for other materials over a narrow frequency range. For problems involving common materials (e.g., water), and when broadband operation is to be studied, the linear relations (1.7) and (1.8) do not hold. A complex flux-field relationship needs to be modeled. This modeling is discussed in time domain in Chapter 8.

1.3 Electrical Properties of the Medium

In general, material properties of the medium ϵ_r, μ_r, and σ are functions of the position, direction, field strength, and frequency of the applied field. However, a majority of the media can be assumed to be homogeneous (material properties independent of position), isotropic (material properties independent of direction), and linear (material properties independent of field strength) over sufficiently large values of field strength and broad range of frequency. These materials or media may be called *simple*. Further, most of the media are nonmagnetic with $\mu_r = 1$. Good conductors are described by a high value of σ, which is constant from dc up to the infrared frequencies. Their permeability and permittivity are approximately equal to that of vacuum. Perfect dielectrics are characterized by $\sigma = 0$. For nonsimple materials, ϵ and μ are described by dielectric and permeability tensors, respectively. For example, sapphire is an anisotropic material, and ferrite is a magnetic material with anisotropy (the value of μ_r varies with direction).

The media employed in the study of electromagnetic phenomena are always nonperfect; that is, some loss is always present in any practical media. The origin of loss in dielectrics is the nonzero conductivity of the dielectric, and gives rise to what is called dielectric loss. If we define the complex dielectric constant of the lossy medium as

$$\hat{\epsilon} = \epsilon - \frac{j\sigma}{\omega} \tag{1.9}$$

we can rewrite (1.5b) for the source-free case as ($\mathbf{J}_s = 0$),

$$\nabla \times \mathbf{H} = j\omega\hat{\epsilon}\mathbf{E} = j\omega\epsilon\left(1 - j\frac{\sigma}{\omega\epsilon}\right)\mathbf{E} \qquad (1.10)$$

or

$$\nabla \times \mathbf{H} = j\omega\epsilon(1 - j\tan\delta)\mathbf{E} \qquad (1.11)$$

where

$$\tan\delta = \frac{\sigma}{\omega\epsilon} \qquad (1.12)$$

is known as the *loss tangent* of the medium. In (1.9), $\epsilon = \epsilon_0\epsilon_r$ is the real part, and $\sigma/\omega = \epsilon\tan\delta$ is the imaginary part of the complex dielectric constant of the medium. The real part of the dielectric constant contributes to the stored electric energy in the medium, and the imaginary part results in loss of energy in the medium. The loss tangent of materials used in practice is found to increase with frequency, although increase is very small at the lower end of microwave frequencies. Table 1.1 lists the relative dielectric constant and loss tangent of dielectric materials commonly used at microwave frequencies.

1.4   Interface and Boundary Conditions

Maxwell's equations along with the constitutive relations may be used to obtain *general* solutions for the electromagnetic problems. To obtain *unique* solutions, we must enforce the boundary conditions at the periphery of the device. Additionally, in a mixed media device, continuity conditions at the interface of two media should be satisfied in order to ensure continuity of fields across the interface.

Interface Conditions
The interface conditions between two media, as shown in Figure 1.1, are given as

$$\hat{n} \times (\mathbf{E}_1 - \mathbf{E}_2) = 0 \qquad (1.13a)$$

Table 1.1   Electronic Properties of Typical Microwave Dielectric Materials

Material	Frequency	ϵ_r	$\tan\delta$ at 25°C
Styrofoam-103.7	3 GHz	1.03	0.0001
Rexolite-1422	3 GHz	2.54	0.0005
FR-4	3 GHz	4.3	0.022
Polystyrene	10 GHz	2.53	0.00047
BeO	10 GHz	6.6	0.0001
GaAs ($\rho = 10^7$ Ωcm)	10 GHz	12.3	0.0016
Sapphire	10 GHz	9.4–11.5	0.0001
Alumina (99.5%)	10 GHz	9.8	0.0003
Quartz (fused)	10 GHz	3.78	0.0001
Teflon	10 GHz	2.1	0.0004
Silicon ($\rho = 10^3$ Ωcm)	10 GHz	11.7	0.005
RT/Duroid 5880	10 GHz	2.2	0.0009
RT/Duroid 6010	10 GHz	102, 10.5, 10.8	0.0028 max.

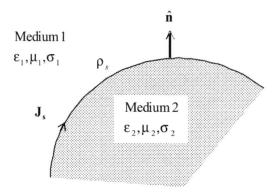

Figure 1.1   Interface between two different media.

$$\hat{n} \times (H_1 - H_2) = J_s \tag{1.13b}$$

$$\hat{n} \cdot (D_1 - D_2) = \rho_s \tag{1.13c}$$

$$\hat{n} \cdot (B_1 - B_2) = 0 \tag{1.13d}$$

where the subscripts 1 and 2 indicate media 1 and 2, respectively. The unit vector $\hat{n}$ is normal to the interface between the media and points into medium 1. The quantities J_s and ρ_s are the (linear) surface current density (in A/m) and surface charge density (in C/m^2) at the interface, respectively. The interface conditions are also applicable to the general time-varying fields. The condition (1.13a) signifies that the tangential component of E is continuous across the interface. It may be noted that the interface conditions for the normal and tangential components of the fields between any two media are not independent of each other; that is, *if the conditions for the tangential components are satisfied, the conditions on the normal components get satisfied*. These conditions are called the natural conditions at the interface.

The interface conditions get simplified for the special cases of interfaces. If the two media are perfect dielectrics $\rho_s = 0$, $J_s = 0$, and the tangential component of H and normal component of D are continuous across the interface. When medium 1 is a perfect dielectric and medium 2 is a perfect conductor, the interface conditions reduce to the following boundary conditions

$$\hat{n} \times E_1 = 0 \tag{1.14a}$$

$$\hat{n} \times H_1 = J_s \tag{1.14b}$$

$$\hat{n} \cdot D_1 = \rho_s \tag{1.14c}$$

$$\hat{n} \cdot B_1 = 0 \tag{1.14d}$$

because the fields inside a perfect conductor are zero. Equation (1.14a) means that the tangential electric field at the surface of a perfect conductor is zero, $E_{\text{tan}} = 0$.

Similarly, (1.14b) signifies that the current induced $\mathbf{J}_s$ on a perfect conductor is equal to the tangential component of $\mathbf{H}_l$. The charge induced on a conductor, in the presence of fields, is given by (1.14c).

Boundary Conditions
The boundary conditions at the periphery of the device must be satisfied to ensure unique solution. These conditions may be classified as Dirichlet type or Neumann type, homogeneous or inhomogeneous. We shall discuss these conditions in relation to the solution of the wave equation

$$\nabla^2 \psi + k^2 \psi = 0 \qquad (1.15)$$

where the function ψ may represent field or potential. The homogeneous Dirichlet condition may be specified as

$$\psi(\mathbf{r}) = 0 \qquad (1.16a)$$

on contour C of a planar surface, Figure 1.2. This condition translates to $E_{\text{tan}} = 0$ on a perfect electric conductor.

The homogeneous Neumann boundary condition is specified as

$$\frac{\partial \psi(\mathbf{r})}{\partial n} = \nabla \psi \cdot \hat{\mathbf{n}} = 0, \text{ on contour C (Figure 1.2)} \qquad (1.16b)$$

that is, directional derivative of ψ along the outward normal $\hat{\mathbf{n}}$ to the boundary equals zero. This condition implies that $\partial H/\partial n = 0$ on a perfect electric conductor, since normal derivative of H is proportional to tangential electric field. The conditions (1.16) are called homogeneous boundary conditions because the right-hand side of the equation is zero. The corresponding inhomogeneous boundary conditions are obtained when the right-hand side is not zero, for example,

$$\psi(\mathbf{r}) = \text{constant} \qquad (1.17a)$$

$$\frac{\partial y(\mathbf{r})}{\partial n} = \text{constant} \qquad (1.17b)$$

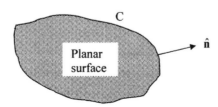

Figure 1.2 A planar surface with contour C and unit outward normal.

Another type of boundary condition called radiation condition or absorbing boundary condition is applied to truncate the infinite region of the problem to the finite region so that computational methods may be employed. The absorbing boundary condition is discussed in Chapter 8.

1.5   Skin Depth

Skin depth, or depth of penetration, is an alternative way to characterize a medium with nonzero conductivity. It is defined as the distance measured from the surface of the lossy medium over which the magnitudes of the fields are reduced to $1/e$, or approximately 37%, of those at the surface of the medium. The skin depth δ of a good conductor ($\sigma/\omega\epsilon \gg 1$) is approximately given by

$$\delta = \sqrt{\frac{2}{\omega\mu\sigma}} \tag{1.18}$$

The skin depth of a good conductor is very small, especially at high frequencies, causing currents to reside near the conductor's surface. The containment of current reduces the effective cross-sectional area of the conductor and therefore increases conduction loss.

1.6   Poynting Vector and Power Flow

When an electromagnetic wave propagates in a medium, it carries power along with it. The instantaneous power density at any location in the medium is given by the *Poynting vector*, defined as

$$\mathbf{S} = \mathbf{E} \times \mathbf{H} \qquad \text{W/m}^2 \tag{1.19}$$

The Poynting vector not only gives the magnitude of power flow but also its direction. The direction of power flow or wave propagation is determined by the right-hand rule of the cross-product and is always perpendicular to both $\mathbf{E}$ and $\mathbf{H}$. For the time-harmonic fields, the Poynting vector in phasor form becomes

$$\mathbf{S} = \mathbf{E} \times \mathbf{H}^* \tag{1.20}$$

where $\mathbf{H}^*$ is the complex conjugate of $\mathbf{H}$. The measurable power is the average power density. It is defined as the time average of the instantaneous power density $\mathbf{S}$, which can be described in terms of the phasor Poynting vector as

$$<\mathbf{S}(\mathbf{r}, t)>_t = \frac{1}{2}\text{Re}(\mathbf{E} \times \mathbf{H}^*) \tag{1.21}$$

The average power crossing a surface s is then given by

$$P_{av} = \frac{1}{2} \operatorname{Re} \iint_{s} (E \times H^*) \cdot ds \tag{1.22}$$

1.7 Image Currents and Equivalence Principle

Image principle is a useful concept in simplifying problems having infinitely long planar conductors. Simplification is introduced by way of replacing the perfect conductor by images of currents in front of it. The image current accounts for the reflection produced by the conductor. Consider the electric and magnetic dipoles placed in front of a perfect electric conductor (PEC) as shown in Figure 1.3(a). The field in the half-space in front of the PEC may be determined by a combination of actual sources and their images. The image sources plus the actual sources, radiating in free space, must produce zero tangential electric field over the perfect conductor. The necessary orientation of image dipoles satisfying this boundary condition is summarized in Figure 1.3(b). This solution is valid in front of the electric conductor only *and not in the image* region where the field should be zero because it was occupied by the perfect conductor. If PEC is replaced by a perfect magnetic conductor (PMC), the orientation of image currents are reversed. A good account of image theory is given in [2–4].

The combination of actual sources and their images, Figure 1.3, is an equivalent problem. The solution to the equivalent problem is supposed to be relatively easier.

The equivalence principle is an important step in setting up the equivalence, and is therefore a useful tool. In this, the surface boundaries are replaced by equivalent sources. The equivalent electric current sheets J_s are determined by the discontinuities in tangential magnetic fields across the boundary as stated by (1.13b),

$$J_s = \hat{n} \times (H_1 - H_2) \tag{1.23a}$$

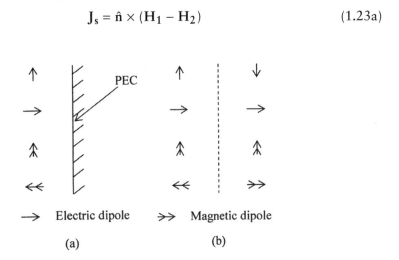

$\rightarrow$ Electric dipole $\gg$ Magnetic dipole

(a)   (b)

Figure 1.3 Image currents produced by a perfect electric conductor (PEC): (a) dipole current sources in front of a PEC; and (b) the electric conductor replaced by image currents for producing the same field in front of the electric conductor.

Similarly, the equivalent magnetic current sheets are determined by the discontinuities in the tangential electric fields across the boundary

$$\mathbf{M_s} = -\hat{n} \times (\mathbf{E_1} - \mathbf{E_2}) \tag{1.23b}$$

To illustrate the equivalence principle let us consider an aperture embedded in a PEC as shown in Figure 1.4. The aperture may represent an antenna or scatterer or may be part of a circuit. Let the fields in the aperture, set up by the excitation, be denoted as $\mathbf{E}$ and $\mathbf{H}$. The side view of the geometry is sketched in Figure 1.5(a) and divides the space into two regions 1 and 2. The actual sources generating $\mathbf{E_i}$, $\mathbf{H_i}$; $i = 1, 2$ are not necessary and can be replaced by equivalent sources according to (1.23) and placed on the boundary between the regions. The equivalent currents are shown in Figure 1.5(b). Only $\mathbf{J_s}$ or $\mathbf{M_s}$ is sufficient to determine $\mathbf{E_l}$ and $\mathbf{H_l}$ [2, p. 108]. Although we have converted the original problem into an

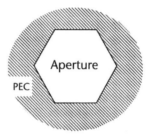

Figure 1.4 An aperture in PEC.

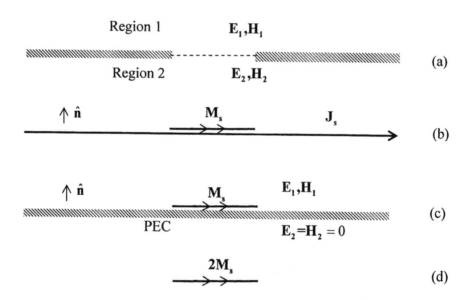

Figure 1.5 Aperture in the PEC and equivalent problems: (a) side view of the aperture; (b) equivalent for region 1 in terms of $\mathbf{M_s}$ and $\mathbf{J_s}$; (c) equivalent for region 1 in terms of $\mathbf{M_s}$ alone; (d) equivalent of (c) in terms of magnetic current in free space.

equivalent one, it is far from simpler. We can simplify the problem if we set $E_2 = H_2 = 0$ without disturbing E_1, H_1 in the region of interest. For this we introduce PEC at the boundary surface and underneath the currents. This is shown in Figure 1.5(c). The equivalent currents are now given by $J_s = \hat{n} \times H_1$, $M_s = -\hat{n} \times E_1$. Since the image of J_s in PEC is $-J_s$, the net effect is zero and we say that the current sheet is shorted. We are now left with M_s on PEC. This equivalence is consistent with the uniqueness concept that only J_s or M_s is sufficient to determine E_1 and H_1 [2, p. 108]. Further simplification is obtained by employing the image principle to obtain the equivalent current in free space as shown in Figure 1.5(d). It may be noted that there are a number of equivalent solutions to the problem. One may choose the easier one.

We next take an example of aperture coupling of waveguide to a cavity, a simple sketch of which is shown in Figure 1.6(a). An equivalent problem is obtained by filling the aperture with PEC as described earlier, thus converting the original problem into two separate problems. The waveguide side and the cavity side of the equivalent problem are drawn in Figure 1.6(b). The current M_s is obtained from the aperture electric field as $M_s = E \times \hat{n}$. This solution is very similar to that shown in Figure 1.5(c).

The equivalence principle is a very useful concept in converting a nonhomogeneous dielectric region problem into two equivalent problems each with homogeneous dielectric [5, p. 14]. The homogeneous Green's function $e^{-jkr}/(4\pi r)$ can then be employed for the two equivalent problems.

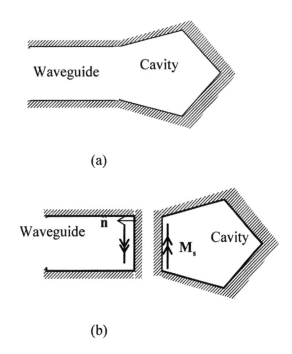

(a)

(b)

Figure 1.6   (a) Aperture coupling between waveguide and a cavity. (b) An equivalent problem of (a).

1.8 Reciprocity Theorem

Reciprocity theorem relates the response at the location of one source due to a second source, to the response at the location of second source due to the first source. Consider two sets of sources $\mathbf{J}^a$, $\mathbf{M}^a$ and $\mathbf{J}^b$, $\mathbf{M}^b$, of the same frequency, existing in the same linear medium. Denote the field produced by the source a alone by $\mathbf{E}^a$, $\mathbf{H}^a$, and that produced by the source b alone by $\mathbf{E}^b$, $\mathbf{H}^b$. Then, reciprocity theorem is stated as

$$<a, b> = <b, a> \tag{1.24a}$$

where $<a, b>$ denotes the reaction of field a on source b, and is given by

$$<a, b> = \iiint (\mathbf{E}^a \cdot \mathbf{J}^b - \mathbf{H}^a \cdot \mathbf{M}^b) \, dv \tag{1.24b}$$

Reciprocity theorem, therefore, states that the reaction produced by the source a on source b is equal to that produced by source b on source a. This theorem is useful in calculating the characteristics of circuits and antennas.

1.9 Differential Equations in Electromagnetics

Electromagnetic fields at a point may be determined using Faraday's law (1.1a), Ampere's law (1.1b), constitutive relations, and boundary conditions directly. This approach is followed in the finite-difference time-domain (FDTD) method (Chapter 8). However, the most convenient approach to determine the steady state fields is to solve the wave equations, which are derived from (1.5) next.

We assume the simplest situation of a source free space ($\rho = 0$, $\mathbf{J}_s = 0$), a medium which is linear, homogeneous, isotropic, and characterized by scalar quantities ϵ and μ. The fields are assumed to be time-harmonic with $\exp(j\omega t)$ variation. To derive the wave equation for $\mathbf{E}$ field, take the curl of (1.5a), make use of (1.5b), and use the constitutive relations (1.7) and (1.8) to obtain

$$\boldsymbol{\nabla} \times \boldsymbol{\nabla} \times \mathbf{E} - k^2 \mathbf{E} = 0 \tag{1.25}$$

where $k = \omega \sqrt{\epsilon \mu}$ is the *wave number* or propagation constant in the medium. Using the vector identity

$$\boldsymbol{\nabla} \times \boldsymbol{\nabla} \times \mathbf{E} = \boldsymbol{\nabla}(\boldsymbol{\nabla} \cdot \mathbf{E}) - \nabla^2 \mathbf{E} \tag{1.26}$$

we may rewrite (1.25) as

$$\nabla^2 \mathbf{E} + k^2 \mathbf{E} = 0 \qquad \text{since } \boldsymbol{\nabla} \cdot \mathbf{E} = 0 \tag{1.27a}$$

where ∇^2 is the *Laplacian operator*. Equation (1.27a) is called the *wave equation* for the electric field. Similarly, the wave equation for the magnetic field can be derived as

$$\nabla^2 \mathbf{H} + k^2 \mathbf{H} = 0 \tag{1.27b}$$

When the electric current excitation $\mathbf{J_s}$ is present, the wave equation becomes inhomogeneous and is given by

$$\nabla^2 \mathbf{E} + k^2 \mathbf{E} = j\omega\mu\mathbf{J_s} \tag{1.28}$$

Each of the expressions (1.27a) and (1.27b) represents three vector field components, each of which satisfies the scalar wave equation or the *Helmholtz equation*

$$\nabla^2 \psi + k^2 \psi = 0, \qquad \psi = E_x, E_y, E_z, H_x, H_y, H_z \tag{1.29}$$

Wave Equation for Guided Waves.   The wave equation for the guided waves can be simplified because the direction of propagation is specified with an assumed propagation constant along this direction. Let us consider guided waves in transmission lines/waveguides and assume that the direction of propagation is z. The ∇ operator may be decomposed into transverse and longitudinal parts as

$$\nabla = \nabla_t + \hat{\mathbf{z}}\frac{\partial}{\partial z} \tag{1.30}$$

where ∇_t is the *transverse* (to the z-axis) *operator*. Assuming that the propagation along the $\pm z$ directions is described by the factor $e^{\mp \gamma z}$, we obtain

$$\frac{\partial}{\partial z} = \mp\gamma \tag{1.31}$$

The Lapacian operator ∇^2 can also be decomposed into transverse and longitudinal parts as

$$\nabla^2 = \nabla \cdot \nabla = \left(\nabla_t + \hat{\mathbf{z}}\frac{\partial}{\partial z}\right) \cdot \left(\nabla_t + \hat{\mathbf{z}}\frac{\partial}{\partial z}\right) = \nabla_t^2 + \left(\frac{\partial}{\partial z}\right)^2 \tag{1.32}$$

$$= \nabla_t^2 + \gamma^2$$

where ∇_t^2 is the *transverse Laplacian operator*. Substituting (1.32) into (1.27) gives

$$\nabla_t^2 \mathbf{E}(x, y) + k_c^2 \mathbf{E}(x, y) = 0 \tag{1.33a}$$

$$\nabla_t^2 \mathbf{H}(x, y) + k_c^2 \mathbf{H}(x, y) = 0 \tag{1.33b}$$

where

$$k_c^2 = k^2 + \gamma^2 = \omega_c^2 \epsilon\mu \tag{1.34}$$

The quantities k_c and ω_c are called the *cutoff wave number* and *cutoff frequency*, respectively, since they reduce to the corresponding parameters at the cutoff ($\gamma = 0$). Equations (1.33) are known as the *wave equations in the transverse plane*, with z as the direction of propagation. Once the cutoff wave number is determined, the propagation constant at any frequency other than the cutoff frequency is obtained from (1.34)

$$\gamma = j\beta = j\sqrt{k^2 - k_c^2} \tag{1.35}$$

for the homogeneous filling of the waveguide/transmission line. In the inhomogeneously filled waveguide, the fields can vary in the z-direction also. The resulting differential equations are not separable and lead to a set of coupled differential equations in the tangential and normal components of the field [6].

Poisson and Laplace Equations
The Poisson equation is obtained from the Gauss law for electrostatics (1.5c) by using the gauge condition

$$\mathbf{E} = -\boldsymbol{\nabla}\varphi(\mathbf{r}) \tag{1.36}$$

Substituting it in (1.5c) gives the Poisson equation

$$\boldsymbol{\nabla}^2\varphi = -\frac{\rho}{\epsilon} \tag{1.37}$$

for the homogeneous case. In the charge free space, (1.37) reduces to

$$\nabla^2\phi = 0 \tag{1.38}$$

This equation is known as *Laplace equation* and is frequently employed in electrostatics. When the frequency is not too high, the electrostatic approach may be used to approximate the results for otherwise full-wave situation, and the approximation is called quasi-static approximation. The quasi-static parameters for microstrip line may be used up to about 3 GHz. We shall use Poisson and Laplace equations to solve problems related to microstrip line, strip line, static charge distribution on a metal wire, and capacitor.

1.10   Electric and Magnetic Vector Potentials

The electromagnetic fields for the source free case may be obtained directly from the wave equations (1.27). In order to determine the field generated by a given source, as in antennas, waveguide and cavity excitations, and so on, the solution is easier to obtain if we introduce auxiliary vector potential functions **A** and **F**. The magnetic vector potential **A** is the solution of

$$\nabla^2\mathbf{A} + k^2\mathbf{A} = -\mu\mathbf{J}_s \tag{1.39a}$$

and the electric vector potential **F** is determined from

$$\nabla^2 \mathbf{F} + k^2 \mathbf{F} = -\epsilon \mathbf{M_s} \tag{1.39b}$$

where $\mathbf{M_s}$ is the magnetic current density. Once the vector potentials are known, the fields produced by these potentials can be obtained as

$$\mathbf{E} = -j\omega \mathbf{A} + \frac{1}{j\omega\mu\epsilon} \nabla(\nabla \cdot \mathbf{A}) - \frac{1}{\epsilon} \nabla \times \mathbf{F} \tag{1.40a}$$

$$\mathbf{H} = -j\omega \mathbf{F} + \frac{1}{j\omega\mu\epsilon} \nabla(\nabla \cdot \mathbf{F}) + \frac{1}{\mu} \nabla \times \mathbf{A} \tag{1.40b}$$

In the source-free space, the differential equations satisfied by the vector potentials are obtained as

$$\nabla^2 \mathbf{A} + k^2 \mathbf{A} = 0 \tag{1.41a}$$

$$\nabla^2 \mathbf{F} + k^2 \mathbf{F} = 0 \tag{1.41b}$$

1.11   Wave Types and Solutions

For the guided-wave propagation, the field configuration assumes a particular form which is dictated by the boundary conditions, medium present, and the frequency of operation. The possible combination of electric and magnetic field allowed to propagate is called a mode type or wave type. In order to classify different *mode* (*or wave*) types, let us assume that the direction of propagation is the z-direction. The various modes are designated as *TE* mode ($E_z = 0$), *TM* mode ($H_z = 0$), *TEM* mode ($E_z = 0$, $H_z = 0$), or hybrid mode ($E_z \neq 0$, $H_z \neq 0$).

Construction of Wave Solutions
Now we shall solve homogeneous wave equations of (1.41) to determine the field distribution in source free space. From (1.41) we observe that each of the rectangular component of the vector potentials **A** and **F** satisfy the scalar wave equation

$$\nabla^2 \psi + k^2 \psi = 0 \tag{1.42}$$

Let us now consider some particular choices of potentials.
     In the first case, let

$$\mathbf{F} = 0, \text{ and } \mathbf{A} = \hat{z}\psi \tag{1.43}$$

Then from (1.40) and (1.43)

$$j\omega\mu\epsilon \mathbf{E} = k^2 \hat{z}\psi + \nabla\nabla \cdot \hat{z}\psi, \text{ and } \mathbf{B} = \nabla \times \hat{z}\psi \tag{1.44}$$

The above expressions can be expanded in rectangular coordinates as

$$E_x = \frac{1}{j\omega\mu\epsilon} \frac{\partial^2 \psi}{\partial x \partial z} \qquad \mu H_x = \frac{\partial \psi}{\partial y} \qquad (1.45a)$$

$$E_y = \frac{1}{j\omega\mu\epsilon} \frac{\partial^2 \psi}{\partial y \partial z} \qquad \mu H_y = -\frac{\partial \psi}{\partial x} \qquad (1.45b)$$

$$E_z = \frac{1}{j\omega\mu\epsilon} \left(\frac{\partial^2}{\partial z^2} + k^2 \right) \psi \qquad H_z = 0 \qquad (1.45c)$$

This field configuration is called *transverse magnetic to z* (TM-to-z). The behavior of function ψ determines the actual field distribution; for example, in the guided wave propagation one may assume $\psi = f(x, y)e^{\pm\gamma z}$.

In the dual case, let

$$\mathbf{A} = 0, \text{ and } \mathbf{F} = \hat{z}\psi \qquad (1.46)$$

Then from (1.40) and (1.46)

$$j\omega\mu\epsilon\mathbf{H} = k^2\hat{z}\psi + \mathbf{\nabla}\mathbf{\nabla} \cdot \hat{z}\psi, \text{ and } \mathbf{D} = \mathbf{\nabla} \times \hat{z}\psi \qquad (1.47)$$

The rectangular components of fields obtained from above are

$$H_x = \frac{1}{j\omega\mu\epsilon} \frac{\partial^2 \psi}{\partial x \partial z}, \qquad \epsilon E_x = -\frac{\partial \psi}{\partial y} \qquad (1.48a)$$

$$H_y = \frac{1}{j\omega\mu\epsilon} \frac{\partial^2 \psi}{\partial y \partial z}, \qquad \epsilon E_y = \frac{\partial \psi}{\partial x} \qquad (1.48b)$$

$$H_z = \frac{1}{j\omega\mu\epsilon} \left(\frac{\partial^2}{\partial z^2} + k^2 \right) \psi, \qquad E_z = 0 \qquad (1.48c)$$

The above field configuration is designated as *transverse electric to z* (TE-to-z) mode. For the guided modes, one may assume $\psi = f(x, y)e^{\pm\gamma z}$.

If both **A** and **F** are nonzero, the field components are again obtained from (1.40).

1.12 Phase Velocity, Dispersion, and Group Velocity

Let us consider *TEM*-wave propagation in free space along the x-direction. Let the plane wave consist of E_y and H_z components. Then, E_y is the solution of

$$\frac{\partial^2 E_y}{\partial x^2} = \frac{1}{c^2} \frac{\partial^2 E_y}{\partial t^2} \qquad (1.49)$$

Consider the following monochromatic, traveling wave solution for the wave equation:

$$E_y(x, t) = e^{j(\omega t - \beta x)} \tag{1.50}$$

This solution when substituted in (1.49) results in

$$\beta = \pm \frac{\omega}{c} \tag{1.51}$$

The phase velocity for the wave is determined by the speed at which a point on the constant phase plane travels. Let the planar phase front coincide with the plane $x = x_0$ at $t = t_0$ as shown in Figure 1.7. The phase value is given by

$$\text{phase} = \omega t_0 - \beta x_0 \tag{1.52}$$

At a later instant $t = t_0 + \Delta t$, this phase surface moves to the point $x = x_0 + \Delta x$. The phase remains the same; that is,

$$\text{phase} = \omega(t_0 + \Delta t) - \beta(x_0 + \Delta x) \tag{1.53}$$

Equating (1.52) and (1.53) gives

$$\omega \Delta t = \beta \Delta x$$

or

$$\text{Phase velocity, } v_{ph} = \frac{\Delta x}{\Delta t} = \frac{\omega}{\beta} = +c \tag{1.54}$$

The phase velocity is therefore independent of the frequency. A narrow-band signal consisting of a number of frequency components will maintain relative phase

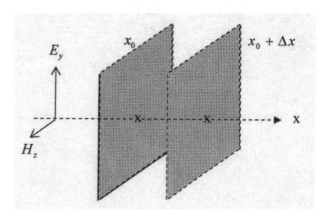

Figure 1.7   Plane wave propagation with phase fronts at two instants.

difference between various components as it propagates. However, if the phase velocity becomes a function of frequency as in waveguides, the relative phase difference will increase with propagation distance and the signal distortion will occur. This phenomenon is called dispersion. In this case, a single phase velocity cannot be assigned to the signal [7, p. 170]. However, group velocity can be assigned if the signal is narrow-band. To determine the group velocity, let us consider a two-tone signal propagating in a dispersive medium. We denote the signal as

$$E_y(x, t) = e^{j((\omega - \Delta\omega)t - (\beta - \Delta\beta)x)} + e^{j((\omega + \Delta\omega)t - (\beta + \Delta\beta)x)}$$

or

$$E_y = e^{j(\omega t - \beta x)} \, 2 \cos(\Delta\omega t - \Delta\beta x) \tag{1.55}$$

where $\beta \pm \Delta\beta$ is the phase constant at $\omega \pm \Delta\omega$. This signal is the product of a carrier wave and an envelope [8, p. 8]. The propagation speed of the envelope $\cos(\Delta\omega t - \Delta\beta x)$ is given by $\Delta\omega/\Delta\beta$. In the limit $\Delta\omega \to 0$, $\Delta\beta \to 0$, the speed is

$$v_{envelope} = \frac{d\omega}{d\beta} \triangleq v_g \tag{1.56}$$

and is called group velocity v_g. The Poynting vector associated with this wave is obtained from (1.20), and is given by

$$\mathbf{S} = \mathbf{E} \times \mathbf{H}^* = \hat{\mathbf{x}} \, 2 \cos^2(\Delta\omega t - \Delta\beta x)/\eta_0, \qquad \eta_0 = 120\pi \tag{1.57}$$

The transportation of energy therefore occurs with the speed of group velocity.
For the *TEM* wave defined by (1.51),

$$v_g = \pm c \tag{1.58}$$

However, phase constant for the modes in an air-filled rectangular waveguide is given by

$$\beta = \sqrt{k_0^2 - k_c^2} \tag{1.59}$$

and therefore

$$v_{ph} = \frac{\omega}{\beta} \quad \text{and} \quad v_g = \frac{c^2\beta}{\omega} \tag{1.60}$$

so that

$$v_{ph} v_g = c^2 \tag{1.61}$$

We shall see later that the *TEM* wave suffers dispersion even in dispersionless medium when the wave equation is discretized for computational methods. This phenomenon is called numerical dispersion and is a characteristic feature of majority of computational methods.

1.13   Characteristics of Transmission Lines

The characteristic impedance Z_0 and the phase velocity v_{ph} of a transmission line are given by

$$Z_0 = \frac{1}{c\sqrt{C_0^a C_0}} \tag{1.62a}$$

$$v_{ph} = \frac{c}{\sqrt{\epsilon_{re}}}, \quad \epsilon_{re} = \frac{C_0}{C_0^a} \qquad c : \text{velocity of light in vacuum} \tag{1.62b}$$

where C_0 is the capacitance per unit length of the line, and C_0^a is the capacitance with the dielectric replaced by air. The value of C_0 is determined from the potential difference V between the conductors of line as

$$C_0 = \frac{Q}{V} \tag{1.63}$$

where Q is the charge per unit length of the line, and is obtained from the voltage distribution about one of the conductors. The voltage distribution can be determined by solving the Laplace equation (1.38) subject to the boundary conditions.

1.14   Charge and Current Singularities

The charge distribution in a finite sized conductor is governed by the Coulomb law of force between the charges, and the charge accumulates at the edges so much so that charge density is infinity at the end of an infinitely thin conductor. Similar distribution has been observed at the corners of thick conductors and at the tip of a wedge. In the presence of time varying fields, the motion of charge carriers generates current. The current density and the charge density in a conductor are related by the continuity condition (1.5e). The current density, therefore, becomes infinity at the end of an infinitely thin conductor and at the tip of a wedge. The order of singularity varies with the conductor thickness and the angle of wedge. The singularity in charge/current gives rise to singularity in field distribution. We derive next the order of singularity at the tip of a wedge [9, p. 202].

Consider a conducting wedge of internal angle α as shown in Figure 1.8(a). Since the behavior of the field near an edge is essentially electrostatic, we solve the Laplace equation in two dimensions to determine the electrostatic potential ϕ,

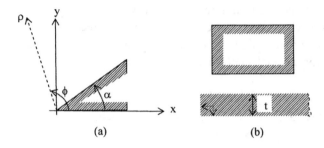

Figure 1.8 (a) Conducting wedge of angle α. (b) Semi-infinite strip of conductor thickness t and a rectangular metal block with $\alpha = \pi/2$.

$$\frac{1}{\rho}\frac{\partial}{\partial\rho}\left(\rho\frac{\partial\phi}{\partial\rho}\right) + \frac{1}{\rho^2}\frac{\partial^2\phi}{\partial\varphi^2} = 0 \tag{1.64}$$

The general solution of this equation near an edge is given by

$$\phi = A\rho^v \sin\left[v(\varphi + \varphi_0)\right] \tag{1.65}$$

where A, v, and φ_0 are the constants to be determined. The electric field can be determined from (1.65) and is given by

$$\mathbf{E} = -\nabla\phi = -\hat{\rho}\frac{\partial\phi}{\partial\rho} - \hat{\varphi}\frac{\partial\phi}{\rho\partial\phi} = -Av\rho^{v-1}\left[\hat{\rho}\sin v(\varphi + \varphi_0) + \hat{\varphi}\cos v(\varphi + \varphi_0)\right] \tag{1.66}$$

Applying the boundary condition that $E_\rho = 0$ at the conducting surfaces defined by $\varphi = 0$ and $\varphi = \alpha$ gives

$$\varphi_0 = -\alpha, \qquad v = \frac{\pi}{2\pi - \alpha} \tag{1.67}$$

The corresponding magnetic field is obtained from (1.5a) to give

$$H_z = j\omega\epsilon A\rho^v \cos\left[v(\varphi - \alpha)\right] \tag{1.68}$$

The charge density near the edge can be obtained from the electric field using (1.13c); that is,

$$\hat{n}.\mathbf{D} = \epsilon_0 E_\varphi = \rho_s$$

or

$$\rho_s = -Av\rho^{v-1}\epsilon_0 \cos v(\varphi - \alpha) \tag{1.69}$$

When the internal angle of the wedge is less than π, ν is less than 1 and E_φ, and therefore, charge density become singular at the edge, whereas E_ρ and E_z remain finite. For the half-plane or semi-infinite strip, $\alpha = 0$, $\nu = 1/2$; and $\rho_s \propto 1/\rho^{1/2}$. For a right-angled wedge and the finite thickness strip [Figure 1.8(b)], $\alpha = \pi/2$, $\nu = 2/3$, and $\rho_s \propto 1/\rho^{1/3}$.

Similar singular behavior of fields is also observed when there is a sharp change in ϵ or μ of the material. The order of singularity depends on the magnitude of change. However, in most of the practical problems, the value of ϵ or μ does not change very much and we do not observe the related singularity of field.

1.15 Classification of Methods of Analysis

Figure 1.9 lists various methods that may be employed for the analysis of boundary value problems in electromagnetics. As shown, the analysis may be based on theoretical approach and/or experimental measurements. The theoretical analysis has the following advantages:

- Analysis may be used to reduce the number of costly tests on prototypes by supporting the design process.
- Analysis may be used to ascertain the advantages as well as limitations of a configuration by carrying out parametric studies.
- Analysis can provide an understanding of the operating principles that could be useful for a new design, for modification of an existing design, and for the development of new configurations.

The theoretical analysis may be divided into analytical methods, model based methods, computational methods, and computational intelligence methods. The well-known analytical methods are listed in Figure 1.9. These methods provide useful design information but have very limited range of applications because of their dependence on closed-form expressions for eigenfunctions and Green's functions.

The use of approximation techniques like perturbation methods and variational methods increases the range of analytical methods at the cost of accuracy. The model-based methods use simplifying assumptions to reduce the complexity of the problem. Physics-based models provide valuable insight into the operation of the device and are helpful in innovations. Some of the well-known models are based on the description of the device in terms of transmission line sections or cavities combined together.

The computational methods are becoming increasingly popular because of their simplicity, versatility, and the availability of software based on them. Computational intelligence methods like artificial neural-network (ANN) method and neuro-fuzzy method may be employed to predict the values from the existing database (created using computational methods) and for optimization.

Experimental measurements may be carried out to characterize the unknown device, and the results may be utilized to develop technology, and for validation of computational methods. The developments in the areas of radar and microwave

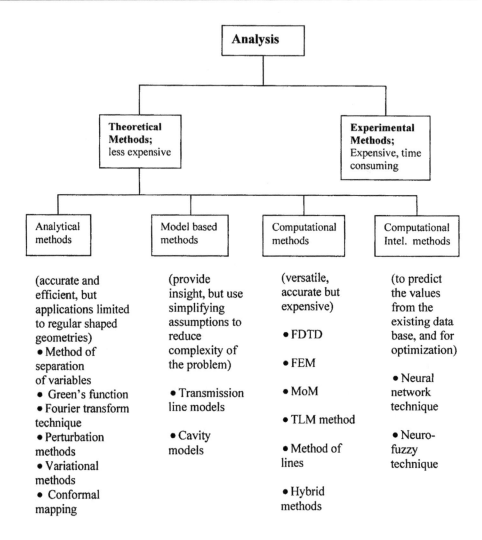

Figure 1.9 Classification of methods for the analysis of a boundary value problem.

circuits before World War II were mainly based on measurements. The experimental methods, however, are expensive, time consuming, and limited in scope because their results are applicable to the set of parameters used in the experiment.

1.16 Mathematical Framework in Electromagnetics

The time-harmonic electric and magnetic fields satisfy the Maxwell's equations (1.5), and may be determined using various approaches, as shown in Figure 1.10. The ab initio method is to solve the Maxwell's equations subject to the boundary conditions imposed by the device. This approach is followed in the FDTD method. An alternative approach, which involves less number of unknowns, is to solve the wave equation or Helmholtz equation, and is derived from the Maxwell's equations. The wave equation for the fields (1.27) or vector potentials (1.39) may be used.

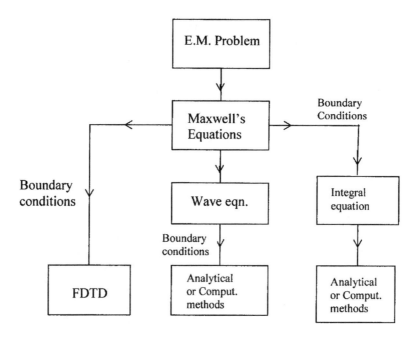

Figure 1.10   Mathematical models for solving an electromagnetic problem.

This is the most popular governing equation. For solving the wave equation one may use various theoretical approaches mentioned in the last section. These include: method of separation of variables, finite difference method, and FDTD. Alternatively, one may derive integral equation or functional form of the Maxwell's equations and find their solution. The differential equations or integral equations for the problem are called governing equations. The common denominator is to determine the unknown fields, charge or potential, or current distribution in the given electromagnetic device as the device dimensions, frequency, and material parameters are varied.

In order to understand the analytical methods, the basic infrastructure, like the knowledge of vector calculus, orthogonal functions, Green's functions, Fourier transform technique, contour integration, and variational methods, are needed. In addition, one needs to know the concepts such as equivalence principle, reaction, polarization currents, image theory, and reciprocity. The computational methods, however, can be *introduced* with the background of electromagnetic theory, microwave engineering, and vector calculus acquired at the undergraduate level. The knowledge of computer programming is desirable for computational methods.

1.17   Overview of Analytical and Computational Methods

There are two broad classes of solution methods employed for the boundary value problems: analytical methods and computational or numerical methods. The most common analytical methods include: method of separation of variables, orthogonal function expansion method, method of images, Green's function, conformal mapping, analysis in spectral or Fourier domain, and variational methods. Most of these methods are covered very well in the literature [10–13]. The computational methods include: finite difference method (FDM), finite-difference time-domain (FDTD) method, finite element method (FEM), method of moments (MoM), boundary integral equation (BIE), mode matching, transmission-line matrix (TLM) method, and method of lines. Some of the useful texts for computational methods include [5, 6, 8, 9, 14–19]. An overview of some of these methods is given next.

Overview of Analytical Methods

Historically, the analytical methods represent the urge of scientists and engineers to solve the mathematical problems with paper and pencil without the use of computers. This was the requirement before World War II and earlier when the computers or calculators were not available. The emphasis was on developing techniques supported by this environment. With the advent of computers and with computing power becoming cheaper, the emphasis is shifting towards computational techniques.

The method of separation of variables is one of the earliest and elegant analytical methods. It is assumed in this method that a function of several variables like $\varphi(x, y, z)$ can be expressed as a product of three functions, each of which depends on one variable only, for example,

$$\varphi(x, y, z) = \varphi_1(x)\,\varphi_2(y)\,\varphi_3(z) \tag{1.70}$$

It is easier to express φ_1, φ_2, φ_3 analytically than the composite function φ. The separability assumption is true when the device geometry can be fitted into one of the several orthogonal coordinate systems.

Each of the unknown functions φ_1, φ_2, φ_3 is expressed in the form of a series to represent a general solution for the corresponding ordinary differential equation. The series expansion may include trigonometric functions, Bessel functions, power series, and polynomials such as Legendre, Chebychev, and so on. The sinusoidal function expansion for $\phi_1(x)$ may be written as

$$\varphi_1(x) = \sum_n A_n \sin(n\pi x) \tag{1.71}$$

where the sinusoidal function is defined over the entire range of x. The expansion functions are chosen to be orthogonal to simplify the algebra. For regular shaped device geometries such as rectangular and circular, the function selected may include eigenfunctions, which leads to eigenfunction expansion of $\varphi(x, y, z)$. The boundary conditions are used to determine the unknowns A_n.

Orthogonal function expansion or series expansion of $\varphi(x, y, z)$ may be used when the method of separation of variables cannot be employed either due to the inseparability of variables or due to the boundary conditions [16, p. 78]. The availability of Green's function is a very significant step in the development of analytical solution. The solution can be determined for various types of excitations once the Green's function is known. For many types of problems, analytical solution cannot be attempted in the space domain, while it is possible to do so in the spectral domain. The most common example is that of layered dielectric configuration of planar circuits and antennas. The inverse transform in the form of an integral provides the solution in space domain. Solving these integrals is tricky because of the singular integrands. The residue calculus is useful in this respect.

The variational concept provides an alternative formulation of the original problem with the advantage that the solution has better accuracy even if the unknown function is modeled crudely. The Schwarz-Christoffel transformations help solve an important class of boundary value problems that involve regions with polygonal boundaries such as planar transmission lines. The analysis results in design equations in the quasi-static limit.

Some of the problems that can be solved analytically include: wave propagation in rectangular, circular, and elliptic waveguides; cavity resonance within rectangular, cylindrical, and spherical cavities; scattering by infinite planes, wedges, circular cylinders, and spheres; and static potential between infinite parallel plates [9, p. 20].

The analytical methods often lead to closed-form solutions and design information. These methods also help advance the computational methods.

Overview of Computational Methods
The analytical methods are biased towards analytical solution of the wave equation. As a result, these methods are applicable to devices with regular shapes and homogeneous dielectric, except for some cases. Most of the device geometries of engineering importance do not conform to these restrictions. Therefore, the range of analytical methods is limited. However, the basic approach of analytical methods can be generalized for application to these geometries. For this, the unknown function $\varphi_1(x)$ may be expanded not in terms of entire domain expansion functions but in terms of subdomain functions as

$$\varphi_1(x) = \sum_m a_m \varphi_{1m}(x) \tag{1.72}$$

where φ_{1m} are called subdomain expansion functions and are defined over a portion of the range of x. These function types could be pulse functions, piecewise linear, or piecewise sinusoidal functions (Chapter 6). The use of subdomain functions makes the solution process almost independent of the device shape. The process described by (1.72) is called discretization of the function, *and is generally accompanied by the corresponding discretization of the device in the form of cells or elements*. A typical discretization in the form of triangular cells is shown in Figure 1.11. The discretization of the device geometry is a common feature of majority of the computational methods and helps to include the effect of dielectric inhomogeneity in the solution.

For the solution based on computational methods, the discretization is carried out such that the dielectric is homogeneous over each cell, as shown in Figure 1.11. The governing equation is solved separately for each of the homogeneous regions, and interface conditions satisfied at the junctions of elements to ensure continuity of the solution over the device geometry. This is a useful feature of computational methods. Including the effect of dielectric inhomogeneity in the analytical methods

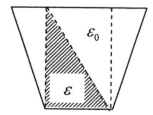

Figure 1.11 Discretization of device geometry in triangular elements with homogeneous dielectric in each element.

requires introducing new concepts such as polarization currents or equivalence principle. The computational methods, therefore, provide general solution procedure and are convenient for the beginners.

The computational methods may be classified as differential equation solvers or integral equation solvers. Some of these methods are listed in Figure 1.12. The most popular of these methods is FEM because it can be efficiently adapted to arbitrary device geometry and dielectric configuration. FDTD method is economical in providing broadband behavior and animation, which is useful in diagnostics of the device. MoM is efficient for analyzing open region geometries. Often computational methods are chosen on the basis of trade-offs between accuracy, speed, storage requirements, versatility, and so on, and are structure dependent.

The solutions based on computational methods are often complete and include various aspects such as internal and external couplings, surface wave effect, and radiation.

Elements of computational methods, their accuracy, and unified formalism for the integral and differential equations are discussed in Chapter 6.

1.18 Summary

This chapter summarizes the basic concepts in electrostatics and wave propagation, and is used as common background material for latter chapters. Of particular

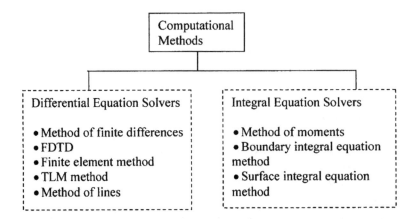

Figure 1.12 Classification of the computational methods.

relevance are the topics such as boundary conditions, differential equations of electromagnetics, modes in a waveguide, dispersion and group velocity, and charge singularity at a wedge.

Diverse approaches for the analysis of boundary value problems in electromagnetics are discussed, bringing out the importance and limitations of theoretical analysis. The analytical tools required in this connection include: orthogonal functions, Green's function, Fourier transform, conformal mapping, and variational methods. Solution in the form of discretization of unknown functions using subdomain basis functions and the associated discretization of device geometry is suggested. Common analytical and computational methods are listed and overviewed, bringing out their advantages and limitations.

References

[1] Kraus, J. D., *Electromagnetics*, 4th ed., New York: McGraw-Hill, 1973.

[2] Harrington, R. F., *Time-Harmonic Electromagnetic Fields*, New York: McGraw-Hill, 1961.

[3] Kong, J. A., *Electromagnetic Wave Theory*, New York: Wiley-Interscience, 1986.

[4] Collin, R. E., *Field Theory of Guided Waves*, 2nd ed., New York: IEEE Press, 1991.

[5] Peterson, A. F., S. L. Ray, and R. Mittra, *Computational Methods for Electromagnetics*, Hyderabad, India: University Press (India) Ltd., 2001.

[6] Volakis, J. L., A. Chatterjee, and L. C. Kempel, *Finite Element Method for Electromagnetics: Antennas, Microwave Circuits and Scattering Applications*, Hyderabad, India: University Press (India) Ltd., 2001.

[7] Pozar, D. M., *Microwave Engineering*, 2nd ed., New York: Wiley, 1998.

[8] Bondeson, A., T. Rylander, and P. Ingelstrom, *Computational Electromagnetics*, Berlin: Springer, 2005.

[9] Jin, J., *The Finite Element Method in Electromagnetics*, 2nd ed., New York: John Wiley, 2002.

[10] Dettman, J. W., *Mathematical Methods in Physics and Engineering*, New York: McGraw-Hill, 1962.

[11] Butkov, E., *Mathematical Physics*, Reading, MA: Addison-Wesley, 1968.

[12] Arfken, G., *Mathematical Methods for Physicists*, 3rd ed., New York: Academic Press, 1985.

[13] Eom, H. J., *Electromagnetic Wave Theory for Boundary Value Problems: An Advanced Course on Analytical Methods*, Berlin: Springer, 2004.

[14] Booton, R. C., Jr., *Computational Methods for Electromagnetics and Microwaves*, New York: John Wiley, 1992.

[15] Chari, M. V. K., and S. J. Salon, *Numerical Methods in Electromagnetism*, San Diego: Academic Press, 2000.

[16] Sadiku, M. N. O., *Numerical Techniques in Electromagnetics*, 2nd ed., Boca Raton, FL: CRC Press, 2001.

[17] Davidson, D. B., *Computational Electromagnetics for RF and Microwave Engineering*, New York: Cambridge University Press, 2005.

[18] Taflove, A., and S. C. Hagness, *Computational Electrodynamics: The Finite-Difference Time-Domain Method*, 3rd ed., Norwood, MA: Artech House, 2005.

[19] Sullivan, D. M., *Electromagnetic Simulation Using the FDTD Method*, New York: IEEE Press, 2000.

CHAPTER 2

Analytical Methods and Orthogonal Functions

2.1 Introduction

A large number of elementary and advanced problems in electromagnetics are formulated in terms of differential equations involving functions of more than one variable. These are known as *partial differential equations* (PDE). The commonly occurring PDEs are:

1. Laplace equation:

$$\nabla^2 \varphi = 0 \tag{2.1}$$

   This is a homogeneous, elliptic PDE equation. It occurs in the studies of electrostatics, dielectrics, static currents, magnetostatics, and so on.

2. Poisson equation:

$$\nabla^2 \varphi = -\rho/\epsilon \tag{2.2}$$

   The Laplace equation gets modified to this form when the charge is present. It is a nonhomogeneous equation due to the source term $-\rho/\epsilon$.

3. The wave equation or Helmholtz equation:

$$\nabla^2 \varphi + k^2 \varphi = 0 \tag{2.3}$$

   This is a hyperbolic PDE and describes the steady state solution of electromagnetic wave propagation, for example, in transmission lines, waveguides, and free space.

4. The time-dependent wave equation:

$$\nabla^2 \varphi - \frac{\partial^2 \varphi}{c^2 \partial t^2} = 0 \tag{2.4}$$

All the above PDE can be written in the operator form as:

$$Lf = g \qquad (2.5)$$

where g is a known source function, f is the unknown scalar function, and L is a differential operator,

$$L : \left(\frac{\partial}{\partial x}, \frac{\partial}{\partial y}, \frac{\partial}{\partial z}, \frac{\partial}{\partial t}, x, y, z, t \right) \qquad (2.6)$$

Two characteristics are common to these equations: (1) The equations are linear in the unknown function φ because ∇^2, $\nabla^2 + k^2$, and $\nabla^2 - \dfrac{\partial^2}{\partial t^2}$ are linear differential operators; and (2) The equations are all second order PDE, because they contain the second order and not higher order derivatives.

The general methods for solving these PDE may be classified into two broad categories: analytical methods and computational methods. Some of these are listed next.

1. *Method of separation of variables*. The PDE is split into ordinary differential equations that may be solved easily. This analytical method may not always work, but it is often the simplest when it does work.
2. *Green's function based integral solutions*. This analytical method is described in Chapter 3 and produces a solution in the form of an integral.
3. *Conformal mapping*. This method is limited to solving the Laplace equation in two dimensions and is discussed in Chapter 4.
4. Integral transforms may be used to solve the PDE. *Fourier transform method* is described in Chapter 5.
5. *Computational methods*. The development of high-speed computing machines has opened up a number of computational methods. Finite difference method (FDM), finite-difference time-domain (FDTD) method, finite element method (FEM), and method of moments (MoM), in this category, are discussed in the text.

The analytical methods provide the most satisfactory solution for the PDE. However, the range of problems which can be solved using the analytical methods are very much limited. The reasons for this limitation are: irregular shape of the structure, dielectric inhomogeneity, and/or inhomogeneous boundary conditions. Therefore, approximation methods, computational methods or any other method is employed in such situations. The analytical solutions, although limited, are useful in validating the results of the computational methods. Also, one is able to appreciate the need for computational methods better after seeing the limitations of other methods. Here, in this chapter we shall summarize the method of separation of variables, orthogonality, Dirac delta function, and eigenfunction based framework of function spaces.

2.2   Method of Separation of Variables

The PDE along with the boundary conditions are to be solved for unique solutions of the problem. The method of *separation of variables* is perhaps the most powerful analytical method employed for this purpose. It is also called the *method of Fourier expansion*, and may also be used as a first step in determining the Green's function (Chapter 3).

Separability of Functions
A function $f(x, y)$ of two variables is said to be separable if we can express it as a product of two functions, each of which is a function of one variable only; that is,

$$f(x, y) = f_1(x) f_2(y) \tag{2.7}$$

The method of separation of variables is based on the property of separability of functions. When this property is applied to the solution of PDE, it means that if $f(x, y)$ is the solution to the PDE, then $f_1(x)$ and $f_2(y)$ are the solutions corresponding to the ordinary differential equations obtained by separating the PDE. It is easier to find solution to each of the ordinary differential equations.

To determine whether the method of separation of variables can be applied to the given problem, we must consider the following: (1) the PDE describing the problem, (2) the solution region, and (3) the boundary conditions. Each of these must satisfy certain conditions. For a problem involving two variables x and y, three things must be considered [1]:

1. The differential operator L must be separable; that is, it must be a function of $f(x, y)$ such that

$$\frac{L\{f_1(x) f_2(y)\}}{f(x, y) f_1(x) f_2(y)} \tag{2.8}$$

   is a sum of a function of x only and a function of y only.
2. All boundary conditions must conform to constant-coordinate surfaces; that is, x = constant, y = constant lines. For a circular region, the constant coordinate surfaces are: ρ = constant, and φ = constant.
3. The boundary conditions at x = constant (or y = constant) line should be either of Dirichlet type ($f = 0$) or Neumann type ($f' = 0$). It should not be Dirichlet type on a part of the boundary and Neumann type on another part.

For the sake of completeness we work out a problem next on the method of separation of variables. We shall come across many examples of the use of this method in Chapters 3 and 5.

Solution of Laplace Equation in a Rectangular Pipe
Consider the cross-section of a rectangular metal pipe of dimensions a and b, as shown in Figure 2.1. The pipe is assumed to be uniform and infinitely long along the

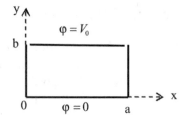

Figure 2.1 A rectangular metal pipe grounded on three sides and maintained at V_0 at the top plate.

z-direction so that the electrostatic potential and the electric field are independent of z. As shown in the figure, three sides of the pipe are maintained at zero potential, while the top plate is maintained at a constant voltage V_0. Insulating gaps on both sides of the top plate prevent short-circuiting.

Assuming that the volume charge density inside the pipe is zero, we wish to find the potential $\varphi(x, y)$ inside the pipe. Mathematically, the problem can be stated as follows: find $\varphi(x, y)$ such that

$$\nabla^2 \varphi(x, y) = 0 \qquad \text{for } 0 \leq x \leq a \text{ and } 0 \leq y \leq b \tag{2.9}$$

and satisfying

$$\varphi(0, y) = 0 \qquad \text{for } 0 \leq y \leq b \tag{2.10a}$$

$$\varphi(a, y) = 0 \qquad \text{for } 0 \leq y \leq b \tag{2.10b}$$

$$\varphi(x, 0) = 0 \qquad \text{for } 0 \leq x \leq a \tag{2.10c}$$

$$\varphi(x, b) = V_0 \qquad \text{for } 0 \leq x \leq a \tag{2.10d}$$

Solution. Because $\varphi(x, y)$ is a function of x and y, (2.9) can be expressed as

$$\frac{\partial^2 \varphi(x, y)}{\partial x^2} + \frac{\partial^2 \varphi(x, y)}{\partial y^2} = 0 \tag{2.11}$$

We now apply the method of separation of variables by letting

$$\varphi(x, y) = X(x) Y(y) \tag{2.12}$$

Substituting (2.12) in (2.11), we obtain

$$Y \frac{d^2 X}{dx^2} + X \frac{d^2 Y}{dy^2} = 0 \tag{2.13}$$

Dividing (2.13) by $\varphi(x, y) = X(x) Y(y)$ yields the following new form:

$$\frac{1}{X}\frac{d^2X}{dx^2} + \frac{1}{Y}\frac{d^2Y}{dy^2} = 0 \tag{2.14}$$

It may be observed that the first term in (2.14) is a function of x only, and the second term is a function of y only, thus satisfying the separability criterion (2.8) of the operator. Also, the boundary conditions conform to constant coordinate surfaces and are homogeneous on a given surface. Therefore, the method of separation of variables can be applied to this problem.

Equation (2.14) may be arranged as

$$\frac{1}{X}\frac{d^2X}{dx^2} = -\frac{1}{Y}\frac{d^2Y}{dy^2} \tag{2.15}$$

Note that the left side of (2.15) is either a function of x only or is a constant. Similarly, the right side is a function of y only or is a constant. Since (2.15) must hold for all x and y in the pipe, it can only hold if both sides are equal to a constant; that is,

$$\frac{1}{X}\frac{d^2X}{dx^2} = C \tag{2.16a}$$

$$\frac{1}{Y}\frac{d^2Y}{dy^2} = -C \tag{2.16b}$$

It may be noted that the original second order PDE (2.11) has been transformed into two second order ordinary differential equations, (2.16). The constant C is called the separation constant. It may be positive, negative, or simply zero. Let us consider these possibilities.

Case 1: $C > 0$.   In this case, C may be expressed as $C = k^2$. Consequently,

$$X(x) = Ae^{kx} + Be^{-kx} \text{ or } X(x) = A_1 \sinh(kx) + B_1 \cosh(kx) \tag{2.17a}$$

and

$$Y(y) = Ae^{jky} + Be^{-jky} \text{ or } Y(y) = A_2 \sin(ky) + B_2 \cos(ky) \tag{2.17b}$$

Case 2: $C < 0$.   For this case, C may be expressed as $C = -k^2$. Therefore,

$$X(x) = Ae^{jkx} + Be^{-jkx} \text{ or } X(x) = A_1 \sin(kx) + B_1 \cos(kx) \tag{2.18a}$$

and

$$Y(y) = Ae^{ky} + Be^{-ky} \text{ or } Y(y) = A_2 \sinh(ky) + B_2 \cosh(ky) \tag{2.18b}$$

Case 3: $C = 0$.   In this case,

$$X(x) = Ax + B \text{ and } Y(y) = A_1 y + B_1 \tag{2.19}$$

Out of the three possible solutions described above, only case 2 can satisfy the boundary conditions at $x = 0$ and $x = a$ because $\sin(kx)$ and $\cos(kx)$ can have infinite number of zeros. Therefore, we must choose (2.18). Consequently, solution of (2.16) yields the following:

$$X(x) = A_1 \sin(kx) + B_1 \cos(kx) \tag{2.20a}$$

$$Y(y) = A_2 \sinh(ky) + B_2 \cosh(ky) \tag{2.20b}$$

or

$$\varphi(x, y) = [A_1 \sin(kx) + B_1 \cos(kx)][A_2 \sinh(ky) + B_2 \cosh(ky)] \tag{2.21}$$

We now apply the boundary condition (2.10a) and obtain

$$B_1 = 0 \tag{2.22}$$

Similarly, (2.10c) is satisfied if $B_2 = 0$. The potential function now takes the following form

$$\varphi(x, y) = A \sin(kx) \sinh(ky) \tag{2.23}$$

where A and k are arbitrary constants. To satisfy (2.10b), we substitute $x = a$ in (2.23) and obtain

$$0 = A \sin(ka) \sinh(ky) \qquad \text{for } 0 \le y \le b \tag{2.24}$$

This equation is satisfied if we choose $ka = m\pi$, where m is an integer. Thus, the function $\varphi(x, y)$ takes the following form:

$$\varphi(x, y) = A \sin\left(\frac{m\pi x}{a}\right) \sinh\left(\frac{m\pi y}{a}\right) \qquad m = 1, 2, 3, \ldots \tag{2.25}$$

It may be noted that (2.25) satisfies the Laplace equation and the first three of the boundary conditions of (2.10). In order to satisfy the last boundary condition we consider a series consisting of terms of (2.25) for different values of m; that is,

$$\varphi(x, y) = \sum_{m=1}^{\infty} A_m \sin\left(\frac{m\pi x}{a}\right) \sinh\left(\frac{m\pi y}{a}\right) \tag{2.26}$$

This series satisfies the Laplace equation and three of the boundary conditions because each term of the series satisfies them. We now impose the fourth boundary condition to obtain

$$V_0 = \sum_{m=1}^{\infty} A_m \sin\left(\frac{m\pi x}{a}\right) \sinh\left(\frac{m\pi b}{a}\right) \qquad \text{for } 0 \le x \le a \qquad (2.27)$$

It may be observed that (2.27) is simply the Fourier series expansion of the constant V_0 in the interval $0 \le x \le a$. We now determine the coefficients of expansion A_m by multiplying both sides of (2.27) by $\sin(n\pi x/a)$ and integrating over the range of x. One obtains

$$V_0 \int_0^a \sin\left(\frac{n\pi x}{a}\right) dx = A_n \sinh\left(\frac{n\pi b}{a}\right) \frac{a}{2} \qquad (2.28)$$

We have employed the following orthogonality condition of functions $\sin(n\pi x/a)$ and $\sin(m\pi x/a)$:

$$\int_0^a \sin\left(\frac{n\pi x}{a}\right) \sin\left(\frac{m\pi x}{a}\right) dx = \begin{cases} 0 & \text{for } m \ne n \\ a/2 & \text{for } m = n \end{cases} \qquad (2.29)$$

Analytical integration of the left side of (2.28) gives the following expression for A_n [2]:

$$A_n = \begin{cases} \dfrac{4V_0}{n\pi \sinh(n\pi b/a)} & \text{for } n \text{ odd} \\ 0 & \text{for } n \text{ even} \end{cases} \qquad (2.30)$$

The final solution for the potential distribution in the pipe is, therefore, given by

$$\varphi(x, y) = \frac{4V_0}{\pi} \sum_{n,odd}^{\infty} \frac{\sin\left(\dfrac{n\pi x}{a}\right) \sinh\left(\dfrac{n\pi y}{a}\right)}{n \sinh(n\pi b/a)} \qquad (2.31)$$

Expression (2.31) is the unique solution to the problem. It satisfies the Laplace equation and all the four boundary conditions. At $y = b$, $\varphi(x, b) = V_0$, and (2.31) yields

$$\sum_{n,odd}^{\infty} \frac{4}{n\pi} \sin\left(\frac{n\pi x}{a}\right) = 1 \qquad (2.32)$$

We can see that the series adds up to unity for $0 \le x \le a$. Figure 2.2 illustrates this result for the sum of the first three terms.

The wave equation, or Helmholtz equation, is separable (similar to the example discussed on Lapalace equation) in a number of orthogonal coordinate systems if the boundary conditions of the problem conform to the constant coordinate surfaces. In

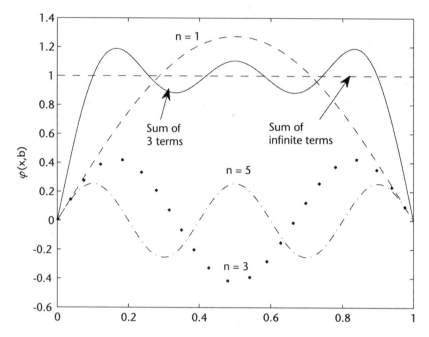

Figure 2.2 $\varphi(x, b)$ given by (2.31) and its first three Fourier components.

the last example, the boundary conditions belong to the $x =$ constant, and $y =$ constant category. If the boundaries are curved lines or surfaces, then it is, in general, convenient to use a coordinate system in which the required surfaces are represented by

$$u = \text{constant}, \qquad v = \text{constant, and so forth}$$

where $u, v, \ldots$ are the coordinates. For a circular region, for example, the problem is separable when expressed in (ρ, φ) coordinate system; and the constant coordinate surfaces are: $\rho =$ constant, and $\varphi =$ constant.

Some of the boundary-value problems involve regular shapes in the form of circular cylinders and spheres (or their portions). In these cases the choice of cylindrical or spherical coordinate systems satisfies the condition of separability and is therefore a natural choice.

The solution of Helmholtz equation in a cylindrical coordinate system is described in terms of Bessel functions $J_n(x)$, Neumann functions $Y_n(x)$, and Hankel functions $H_n^{(1)}(x)$ and $H_n^{(2)}(x)$; and the solution in a spherical coordinate system is described in terms of spherical Bessel functions $j_n(x)$, $y_n(x)$, $h_n^{(1)}(x)$, $h_n^{(2)}(x)$ [3]. The solution of Laplace equation in a spherical coordinate system generates Legendre functions $P_n(x)$ [3].

We will come across the use of method of separation of variables in Chapter 3 on Green's functions and Chapter 5 on Fourier transform method. If the method of separation of variables cannot be applied, we may use Green's function method, integral transform technique, or computational methods.

The method of separation of variables utilizes the orthogonality of functions to determine the expansion coefficients in a Fourier series expansion. Because of this useful property, the orthogonal functions can be employed to expand any arbitrary continuous function or piecewise continuous function; that is, they serve as basis functions in the expansion of arbitrary functions.

2.3   Orthogonality Condition

Orthogonality of functions can be described in the same way as the orthogonality of vectors. For instance, let us define a vector

$$\mathbf{f} = \sum_{i=1}^{N} a_i \hat{\mathbf{x}}_i \qquad (2.33)$$

where $\hat{\mathbf{x}}_i$ is a unit vector in the ith direction, and a_i is the amplitude of $\mathbf{f}$ in the ith direction. In the three-dimensional Euclidean space, $N = 3$. The value of a_i can be determined easily by using the orthogonality constraint of vectors defined as

$$\hat{\mathbf{x}}_i \cdot \hat{\mathbf{x}}_j = \delta_{ij} \qquad (2.34)$$

where the kronecker delta is defined as

$$\delta_{ij} = \left\{ \begin{array}{ll} 1 & \text{for } i = j \\ 0 & \text{for } i \neq j \end{array} \right\}$$

Thus, taking the dot product of (2.33) with $\hat{\mathbf{x}}_j$ one obtains

$$\mathbf{f} \cdot \hat{\mathbf{x}}_j = \sum_{i=1}^{N} a_i \hat{\mathbf{x}}_i \cdot \hat{\mathbf{x}}_j \qquad (2.35)$$

Use of the orthogonality constraint (2.34) gives

$$\mathbf{f} \cdot \hat{\mathbf{x}}_j = \sum_{i=1}^{n} a_i \delta_{ij} = a_j \qquad (2.36)$$

Equation (2.36) can be interpreted to mean that the coefficient a_j is given by the projection of vector $\mathbf{f}$ on the jth coordinate axes. This example illustrates the simplicity with which we can determine a particular vector component. Unit vectors $\hat{\mathbf{x}}_i$ are also called the basis vectors. Figure 2.3 illustrates the meaning of projection graphically. Here, vector $\mathbf{f}$ is projected on x-axis to obtain its x-component $\mathbf{f} \cdot \hat{\mathbf{x}}$.

The idea of the expansion of an arbitrary vector in terms of basis vectors can be extended to functions also. Let us consider a function $f(x)$ specified in the interval (a, b). We can expand this function in a set of orthogonal functions $u_i(x)$, $i = 1, 2, 3, \ldots, \infty$ as follows:

$$f(x) = \sum_{i=1}^{\infty} c_i u_i(x) \qquad (2.37)$$

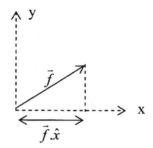

Figure 2.3 Projection of $\bar{f}$ on x-axis to determine its x-component.

Here coefficients c_i are analogous to a_i, and $u_i(x)$ analogous to $\hat{x}_i$. However, the summation now extends over an infinite number of terms. The functions $u_i(x)$ are called expansion functions or basis functions. The orthogonality condition of functions $u_i(x)$ is expressed as follows (this condition is derived in Section 2.4.1):

$$\int_a^b u_i(x)\, u_j(x)\ dx = N_i\, \delta_{ij} \tag{2.38}$$

where N_i is the norm of $u_i(x)$, and is defined as

$$N_i = \int_a^b u_i^2(x)\ dx \tag{2.39}$$

The norms are used to normalize the functions $u_i(x)$, and simply means dividing $u_i(x)$ by $\sqrt{N_i}$. The resulting functions $u_i(x)/\sqrt{N_i}$ are called normalized orthogonal functions, or *orthonormal functions* for brevity.

We can determine c_i of (2.37) in a manner similar to that used to determine a_i. For this, we multiply both sides of (2.37) by $u_j(x)$ and integrate over the range of x; that is,

$$\int_a^b u_j(x)\, f(x)\ dx = \int_a^b \sum_i c_i u_i(x) u_j(x)\ dx$$

$$= \sum_i c_i \int_a^b u_i(x)\, u_j(x)\ dx$$

$$= \sum_i c_i N_j \delta_{ij} \qquad \text{by virtue of (2.38)}$$

$$= N_j c_j$$

or

$$c_j = \frac{\int\limits_a^b f(x)u_j(x)\,dx}{N_j} \tag{2.40}$$

It may be pointed out that, while determining c_j, it is necessary to integrate over the interval (a, b) so that the orthogonality property of the basis functions $u_i(x)$ can be utilized. Analogous to a_j of (2.36), the coefficient c_j is called projection of a function. This point is graphically illustrated in Figure 2.4 where the function $f(x) = \frac{5x}{6} - \frac{x^2}{2} - \frac{x^4}{3}$ and the normalized basis function $u(x) = \sqrt{2}\,\sin(\pi x)$ are plotted over the interval $(0, 1)$. The projection c is defined as

$$(f, u) = \sqrt{2} \int\limits_0^1 f(x)\,\sin(\pi x)\,dx = 0.1997 \tag{2.41}$$

and is given by the area under the curves as shown in the figure. It may be noted that the length of $f(x)$, defined by the norm, is 0.2012. The notation (f, u) is called the inner product of two functions, analogous to the dot product of vectors. Inner product of functions is further illustrated in Section 2.5.

We shall see later that the integral in (2.40) should include a weight function $w(x)$. We have taken the weight function $w(x) = 1$ in (2.40) to simplify the

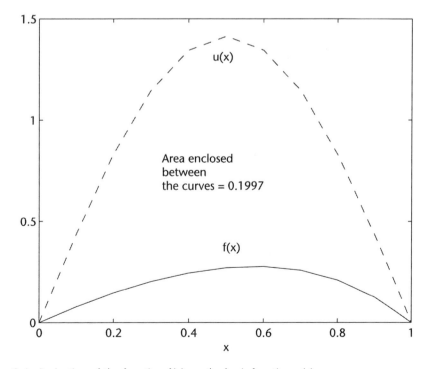

Figure 2.4   Projection of the function $f(x)$ on the basis function $u(x)$.

situation. Orthogonality implies independence. Orthogonal functions are frequently employed in analytical and computational methods.

Examples of Orthogonal Function Expansion

Fourier Series Expansion. Any periodic, differentiable function $f(\theta)$ with period 2π can be expanded in terms of orthogonal functions $\cos(n\theta)$ and $\sin(n\theta)$ as

$$f(\theta) = \frac{a_0}{2} + \sum_{n=1}^{\infty} [a_n \cos(n\theta) + b_n \sin(n\theta)] \qquad -\pi \le \theta \le \pi \qquad (2.42)$$

The series converges uniformly in the interval $-\pi \le \theta \le \pi$. Uniform convergence means convergence for all values of θ. The Fourier coefficients a_n, b_n are evaluated from the following formulae (utilizing the orthogonality properties of expansion functions):

$$a_n = \frac{1}{\pi} \int_{-\pi}^{\pi} f(\theta) \cos(n\theta) \, d\theta \qquad n \ge 0 \qquad (2.43)$$

$$b_n = \frac{1}{\pi} \int_{-\pi}^{\pi} f(\theta) \sin(n\theta) \, d\theta \qquad n \ge 1 \qquad (2.44)$$

First, let

$$f(\theta) = \theta^2 \qquad \text{for } -\pi \le \theta \le \pi \qquad (2.45)$$

Then,

$$\theta^2 = \frac{a_0}{2} + \sum_{n=1}^{\infty} [a_n \cos(n\theta) + b_n \sin(n\theta)] \qquad (2.46)$$

From (2.43) and (2.44) we obtain

$$a_0 = \frac{1}{\pi} \int_{-\pi}^{\pi} \theta^2 \, d\theta = \frac{2}{3} \pi^2 \qquad (2.47a)$$

$$a_n = \frac{1}{\pi} \int_{-\pi}^{\pi} \theta^2 \cos(n\theta) \, d\theta = (-1)^n \frac{4}{n^2} \qquad n \ge 1 \qquad (2.47b)$$

$$b_n = \frac{1}{\pi} \int_{-\pi}^{\pi} \theta^2 \sin(n\theta) \, d\theta = 0 \qquad (2.47c)$$

The coefficients b_n are expected to be zero because they are coefficients of odd symmetry terms $\sin(n\theta)$, and $f(\theta) = \theta^2$ is an even function. Finally,

$$\theta^2 = \frac{a_0}{2} + \sum_{n=1}^{\infty} a_n \cos(n\theta) \tag{2.48}$$

with a_0 and a_n given by (2.47).

Next, consider a periodic square wave signal defined as

$$f(\theta) = \begin{cases} -1 & \text{for } -\pi < \theta < 0 \\ 1 & \text{for } 0 < \theta < \pi \end{cases} \tag{2.49}$$

The Fourier coefficients for this case are

$$a_0 = \frac{1}{\pi} \int_{-\pi}^{\pi} f(\theta)\, d\theta = \frac{1}{\pi} \int_{-\pi}^{0} (-1)\, d\theta + \frac{1}{\pi} \int_{0}^{\pi} (+1)\, d\theta = 0 \tag{2.50a}$$

$$a_n = \frac{1}{\pi} \int_{-\pi}^{0} (-\cos(n\theta))\, d\theta + \frac{1}{\pi} \int_{0}^{\pi} \cos(n\theta)\, d\theta = 0 \tag{2.50b}$$

These coefficients are zero because $f(\theta)$ is an odd function. Similarly,

$$b_n = \frac{1}{\pi} \int_{-\pi}^{0} (-\sin(n\theta))\, d\theta + \frac{1}{\pi} \int_{0}^{\pi} \sin(n\theta)\, d\theta \tag{2.51}$$

$$= \begin{cases} \dfrac{4}{n\pi}, & \text{for } n \text{ odd} \\ 0, & \text{for } n \text{ even} \end{cases}$$

Therefore,

$$f(\theta) = \frac{4}{\pi} \sum_{n=odd}^{\infty} \frac{1}{n} \sin(n\theta) \tag{2.52}$$

The above example shows that a periodic square wave signal consists of suitably weighted odd harmonics, similar to that shown in Figure 2.2 for $\varphi(x, b)$. Other examples of orthogonal functions include Legendre and Chebyshev polynomials.

Orthogonal functions are the solutions of a particular type of differential equation. The origin of orthogonal functions is described next. The emphasis now shifts from solving the PDE to understanding the general properties of the solutions.

2.4   Sturm-Liouville Differential Equation

Orthogonal functions are the solutions of Sturm-Liouville differential equation (S-LDE) with proper boundary conditions. The S-LDE represents most of the useful second order differential equations of physics and engineering. The operator form of eigenvalue S-LDE is given by, for one independent variable,

$$L(y) + \lambda w(x)y(x) = 0 \qquad a \leq x \leq b \qquad (2.53a)$$

where L is the differential operator, the function $y(x)$ is the eigenfunction of operator L, λ is the corresponding eigenvalue, and $w(x)$ is the weight function with $w(x) \geq 0$. The operator L is defined as

$$L(y) = [p(x)y'(x)]' - q(x)y(x) \qquad (2.53b)$$

where ($'$) stands for d/dx. Substituting (2.53b) in (2.53a) yields the following form of S-LDE:

$$[p(x)y'(x)]' - q(x)y(x) + \lambda w(x)y(x) = 0 \qquad (2.54)$$

The coefficients p and q are functions of x. Different expressions for $p(x)$ and $q(x)$ lead to different interesting differential equations. An example is the wave equation or Helmholtz equation

$$\frac{d^2\phi}{dx^2} + k^2\phi = 0 \qquad (2.55)$$

Comparison of (2.55) and (2.53) shows that

$$L \equiv \frac{d^2}{dx^2} + k^2, \; y(x) \equiv \phi, \; w(x) = 1 \qquad (2.56)$$

The eigenvalue equation corresponding to (2.55) may then be written as

$$\frac{d^2y}{dx^2} + k^2y + \lambda w(x)y(x) = 0 \qquad (2.57)$$

Generalizing the wave equation to three-dimensions, d^2/dx^2 of (2.55) is replaced by ∇^2 which may be expressed in rectangular, cylindrical, spherical, or any other separable coordinate system, and the corresponding expressions for the coefficient functions p, q, and w obtained from comparison.

2.4.1   Orthogonality of Eigenfunctions

Let us consider two different solutions of S-LDE (2.54) with eigenfunctions $y_m(x)$ and $y_n(x)$ and the corresponding eigenvalues λ_m and λ_n. Since $y_m(x)$ and $y_n(x)$ are eigenfunctions, they satisfy (2.54); that is,

$$\frac{d}{dx}\left[p(x)\frac{dy_m(x)}{dx}\right] - q(x)y_m(x) + \lambda_m w(x)y_m(x) = 0 \qquad (2.58a)$$

$$\frac{d}{dx}\left[p(x)\frac{dy_n(x)}{dx}\right] - q(x)y_n(x) + \lambda_n w(x)y_n(x) = 0 \qquad (2.58b)$$

We now use a standard procedure to derive the orthogonality condition for the eigenfunctions $y_m(x)$ and $y_n(x)$. In this procedure, we multiply (2.58a) by $y_n(x)$ and (2.58b) by $y_m(x)$, subtract the resulting equations, and integrate over the range (a, b) to obtain

$$\int_a^b \left\{y_n\frac{d}{dx}\left[p(x)\frac{dy_m}{dx}\right] - y_m\frac{d}{dx}\left[p(x)\frac{dy_n}{dx}\right]\right\} dx = [\lambda_m - \lambda_n]\int_a^b w(x)y_m y_n \, dx \qquad (2.59)$$

Integration by parts of the left side yields

$$LHS = y_n p(x)\frac{dy_m}{dx}\bigg|_a^b - y_m p(x)\frac{dy_n}{dx}\bigg|_a^b$$

or

$$LHS = p(x)\left[y_n(x)y_m' - y_m(x)y_n'\right]_a^b \qquad (2.60)$$

The value of this expression depends on the eigenfunctions and their derivatives. If this value happens to vanish due to the boundary conditions on $y_m(x)$ and $y_n(x)$ or their derivatives, then (2.59) becomes

$$[\lambda_m - \lambda_n]\int_a^b y_m w(x)y_n \, dx = 0 \qquad (2.61)$$

Since $\lambda_m \neq \lambda_n$, in general, (2.61) can be satisfied if

$$\int_a^b y_m w(x)y_n \, dx = 0 \qquad (2.62)$$

Equation (2.62) *is the orthogonality condition of eigenfunctions* $y_m(x)$ *and* $y_n(x)$ *with respect to the weight function* $w(x)$.

2.4.2 Boundary Conditions for Orthogonal Functions

It is clear from the above presentation that the orthogonality of eigenfunctions of S-LDE is subject to vanishing of (2.60); that is,

$$p(x)\left[y_n(x)y'_m - y_m(x)y'_n\right]_a^b = 0 \tag{2.63}$$

The above equation can be satisfied by a variety of conditions, namely:

(a) The functions $y_m(x)$ and $y_n(x)$ vanish at $x = a$, and $x = b$. These boundary conditions are called the Dirichlet conditions. An example of the Dirichlet boundary condition is $y_m(x) = 0$, and it may be interpreted as

$$E_{\tan} = 0, \qquad \text{on a perfect electric conductor} \tag{2.64a}$$

$$H_{\tan} = 0, \qquad \text{on a perfect magnetic conductor} \tag{2.64b}$$

(b) The functions $y'_m(x)$ and $y'_n(x)$ vanish at $x = a$, and $x = b$. These boundary conditions are called the Neumann conditions. The Neumann boundary condition $dy_m/dn = 0$ can be satisfied by requiring that at a perfect electric conductor,

$$\frac{dH_{\tan}}{dn} = 0 \tag{2.65}$$

where $\hat{n}$ is the unit vector normal to the boundary surface.

(c) Dirichlet condition specified, say for $n = a$ and Neumann condition for $y = b$. Let $y_m(x)$ and $y_n(x)$ be zero at $x = a$, and $y'_m(x)$ and $y'_n(x)$ are zero at $x = b$.

(d) A linear combination of $y_m(x)$ and its derivative $y'_m(x)$ vanish at $x = a$, and $x = b$; that is,

$$y_m(a) + \alpha y'_m(a) = 0 \qquad m = 1, 2, 3, \ldots \tag{2.66a}$$

$$y_m(b) + \beta y'_m(b) = 0 \qquad m = 1, 2, 3, \ldots \tag{2.66b}$$

where α and β are constants. These boundary conditions are called mixed type or Cauchy boundary conditions. We come across these boundary conditions at the surface of an imperfect electric conductor. It is expressed as

$$\frac{dy}{dn} + \alpha y \equiv \frac{dE}{dn} + \alpha E = 0 \tag{2.67}$$

There are other boundary conditions which satisfy (2.63). For these the reader is referred to [4].

2.4.3   Examples of Sturm-Liouville Type of Differential Equations

The differential equations of electromagnetics are the special cases of Sturm-Liouville differential equation. Some of these are described next.

Voltage V(z) on a Transmission Line.   Consider a transmission line of finite length with propagation constant γ. The differential equation satisfied by $V(z)$ is

$$\frac{d^2V}{dz^2} + \gamma^2 V(z) = 0 \qquad (2.68)$$

It is a S-LDE with

$$L = \frac{d^2}{dz^2}, \ y = V(z), \ p = 1, \ q = 0, \ w = 1, \text{ and } \lambda = \gamma^2 \qquad (2.69)$$

The transmission line may be terminated at its ends, and the terminations will determine the boundary conditions leading to unique solutions. If the line is terminated, in short-circuits at its ends at $z = a$ and $z = b$; the corresponding boundary conditions are

$$V(a) = 0, \ V(b) = 0 \qquad \text{Dirichlet type} \qquad (2.70a)$$

When the line is terminated in open circuits, the boundary conditions become

$$\left.\frac{dV}{dz}\right|_{z=a} = 0, \left.\frac{dV}{dz}\right|_{z=b} = 0 \qquad \text{Neumann type} \qquad (2.70b)$$

For the load impedance at the ends, the boundary conditions can be expressed as

$$\frac{dV}{dz} + \gamma V = 0 \qquad (2.70c)$$

For example, let the terminating impedance at $z = b$ be Z_L. Then,

$$Z_L = \frac{V(b)}{I(b)}, \text{ and } I(b) = -\frac{dV}{dz} \qquad (2.71)$$

Rearranging the expression for Z_L gives

$$V(b) + Z_L \frac{dV}{dz} = 0 \qquad (2.72)$$

and is in a form similar to (2.70c). The solutions $V(z)$ of (2.68) are orthogonal functions.

Current $I(z)$ on a Transmission Line.   The differential equation satisfied by $I(z)$ on a transmission line is

$$\frac{d^2I}{dz^2} + \gamma^2 I(z) = 0 \qquad (2.73)$$

The boundary conditions imposed by the terminations in this case are

$$I(z) = 0 \qquad \text{at the open circuit} \qquad (2.74a)$$

$$\frac{dI}{dz} = 0 \qquad \text{at the short circuit, and} \qquad (2.74b)$$

$$\frac{dI}{dz} + \gamma I = 0 \qquad \text{at the terminal admittances} \qquad (2.74c)$$

Again the solutions $I(z)$ are orthogonal functions.

Plane Wave Propagation Normal to a Boundary.   Let us consider a plane wave incident normally on an interface as shown in Figure 2.5. The propagation phenomenon is described by the differential equation

$$\nabla^2 f(r) + k^2 f(r) = 0 \qquad (2.75)$$

This is an S-LDE with

$$L \equiv \nabla^2, \; y = f(r), \; p = 1, \; q = 0, \; w = 1, \text{ and } \lambda = k^2$$

The function $f(r)$ may represent electric or magnetic field vector. The boundary condition, in this case, is decided by the interface. For simplicity, let us consider the interface as air-metal boundary. Then, $E_{\tan}$ should be zero at the metal surface, and implies that,

$$f = 0, \text{ or } E_{\tan} = 0, \text{ if } f \text{ represents the electric field} \qquad (2.76a)$$

$$\frac{df}{dz} = 0, \text{ or } \frac{dH_{\tan}}{dz} = 0, \text{ if } f \text{ represents the magnetic field} \qquad (2.76b)$$

Waves Guided by Perfect Electric Conductors [5].   Consider a waveguide of perfect electric conductor as shown in Figure 2.6. The modes guided by this waveguide can be classified as *TM-to-z* and *TE-to-z*. These are described next in context to the boundary conditions.

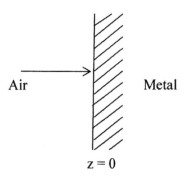

Air   Metal

$z = 0$

Figure 2.5   Wave incident normal to a metal-air interface.

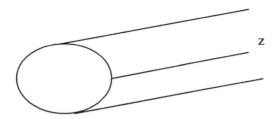

Figure 2.6   Configuration of a cylindrical waveguide.

TM-to-z modes.   For wave propagation along the z-direction and described by $e^{\pm jk_z z}$, the differential equation is given by

$$\left(\nabla_t^2 + k^2 - k_z^2\right)A_z = 0 \tag{2.77a}$$

where A_z is the z-component of the magnetic vector potential. Here, $L \equiv \nabla_t^2$ and $\lambda = k^2 - k_z^2$. The boundary conditions satisfied by A_z are

$$E_z = 0 \text{ or } A_z = 0 \text{ on the metal surface} \tag{2.77b}$$

TE-to-z modes.   The differential equation for these modes is

$$\left(\nabla_t^2 + k^2 - k_z^2\right)F_z = 0 \tag{2.78a}$$

where F_z is the z-component of the electric vector potential. Again, $L \equiv \nabla_t^2$ and $\lambda = k^2 - k_z^2$. The boundary conditions to be satisfied on the metal surface are

$$\frac{\partial F_z}{\partial n} = 0 \tag{2.78b}$$

where $\hat{n}$ is the unit normal vector to the waveguide surface. Similar wave equations and boundary conditions can be written for the waves guided by a dielectric slab.

2.5   Eigenfunction Expansion Method

The homogeneous and nonhomogeneous differential equations may be solved by using the eigenfunction expansion method. In this method we expand the unknown function and the excitation in terms of the eigenfunctions of the operator and the orthogonality property of the eigenfunctions is used to determine the expansion coefficients. We illustrate this procedure next by an example.

Linear Resonator Problem

Consider a linear resonator of length l subjected to excitation $g(x)$. The amplitude distribution on the resonator is described by the following differential equation:

$$\frac{d^2y}{dx^2} + k^2 y = -g(x), \qquad 0 \le x \le l \tag{2.79}$$

Let the resonator be terminated at the ends such that the amplitude at these points is zero, giving rise to the boundary conditions $y(0) = y(l) = 0$. To solve the differential equation we use the method of eigenfunction expansion. For this, we expand $y(x)$ and $g(x)$ in terms of the eigenfunctions of the operator d^2/dx^2. The eigenfunctions are obtained by solving the corresponding eigenvalue differential equation

$$\frac{d^2 \psi_n}{dx^2} + k^2 \psi_n = \lambda_n \psi_n \qquad \psi_n(0) = \psi_n(l) = 0 \tag{2.80}$$

where λ_n is the eigenvalue corresponding to the eigenfunction $\psi_n(x)$. One obtains

$$\psi_n(x) = \sqrt{\frac{2}{l}} \sin\left(\frac{n\pi x}{l}\right) \qquad n = 1, 2, 3, \ldots \tag{2.81}$$

Choosing this set as a basis function set, we write

$$y(x) = \sum_{n=1}^{\infty} y_n \psi_n(x) \tag{2.82}$$

and

$$g(x) = \sum_{n=1}^{\infty} g_n \psi_n(x) \tag{2.83}$$

Substituting these expansions in (2.79) yields

$$\sum_{n=1}^{\infty} \left[k^2 - \left(\frac{n\pi}{l}\right)^2 \right] y_n \psi_n(x) = -\sum_{n=1}^{\infty} g_n \psi_n(x) \tag{2.84}$$

Comparing both sides term by term we obtain the following relation between the coefficients y_n and g_n:

$$\left[k^2 - \left(\frac{n\pi}{l}\right)^2 \right] y_n = -g_n \tag{2.85}$$

Alternatively, one can use orthogonality property of $\psi_n(x)$ to obtain the above expression. Now substitute the value of y_n from (2.85) in (2.82) to complete the solution:

$$y(x) = \sum_{n=1}^{\infty} \frac{g_n}{[n\pi/l]^2 - k^2} \sqrt{\frac{2}{l}} \sin\left(\frac{n\pi x}{l}\right) \tag{2.86}$$

The eigenvalues corresponding to the eigenfunctions can be obtained by substituting (2.86) in (2.80):

$$\lambda_n = k^2 - \left(\frac{n\pi}{l}\right)^2 \qquad n = 1, 2, 3, \ldots \tag{2.87}$$

The eigenvalues are such that $\lambda_1 > \lambda_2 > \ldots \lambda_n$.

Completeness and Convergence
The set of eigenfunctions (2.81) forms a complete set, that is, any function such as $y(x)$ which can be composed of eigenfunctions, can be constructed to the desired accuracy [6, p. 26]. The series (2.86) must therefore *converge uniformly* by the completeness of eigenfunctions set. By uniform convergence we mean that $y(x)$ of (2.86) should approach exact solution $y_{exact}(x)$ for each point x as the value of integer n approaches infinity.

Example 2.1. Let us test the convergence property of the solution of the differential equation

$$-\frac{d^2y}{dx^2} = 1 + 4x^2, \qquad 0 \le x \le 1 \tag{2.88}$$

subject to $y(0) = y(1) = 0$. The exact solution obtained by direct integration is

$$y(x) = \frac{5x}{6} - \frac{x^2}{2} - \frac{x^4}{3}$$

Solution. Comparing the differential equations (2.79) and (2.88) we find that $k = 0, f = 1 + 4x^2, l = 1$. Although the operators in these differential equations are different; $-d^2/dx^2$ and $-d^2/dx^2 - k^2$, their eigenfunctons are same (being governed by d^2/dx^2 and the boundary conditions) and the eigenvalues differ by constant k^2. The solution of (2.88) is, therefore, given by

$$y(x) = \sum_{n=1}^{\infty} \frac{g_n}{(n\pi)^2} \sqrt{2} \sin(n\pi x) \tag{2.89}$$

where the coefficients g_n are obtained by expanding the excitation function as

$$1 + 4x^2 = \sum_{n=1}^{\infty} g_n \psi_n(x)$$

or

$$g_n = \sqrt{2} \int_0^1 (1 + 4x^2) \sin(n\pi x) \, dx$$

or

$$g_n = \sqrt{2} \begin{cases} -\dfrac{4}{n\pi} & \text{for } n \text{ even} \\[2ex] \dfrac{2}{n\pi}\left(3 - \dfrac{8}{(n\pi)^2}\right) & \text{for } n \text{ odd} \end{cases} \tag{2.90}$$

Let us determine if the series solution (2.89) converges uniformly by comparing it to the exact value $y(x) = \dfrac{5x}{6} - \dfrac{x^2}{2} - \dfrac{x^4}{3}$. The solutions are compared in Figure 2.7 for $n = 6$. The convergence with n in this case is so fast that $n = 6$ is sufficient to prove convergence of the series. Therefore, we have not plotted the solution for higher values of n. The agreement is seen to be very good for all values of x, and the series *converges uniformly* as predicted by the completeness of the basis set $\{\psi_n, n = 1, 2, 3, \ldots \}$. A less restrictive criterion for convergence of a series is called *convergence in the mean or average sense*. It is defined as

$$\lim_{m \to n} \int_0^1 \left[y(x) - \sum_{i=1}^m c_i \psi_i(x) \right]^2 dx = 0 \tag{2.91}$$

The inhomogeneous differential equations discussed above can be written as the operator equation $L(f) = g$. Its solutions were obtained by expanding the

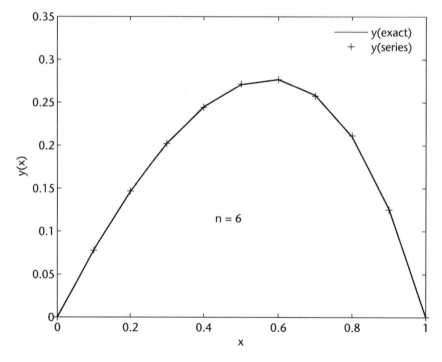

Figure 2.7 Comparison of the exact solution with the series solution.

functions f and g in eigenfunction set $\{\psi_n\}$ of the operator L. The solutions other than eigenfunction set are also possible. The set $\{\psi_n\}$ may be thought of as a specific case of vector space or function space for L. The set of functions are said to constitute a function space if its elements satisfy certain properties to be discussed shortly.

The solution of operator equation is best described in terms of functional analysis [7, 8]. The concepts involved are abstract in nature, and we shall try to bring physical meaning to these concepts by interpreting the function space in terms of eigenfunction set. Example 2.1 will be used as a case study to make this exercise interesting.

2.6 Vector Space/Function Space

Associated with each operator L is a function space or vector space or a set of functions. This space is called the domain of L, the admissible elements of which satisfy the properties of the operator. For example, the domain of operator equation $L(f) = -d^2f/dx^2$ with boundary conditions $f(0) = f(1) = 0$ over the interval $0 \leq x \leq 1$ consists of functions which are differentiable at least twice and are zero at $x = 0$ and $x = 1$. These functions, if linearly independent, are called the basis functions and are said to *span* the space. The basis functions are analogous to the unit vectors of the three-dimensional Euclidean space using which any three-dimensional vector $\vec{f}$ can be configured as $\vec{f} = a\hat{x} + b\hat{y} + c\hat{z}$. The eigenfunction set $\{\psi_n\}$ of (2.81) meets the requirements stated above, and are linearly independent. This set can be used as the basis set for the domain of L. In order to help visualize function space we have plotted the first four basis functions of the set $\left\{\sqrt{2}\,\sin(n\pi x),\ n = 1, 2, 3, \ldots\right\}$ in Figure 2.8. This figure also shows an arbitrary function $y(x)$ which is a linear combination of this basis set. The function space resulting from the operation $L\psi_n(x)$ is called the *range* of operator L. The range space may or may not be the same as the domain. It depends on the operator and the choice of basis functions for the domain. For the operator $L = -d^2/dx^2$ and the domain described by $\{\sin(m\pi x)\}$, the range space is also $\{\sin(n\pi x)\}$. The function space is frequently referred to as vector space in the literature because of the similarities between the basis functions and basis vectors and mathematical operations on them.

The function space $\{\psi_n\}$ is characterized by some more properties which are defined next. We shall restrict ourselves to linear function spaces because the operators we are using are linear.

Dimensionality of Function Space.   The dimensionality of a linear function space is given by the maximum number of linearly independent elements in it. The set $\{\sin(m\pi x),\ m = 1, 2, 3, \ldots, N\}$ is N-dimensional because all the elements are independent of each other over $(0, 1)$. The dimensionality of the set $\{1, x, x^2, x^3, \ldots x^n\}$ is also n.

Completeness of a Set.   A set of elements is said to be complete if any well-behaved function $y(x)$ in the space can be approximated by a series *to any desired*

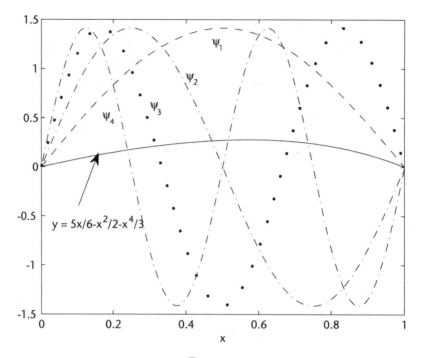

Figure 2.8 Plot of the basis functions $\psi_i = \sqrt{2}\,\sin(i\pi x)$ for the four-dimensional function space. An arbitrary function $y(x)$ is also plotted in this function space.

accuracy; that is, $y(x) = \sum\limits_{i=1}^{n} a_i\psi_i(x)$. More precisely, the set $\{\psi_i\}$ is complete if the mean square error vanishes in a limiting sense,

$$\lim_{m \to n} \int \left[y(x) - \sum_{i=1}^{m} a_i\psi_i(x) \right]^2 dx = 0 \qquad (2.92)$$

Orthogonality of Functions.   The orthogonality of functions was illustrated in Section 2.3. Any pair of functions are said to be orthogonal over the interval (a, b) if

$$\int_{a}^{b} f_i(x)g_j(x)\,dx = c\delta_{ij} = \begin{cases} c & \text{for } i = j \\ 0 & \text{for } i \neq j \end{cases} \qquad (2.93)$$

where c is a constant, and equals 1 for normalized functions. The basis set $\{e^{jnx},\ -\infty \leq n \leq \infty,\ 0 \leq x \leq 2\pi\}$ is orthogonal with $c = 2\pi$. Orthogonality of functions implies their linear independence, but not vice versa. For example, the set $\{1, x, x^2, x^3, \ldots x^n\}$ is linearly independent but not orthogonal over the interval $(0, 1)$. Any nonorthogonal set can be made orthogonal using the *Gram-Schmidt* orthogonalization procedure [9].

Vector Representation of a Function. Any well-behaved function $y(x)$ in the function space can be decomposed as

$$y(x) = \sum_{i=1}^{n} a_i \psi_i(x) \qquad (2.94)$$

This expression may be written in the vector form as

$$y = [a_1, a_2, a_3, \ldots a_n][\psi_1, \psi_2, \psi_3, \ldots \psi_n]^t \qquad (2.95)$$

where $a_1, a_2, a_3, \ldots a_n$ are called the components of function y in the space spanned by ψ_i. The component values are obtained from (2.94) by using orthogonality of functions. The vector representation of function is very convenient for use with computers. One may use nonorthogonal basis functions and determine the coefficient vector by taking the inner product of $y(x)$ with the basis functions. The inner product of two functions is discussed next.

Having described the structure of a function space in the form of basis vectors, we now define some of the important operations on functions. These operations are needed in the course of their usage. The inner product between a pair of functions is of central importance. The inner product and the various properties based on this are discussed next.

Inner Product. The inner product or scalar product of any pair of functions f and g in a space is denoted by (f, g) and is defined as

$$(f(x), g(x)) = \int_a^b f(x)g(x)\, dx \qquad (2.96)$$

As illustrated in Section 2.3, the inner product describes the projection of one function on the other. The physical meaning of projection of a function was also discussed there. The projection of a function onto itself is unity if the function is normalized.

The inner product operation makes it possible to decompose an arbitrary function into basis functions by taking projections. For example, the inner product has been used to determine the coefficients when $y(x) = \dfrac{5x}{6} - \dfrac{x^2}{2} - \dfrac{x^4}{3}$ is projected onto the basis functions $\psi_1 = \sqrt{2}\,\sin(\pi x)$ and $\psi_2 = \sqrt{2}\,\sin(2\pi x)$ over the interval $(0, 1)$. The result is

$$y_2(x) = 0.1997 \sin(\pi x) - 0.0228 \sin(2\pi x) \qquad (2.97)$$

where $y_2(x)$ denotes two-term approximation of $y(x)$. The error, determined by the remainder terms, is orthogonal to both ψ_1 and ψ_2 since the remainder terms correspond to the projections of $y(x)$ on ψ_3, ψ_4, ... Because of the orthogonal nature of projections, the error is of second order [9, p. 18].

Expressing the functions in basis set, the inner product may be written in the form of vector product as

$$(f(x), g(x)) = \int_a^b f(x)g(x)\, dx = \int_a^b \sum_i a_i \psi_i \sum_j b_j \psi_j\, dx$$

or

$$= \sum_{i,j} a_i b_j \int_a^b \psi_i \psi_j\, dx = \sum_i a_i b_i$$

or

$$(f, g) = [a][b]^t \tag{2.98}$$

where $[a]$ and $[b]$ are coefficients vectors corresponding to the functions f and g, respectively, and the superscript t stands for transpose.

Norm of a Function.   The norm of a function is a scalar quantity and is a measure of its length. By taking the inner product of a function with itself, we obtain its norm as

$$\|f\| = \sqrt{(f, f)} = \sqrt{[a][a]^t} \tag{2.99}$$

$$= \sqrt{\sum_{i=1}^n |a_i|^2}$$

The expression (2.99) is obtained by replacing g by f in (2.98). The norm of the function $y(x) = \dfrac{5x}{6} - \dfrac{x^2}{2} - \dfrac{x^4}{3}$ over $(0, 1)$ is found to be 0.2012.

Metric.   The norm of a function may be used to measure the distance between any pair of functions and is called a *metric*. It is defined as

$$d(f, g) = \|f - g\| = \sqrt{(f - g, f - g)} \tag{2.100}$$

Let f represent the exact function and g represent its approximation, then metric $d(f, g)$ determines the accuracy of this approximation. We now use metric to determine the accuracy of approximating $y(x) = \dfrac{5x}{6} - \dfrac{x^2}{2} - \dfrac{x^4}{3}$ by the series $y_{approx}(x) = \displaystyle\sum_{n=1}^{\infty} \dfrac{g_n}{(n\pi)^2} \sqrt{2}\, \sin(n\pi x)$, where g_n is given by (2.90). The series approximation of $y(x)$ for $n = 4$ is given by

$$y_4(x) = 0.1997 \sin(\pi x) - 0.0228 \sin(2\pi x) + 0.0098 \sin(3\pi x) \quad (2.101)$$
$$- 0.00285 \sin(4\pi x)$$

The metric $d(y - y_4)$ is found to be 0.0590253, and $d(y - y_3) = 0.0590882$. Since the metric is hardly changing with the increase in the value of n, we may say that the series approximation is very good for $n = 4$. Another criterion to measure solution accuracy is to determine $\|f - g\|^2$. We obtain $\|f - g\|^2 = (0.0590882)^2 = 0.0035$ for $n = 3$.

Schwarz Inequality.   It is defined as

$$\|fg\| \leq \|f\| \cdot \|g\| \quad (2.102)$$

The equality sign holds when the function f is a multiple of function g.

2.6.1   Operators

We had introduced differential operators as part of the differential equations. Some specific differential operators are: ∇^2 and $\nabla^2 + k^2$ for scalar fields, and $L = \nabla \times \nabla - k_0^2 \mu_r \epsilon_r$ for the vector fields in homogeneous media. The operators are called linear if they satisfy the linearity conditions. The integral operators are derived from either Green's functions or functionals. Common examples of integral operators are the integral equations and integral transforms of the following type:

$$f(\alpha) = \int_a^b g(t) K(\alpha, t) \, dt \quad (2.103)$$

where $K(\alpha, t)$ is called the kernel and may be replaced by the Green's function. The integral operators may also be called the inverse operator L^{-1} of the corresponding differential operator L. We shall come across integral operators in Chapter 5 on Fourier transform method, solution of integral equations using method of moments (Chapter 11), and variational methods (Chapter 9). Another class of operators is the integro-differential operators. As the name implies, they combine both the integral and differential operations as in (see Chapter 11)

$$E_z = \frac{1}{j\omega\epsilon_0} \int_{-\ell/2}^{\ell/2} \left[\frac{\partial^2 G(z, z')}{\partial z^2} + k^2 G(z, z') \right] I(z') \, dz' \quad (2.104)$$

Properties of Operators
An operator defines mapping of function in one space to the mapping in another space. The space of functions on which the operator operates is called the domain of the operator, and the space of functions resulting from the operation is called the range of the operator. In the operator equation $Lf = g$, the domain of L is the space F of functions f and the range of L is the space G of functions g. The functions

f should satisfy the boundary conditions and should be differentiable if L is a differential operator. However, no such conditions are imposed on functions g.

Linear Operators.   An operator is said to be linear if it satisfies the linearity properties such as

$$L(f + g) = Lf + Lg \tag{2.105a}$$

$$L(\alpha f) = \alpha Lf \tag{2.105b}$$

The differential operators ∇^2 and $\nabla^2 + k^2$ satisfy the above properties and are linear. The operation $\sin(.)$ is not linear because $\sin(\theta + \varphi) \neq \sin\theta + \sin\varphi$.

Real Operators.   An operator is said to be real if Lf is real whenever f is real. If

$$(f^*, Lf) > 0, \text{ the operator is positive definite} \tag{2.106a}$$

$$\geq 0, \text{ the operator is positive semi-definite} \tag{2.106b}$$

$$< 0, \text{ the operator is negative definite} \tag{2.106c}$$

Adjoint Operators.   The adjoint of an operator, L^a defined as

$$(Lf, g) = (f, L^a g) \tag{2.107}$$

for all functions f in the domain of L. The adjoint operator corresponding to the Sturm-Liouville type operator

$$L = p_0(x)\frac{d^2}{dx^2} + p_1(x)\frac{d}{dx} + p_2(x) \tag{2.108}$$

is given by [10, p. 498]

$$L^a = p_0\frac{d^2}{dx^2} + \left(2p_0' - p_1\right)\frac{d}{dx} + \left(p_0'' - p_1' + p_2\right) \tag{2.109}$$

An operator is said to be self-adjoint if $L^a = L$, and naturally the domain of L^a is that of L. Comparing (2.108) and (2.109), one finds that the operator L is self-adjoint if the following condition is satisfied:

$$p_0'(x) = p_1(x) \tag{2.110}$$

We illustrate these properties next.

Example 2.2.   Consider the differential equation

$$-\frac{d^2 f}{dx^2} = g, \qquad 0 \leq x \leq 1 \tag{2.111}$$

with the boundary conditions $f(0) = f(1) = 0$ [9].

The operator in the given differential equation is $L = -\dfrac{d^2}{dx^2}$. The eigenfunctions

of the operator are obtained by solving the eigenvalue equation $-\dfrac{d^2 \psi_n}{dx^2} = \lambda \psi_n$

subject to the boundary conditions stated above. The normalized eigenfunctions are found to be $\{\sqrt{2} \sin(n\pi x), n = 1, 2, 3, \ldots \}$. These eigenfunctions are orthogonal, span the space, and can be used as the basis functions to express any arbitrary function in the space. This set is complete also.

The operator is a real operator, since $Lf = -\dfrac{d^2}{dx^2}(\sin(n\pi x)) = n^2 \pi^2 f$ is real.

The set of eigenfunctions given above describes the domain as well as the range of L. Also, L is a positive definite operator, since

$$(f^*, Lf) = -\int_0^1 f^* \frac{d^2 f}{dx^2} \, dx$$

$$= -f^* \left.\frac{df}{dx}\right|_0^1 + \int_0^1 \frac{df}{dx} \frac{df^*}{dx} \, dx \tag{2.112}$$

$$= \int_0^1 \left|\frac{df}{dx}\right|^2 \, dx \text{ by virtue of the boundary conditions on } f$$

$$> 0$$

Adjoint Operator of L.

$$(Lf, g) = \left(-\frac{d^2 f}{dx^2}, g\right)$$

$$= -\int_0^1 \frac{d^2 f}{dx^2} g(x) \, dx$$

$$= -g \left.\frac{df}{dx}\right|_0^1 + \int_0^1 \frac{df}{dx} \frac{dg}{dx} \, dx$$

$$= -g \left.\frac{df}{dx}\right|_0^1 + f \left.\frac{dg}{dx}\right|_0^1 - \int_0^1 f \frac{d^2 g}{dx^2} \, dx$$

$$= -g \left.\frac{df}{dx}\right|_0^1 - \int_0^1 f \frac{d^2 g}{dx^2} \, dx$$

by virtue of the boundary conditions on f. If we choose $g(0) = g(1) = 0$, then

$$(Lf, g) = -\int_0^1 f \frac{d^2 g}{dx^2} \, dx = \left(f, -\frac{d^2 g}{dx^2} \right) = (f, Lg) \tag{2.113}$$

Comparing this expression with the definition of adjoint operator, (2.107), we find that in this case

$$L^a = L = -\frac{d^2}{dx^2} \tag{2.114}$$

The conditions $g(0) = g(1) = 0$ imply that the domain of adjoint operator is the same as that of L.

The self-adjoint nature of this operator is also evident from the fact that the condition (2.110) is satisfied by $L = -d^2/dx^2$ ($p_0 = 1$ and $p_1 = 0$).

A self-adjoint operator is also called Hermitian operator with the important property that the eigenfunctions of a Hermitian operator are orthogonal and they form a complete set. Since the operator $L = -d^2/dx^2$ is Hermitian, its eigenfunction set $\{ \sqrt{2} \sin(n\pi x), n = 1, 2, 3, \dots \}$ is orthogonal over $(0, 1)$ and is a complete set. This has been proved independently earlier.

Integral Operator or Inverse Operator of L. By definition, inverse operator for $Lf = g$ is given by $f = L^{-1}g$. The inverse operator in electromagnetics can be described as the Green's function G for the operator L and is denoted as

$$L^{-1} = \int_0^1 dx' G(x, x') \tag{2.115}$$

Sometimes, Green's function is also called the Green's operator. Methods to obtain Green's function for a given differential equation are discussed in Chapter 3. For $L = -d^2/dx^2$, the Green's function is given by (3.17)

$$G(x; x') = \begin{cases} x(1 - x'), & \text{for } x < x' \\ x'(1 - x), & \text{for } x > x' \end{cases} \tag{2.116}$$

Boundary conditions are not required for the domain of integral operators, because these are built into the Green's function. The inverse operator is self-adjoint and positive definite whenever L is self-adjoint and positive definite [9, p. 5]. The eigenfunctions of the differential operators are also the eigenfunctions of the integral operators and vice versa [7, p. 23]. If the eigenvalues of differential operators are λ_n, the eigenvalues of the integral operators are $1/\lambda_n$ [7, p. 23].

The Poisson equation $-\epsilon \nabla^2 \varphi = \rho$ is a well-known example of the three-dimensional version of the differential equation (2.88). The operator here is $L = -\epsilon \nabla^2$.

This operator is positive definite and self-adjoint [9]. The inverse problem corresponding to the Poisson equation is the integral equation

$$\varphi(x, y, z) = \iiint \frac{\rho(x', y', z')}{4\pi\epsilon R}\, dx'\, dy'\, dz' \tag{2.117}$$

where $R = \sqrt{(x - x')^2 + (y - y')^2 + (z - z')^2}$, and $1/(4\pi\epsilon R)$ is the Green's function. The inverse operator to $L = -\epsilon\nabla^2$ is therefore

$$L^{-1} = \iiint dx'\, dy'\, dz'\, \frac{1}{4\pi\epsilon R} \tag{2.118}$$

2.6.2   Matrix Representation of Operators

We have discussed the vector representation of a function, (2.95). If the operator L can be represented by a matrix, the operation Lf may be carried out using matrix-vector product. Using (2.94) for f, the operation Lf may be written as

$$Lf = L\left(\sum_{j=1}^{n} \alpha_j \psi_j\right) \tag{2.119}$$

$$= \sum_{j=1}^{n} \alpha_j L\psi_j$$

assuming the operator to be linear. This expression implies that the operation L on any function is completely specified by the effect of L on the basis. Let the operator L transform the basis ψ_i to the function h_i; that is,

$$L(\psi_i) = h_i \tag{2.120}$$

Also, the function h_i can be expanded similar to (2.94) as

$$h_i = \sum_{j=1}^{n} a_{ji}\psi_j \qquad i = 1, 2, 3, \ldots, n$$

or

$$L(\psi_i) = \sum_{j=1}^{n} a_{ji}\psi_j \qquad i = 1, 2, 3, \ldots, n \tag{2.121}$$

The above expression signifies that the operator L can be described in terms of n^2 numbers a_{ji}; $i, j = 1, 2, 3, \ldots n$, which can be arranged in the form of a matrix as

$$[A] = \begin{bmatrix} a_{11} & a_{12} & . & . & a_{1n} \\ a_{21} & a_{22} & . & . & a_{2n} \\ . & . & . & . & . \\ . & . & . & . & . \\ a_{n1} & a_{n2} & . & . & a_{nn} \end{bmatrix}$$

$$\uparrow$$
$$\text{vector } a_2 = Lu_2$$

(2.122)

The matrix A is the discrete form of operator with matrix elements a_{ji}.

Employing the matrix representation of the operator L, we can now determine $g = Lf$ in the vector form as

$$[g] = [A][\alpha]^t$$

(2.123)

With the choice of eigenfunctions as basis functions, the functions f and g can be represented by column vectors as

$$f = [\alpha_1 \quad \alpha_2 \quad \alpha_3 \quad \ldots \quad \alpha_n]^t$$

(2.124a)

$$g = [\beta_1 \quad \beta_2 \quad \beta_3 \quad \ldots \quad \beta_n]^t$$

(2.124b)

and the operator L now becomes a diagonal matrix A

$$A = \begin{bmatrix} \lambda_1 & 0 & 0 & . & . & 0 \\ 0 & \lambda_2 & 0 & . & . & 0 \\ 0 & 0 & \lambda_3 & 0 & . & 0 \\ . & . & . & . & . & . \\ . & . & . & . & . & . \\ 0 & 0 & 0 & . & . & \lambda_n \end{bmatrix}$$

(2.124c)

with λ_n ($n = 1, 2, 3, \ldots$) as the eigenvalues.

The operator equation $Lf = g$ may now be written in matrix form as

$$[A][\alpha] = [\beta]$$

(2.125)

For a given problem $[A]$ and $[\beta]$ are known, and the unknown $[\alpha]$ is obtained as

$$[\alpha] = [A]^{-1}[\beta]$$

(2.126)

Specifically, if the eigenfunction set is employed as the function space, the matrix A is diagonal as described by (2.124c). The inverse of a diagonal matrix is also diagonal with nonzero elements $a_{ii} = 1/\lambda_i$. The vector $[\alpha]$ is therefore given by

$$\alpha_i = \beta_i/\lambda_i$$

(2.127)

and the solution

$$f(x) = \sum_{i=1}^{n} \alpha_i \psi_i(x) \tag{2.128}$$

This expression is identical with (2.86). The vector-matrix form of the present solution is very convenient for computer implementation.

The use of eigenfunctions as the basis for the function space limits the application of this concept to only those problems for which eigenfunctions can be determined. This leaves out a large number of useful geometries and requires that the eigenfunctions be determined first. The expression (2.122) is general and holds for any type of basis functions. The coefficients a_{ji} in (2.121) can be determined by employing inner product and this is discussed in Chapter 6. The matrix representation of operators based on subdomain function space is discussed in Chapter 6.

Relationship Between Operators, Eigenvalues and Matrices.   The operators are characterized by eigenfunctions and the corresponding eigenvalues. Some of the important operators and the properties of their eigenvalues (λ) are given in Table 2.1 [11, p. 32].

An operator can be described by its matrix, and is diagonal if eigenfunctions are employed as basis. Otherwise, the matrix is nondiagonal. Some of the common matrix types characterizing the operators are listed in Table 2.2.

Table 2.1   Operators and the Properties of Their Eigenvalues

Operator Type	Properties of Eigenvalues
Hermitian	Real eigenvalues and > 0
Positive semi-definite	Eigenvalues ≥ 0
Positive definite	Eigenvalues > 0
Negative definite	Eigenvalues < 0
Nonsingular	$\lambda = 0$ is not an eigenvalue

Table 2.2   Matrix Types

$[A]^t = [A]$	Symmetric matrix
$[A] = [A^*]^t$	Hermitian (or self-adjoint) matrix
$[A]^t[A] = [I]$	Orthogonal matrix
$A_{ij} > 0$	Positive definite matrix
$A_{ij} \geq 0$	Positive semi-definite matrix
$A_{ii} > \sum_{j=1,\, i \neq j}^{N} |A_{ij}|$ for all i	Diagonally dominant matrix
$A_{ij} = 0$ if $i \neq j$	Diagonal matrix
$A_{ij} = A_{ji}$	Toeplitz matrix

2.6.3   Generic Solution of Sturm-Liouville Type Differential Equations

The homogeneous differential equations of SL type were solved in Sections 2.4 and 2.5, and the solutions consisted of eigenfunctions ψ_n and the corresponding eigenvalues λ_n. The eigenfunctions are orthogonal and most of the differential equations of electromagnetics belong to the SL type. The eigenfunctions may be employed in the solution of corresponding inhomogeneous differential equations denoted as $L(f) = g$. However, availability of eigenfunctions limits its range of applications to problem geometries which can be described by separable coordinate systems. Harrington has determined the eigenfunctions for irregular shaped geometries also [12, p. 52]. This extra effort is not necessary because one can solve the operator equation without using eigenfunctions as basis functions for f, g, and L. The methods employed are Raleigh-Ritz method and the method of weighted-residuals, and these form the basis of computational methods. The governing equations are in the form of functionals for the Raleigh-Ritz method, and the integral or differential equation for the method of weighted residuals. The expansion functions are generalized to include subdomain piecewise continuous functions. The computational methods are discussed later.

2.7   Delta-Function and Source Representations

We come across Dirac δ-function as a part of many inhomogeneous differential equations such as Green's function equation and impulse-response equation. The Green's function equation corresponding to (2.79) is

$$\frac{d^2 G(x; x')}{dx^2} + k^2 G(x; x') = -\delta(x - x') \tag{2.129}$$

where x' is the source coordinate and x is the field coordinate. Equation (2.129) can be solved in a manner similar to (2.79) by expanding the δ-function in terms of complete set of eigenfunctions of the operator. We define the δ-function first.

The Dirac delta function is defined as

$$\delta(x - a) = 0, \qquad \text{for } x \neq a \tag{2.130}$$

such that

$$\int_{-\infty}^{\infty} f(x)\, \delta(x - a)\, dx = f(a) \tag{2.131}$$

where $f(x)$ is any continuous function. The property (2.130) specifies that δ-function is a highly localized or peaked representation of a distribution. Equation (2.131) defines the sifting property; that is, δ-function acts like a sieve and samples the value of $f(x)$ at the point $x = a$. It follows from (2.131) that, for $f(x) = 1$,

$$\int_R \delta(x - a)\, dx = 1 \qquad \text{when the range } R \text{ contains the point } a \quad (2.132)$$

Dirac δ-function is not physically rigorous. It is an idealized version of many physical functions like a pulse, physical blow, and so on. Because of the idealization, δ-function is a good mathematical concept and is used to represent highly peaked sources in Green's function determination, impulse response, and so on. Generally speaking, there are three requirements that must be satisfied when representing volume source densities (charges or current) that are singular in nature [13]. These representations must: (1) be dimensionally correct, (2) locate the source exactly, and (3) give the correct total source strength (charge or current) after integration.

The one-dimensional representation of δ-function given in (2.130) may be generalized to vector representation and written as $\delta(\mathbf{r} - \mathbf{r}')$, where $\mathbf{r}$ and $\mathbf{r}'$ are vectors. In two- and three-dimensional rectangular coordinates we may write

$$\delta(\mathbf{r} - \mathbf{r}') = \delta(x - x')\,\delta(y - y') \tag{2.133a}$$

$$\delta(\mathbf{r} - \mathbf{r}') = \delta(x - x')\,\delta(y - y')\,\delta(z - z') \tag{2.133b}$$

Delta-function expressions in polar coordinates (ρ, ϕ, z) may be derived as follows. In two-dimensional polar coordinates (ρ, ϕ), let us write

$$\delta(\mathbf{r} - \mathbf{r}') = C_1\,\delta(\rho - \rho')\,\delta(\phi - \phi') \tag{2.134}$$

The constant C_1 is determined by using the property (2.132) and the surface element $ds = \rho d\rho d\phi$; that is,

$$\int_0^{2\pi}\int_0^{\infty} C_1\,\delta(\rho - \rho')\,\delta(\phi - \phi')\rho d\rho d\phi = 1$$

One obtains $C_1 = 1/\rho$ and, therefore,

$$\delta(\mathbf{r} - \mathbf{r}') = \frac{1}{\rho}\,\delta(\rho - \rho')\,\delta(\phi - \phi') \tag{2.135a}$$

This may be extended to three-dimensional cylindrical coordinates (ρ, ϕ, z) as

$$\delta(\mathbf{r} - \mathbf{r}') = \frac{1}{\rho}\,\delta(\rho - \rho')\,\delta(\phi - \phi')\,\delta(z - z') \tag{2.135b}$$

The delta function in spherical coordinates is given by $(dv = rd\phi r \sin\theta\, d\theta\, dr)$

$$\delta(\mathbf{r} - \mathbf{r}') = \frac{1}{r^2 \sin\theta}\,\delta(r - r')\,\delta(\theta - \theta')\,\delta(\phi - \phi') \tag{2.136}$$

Representation of Surface/Sheet Sources in Terms of δ-Function: Delta function, as defined by $\delta(z - z')$, is a highly peaked function at the point $z = z'$. In three-dimensional space, the point $z = z'$ represents a plane perpendicular to the z-axis. The delta function $\delta(z - z')$ can therefore be used to represent a highly peaked function in the z-direction. This may be the surface density of charge or current, which will have no thickness along z. An example of this representation is shown in Figure 2.9. Mathematically, if ρ_s is the surface charge density at the plane $z = z'$, the corresponding volume charge density is given by

$$\rho_v(x, y, z) = \rho_s(x, y)\,\delta(z - z') \qquad \text{C/m}^3 \qquad (2.137a)$$

For the charge distribution on a spherical surface $r = a$ with surface charge density $\rho_s(\theta, \varphi)$, the volume charge density is expressed as

$$\rho_v(r, \theta, \varphi) = \rho_s(\theta, \varphi)\,\delta(r - a) \qquad \text{C/m}^3 \qquad (2.137b)$$

Similarly, if the surface charge is distributed on an infinite cylindrical surface $\rho = a$, then the correct expression for the volume charge density is

$$\rho_v(\rho, \varphi, z) = \rho_s(\varphi, z)\,\delta(\rho - a) \qquad \text{C/m}^3 \qquad (2.137c)$$

It may be noted that although the surface charge density is finite, the volume charge density is infinite because of the δ-function in ρ_v. The total surface charge, which is identical to the total volume charge in this case, is obtained as [13]

$$Q = \int_v \rho_v\, dv = \iint_s \rho_s(x, y)\, dx\, dy \int_z \delta(z - z')\, dz$$

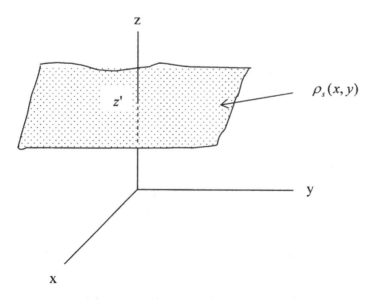

Figure 2.9 Sheet source representation as a δ-function, $\rho_v = \rho_s(x, y)\,\delta(z - z')$.

or

$$Q = \iint\limits_{s} \rho_s(x, y) \, dx \, dy, \text{ because } \int \delta(z - z') \, dz = 1 \qquad (2.138)$$

Representation of Line Sources in Terms of δ-Functions. The line charge density is a filament of charge having a length but no thickness. In certain cases, the charge distribution on a finite diameter conducting wire can be approximated by a filamentary line charge density. The wire must look like a filament to an observer who is located several meters away from it. Since geometrically a line is the result of intersection of two sheets or surfaces, the line source is a distribution which is highly peaked in two directions mutually perpendicular to the line. Therefore, its representation should be in the form of product of two δ-functions along these directions. For the line charge drawn in Figure 2.10, the volume charge density is given by

$$\rho_v(x, y, z) = \rho_l(z) \, \delta(y - y') \, \delta(x - x') \qquad \text{C/m}^3 \qquad (2.139\text{a})$$

where ρ_l is the linear density of the source. The volume density is again infinite. For a line charge running parallel to the z axis and passing through the point $(\rho', \varphi', 0)$ in the $z = 0$ plane with line charge density $\rho_l(z)$, the volume charge density is given by

$$\rho_v(\rho, \varphi, z) = \frac{1}{\rho} \rho_l(z) \, \delta(\rho - \rho') \, \delta(\varphi - \varphi') \qquad \text{C/m}^3 \qquad (2.139\text{b})$$

Representation of Point Sources in Terms of δ-Functions: Let us consider a point charge located at $P(x', y', z')$. The point charge has zero volume but finite charge Q. Therefore, its volume charge density is infinite. The charge density associated with it can be expressed as

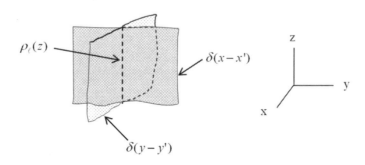

Figure 2.10 Line source representation as product of two δ-functions, $\rho_v = \rho_\ell(x, y) \, \delta(z - z') \, \delta(y - y')$.

$$\rho_v(x, y, z) = Q\delta(x - x')\,\delta(y - y')\,\delta(z - z') \qquad \text{C/m}^3 \qquad (2.140\text{a})$$

In cylindrical coordinates we have,

$$\rho_v(\rho, \varphi, z) = \frac{1}{\rho}\,Q\delta(\rho - \rho')\,\delta(\varphi - \varphi')\,\delta(z - z') \qquad \text{C/m}^3 \qquad (2.140\text{b})$$

and in spherical coordinates,

$$\rho_v(r, \theta, \varphi) = \frac{1}{(r)^2\,\sin\,\theta}\,Q\delta(r - r')\,\delta(\theta - \theta')\,\delta(\varphi - \varphi') \qquad \text{C/m}^3$$

$$(2.140\text{c})$$

Similarly, a current element of infinitesimal length directed along the z-direction, of unit strength, and located at $P(x', y', z')$ as shown in Figure 2.11 can be represented by the following current density function:

$$\mathbf{J} = \hat{\mathbf{z}}\delta(x - x')\,\delta(y - y')\,\delta(z - z') \qquad \text{A/m}^3 \qquad (2.141)$$

The dimension of δ-function is reciprocal of its argument because of the unit strength property of (2.132). That is, if x is measured in meters, then $\delta(x)$ is expressed in m^{-1}.

Eigenfunction Expansion of Delta-Function
Let us assume that it is possible to find a complete set of orthonormal eigenfunctions $\psi_n(x)$ of the homogeneous differential equation corresponding to (2.129). These functions are the natural modes of the system under consideration (e.g., modes of a rectangular waveguide, modes of oscillations of a vibrating string, and plane waves). We can use this set to expand the delta function of (2.129). We assume an expansion of the form

$$\delta(x - x') = \sum_{n=1}^{\infty} a_n(x')\,\psi_n(x) \qquad (2.142)$$

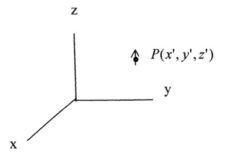

Figure 2.11 Representation of an infinitesimal current source as product of three δ-functions, $\mathbf{J} = \hat{\mathbf{z}}\delta(x - x')\,\delta(y - y')\,\delta(z - z')$.

The value of the coefficients a_n is obtained by multiplying both sides of (2.142) by $\psi_n(x)$ and integrating over the domain of x, say (a, b),

$$a_m(x') = \int_a^b \delta(x - x')\,\psi_m(x)\,dx \tag{2.143}$$

by virtue of the orthonormality property of $\psi_n(x)$. Using the sifting property of δ-function, (2.143) becomes

$$a_m(x') = \psi_m(x') \tag{2.144}$$

Therefore,

$$\delta(x - x') = \sum_{n=1}^{\infty} \psi_n(x')\,\psi_n(x) \tag{2.145}$$

It may be verified that the above expansion satisfies the properties of the δ-function; for example,

$$\int f(x)\,\delta(x - x')\,dx = \int \sum_n a_n \psi_n(x) \sum_m \psi_m(x')\,\psi_m(x)\,dx \tag{2.146}$$

by expanding the functions $f(x)$ and $\delta(x - x')$ in terms of the set $\psi_n(x)$. The above expression can be rearranged as follows:

$$\int f(x)\,\delta(x - x')\,dx = \sum_n \sum_m a_n \psi_m(x') \int \psi_m(x)\,\psi_n(x)\,dx$$

$$= \sum_n \sum_m a_n \psi_m(x')\,\delta_{mn} \tag{2.147}$$

$$= \sum_n a_n \psi_n(x')$$

$$= f(x')$$

Following the above procedure one can obtain the series expansion for the δ-function as

$$\delta(x) = \frac{1}{2l} + \frac{1}{l} \sum_{n=0}^{\infty} \cos\left(\frac{n\pi x}{l}\right) \qquad \text{for } -l \leq x \leq l \tag{2.148a}$$

$$\delta(\varphi - \varphi') = \frac{1}{2\pi} \sum_{n=-\infty}^{\infty} e^{jn(\varphi - \varphi')} \tag{2.148b}$$

Some of the useful representations of δ-function are listed next.

Plane Wave Representation or Integral Representation.

$$\delta(x - x') = \frac{1}{2\pi} \int\limits_{-\infty}^{\infty} e^{jk(x-x')} \, dk \qquad (2.149a)$$

that is, δ-function is composed of an infinite number of plane waves with different k-values but same amplitude. The sine and cosine representations equivalent to the plane wave representation of (2.149a) are given as

$$\delta(x - x') = \frac{2}{\pi} \int\limits_{0}^{\infty} \sin(kx) \sin(kx') \, dk \qquad (2.149b)$$

$$\delta(x - x') = \frac{2}{\pi} \int\limits_{0}^{\infty} \cos(kx) \cos(kx') \, dk \qquad (2.149c)$$

The Fourier transform of the δ-function is defined as

$$F\{\delta(x)\} = \int\limits_{-\infty}^{\infty} \delta(x) e^{jkx} \, dx = e^{jkx}\big|_{x=0} = 1 \qquad (2.150)$$

That is, all the Fourier components have the same amplitude.

Bessel Integral Representation [10, 14].

$$\delta(x - x') = x \int\limits_{0}^{\infty} J_n(\alpha x) J_n(\alpha x') \, \alpha d\alpha \qquad (2.151)$$

where $J_n(.)$ is nth order Bessel function. This expansion is useful in developing Green's function in terms of Bessel functions.

There are other types of expansions and representations of δ-function. For these, the reader is referred to [10, 14].

2.8 Summary

This chapter summarizes the popular method of separation of variables, and emphasizes the properties of orthogonal functions. The concept of orthogonality is central to the analytical and computational methods. Orthogonality of functions is introduced as an extension of orthogonality of vectors. Fourier series expansion is used to illustrate orthogonal function expansion. The solution of Sturm-Liouville differential equation is discussed in terms of orthogonal function expansion. Various types of differential equations that can be represented by Sturm-Liouville type are enumerated. Examples of orthogonal functions in the form of trigonometric functions, polynomials, series expansion, and so on, are given. Orthogonal function expansion of the unknown function is discussed and applied to the solution of nonhomogeneous differential equations in a very general manner.

The notable part of this chapter is the eigenfunction-based framework of function spaces. The eigenfunction approach is used to make the subject interesting and simple. Concepts like completeness of a set, convergence of series, eigenfunctions of the operator, vector representation of function, matrix representation of an operator are illustrated through the solution of one-dimensional differential equation. Limitations of the eigenfunction-based function space are brought out and the generic solution in terms of subdomain basis functions is suggested. The delta function is essential to the development of Green's function. Various expansions of Dirac delta functions such as Fourier cosine series expansion, plane wave expansion, and Bessel function expansion are listed. The last section describes delta function in terms of eigenfunction expansion and prepares the background for the next chapter on Green's function.

References

[1]   Weinberger, H. F., *A First Course in Partial Differential Equations*, New York: John Wiley, 1965, Chapter 4.

[2]   Shen, L. C., and J. A. Kong, *Applied Electromagnetism*, Monterey, CA: Brooks/Cole, 1983.

[3]   Sadiku, M. N. O., *Numerical Techniques in Electromagnetics*, 2nd ed., Boca Raton, FL: CRC Press, 2001.

[4]   Stinson, D. C., *Intermediate Mathematics of Electromagnetics*, Englewood Cliffs, NJ: Prentice-Hall, 1976.

[5]   Harrington, R. F., *Time Harmonic Electromagnetic Fields*, New York: McGraw-Hill, 1961.

[6]   Van Bladel, J., *Electromagnetic Fields*, Washington, D.C.: Hemisphere, 1985.

[7]   Dettman, J. W., *Mathematical Methods in Physics and Engineering*, New York: McGraw-Hill, 1962.

[8]   Friedman, B., *Principles and Techniques of Applied Mathematics*, New York: John Wiley, 1956.

[9]   Harrington, R. F., *Field Computations by Moment Methods*, Malabar, FL: R. E. Krieger, 1968.

[10] Arfken, G., *Mathematical Methods for Physicists*, 3rd ed., New York: Academic Press, 1985.

[11] Volakis, J. L., A. Chatterjee, and L. C. Kempel, *Finite Element Method for Electromagnetics*, Hyderabad (India): Universities Press, 2001.

[12] Harrington, R. F., "Characteristic Modes for Antennas and Scatterers," in R. Mittra, (ed.), *Numerical and Asymptotic Techniques in Electromagnetics*, Berlin: Springer, 1975.

[13] Neff, H. P., Jr., *Basic Electromagnetic Fields*, 2nd ed., New York: Harper & Row, 1987.

[14] Butkov, E., *Mathematical Physics*, Reading, MA: Addison-Wesley, 1968.

Problems

P2.1. Write down the periodic function $f(x)$ satisfying the following boundary conditions:

1. $f(x) = 0$ at $x = 0, a$.
2. $f(x) = 0$ at $x = \pm a/2$.
3. $f(x) = 0$ at $x = 0, a$, and $\partial f/\partial x = 0$ at $x = 0$.
4. $f(x) = 0$ at $x = 0, a/2$, and $\partial f/\partial x = 0$ at $x = -a/2$.

Normalize the functions for the cases (1) and (2) above.

P2.2. Determine the Fourier cosine series (or Fourier sine series) expansion of the function $\delta(x - x_0)$ over the interval $-a \le x \le a, -a \le x_0 \le a$ and use this expansion to show that

$$\int_{-a}^{a} \delta(x - x_0)\, dx = 1$$

Hint: Expand 1 also in Fourier sine series.

P2.3. Consider the mode functions

$$\psi_{mn}(x, y) = \sin\left(\frac{m\pi x}{a}\right) \sin\left(\frac{n\pi y}{b}\right) \qquad m, n = 1, 2, 3, \ldots$$

defined over the interval $0 \le x \le a, 0 \le y \le b$.

1. Show that the mode functions are orthogonal.
2. Normalize the mode functions.
3. Use the normalized mode functions to expand the function $f(x, y) = c$ in a series.

P2.4. Express the following functions in terms of orthogonal functions given against each:

1. $f(x) = 2x/a$, $0 \le x \le a/2$ in terms of $\sin(n\pi x/a)$.
2. $f(x) = \sin(n\pi x/a)$, $0 \le x \le a$ in terms of $\cos(n\pi x/a)$.

P2.5. Determine the solution of differential equation

$$\frac{d^2 y}{dx^2} = -1 \qquad 0 \le x \le 1$$

subject to the boundary conditions $y(0) = 0 = y(1)$ by direct integration. Express the resulting solution in Fourier sine series.

CHAPTER 3

Green's Function

3.1   Introduction

Green's function is one of the very useful tools for solving problems in electromagnetics. It has the same meaning and importance for electromagnetics as the impulse response for circuits. Green's function describes the response of an electromagnetic system to a delta function source. Once the Green's function is known, the response to an arbitrary excitation can be obtained using superposition. Green's function is an important part of analytical tools in electromagnetics. The efficiency of the method of moments solution of open region problems such as in radiation, scattering, and planar lines is mainly due to the Green's function. We shall first illustrate the meaning of and than describe methods to construct Green's functions.

Let us consider the Poisson equation of electrostatics:

$$\nabla^2 \phi(\mathbf{r}) = \frac{-\rho(\mathbf{r})}{\epsilon} \tag{3.1}$$

Its solution in the form of an integral equation is given by

$$\phi(\mathbf{r}_1) = \int d\phi = \frac{1}{4\pi\epsilon} \int_\tau \frac{\rho(\mathbf{r}_2)}{|\mathbf{r}_1 - \mathbf{r}_2|} \, d\tau \tag{3.2}$$

According to (3.2), the net potential ϕ at $\mathbf{r}_1$ is due to the superposition of $d\phi$, produced by the elemental charge $\rho(\mathbf{r}_2)d\tau$ located at $\mathbf{r}_1 - \mathbf{r}_2$. This is shown in Figure 3.1. *The effectiveness of the charge $\rho(\mathbf{r}_2)d\tau$ depends on the factor $\{4\pi|\mathbf{r}_1 - \mathbf{r}_2|\}^{-1}$. For this reason, $\{4\pi|\mathbf{r}_1 - \mathbf{r}_2|\}^{-1}$ is often called the influence function. This is also called the Green's function* $G(\mathbf{r}_1; \mathbf{r}_2)$; that is, it represents the influence or effect produced by unit charge and we write (3.2) as

$$\phi(\mathbf{r}_1) = \frac{1}{\epsilon} \int_\tau G(\mathbf{r}_1; \mathbf{r}_2)\rho(\mathbf{r}_2) \, d\tau \tag{3.3}$$

where

$$G(\mathbf{r}_1; \mathbf{r}_2) = \frac{1}{4\pi|\mathbf{r}_1 - \mathbf{r}_2|} \tag{3.4}$$

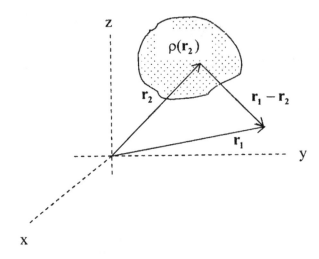

Figure 3.1   Illustration showing the region with charge and the potential produced at a distance $r_1 - r_2$.

In general, we may define the Green's function as the effect produced by a source of unit intensity. It has the same significance in electromagnetics as the impulse response in circuit theory. We shall denote source position by primed coordinates and field position by unprimed coordinates.

The Green's function can be expressed in the closed form, in the form of a series expansion, or in the form of an integral depending upon the boundary conditions. For example, if the boundary conditions are either Neumann or Dirichlet or mixed type it is possible to attempt closed form and series solutions. However, the open boundary case is best solved with an integral form of Green's function. We shall describe the closed form solution using the *direct construction approach* and the series form of solution based on eigenfunction expansion. The integral form of Green's function is described in Chapters 5 and 11.

3.2   Direct Construction Approach for Green's Function

We shall illustrate this method through an example. Consider a stretched string of length L tied with the nails at the two ends as shown in Figure 3.2(a). The string is at rest under an external distributed load given by $F(x)$ (force per unit length). The deflection y of the string is described mathematically by the following differential equation:

$$\frac{d^2 y}{dx^2} = \frac{F(x)}{T} = f(x) \qquad \text{for } 0 \leq x \leq L \tag{3.5}$$

Here, T is the tension in the string. The boundary conditions to be satisfied by (3.5) are: $y(0) = y(L) = 0$, the deflection at the end points is zero because the string is tied there.

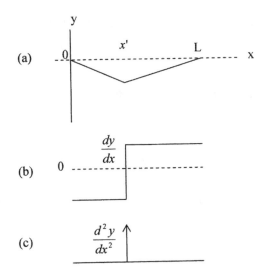

Figure 3.2 Illustration of a stretched string under concentrated load: (a) deflection $y(x)$ of the string; (b) slope of the string, dy/dx; and (c) derivative of the slope, d^2y/dx^2.

Before solving (3.5) let us first solve the corresponding problem for a *concentrated* unit load at the point $x = x'$. The solution to the concentrated load case is the Green's function, and the associated differential equation is obtained by replacing $y(x)$ by $G(x, x')$ and $f(x)$ in (3.5) by the Dirac delta function,

$$\frac{d^2 G(x; x')}{dx^2} = \delta(x - x') \tag{3.6}$$

The boundary conditions on G are the same as that on $y(x)$; that is,

$$G(0; x') = G(L; x') = 0 \tag{3.7}$$

We expect the shape of the string to look like the one shown in Figure 3.2(a).

Solution. First separate the problem region or domain $0 \le x \le L$ into two subregions, isolating the singularity at $x = x'$. Let us call $0 \le x < x'$ as region I, and $x' < x \le L$ as region II. The Green's function in these regions satisfies the homogeneous differential equation corresponding to (3.6) because the source point $x = x'$ is not included in either region; that is,

$$\frac{d^2 G(x; x')}{dx^2} = 0, \qquad \text{for } x \ne x' \tag{3.8}$$

Solution in Region I. The solution of (3.8) must have the form

$$G_I(x; x') = Ax + B, \qquad \text{for } 0 \le x < x' \tag{3.9}$$

Also, the boundary condition at $G(0; x') = 0$ implies $B = 0$. Therefore,

$$G_I(x; x') = Ax, \qquad \text{for } 0 \le x < x' \tag{3.10}$$

Solution in Region II.   Let us write

$$G_{II}(x; x') = Cx + D, \qquad \text{for } x' < x \le L \tag{3.11}$$

The boundary condition $G(L; x') = 0$ implies $CL + D = 0$. Therefore,

$$G_{II}(x; x') = C(x - L), \qquad \text{for } x' < x \le L \tag{3.12}$$

The unknowns A and C of (3.10) and (3.12), respectively, can be determined using two conditions:

1. Since $G(x; x')$ represents the shape of the string, it must be continuous at $x = x'$; that is, $G_I = G_{II}$ at $x = x'$, or

$$Ax' = C(x' - L) \text{ or } C = \frac{Ax'}{(x' - L)} \tag{3.13}$$

   and

$$G_{II}(x; x') = Ax' \frac{x - L}{x' - L} \tag{3.14}$$

2. There is a discontinuity in the slope at $x = x'$, see Figure 3.2(b). To determine the magnitude of this discontinuity we integrate (3.6) over the range of x:

$$\int\limits_0^L \frac{d^2 G(x; x')}{dx^2} \, dx = \int\limits_0^L \delta(x - x') \, dx \tag{3.15}$$

Since $\dfrac{d^2 G}{dx^2} = 0$ and $\delta(x - x') = 0$ except at $x = x'$, from (3.8) and (2.130), respectively, the contribution to the integrals arises near the point $x = x'$. Equation (3.15) may therefore be written as [5, p. 9]

$$\int\limits_{x'-\epsilon}^{x'+\epsilon} \frac{d^2 G(x; x')}{dx^2} \, dx = \int\limits_{x'-\epsilon}^{x'+\epsilon} \delta(x - x') \, dx$$

or

$$\left. \frac{dG_{II}}{dx} \right|_{x'+\epsilon} - \left. \frac{dG_I}{dx} \right|_{x'-\epsilon} = 1, \text{ using (2.131) for } \delta\text{-function}$$

or

$$\frac{Ax'}{x' - L} - A = 1, \Rightarrow A = \frac{x' - L}{L} \tag{3.16}$$

Substitute for A in G_I and G_{II} to obtain

$$G(x; x') = \begin{cases} x(x' - L)/L & \text{for } x \le x' \\ x'(x - L)/L & \text{for } x \ge x' \end{cases} \tag{3.17}$$

Using the superposition principle and the Green's function, we now obtain the solution for (3.5) as

$$y(x) = \int_0^L G(x; x') f(x') \, dx' \tag{3.18}$$

From the above example we can list the following properties of the Green's function:

1. It is a continuous function of variable x.
2. It satisfies the same boundary conditions as $y(x)$.
3. Green's function is symmetrical in the source point x' and field point x; that is, $G(x; x') = G(x'; x)$.
4. dG/dx and d^2G/dx^2 are continuous in the interval $(0, L)$ except at the source point $x = x'$.

3.2.1 Green's Function for the Sturm-Liouville Differential Equation

Sturm-Liouville differential equation is a generalized version of the second-order differential equations employed in electromagnetics. This differential equation in one variable was defined in Chapter 2 and in operator form is given by

$$L\{y\} + \lambda w(x) y(x) = \pm f(x) \tag{3.19}$$

where $f(x)$ is the source function and the operator L is defined as

$$L\{y\} = \frac{d}{dx} [p(x) y'(x)] - q(x) y(x) \tag{3.20}$$

Here $y'(x)$ stands for dy/dx. The boundary conditions ensure unique solutions.

The Green's function differential equation corresponding to (3.19) is obtained by replacing the source term $f(x)$ by the δ-function and $y(x)$ by $G(x; x')$,

$$L\{G\} + \lambda w(x) G(x; x') = \pm \delta(x - x') \tag{3.21}$$

where $L\{G(x; x')\}$ is obtained from (3.20) by a similar substitution,

$$L\{G(x; x')\} = \frac{d}{dx}[p(x)G'(x; x')] - q(x)G(x; x') \qquad (3.22)$$

The *direct construction approach* can be used to derive Green's function for (3.21), after $p(x)$ and $q(x)$ have been specified. We apply continuity of G and discontinuity of dG/dx at $x = x'$ to determine the unknowns in the expansion of $G(x; x')$ [1, p. 508]

$$G_I|_{x'} = G_{II}|_{x'} \qquad (3.23a)$$

$$\left.\frac{dG_{II}}{dx}\right|_{x'} - \left.\frac{dG_I}{dx}\right|_{x'} = \pm\frac{1}{p(x')} \qquad (3.23b)$$

Procedure for Finding G(x; x')
   1. Divide the domain of the variable x, say, (a, b) into two regions $a \leq x < x'$ and $x' < x \leq b$, leaving out the source point $x = x'$. Call these regions I and II.
   2. Find two linearly independent solutions of the homogeneous differential equation $L\{G\} + \lambda w(x)G(x; x')$ subject to the given boundary conditions, and call them G_I and G_{II}. Linear independence is ensured if the wronskian $W = G_I G_{II}' - G_I' G_{II} \neq 0$. Due to the requirement of linear independence, we cannot choose the solution $G_I = G_{II} = \sin(n\pi x/L)$ for the string problem.
   3. Apply the conditions (3.23) at the source point $x = x'$ to determine the unknowns of the solutions.

Exercise.   Show that the solution to

$$\frac{d^2 G}{dx^2} + k^2 G = \delta(x - x') \qquad 0 \leq x \leq L$$

subject to $G(0; x') = G(L; x') = 0$ is given by

$$G(x; x') = \begin{cases} \dfrac{\sin k(L - x')\sin(kx)}{k\sin(kL)} & \text{for } x \leq x' \\[4mm] \dfrac{\sin k(L - x)\sin(kx')}{k\sin(kL)} & \text{for } x \geq x' \end{cases} \qquad (3.24)$$

3.2.2   Green's Function for a Loaded Transmission Line

Consider a transmission line of characteristic impedance Z_0, propagation constant γ, and length L. The voltage $v(z)$ and the current $i(z)$ in the transmission line are governed by the following telegraphist's equations:

$$\frac{dv}{dz} = -Zi \qquad (3.25a)$$

and

$$\frac{di}{dz} = -Yv \qquad (3.25b)$$

where $Z = R + j\omega L$ and $Y = G + j\omega C$ are the impedance per unit length and admittance per unit length, respectively, for the transmission line. R, L, G, and C are the resistance, inductance, conductance, and capacitance per unit length, respectively, of the line. They are also called the line parameters because they define the characteristics of the transmission line, for example,

$$\text{Characteristic impedance } Z_0 = \sqrt{(R + j\omega L)/(G + j\omega C)} = \sqrt{Z/Y} \quad (3.26)$$

$$\text{Propagation constant } \gamma = \sqrt{(R + j\omega L)(G + j\omega C)} = \sqrt{ZY} \quad (3.27)$$

Let the transmission line be excited by a current source $i_0(z')$ as shown in Figure 3.3. Due to the presence of source, (3.25b) will now get modified as

$$\frac{dv}{dz} = -Zi \qquad (3.28a)$$

and

$$\frac{di}{dz} = -Yv + i_0(z') \qquad (3.28b)$$

One of the ways to solve the above equations is the Green's function method. For this, equations (3.28) are reduced to a second-order differential equation in a single variable v or i first. It means that one of the variables from (3.28) should be eliminated, which requires (3.28a) or (3.28b) to be differentiated. At times, however, the source distribution is discontinuous or has discontinuous derivatives. In the present case, this can be avoided by differentiating (3.28a) with respect to z and eliminating di/dz; one obtains

$$\frac{d^2v}{dz^2} - \gamma^2 v = -Zi_0(z') \qquad \text{since } YZ = \gamma^2 \qquad (3.29)$$

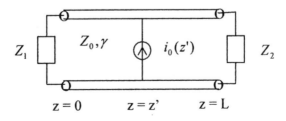

Figure 3.3   Current source excitation of a transmission line terminated at the ends.

Remark. If both the voltage and current sources are used for exciting the transmission line, (3.28a) will have a voltage source term $v_0(z')$ on the right side. So, we cannot differentiate this equation without running into the problem of source discontinuity. A solution to this problem is to invoke the linearity of telegraphists/ transmission line equations and use superposition; that is, we work out two problems; one with the current source as above, and the other with the voltage source alone. The voltage and current distributions obtained from the two solutions are added to obtain the final result [2].

The Green's function is used here to solve (3.29). The first step is to replace the excitation $i_0(z')$ by the delta function, and (3.29) becomes for Green's function

$$\frac{d^2 v}{dz^2} - \gamma^2 v = -Z\delta(z - z') \tag{3.30}$$

For simplicity we assume that the transmission line is terminated in short circuits at $z = 0$, and $z = L$; that is, $Z_1 = Z_2 = 0$. We shall determine v for this problem using the *direct construction approach*. Next we divide the region $0 \leq z \leq L$ into two subregions I and II such that

Region I: $0 \leq z < z'$ Region II: $z' < z \leq L$

Now write down the solution of source free differential equation,

$$\frac{d^2 v}{dz^2} - \gamma^2 v = 0 \tag{3.31}$$

in the two regions. The general solution is

$$v = Ae^{\gamma z} + Be^{-\gamma z} \tag{3.32}$$

where $\pm\gamma$ are the solutions of (3.31). For region I, we can express (3.32) as

$$v_1 = v_{11}e^{\gamma z} + v_{12}e^{-\gamma z} \tag{3.33a}$$

Similarly, for region II,

$$v_2 = v_{21}e^{\gamma z} + v_{22}e^{-\gamma z} \tag{3.33b}$$

The four unknowns v_{11}, v_{12}, v_{21}, and v_{22} are determined from the boundary conditions and the conditions at the source point.

The boundary condition $v_1 = 0$ at $z = 0$ gives rise to $v_{12} = -v_{11}$, and therefore

$$v_1 = 2v_{11}\sinh(\gamma z) \tag{3.34a}$$

Similarly, the use of boundary condition at $z = L$ yields

$$v_2 = v_{21}(e^{\gamma z} - e^{-\gamma z + 2\gamma L}) \tag{3.34b}$$

The continuity of voltage across the current source at $z = z'$ means

$$v_1(z') = v_2(z') \tag{3.35a}$$

The discontinuity of current at $z = z'$ implies

$$\left.\frac{dv_2}{dz}\right|_{z'} - \left.\frac{dv_1}{dz}\right|_{z'} = -Z$$

or

$$\left.\frac{-1}{Z}\frac{dv_2}{dz}\right|_{z'} + \left.\frac{1}{Z}\frac{dv_1}{dz}\right|_{z'} = 1$$

or

$$i_2(z') - i_1(z') = 1 \tag{3.35b}$$

Use of conditions (3.35) in (3.34) leads to the simultaneous equations:

$$v_{21}(e^{\gamma z'} - e^{2\gamma L - \gamma z'}) - 2v_{11}\sinh(\gamma z') = 0 \tag{3.36a}$$

$$v_{21}(e^{\gamma z'} + e^{2\gamma L - \gamma z'}) - 2v_{11}\cosh(\gamma z') = -\frac{Z}{\gamma} = -Z_0 \tag{3.36b}$$

solution of which yields

$$v_{11} = \frac{Z_0}{2}\frac{1 - e^{2\gamma(z'-L)}}{1 - e^{-2\gamma L}}e^{-\gamma z'}, \text{ and } v_{21} = -\frac{Z_0}{2}\frac{e^{\gamma z'} - e^{-\gamma z'}}{1 - e^{-2\gamma L}}e^{-2\gamma L} \tag{3.37}$$

Substituting for v_{11} and v_{21} in (3.34) produces the following Green's function for the current excited voltage on the transmission line:

$$v_1 = Z_0\sinh(\gamma z)\frac{e^{-\gamma z'} - e^{\gamma z' - 2\gamma L}}{1 - e^{-2\gamma L}} \qquad 0 \le z \le z' \tag{3.38a}$$

$$v_2 = Z_0\sinh(\gamma z')\frac{e^{-\gamma z} - e^{\gamma z - 2\gamma L}}{1 - e^{-2\gamma L}} \qquad z' \le z \le L \tag{3.38b}$$

It may be noted that the voltage is symmetric in z and z'. It is due to the reciprocity property of the passive network.

3.3   Eigenfunction Expansion of Green's Function

This is the most popular method to determine the Green's function. Due to the analytical difficulties, however, this method is applicable to regular shaped geometries for which the eigenfunctions can be obtained analytically. We shall illustrate this method by an example.

Determine the solution of

$$\frac{d^2 G(x; x')}{dx^2} = \delta(x - x') \text{ subject to } G(0; x') = G(L; x') = 0 \qquad (3.39)$$

employing the method of eigenfunction expansion. For this, we first determine the eigenfunctions (of the operator d^2/dx^2) by solving the eigenvalue differential equation corresponding to (3.39). The eigenvalue equation is

$$\frac{d^2 y}{dx^2} = \lambda y \text{ subject to } y(0) = y(L) = 0 \qquad (3.40)$$

The eigenfunctions $y_n(x)$ for this problem are: $y_n(x) = \sin(n\pi x/L)$, $n = 1, 2, 3,$ $\ldots$, with the eigenvalues $\lambda_n = -(n\pi/L)^2$.

Since the eigenfunctions satisfy the same boundary conditions as the original problem they can be used to expand the unknown Green's function. We therefore write

$$G(x; x') = \sum_{n=1}^{\infty} a_n(x') y_n(x) \qquad (3.41a)$$

Substituting this expansion in the Green's function equation (3.39) gives

$$-\sum_{n=1}^{\infty} a_n(x') \left(\frac{n\pi}{L}\right)^2 \sin\left(\frac{n\pi x}{L}\right) = \delta(x - x') \qquad (3.41b)$$

The coefficients a_n are determined by using orthogonality constraint of eigenfunctions; that is, multiply both sides of (3.41b) by $\sin(m\pi x/L)$ and integrate over the range of x to obtain

$$-a_n(x') \left(\frac{n\pi}{L}\right)^2 \int_0^L \sin^2\left(\frac{n\pi x}{L}\right) dx = \int_0^L \sin\left(\frac{n\pi x}{L}\right) \delta(x - x') dx$$

or

$$a_n = \frac{-2L}{(n\pi)^2} \sin\left(\frac{n\pi x'}{L}\right) \qquad (3.42)$$

Therefore,

$$G(x; x') = -2L \sum_{n=1}^{\infty} \frac{1}{(n\pi)^2} \sin\left(\frac{n\pi x}{L}\right) \sin\left(\frac{n\pi x'}{L}\right) \tag{3.43}$$

As an extension of (3.43), it can be shown that the solution to

$$\frac{d^2 G}{dx^2} + k^2 G = \delta(x - x') \tag{3.44}$$

subject to $G(0; x') = G(L; x') = 0$, is given by

$$G(x; x') = \frac{2}{L} \sum_{n=1}^{\infty} \frac{\sin\left(\frac{n\pi x}{L}\right) \sin\left(\frac{n\pi x'}{L}\right)}{k^2 - \left(\frac{n\pi}{L}\right)^2} \tag{3.45}$$

Out of the two alternate solutions (3.24) and (3.45), the solution (3.24) does not involve any summation, and is therefore computationally efficient. The solution (3.45) reduces to the solution of $\dfrac{d^2 G(x; x')}{dx^2} = \delta(x - x')$ in the limit $k \to 0$. This is because the operator d^2/dx^2 is the same in both the differential equations. This observation is true only for the solutions of inhomogeneous equations.

Exercise.   Show that the alternate solutions (3.24) and (3.45) to the differential equation $\dfrac{d^2 G(x; x')}{dx^2} + k^2 G = \delta(x - x')$ are equivalent.

3.4   Green's Function in Two Dimensions

Let us consider a line charge in an infinitely long rectangular metal pipe. Cross-section of the geometry is shown in Figure 3.4. The distribution of static potential ϕ in the pipe is governed by the Poisson equation

$$\nabla^2 \phi = -\frac{\rho}{\epsilon} \tag{3.46}$$

where ρ is the charge density. The metal pipe is infinitely long and uniform along the z-direction. The boundary conditions are also invariant with z. Therefore, the potential V is invariant with respect to z, that is, $\partial\phi/\partial z = 0$ and the Laplacian ∇^2 reduces to

$$\nabla_t^2 = \frac{\partial^2}{\partial x^2} + \frac{\partial^2}{\partial y^2} \tag{3.47}$$

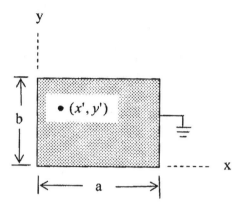

Figure 3.4 Cross-section of a rectangular metal pipe of size $a \times b$ excited by a line source of charge.

where ∇_t^2 is called the transverse Laplacian operator. For the two-dimensional problem, the line charge becomes a point charge located at (x', y') (Figure 3.4). The solution to the original problem can be obtained if we determine the corresponding Green's function, which is defined as follows:

$$\frac{\partial^2 G}{\partial x^2} + \frac{\partial^2 G}{\partial y^2} = -\frac{1}{\epsilon}\,\delta(x - x')\,\delta(y - y') \tag{3.48}$$

subject to the boundary conditions

$$G(0, y; x', y') = G(a, y; x', y') = 0 \tag{3.49}$$

$$G(x, 0; x', y') = G(x, b; x', y') = 0 \tag{3.50}$$

Here we have fixed the potential at the four walls to be zero. For an arbitrary potential V_0 on the walls the solution is given by adding V_0 to the solution of the above problem. The partial differential equation (3.48) is next solved employing double series and single series expansions of Green's function.

3.4.1 Double Series Expansion Method

Let us expand the Green's function in terms of the eigenfunctions $\phi_m(x, y)$ of the operator ∇_t^2; that is,

$$G(x, y; x', y') = \sum_m \sum_n a_{mn}\,\phi_{mn}(x, y) \tag{3.51}$$

The eigenfunctions are the solutions of the following eigenvalue equation,

$$\nabla^2 \phi_{mn}(x, y) = \lambda_{mn}\,\phi_{mn}(x, y) \tag{3.52}$$

and the same boundary conditions as for the original problem,

$$\phi_{mn}(x, y) = 0 \text{ at } x = 0; \ x = a; \ y = 0; \ y = b \tag{3.53}$$

The orthonormal eigenfunctions are obtained as

$$\phi_{mn}(x, y) = \frac{2}{\sqrt{ab}} \sin\left(\frac{m\pi x}{a}\right) \sin\left(\frac{n\pi y}{b}\right) \qquad n, m = 1, 2, 3, \ldots \tag{3.54}$$

with eigenvalues

$$\lambda_{mn}(x, y) = -\left[\left(\frac{m\pi}{a}\right)^2 + \left(\frac{n\pi}{b}\right)^2\right] \tag{3.55}$$

Therefore, (3.51) can be written as

$$G(x, y; x', y') = \frac{2}{\sqrt{ab}} \sum_{m=1}^{\infty} \sum_{n=1}^{\infty} a_{mn} \sin\left(\frac{m\pi x}{a}\right) \sin\left(\frac{n\pi y}{b}\right) \tag{3.56}$$

Each term of G satisfies the boundary conditions at the four walls. To determine the expansion coefficients a_{mn} we substitute (3.56) in (3.48) and obtain

$$\frac{2}{\sqrt{ab}} \sum_{m=1}^{\infty} \sum_{n=1}^{\infty} \left[\left(\frac{m\pi}{a}\right)^2 + \left(\frac{n\pi}{b}\right)^2\right] a_{mn} \sin\left(\frac{m\pi x}{a}\right) \sin\left(\frac{n\pi y}{b}\right) = \frac{1}{\epsilon} \delta(x - x')\delta(y - y')$$
$$\tag{3.57}$$

To remove the double summation from the left side and hence to obtain an expression for a_{mn}, we multiply both sides of (3.57) by $\sin(m'\pi x/a)\sin(n'\pi y/b) \, dx \, dy$ and integrate over the range of x and y. Use of the orthogonality condition of the functions $\sin(m\pi x/a)$ and $\sin(m'\pi x/a)$, and $\sin(n\pi y/b)$ and $\sin(n'\pi y/b)$ yield

$$\frac{\sqrt{ab}}{2}\left[\left(\frac{m\pi}{a}\right)^2 + \left(\frac{n\pi}{b}\right)^2\right] a_{mn} = \frac{1}{\epsilon} \sin\left(\frac{m\pi x'}{a}\right) \sin\left(\frac{n\pi y'}{b}\right)$$

or

$$a_{mn} = \frac{2}{\epsilon\sqrt{ab}} \frac{\sin\left(\dfrac{m\pi x'}{a}\right) \sin\left(\dfrac{n\pi y'}{b}\right)}{\left[\left(\dfrac{m\pi}{a}\right)^2 + \left(\dfrac{n\pi}{b}\right)^2\right]} \tag{3.58}$$

Now substitute this expression in (3.56) to obtain

$$G(x, y; x', y') = \frac{4}{\epsilon ab} \sum_m \sum_n \frac{\sin\left(\frac{m\pi x}{a}\right)\sin\left(\frac{n\pi y}{b}\right)}{\left(\frac{m\pi}{a}\right)^2 + \left(\frac{n\pi}{b}\right)^2} \sin\left(\frac{m\pi x'}{a}\right)\sin\left(\frac{n\pi y'}{b}\right)$$

$$(3.59)$$

Corr.: Green's function for the Helmholtz equation

$$\frac{\partial^2 G}{\partial x^2} + \frac{\partial^2 G}{\partial y^2} + k^2 G = \delta(x - x')\,\delta(y - y') \tag{3.60}$$

may be obtained in a similar manner. The final result can also be obtained by replacing $\left(-\left[\left(\frac{m\pi}{a}\right)^2 + \left(\frac{n\pi}{b}\right)^2\right]\right)$ by $k^2 - \left[\left(\frac{m\pi}{a}\right)^2 + \left(\frac{n\pi}{b}\right)^2\right]$ and ϵ by (-1) in (3.59). One obtains

$$G(x, y; x', y') = \frac{4}{ab} \sum_m \sum_n \frac{\sin\left(\frac{m\pi x}{a}\right)\sin\left(\frac{n\pi y}{b}\right)}{k^2 - \left[\left(\frac{m\pi}{a}\right)^2 + \left(\frac{n\pi}{b}\right)^2\right]} \sin\left(\frac{m\pi x'}{a}\right)\sin\left(\frac{n\pi y'}{b}\right)$$

$$(3.61)$$

3.4.2 Single Series Expansion Method

The eigenfunction expansion (3.61) is in the form of double Fourier series, computation of which requires more computer resources. We next show that the solution for the same problem can be attempted in the form of a single series, and is therefore less computational intensive. For this, we shall use *direct construction method* of Section 3.2 along one of the directions.

Let us start by satisfying only two boundary conditions, namely, those at the edges $y = 0$ and $y = b$. This can be done by representing the Green's function as a single Fourier series with respect to y:

$$G(x, y; x', y') = \sum_n g_n \sin\left(\frac{n\pi y}{b}\right) \tag{3.62}$$

The coefficients g_n are therefore functions of x, x' and y'. *An alternative to (3.62) is obtained by satisfying the boundary conditions at the planes $x = 0$ and $x = a$.* Now, substitute the above series in Green's function partial differential equation (3.48), multiply by the factor $\sin(m\pi y/b)$, and integrate over the y variable to obtain

$$\frac{d^2 g_m}{dx^2} - \left(\frac{m\pi}{b}\right)^2 g_m(x, x', y') = \frac{-2}{\epsilon b} \sin\left(\frac{m\pi y'}{b}\right)\delta(x - x') \tag{3.63}$$

This ordinary differential equation shows that the expansion functions g_m are one-dimensional Green's functions in their own right. We can obtain these functions using the *method of direct construction*. For this, we divide the range of x into two regions, and find the solution of

$$\frac{d^2 g_m}{dx^2} - \left(\frac{m\pi}{b}\right)^2 g_m(x) = 0 \qquad (3.64)$$

subject to $g_m(0) = g_m(a) = 0$. The solutions of (3.64) are described in the form of exponentials or hyperbolic functions because of the second term with negative sign. In order to satisfy the boundary condition at $x = 0$, we may choose the following form:

$$g_m^I(x) = A_m \sinh\left(\frac{m\pi x}{b}\right) \qquad \text{for } x < x' \qquad (3.65a)$$

Similarly, boundary condition at $x = a$ leads to the following form:

$$g_m^{II}(x) = B_m \sinh\left(\frac{m\pi(a - x)}{b}\right) \qquad \text{for } x > x' \qquad (3.65b)$$

Continuity condition at the source, $g_m^I(x') = g_m^{II}(x')$, gives

$$A_m \sinh\left(\frac{m\pi x'}{b}\right) = B_m \sinh\left(\frac{m\pi(a - x')}{b}\right) \qquad (3.66)$$

The following condition derived from integration of (3.63) gives

$$\left.\frac{dg_m^{II}}{dx}\right|_{x'} - \left.\frac{dg_m^I}{dx}\right|_{x'} = \frac{-2}{\epsilon b} \sin\left(\frac{m\pi y'}{b}\right)$$

or

$$-\frac{m\pi}{b}\left[B_m \cosh\left(\frac{m\pi(a - x')}{b}\right) + A_m \cosh\left(\frac{m\pi x'}{b}\right)\right] = \frac{-2}{\epsilon b} \sin\left(\frac{m\pi y'}{b}\right) \qquad (3.67)$$

Solving (3.66) and (3.67) for A_m and B_m yields

$$A_m = \frac{\frac{2}{\epsilon m\pi} \sin\left(\frac{m\pi y'}{b}\right) \sinh\left(\frac{m\pi(a - x')}{b}\right)}{\sinh\left(\frac{m\pi a}{b}\right)} \qquad (3.68a)$$

$$B_m = \frac{\frac{2}{\epsilon m \pi} \sin\left(\frac{m\pi y'}{b}\right) \sinh\left(\frac{m\pi x'}{b}\right)}{\sinh\left(\frac{m\pi a}{b}\right)} \tag{3.68b}$$

Use of these constants in (3.65) provides the following expression for the Green's function:

$$G(x, y; x', y') =$$

$$
\begin{cases}
\displaystyle\sum_{m=1}^{\infty} \frac{2}{m\pi\epsilon} \sin\left(\frac{m\pi y'}{b}\right) \sin\left(\frac{m\pi y}{b}\right) \frac{\sinh\left(\frac{m\pi(a - x')}{b}\right) \sinh\left(\frac{m\pi x}{b}\right)}{\sinh\left(\frac{m\pi a}{b}\right)} & \text{for } (x \leq x') \\[3em]
\displaystyle\sum_{m=1}^{\infty} \frac{2}{m\pi\epsilon} \sin\left(\frac{m\pi y'}{b}\right) \sin\left(\frac{m\pi y}{b}\right) \frac{\sinh\left(\frac{m\pi(a - x)}{b}\right) \sinh\left(\frac{m\pi x'}{b}\right)}{\sinh\left(\frac{m\pi a}{b}\right)} & \text{for } (x \geq x')
\end{cases}
$$

$$\tag{3.69a}$$

It may be observed that the Green's function is symmetric in x and x'; and y and y'. The advantage of single series expansion of (3.69a) over the double series expansion of (3.61) lies in the computational speed for the single series. Also, expression (3.69a) can be easily generalized to the parallel plate case. The given rectangular pipe geometry, Figure 3.4, will reduce to the parallel plate geometry as $a \to \infty$. In the limit, (3.69a) reduces to

$$G(x, y; x', y') =$$

$$
\begin{cases}
\displaystyle\sum_{m=1}^{\infty} \frac{2}{m\pi\epsilon} \sin\left(\frac{m\pi y'}{b}\right) \sin\left(\frac{m\pi y}{b}\right) \sinh\left(\frac{m\pi x}{b}\right) \exp\left(\frac{-m\pi x'}{b}\right) & \text{for } (x \leq x') \\[2em]
\displaystyle\sum_{m=1}^{\infty} \frac{2}{m\pi\epsilon} \sin\left(\frac{m\pi y'}{b}\right) \sin\left(\frac{m\pi y}{b}\right) \sinh\left(\frac{m\pi x'}{b}\right) \exp\left(\frac{-m\pi x}{b}\right) & \text{for } (x \geq x')
\end{cases}
$$

$$\tag{3.69b}$$

A formula similar to (3.69a) can be obtained by representing the Green's function as a single Fourier sine series with respect to x. The solution is

$$G(x, y; x', y') =$$

$$
\begin{cases}
\displaystyle\sum_{m=1}^{\infty} \frac{2}{m\pi\epsilon} \sin\left(\frac{m\pi x'}{a}\right) \sin\left(\frac{m\pi x}{a}\right) \frac{\sinh\left(\dfrac{m\pi(b-y')}{a}\right) \sinh\left(\dfrac{m\pi y}{a}\right)}{\sinh\left(\dfrac{m\pi b}{a}\right)} & \text{for } (y \leq y') \\[4em]
\displaystyle\sum_{m=1}^{\infty} \frac{2}{m\pi\epsilon} \sin\left(\frac{m\pi x'}{a}\right) \sin\left(\frac{m\pi x}{a}\right) \frac{\sinh\left(\dfrac{m\pi(b-y)}{a}\right) \sinh\left(\dfrac{m\pi y'}{a}\right)}{\sinh\left(\dfrac{m\pi b}{a}\right)} & \text{for } (y \geq y')
\end{cases}
$$

$$(3.70a)$$

Generalizing (3.70a) to a parallel plate case with plate separation a $(b \to \infty)$, one obtains

$$G(x, y; x', y') =$$

$$
\begin{cases}
\displaystyle\sum_{m=1}^{\infty} \frac{2}{m\pi\epsilon} \sin\left(\frac{m\pi x'}{a}\right) \sin\left(\frac{m\pi x}{a}\right) \sinh\left(\frac{m\pi y}{a}\right) \exp\left(\frac{-m\pi y'}{a}\right) & \text{for } (y \leq y') \\[2em]
\displaystyle\sum_{m=1}^{\infty} \frac{2}{m\pi\epsilon} \sin\left(\frac{m\pi x'}{a}\right) \sin\left(\frac{m\pi x}{a}\right) \sinh\left(\frac{m\pi y'}{a}\right) \exp\left(\frac{-m\pi y}{a}\right) & \text{for } (y \geq y')
\end{cases}
$$

$$(3.70b)$$

The Green's functions derived above can be used to determine the potential distribution in the pipe for an arbitrary excitation. The principle of superposition may be used for this purpose. The free space Green's functions for the Laplace and Helmholtz equations are listed in Section 3.6.

3.4.3   Green's Function in Spectral Domain

For geometries with layered dielectrics (e.g., planar transmission lines), it is not possible to obtain the Green's function in closed form in the space domain. However, when the analysis is carried out in spectral domain one can determine the Green's function in a closed form. The procedure is described for microstrip line in Chapter 5.

3.5   Green's Function for Probe Excitation of *TE*-Modes in Rectangular Waveguide

The Green's function may be used to determine the distribution for an arbitrary excitation using superposition such as (3.3). We shall illustrate the procedure for probe excitation of modes in rectangular waveguide. A waveguide may support a large number of modes but the actual excitation of the modes is decided by the distribution of excitation current. Depending on the source function (concentrated or distributed) some of the modes may be excited with large amplitudes whereas the other modes may not be excited at all [3, 4]. We analyze one such problem next.

Consider a uniform, *unit* strength line current extending across the rectangular waveguide, parallel with the y-axis and located at (x', z') as shown in Figure 3.5. The line current is an approximation of the excitation produced by a coaxial-to-waveguide adapter. The current is assumed to be uniform along the y-coordinate, that is, it does not vary with y. The field produced by this current does not vary with y because the excitation current and the waveguide dimensions are uniform along the y-direction. Mathematically, it means $\partial/\partial y \equiv 0$; and the problem may be treated as two-dimensional. The operator ∇^2, therefore, becomes

$$\nabla^2 = \nabla_t^2 = \frac{\partial^2}{\partial x^2} + \frac{\partial^2}{\partial z^2} \tag{3.71}$$

For simplicity, we shall follow the vector potential approach to determine the field components. The potential **A** and the excitation current **J** are related through (1.39a); that is,

$$\left(\nabla_t^2 + k^2\right)\mathbf{A} = -\mu\mathbf{J}(x', z') \tag{3.72}$$

Since $\mathbf{J} = \hat{y}J_y$, (3.72) reduces to the following scalar equation:

$$\left(\nabla_t^2 + k^2\right)A_y = -\mu J_y(x', z') \tag{3.73}$$

The unit strength excitation current, located at (x', z'), is represented by the delta function as

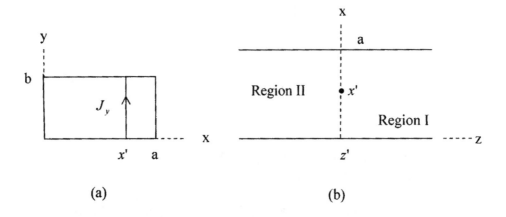

Figure 3.5 Line source excitation of *TE*-modes in a rectangular waveguide: (a) cross-sectional view; and (b) top view.

$$J_y = \delta(x - x')\,\delta(z - z') \qquad \text{amp/m} \tag{3.74}$$

Substituting for ∇_t^2 and J_y in (3.72) one obtains the PDE for the Green's function as

$$\left(\frac{\partial^2}{\partial x^2} + \frac{\partial^2}{\partial z^2} + k^2\right) A_y = -\mu\delta(x - x')\,\delta(z - z') \tag{3.75}$$

Field Components Produced by A_y
From (1.40a), we know that $\mathbf{E} = -j\omega\mathbf{A}$; therefore, $E_y = -j\omega A_y$. Also, $\mu\mathbf{H} = \nabla \times \mathbf{A}$ implies

$$\mu H_x = \frac{-\partial A_y}{\partial z}, \qquad \mu H_z = \frac{\partial A_y}{\partial x} \tag{3.76}$$

All other field components are zero because $\partial/\partial y \equiv 0$. The modes described by E_y, H_x and H_z can be classified as *TE-to-z* or *TM-to-y* modes.

Boundary Conditions on A_y
Because E_y is proportional to A_y, and E_y is tangential to the waveguide walls at $x = 0$ and $x = a$, the boundary conditions on A_y are

$$A_y = 0 \text{ at } x = 0, a \tag{3.77}$$

Conditions on A_y for Large Values of $|z|$
The fields propagate away from the excitation source along the z-direction. Therefore, the solution of (3.75) for large values of z can be written as (using single series expansion of Section 3.4.2)

$$A_y^I = \sum_{m=1}^{\infty} B_m \sin\left(\frac{m\pi x}{a}\right) e^{-k_3 z} \qquad \text{for } z > z' \tag{3.78a}$$

$$A_y^{II} = \sum_{m=1}^{\infty} C_m \sin\left(\frac{m\pi x}{a}\right) e^{+k_3 z} \qquad \text{for } z < z' \tag{3.78b}$$

The factor $\sin(m\pi x/a)$ accounts for the boundary conditions (3.77). The relationship between k, k_3 and $m\pi/a$ can be determined by substituting (3.78) in the source-free PDE corresponding to (3.75). One obtains

$$k_3 = \sqrt{(m\pi/a)^2 - k^2} \tag{3.79}$$

The coefficients B_m and C_m of (3.78) may be determined by satisfying the conditions at the source. These are as follows:

1. The magnetic field is discontinuous at the current source. This discontinuity is

$$\hat{\mathbf{n}} \times (\mathbf{H}^{\mathrm{I}} - \mathbf{H}^{\mathrm{II}}) = \mathbf{J} \qquad (3.80\mathrm{a})$$

2. The electric field is continuous across the current source; that is,

$$\hat{\mathbf{n}} \times (\mathbf{E}^{\mathrm{I}} - \mathbf{E}^{\mathrm{II}}) = 0 \qquad (3.80\mathrm{b})$$

Using $\hat{\mathbf{n}} = \hat{\mathbf{z}}$ we obtain at $z = z'$,

$$\hat{\mathbf{z}} \times \mathbf{E}^{\mathrm{I}} = \hat{\mathbf{z}} \times \mathbf{E}^{\mathrm{II}} \Rightarrow E_y^{I} = E_y^{II} \text{ or } A_y^{I} = A_y^{II} \qquad (3.81)$$

and

$$\hat{\mathbf{z}} \times (\mathbf{H}^{\mathrm{I}} - \mathbf{H}^{\mathrm{II}}) = \hat{\mathbf{y}}\mu\delta(x - x') \Rightarrow H_x^{I} - H_x^{II} = \mu\delta(x - x')$$

or

$$\frac{\partial A_y^{I}}{\partial z} - \frac{\partial A_y^{II}}{\partial z} = -\mu\delta(x - x') \qquad (3.82)$$

These source conditions can also be obtained by following (3.66) and (3.67), that is, without invoking electromagnetics. Substituting for A_y^{I} and A_y^{II} in (3.81) and (3.82) gives

$$B_m e^{-k_3 z'} = C_m e^{k_3 z'} \qquad (3.83\mathrm{a})$$

and

$$-\sum_{m=1}^{\infty} k_3 B_m \sin\left(\frac{m\pi x}{a}\right) e^{-k_3 z'} - \sum_{m=1}^{\infty} k_3 C_m \sin\left(\frac{m\pi x}{a}\right) e^{k_3 z'} = -\mu\delta(x - x') \qquad (3.83\mathrm{b})$$

We multiply (3.83b) by $\sin(m'\pi x/a)$ and integrate over $(0, a)$ to remove the summation sign

$$-k_3 \frac{a}{2}(B_m e^{-k_3 z'} + C_m e^{k_3 z'}) = -\mu \sin\left(\frac{m\pi x'}{a}\right) \qquad (3.84)$$

From (3.83a) and (3.84) one obtains

$$B_m e^{-k_3 z'} = \frac{\mu}{a k_3} \sin\left(\frac{m\pi x'}{a}\right) \qquad (3.85)$$

The Green's function for the probe excitation is therefore obtained as

$$A_y(x, z; x', z') = \frac{\mu}{a} \sum_{m=1}^{\infty} \frac{\sin\left(\dfrac{m\pi x}{a}\right) \sin\left(\dfrac{m\pi x'}{a}\right)}{k_3} e^{-k_3|z-z'|} \qquad (3.86)$$

The Fourier transform solution to the above problem is discussed in Section 5.2.

Special Case.   Let us consider the case in which the waveguide is short-circuited at $z = 0$. This situation is very similar to that which occurs in a coaxial-to-waveguide adapter. Due to the boundary condition imposed by the short, the expression for (3.78b) is modified and we write

$$A_y^I = \sum_{m=1}^{\infty} B_m \sin\left(\frac{m\pi x}{a}\right) e^{-k_3 z} \qquad \text{for } z > z' \qquad (3.87a)$$

$$A_y^{II} = \sum_{m=1}^{\infty} C_m \sin\left(\frac{m\pi x}{a}\right) \sinh(k_3 z) \qquad \text{for } z < z' \qquad (3.87b)$$

Applying the conditions at the source point, one obtains

$$B_m = \mu \frac{\sinh(k_3 z')}{a k_3} \sin\left(\frac{m\pi x'}{a}\right) \qquad (3.88a)$$

$$C_m = \mu \frac{e^{-k_3 z'}}{a k_3} \sin\left(\frac{m\pi x'}{a}\right) \qquad (3.88b)$$

Therefore, vector potential Green's function for a waveguide short-circuited at one end and excited by a filamentary current parallel to the height of the waveguide can be written as

$$A_y(x, z; x', z') = \begin{cases} \dfrac{\mu}{a} \displaystyle\sum_{m=1}^{\infty} \dfrac{\sin\left(\dfrac{m\pi x}{a}\right) \sin\left(\dfrac{m\pi x'}{a}\right)}{k_3} e^{-k_3 z} \sinh(k_3 z') & \text{for } z \geq z' \\[4mm] \dfrac{\mu}{a} \displaystyle\sum_{m=1}^{\infty} \dfrac{\sin\left(\dfrac{m\pi x}{a}\right) \sin\left(\dfrac{m\pi x'}{a}\right)}{k_3} e^{-k_3 z'} \sinh(k_3 z) & \text{for } z \leq z' \end{cases}$$

$$\qquad (3.89)$$

The electric field Green's function is obtained by using $E_y = -j\omega A_y$ and the expression for $A_y(x, z; x', z')$.

Excitation Coefficients for Various Modes

The expansion coefficient B_m represents the coefficient of excitation for the various modes at $z \geq z'$. It is given by (3.88a) with

$$k_3 = \sqrt{(m\pi/a)^2 - k^2} \tag{3.90}$$

Let us select $x' = a/2$. Then,

$$B_m = \frac{\sinh(k_3 z')}{a k_3} \sin(m\pi/2) \tag{3.91}$$

Equation (3.91) suggests that only modes with odd-m are getting excited. To maximize the excitation of a particular mode we should choose z' suitably, and to determine this value let us calculate $\sinh(k_3 z')$,

$$\sinh(k_3 z') = \sinh\left\{z'\sqrt{(m\pi/a)^2 - k^2}\right\}$$

or

$$\sinh(k_3 z') = \sinh\left\{kz'\sqrt{(m\lambda/2a)^2 - 1}\right\} \qquad \text{for } k = 2\pi/\lambda \tag{3.92}$$

For $m = 1$ and $\lambda < 2a$,

$$\sinh(k_3 z') = j \sin\left\{kz'\sqrt{1 - (\lambda/2a)^2}\right\} \tag{3.93}$$

The factor $\sinh(k_3 z')$ is maximum when $kz'\sqrt{1 - (\lambda/2a)^2} = \pi/2$.

For $m = 3$, k_3 is positive and $e^{-k_3 z'}$ represents a decaying wave. Therefore, when the current source is located at $x' = a/2$ it excites only TE_{10}-mode. All other modes are evanescent for $\lambda < 2a$.

We have determined Green's function for the problems when the domain is bounded and the solution is in the form of series expansion. It provides an alternative to the eigenfunction expansion method of Section 2.5 for the solution of inhomogeneous differential equations. In the language of operators, the inhomogeneous PDE may be written as

$$L(f) = g \tag{3.94}$$

The corresponding Green's function PDE is obtained by replacing the excitation term by the delta function

$$LG(\mathbf{r};\mathbf{r}') = \delta(\mathbf{r} - \mathbf{r}') \tag{3.95}$$

where $\mathbf{r}$ is the field position vector and $\mathbf{r}'$ is the source position vector. Once the Green's function has been obtained, the solution to the inhomogeneous PDE of (3.94) in terms of Green's function is expressed as

$$f(\mathbf{r}) = \int_\tau G(\mathbf{r};\mathbf{r}') g(\mathbf{r}')\, d\tau' \tag{3.96}$$

This is an integral equation. Green's function, therefore, provides a useful link between the differential and integral equations.

3.6   Green's Function for Unbounded Region

The real advantage of Green's function lies in solving problems with unbounded regions, and as a consequence the literature on radiation and scattering problems is replete with examples of Green's functions. Due to the constraint on computer resources, the unbounded region in computational methods is handled by imposing either absorbing boundary conditions or radiation conditions. This approach requires extra computing resources. The use of Green's function does not require any such termination and provides the most efficient solution. However, constructing Green's function for inhomogeneous dielectric configurations is a limitation. The method of moments (MoM) analysis, Chapter 11, is based on the availability of Green's function.

Free space Green's functions for the Laplace and Helmholz PDE are given next, and are derived in Chapter 5 employing the Fourier transform method.

One-Dimensional Cases
For the one-dimensional Laplace equation defined as

$$\frac{d^2G}{dx^2} = \delta(x - x') \tag{3.97a}$$

Green's function does not exist for the open region problem [7, p. 295]. For finite values of x and x', the Green's function is given by

$$G(x; x') = |x - x'|/2 \tag{3.97b}$$

For the one-dimensional Helmholtz equation defined as

$$\frac{d^2G}{dx^2} + k^2G = -\delta(x - x') \tag{3.98a}$$

the Green's function is given by

$$G(x; x') = \frac{j}{2k} \exp\left(jk|x - x'|\right) \tag{3.98b}$$

Two-Dimensional Cases (Line Sources)
For the two-dimensional Laplace equation defined as

$$\frac{\partial^2G}{\partial x^2} + \frac{\partial^2G}{\partial y^2} = \delta(\rho - \rho'), \qquad \text{where } \rho = \sqrt{x^2 + y^2} \tag{3.99a}$$

the Green's function is given by

$$G(\rho; \rho') = \frac{1}{2\pi} \ln|\rho - \rho'| \tag{3.99b}$$

For the two-dimensional Helmholtz equation defined as

$$\left[\frac{1}{\rho}\frac{d}{d\rho}\left(\rho\frac{d}{d\rho}\right) + k^2\right] G(\rho; \rho') = \frac{\delta(\rho - \rho')}{2\pi\rho} \tag{3.100}$$

the Green's function for outward propagation waves is given by

$$G(\rho; \rho') = \frac{-j}{4} H_0^{(2)}\left(k|\rho - \rho'|\right) \tag{3.101}$$

where $H_0^{(2)}(.)$ is a Hankel function of the second kind. The asymptotic value of the Green's function for large $|\rho - \rho'|$ is

$$\lim_{|\rho-\rho'|\to\infty} G(\rho; \rho') = \frac{1}{4j}\sqrt{\frac{2}{j\pi k|\rho - \rho'|}}\, e^{jk|\rho-\rho'|} \tag{3.102a}$$

For small values of $|\rho - \rho'|$,

$$\lim_{|\rho-\rho'|\to 0} G(\rho; \rho') = \frac{j}{4}\left(1 + j\frac{2}{\pi}\ln\frac{\gamma k|\rho - \rho'|}{2}\right) \tag{3.102b}$$

where $\gamma = 1.781\ldots$ is Euler's constant.

Three-Dimensional Cases (Point Sources)
For the Laplace equation in three dimensions defined as

$$\nabla^2 G = \delta(\mathbf{r} - \mathbf{r}') \tag{3.103}$$

the Green's function is given by

$$G(\mathbf{r};\mathbf{r}') = \frac{-1}{4\pi(\mathbf{r} - \mathbf{r}')} \tag{3.104}$$

For the Helmholtz equation in three dimensions defined as

$$\nabla^2 G + k^2 G = \delta(\mathbf{r} - \mathbf{r}') \tag{3.105}$$

the Green's function is given by

$$G(\mathbf{r};\mathbf{r}') = \frac{-1}{4\pi|\mathbf{r} - \mathbf{r}'|}\exp\left(jk|\mathbf{r} - \mathbf{r}'|\right) \tag{3.106}$$

3.7 Summary

Green's function is one of the very useful tools for solving problems in electromagnetics. It has the same meaning and importance as the impulse response for circuits. Green's function describes the response of an electromagnetic system to a delta function source. Once the Green's function is known, the response to an arbitrary excitation can be obtained using superposition. Two well-known techniques for determining the Green's function are illustrated through a number of examples. These techniques are direct construction approach and eigenfunction expansion. Green's functions for the unbounded region are listed at the end. These Green's functions find application in the method of moments solution of boundary value problems.

References

[1] Butkov, E., *Mathematical Physics*, Reading, MA: Addison-Wesley, 1968.

[2] Weeks, W. L., *Electromagnetic Theory for Engineering Applications*, New York: John Wiley, 1964.

[3] Harrington, R. F., *Time-Harmonic Electromagnetic Fields*, New York: McGraw-Hill, 1961.

[4] Collin, R. E., *Field Theory of Guided Waves*, 2nd ed., New York: IEEE Press, 1991, Chapter 2.

[5] Van Bladel, J., *Electromagnetic Fields*, Washington, D.C.: Hemisphere Publishing Corp., 1985.

[6] Kong, J. A., *Electromagnetic Wave Theory*, New York: John Wiley, 1986.

[7] Sadiku, M. N. O., *Numerical Techniques in Electromagnetics*, 2nd ed., Boca Raton, FL: CRC Press, 2001.

[8] Eom, H. J., *Electromagnetic Wave Theory for Boundary Value Problems*, Berlin: Springer, 2004.

Problems

P3.1. Use the Green's function method to solve the following differential equation:

$$\frac{d^2\phi}{dx^2} = -S(x)$$

with

$$S(x) = \begin{cases} 1 & \text{for } 0 \leq x \leq 1 \\ 0 & \text{otherwise} \end{cases}$$

subject to the boundary conditions $\phi(0) = 0$; $\phi(1) = 2$.

P3.2. Consider a transmission line of characteristic impedance Z_0, and propagation constant γ terminated by loads at its two ends. This line is excited by a point-current-source at $z = z'$ as shown in Figure 3.3. Show that the line voltage is given by

$$v(z) = \begin{cases} \dfrac{Z_0}{2}(e^{\gamma z} + \rho_1 e^{-\gamma z})\dfrac{e^{-\gamma z'} + \rho_2 e^{\gamma(z' - 2L)}}{1 - \rho_1 \rho_2 e^{-2\gamma L}} & \text{for } z \leq z' \\[4mm] \dfrac{Z_0}{2}(e^{\gamma z'} + \rho_1 e^{-\gamma z'})\dfrac{e^{-\gamma z} + \rho_2 e^{\gamma(z - 2L)}}{1 - \rho_1 \rho_2 e^{-2\gamma L}} & \text{for } z \geq z' \end{cases}$$

where ρ_1 and ρ_2 are the reflection coefficients of the loads Z_1 and Z_2, respectively.

P3.3. A two-section transmission line with a unit-current-source at the plane $z = h$ and short-circuits at the two ends is shown in Figure 3.6. The lengths and the characteristic impedances of these sections are also shown there.

1. Write down the differential equation and boundary conditions satisfied by the voltage Green's function.
2. Derive an expression for the current at the plane $z = a$.

P3.4. Consider a three-section transmission line with a current source I_s at the plane $z = h_1$ and the short-circuits at the two ends as shown in Figure 3.7. The lengths and the characteristic impedances of the three sections are: Z_{01}, h_1; Z_{02}, h_2; Z_{03}, h_3.

1. Write down the differential equation and the boundary conditions satisfied by the voltage Green's function.
2. Derive the expression for the Green's function at the plane $z = h_1$.
3. Use the result of (2) to show that the Green's function at the plane $z = h_1 + h_2$ is given by

$$G(h_1 + h_2) = G(h_1)\left[\frac{e^{-\gamma_2 h_2} + \dfrac{Z_{03}\tanh(\gamma_3 h_3) - Z_{02}}{Z_{03}\tanh(\gamma_3 h_3) + Z_{02}}e^{\gamma_2 h_2}}{1 + \dfrac{Z_{03}\tanh(\gamma_3 h_3) - Z_{02}}{Z_{03}\tanh(\gamma_3 h_3) + Z_{02}}}\right]$$

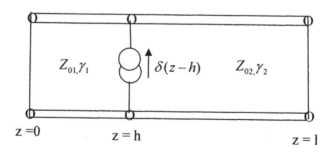

Figure 3.6 Current-source excitation of two-section transmission line shorted at the ends.

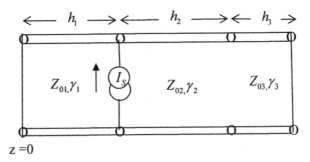

Figure 3.7 Current-source excitation of three-section transmission line shorted at the ends.

where $G(h_1)$ is the Green's function at the plane $z = h_1$. The result of this problem is useful in a three-layered planar transmission line when the spectral domain approach is used to determine the Green's function.

P3.5. Consider a transmission line of characteristic impedance Z_0, and phase constant β terminated by source impedance Z_0 at one end and load impedance Z_L at the other end. This line is excited by a point-voltage-source V_0 at $z = 0$ as shown in Figure 3.8. Show that the line current for $z > 0$ is given by

$$I(z) = -\frac{V_0}{2Z_0}\left[e^{j\beta z} + \frac{Z_0 + Z_L}{Z_0 - Z_L}e^{2j\beta\ell - j\beta z}\right]$$

P3.6. Consider an electric line source radiating in free space above a grounded dielectric sheet of thickness h, and as shown in Figure 3.9. Assume the line source to be time harmonic and uniform in the z-direction.

1. Write down the Helmholtz equation satisfied by the magnetic vector potential **A**.
2. Derive an expression for **A**.

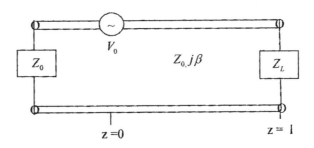

Figure 3.8 Voltage-source excitation of a transmission line loaded at one end and terminated in matched load at the other end.

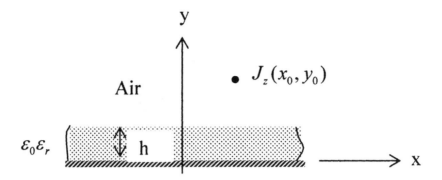

Figure 3.9 Line source excitation of a grounded dielectric slab.

P3.7. Consider a uniform transmission line of characteristic impedance Z_0 and propagation constant γ terminated in matched loads at both ends. The transmission line has a unit-point-voltage source at $z = a$ (i.e., $V_a = \delta(z - a)$), and a unit-point-current source at $z = b$ (i.e., $I_b = \delta(z - b)$), as shown in Figure 3.10. Show that

$$I_b(a) = -V_a(b)$$

that is, the current at the point $z = a$ due to a unit-point-current source at $z = b$ is the negative of the voltage at $z = b$ due to a unit-point-voltage source at $z = a$ (*Reciprocity Theorem*).

P3.8. *Excitation of modes in a parallel plate waveguide* [3, p. 194]. Let there be a sheet of x-directed current J_x over $z = 0$ plane of a parallel plate waveguide formed by conductors over the $y = 0$ and $y = b$ planes as shown in Figure 3.11. The guide is matched in both the $+z$ and $-z$ directions. Show that the field produced by the current sheet is

$$E_x = \sum_{n=1}^{\infty} B_n \sin(n\pi y/b)e^{-\gamma_n|z|} \qquad \gamma_n^2 = (n\pi/b)^2 - k_0^2$$

where

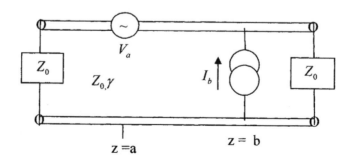

Figure 3.10 Voltage and current source excitation of a transmission line.

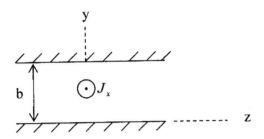

Figure 3.11 Current source excitation of modes in a parallel plate waveguide.

$$B_n = \frac{j\omega\mu}{\gamma_n b} \int\limits_0^b J_x(y) \sin(n\pi y/b) \, dy$$

Use $J_x(y) = I_0 \delta(y - d)$, and determine the value of d/b for which B_1, the amplitude of TE_{10} mode, is maximum.

P3.9. Consider the mode functions $\psi_{mn} = \cos(m\pi x/a)\cos(n\pi y/b)$; $m, n = 0, 1, 2, 3 \ldots$ defined over the interval $0 \le x \le a$, $0 \le y \le b$, $0 \le z \le d$ in a microstrip rectangular cavity with magnetic walls at the four sides and electric walls at the top and bottom.

1. Show that the mode functions are orthogonal.
2. Normalize these mode functions.
3. Use the normalized mode functions to expand the function

    $$f(x, y) = j\omega\mu_0 d\delta(x - x')\delta(y - y')$$

4. Employ this expansion to show that

    $$\int\limits_0^b \int\limits_0^a \delta(x - x')\delta(y - y') \, dx' \, dy' = 1$$

P3.10. Write down the eigenfunctions for the Laplace's equation

$$\nabla^2 \phi(x, y) = 0 \qquad 0 \le x \le a, 0 \le y \le b$$

subject to the following boundary conditions:

$$\phi = 0 \text{ at } y = 0, b$$
$$\phi = 0 \text{ at } x = a$$
$$\partial\phi/\partial x = 0 \text{ at } x = 0$$

1. Use these eigenfunctions to obtain the solution of

$$-\nabla^2 \phi(x, y) = \sin(2\pi y/b)$$

2. Verify that the solution satisfies the differential equation and the boundary conditions.
3. Solve the differential equation in (1) above using the Green's function approach, and compare the two solutions.

P3.11. Solve the following Poisson's equation:

$$\nabla^2 \phi(x, y) = -1$$

in the range $0 \le x \le a$, $0 \le y \le b$, subject to the condition $\phi = 0$ on the boundary.

P3.12. Consider a rectangular waveguide with a line source of magnetic current $\mathbf{J}^m$ located at (x', y'). The magnetic current is z-directed and uniform along z (i.e., $\partial/\partial z \equiv 0$). The vector potential $\mathbf{A}^m$ will have z-component only and satisfies the following differential equation:

$$\nabla^2 A_z^m(x, y) + k_0^2 A_z^m(x, y) = -J_z^m$$

and the boundary conditions

$$\frac{\partial A_z^m}{\partial x} = 0 \text{ at } x = 0, a; \qquad \frac{\partial A_z^m}{\partial y} = 0 \text{ at } y = 0, b$$

Assuming the current density to be

$$J_z^m(x', y') = J_0 \delta(x - x') \delta(y - y')$$

obtain the expression for A_z^m using the *direct construction approach*.

P3.13. Consider a rectangular waveguide with a line source of electric current $\mathbf{J}$ located at (x', y'). The electric current is z-directed. The vector potential $\mathbf{A}$ will have z-component only and satisfies the following differential equation:

$$\nabla^2 A_z + k_0^2 A_z = -\mu_0 J_z$$

The boundary conditions to be satisfied by A_z are: $A_z = 0$ at $x = 0, a$; $y = 0$, b. Assuming the current density to be

$$J_z(x', y', z) = J_0 \delta(x - x') \delta(y - y') e^{-j\beta z}$$

obtain the expression for A_z.

P3.14. Consider the excitation of *TE*-modes by a probe in a rectangular waveguide of Figure 3.5. Assume the current on the probe as

$$J_y(x, y, z) = \delta(x - x')\,\delta(z - z')\,\cos(qy)$$

What are the mode amplitudes that are excited by this probe? Where should one place the probe to achieve maximum excitation for the TE_{10}-mode [6, p. 210]?

P3.15. A strip transmission line consists of a metal strip sandwiched between two parallel plates as shown in Figure 3.12. Assume that the strip is infinitely thin of width W and the plate separation is b. The space between the parallel plates is filled with a homogeneous medium characterized by $\epsilon_0\,\epsilon_r$ and μ_0. Solve the Poisson equation to determine the potential Green's function and show that it is given by

$$G(x, y; x', y') = \frac{1}{\pi\epsilon_0\,\epsilon_r}\sum_{n=1}^{\infty}\frac{1}{n}\sin\left(\frac{n\pi y}{b}\right)\sin\left(\frac{n\pi y'}{b}\right)e^{-n\pi|x - x'|/b}$$

P3.16. The free space one-dimensional Green's function for the Helmholtz equation is the solution of

$$\frac{d^2 G}{dx^2} + k_0^2 G = -\delta(x - x')$$

subject to the radiation condition $G(\pm\infty, x') = 0$. Use direct construction method and

$$G(x; x') = \begin{cases} Ae^{jk_0 x} & \text{for } x > x' \\ Be^{-jk_0 x} & \text{for } x < x' \end{cases}$$

to show that [8, p. 189]

$$G(x; x') = \frac{j}{2k}\exp\left(jk|x - x'|\right)$$

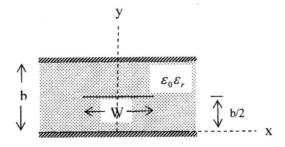

Figure 3.12 Geometry of a strip line.

CHAPTER 4

Contour Integration and Conformal Mapping

Our interest in the functions of a complex variable is because of their applications to solving improper integrals and the solution of Laplace equation based on conformal transformation. In this chapter we first review the fundamentals of functions of a complex variable, including analytic functions, calculus of residues, Cauchy's integral theorem, and improper integrals. Then we discuss conformal mapping and Schwarz-Christoffel transformation. Applications to planar transmission lines are included.

4.1 Introduction

The lossless systems are idealizations of real-world systems which are lossy in nature. While the characteristics of the lossless system are described by real numbers, the losses are accounted for by adding an imaginary part to the characteristics and therefore represented by complex numbers. For example, a lossy transmission line is described by characteristic impedance and propagation constant which are complex in nature. When the losses are small, perturbation approach may be employed to include the effect of losses. Complex numbers may be considered as variables if either its real part or the imaginary part, or both, vary.

A complex variable z may be expressed in the Cartesian form as

$$z = x + jy \tag{4.1}$$

where x and y are real variables. Polar form of the complex variable is given by

$$z = re^{j\theta} \tag{4.2}$$

These representations are shown in Figure 4.1.

Representation of a Function of Complex Variable
A function of a complex variable z may be represented as

$$f(z) = f(x, y) \tag{4.3}$$

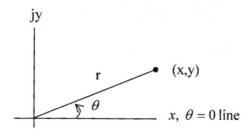

Figure 4.1   Cartesian and polar representations of a complex number.

Analogous to the complex variable (4.1), the function $f(z)$ may be written as a sum of two functions $u(x, y)$ and $v(x, y)$, each of which is a function of two real variables x and y, and is defined as

$$f(z) = u(x, y) + jv(x, y) \qquad (4.4)$$

Applications of Complex Variables
(i) For an analytic function $f(z)$ (defined later), the functions u and v satisfy the Laplace equation,

$$\nabla_t^2 u = 0, \text{ and } \nabla_t^2 v = 0 \qquad (4.5)$$

Let the function u describe a two-dimensional electrostatic potential. The function v, which describes a family of curves orthogonal to the function u, may then be used to describe the electric field E.

(ii) In many problems, mapping in the complex plane permits us to create a geometry that is easier to analyze. For example, conformal transformation has been used to transform coaxial line, and planar lines to parallel plate geometry. It is easier to determine the capacitance for a parallel plate line.

(iii) Some second order differential equations may be solved by the power series expansion. The same power series may be used in the complex plane. Taylor series expansion and analytic continuation can be used for extending the region in which the solution is valid.

(iv) The integration in the complex plane has a number of useful applications. These include: evaluation of definite improper integrals, obtaining asymptotic solutions of differential equations, and inverting integral transforms.

Next, we define some of the important properties of $f(z)$.

4.1.1   Analytic Function

A function $f(z)$ is said to be analytic at a point $z = z_0$ if the derivative $f'(z)$ exists at z_0 and in some small region around z_0. The definition for $f'(z)$ is identical in form to that of the derivative of a function of a real variable; that is,

$$f'(z_0) = \lim_{\Delta z \to 0} \frac{f(z_0 + \Delta z) - f(z_0)}{\Delta z} \tag{4.6}$$

$$= \lim_{\substack{\Delta x \to 0 \\ \Delta y \to 0}} \frac{u(x_0 + \Delta x, y_0 + \Delta y) + jv(x_0 + \Delta x, y_0 + \Delta y) - u(x_0, y_0) - jv(x_0, y_0)}{\Delta x + j\Delta y}$$

The limit involved is two-dimensional in nature because $\Delta z = \Delta x + j\Delta y$. Therefore, many results from the calculus of real variables do not carry over to the calculus of complex variables [1]. Use of Taylor series expansion in (4.6) shows that $f'(z)$ exists if

$$\frac{\partial u}{\partial x} = \frac{\partial v}{\partial y} \tag{4.7a}$$

and

$$\frac{\partial u}{\partial y} = -\frac{\partial v}{\partial x} \tag{4.7b}$$

The expressions (4.7) are called *Cauchy-Riemann conditions* for a function to be analytic [1–5]. In words, a function $f(z)$ is analytic at $z = z_0$ provided: (1) the first-order partial derivatives of u and v exist in the neighborhood of z_0, (2) these partial derivatives are continuous, and (3) satisfy (4.7) at z_0. A function $f(z)$ is not analytic at z_0 if: (1) $z = z_0$ is a singular point, or (2) $z = z_0$ is a branch point or if $f(z)$ is multivalued at z_0, or (3) the derivatives of all orders of $f(z)$ do not exist at z_0. Henceforth, it will be assumed that we are working with analytic functions.

4.1.2 Analytic Continuation

Analytic continuation is a process of extending the region in which the function is defined. The function can be extended indefinitely, through analytic continuation, if the function has isolated singularities only [2]. Let D_1 and D_2 be the two regions that overlap (i.e., $D_1 \cap D_2 \neq 0$ as shown in Figure 4.2). If f_1 is analytic on D_1, f_2 is analytic on D_2, and $f_1(z) \equiv f_2(z)$ on $D_1 \cap D_2$, then f_2 is called the analytic continuation of f_1 from D_1 into D_2. In practice, we determine a single function $F(z)$ such that

$$F(z) = f_1(z) \qquad \text{when } z \text{ is in } D_1 \tag{4.8a}$$

$$= f_2(z) \qquad \text{when } z \text{ is in } D_2 \tag{4.8b}$$

and is analytic in the domain $D_1 \cap D_2$. We shall use analytic continuation to determine improper integrals along the real axis.

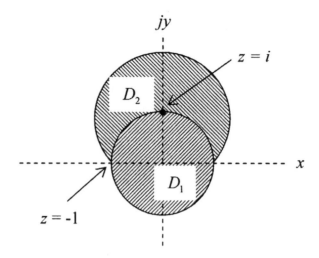

Figure 4.2 Definition of analytic continuation.

4.2 Calculus of Residues

4.2.1 Poles and Branch-Point Singularities

A function of a complex variable may have two types of singularities: pole singularity and branch-point singularity. The pole singularity is characterized by the value of the function blowing to infinity at the singular point. Mathematically, a function $f(z)$ has pole at $z = z_0$ if $\underset{z \to z_0}{lt} \; |f(z)| \to \infty$. A pole is of order m if $f(z)$ is of the form $f(z) = b/(z - z_0)^m$. A pole of order $m = 1$ is called a simple pole.

The branch-point singularity is associated with the ambiguity produced by the multivalued nature of the functions such as $f(z) = (z - z_0)^\alpha$, where $|\alpha| < 1$. The branch-point is determined by setting $f(z)$ to zero and the number of values of the function is given by $1/|\alpha|$. For example, the function $f(z) = z^{1/2}$ has branch-point at $z = 0$. The function is double-valued because for any value of z defined as $z = (\rho, \varphi + 2n\pi)$ in the neighborhood of $z = 0$, there are two distinct values of $f(z)$. The double value behavior is a complex analogue of $y^2 = x$ in which two values of y, $\pm y$ correspond to each value of x. Similarly, the function $f(z) = (z^2 - 1)^{1/3}$ is triple-valued with branch points at $z = \pm 1$.

4.2.2 Cauchy Integral Theorem

If a function $f(z)$ is analytic in some region R (see Figure 4.3), then for every closed path C in R the line integral of $f(z)$ around C is zero; that is,

$$\oint_c f(z) \, dz = 0 \tag{4.9}$$

Conventionally, the positive direction of the line integral is taken to be counter-clockwise direction, as shown in the figure.

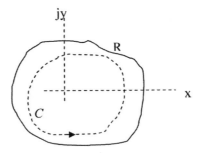

Figure 4.3 Closed contour C within a region R in which the function $f(z)$ is analytic.

Corollary: Cauchy Integral Formula
If a function $f(z)$ is analytic on a closed contour C and within the interior region bounded by C, then

$$\oint_c \frac{f(z)}{z - z_0}\, dz = \begin{cases} 2\pi j f(z_0) & \text{if } z_0 \text{ is interior to } C \\ 0 & \text{if } z_0 \text{ is exterior to } C \end{cases} \tag{4.10}$$

For the point z_0 interior to C, (4.10) gives

$$f(z_0) = \frac{1}{2\pi j} \oint_c \frac{f(z)}{z - z_0}\, dz \tag{4.11}$$

This expression may be interpreted to mean that the value of an analytic function at an interior point z_0 can be determined once the value of the function on the contour is specified.

Proof. Although $f(z)$ is assumed analytic, the integrand $f(z)/(z - z_0)$ is not analytic at z_0 because $z = z_0$ is a singular point. The contour may be deformed as shown in Figure 4.4 to avoid the singularity. It need not be circular as shown. The circle C_2 is of radius ρ. In the limiting case of $\rho \to 0$, the modified contour will approach the original contour. The function $f(z)/(z - z_0)$ is analytic over the deformed contour and the Cauchy integral theorem can be applied. The contour integral may be split into a number of line integrals as

$$\oint_c \frac{f(z)}{z - z_0}\, dz = \int_{C_1} \frac{f(z)}{z - z_0}\, dz + \int_D^C \frac{f(z)}{z - z_0}\, dz - \int_{C_2} \frac{f(z)}{z - z_0}\, dz + \int_B^A \frac{f(z)}{z - z_0}\, dz = 0 \tag{4.12}$$

The value of the integral is set to zero because the point z_0 is outside the contour. The line integrations over DC and BA, being parallel to each other and oppositely directed, cancel each other. Therefore,

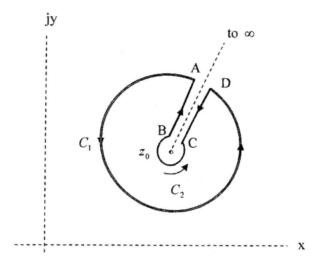

Figure 4.4 A possible construction of contour for evaluating a contour integral with pole at $z = z_0$.

$$\int_{C_1} \frac{f(z)}{z - z_0} \, dz - \int_{C_2} \frac{f(z)}{z - z_0} \, dz = 0 \tag{4.13}$$

In the limiting case when the lines CD and AB merge, the contour C_1 becomes the original contour c, and C_2 becomes a circle. Therefore, (4.13) may be written as

$$\oint_c \frac{f(z)}{z - z_0} \, dz - \int_{C_2} \frac{f(z)}{z - z_0} \, dz = 0 \tag{4.14}$$

The second integral can be evaluated easily because contour C_2 is a circle by construction. Let the points on the circle be defined as

$$z = z_0 + \rho e^{j\theta} \qquad dz = j\rho e^{j\theta} \, d\theta \tag{4.15}$$

Polar representation for z is used here because of the circular shape of the contour. Here ρ is small and will eventually be made to approach zero. We have,

$$\oint_{C_2} \frac{f(z)}{z - z_0} \, dz = \underset{\rho \to 0}{Lt} \; j \int_0^{2\pi} \frac{f(z_0 + \rho e^{j\theta})}{\rho e^{j\theta}} \rho e^{j\theta} \, d\theta$$

$$= jf(z_0) \int_0^{2\pi} d\theta \tag{4.16}$$

$$= 2\pi j f(z_0)$$

Substitution in (4.14) yields

$$\oint_C \frac{f(z)}{z - z_0}\, dz = 2\pi j f(z_0) \tag{4.17}$$

In the Cauchy integral formula, the integrand has a simple pole at $z = z_0$, and $f(z)$ was analytic over the contour. What about those integrals where the integrand cannot be expressed in such a simple form? Residue theorem provides the solution in such cases.

4.2.3   Residue Theorem

Let C be a closed contour within and on which a function $f(z)$ is analytic except for a finite number of poles at $z_1, z_2, z_3, \ldots$ interior to C as shown in Figure 4.5. The shape of the contour is arbitrary so long as it passes about the poles as shown. If $K_1, K_2, K_3, \ldots K_n$ denote the residues of $f(z)$ at these poles, then

$$\oint_C f(z)\, dz = 2\pi j (K_1 + K_2 + K_3 + \ldots + K_n) \tag{4.18}$$

where the integral is taken counter-clockwise around C. To apply the residue theorem to a given problem the contour should be drawn to avoid the poles as in Figure 4.4.

Determination of Residues Corresponding to the Simple Poles
For this purpose, the given function $f(z)$ is expanded in Laurent series about the pole. For a pole at $z = z_0$, the Laurent series can be written as

$$f(z) = \sum_{n=-\infty}^{\infty} a_n (z - z_0)^n \tag{4.19}$$

For a simple pole the coefficients for $n < -1$ are zero. Therefore,

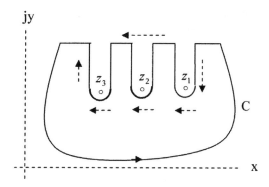

Figure 4.5   Modification of the contour of Figure 4.4 when there are a number of poles.

$$f(z) = \frac{a_{-1}}{z - z_0} + a_0 + \sum_{n=1}^{\infty} a_n(z - z_0)^n \qquad (4.20)$$

The coefficient a_{-1} is called the residue corresponding to the simple pole, and is determined as

$$a_{-1} = f(z)(z - z_0)|_{z=z_0} \qquad (4.21)$$

since all other terms of $f(z)(z - z_0)$ become zero at $z = z_0$. Residue for a simple pole at z_0 is therefore given by

$$\text{Residue} = f(z)(z - z_0)|_{z=z_0} \qquad (4.22)$$

Example 4.1.   Determine the poles and corresponding residues for the function

$$f(z) = \frac{1}{(z^2 + 1)}$$

Solution.   The function $f(z)$ has simple poles at $z = \pm j$. The corresponding residues are:

$$\text{Residue at } z = j: f(z)(z - j)|_{z=j} = \frac{1}{z + j}\bigg|_{z=j} = \frac{1}{2j} \qquad (4.23a)$$

$$\text{Residue at } z = -j: f(z)(z + j)|_{z=-j} = \frac{1}{z - j}\bigg|_{z=-j} = -\frac{1}{2j} \qquad (4.23b)$$

We now use the Cauchy integral formula and the residue theorem to determine improper integrals.

4.3   Evaluation of Definite Improper Integrals

We come across improper integrals in the solution of partial differential equations (PDE). For example, the solution of inhomogeneous PDE using the Green's function method (Chapter 3) may be expressed as

$$\phi(x) = \int G(x; x')f(x')\, dx' \qquad (4.24)$$

Similarly, the use of Fourier transform method (Chapter 5) gives rise to a solution in the form of inverse Fourier integral:

$$f(x) = \frac{1}{\sqrt{2\pi}} \int_{-\infty}^{\infty} \tilde{f}(\alpha)e^{-j\alpha x}\, d\alpha \qquad (4.25)$$

In the above cases, the integration is carried out along the real axis, and some of these integrals are improper integrals in the sense that the integrands may have poles and/or branch-point singularities over the range of integration. Such integrals can be evaluated by means of contour integration in the complex plane.

4.3.1   Improper Integral Along the Real Axis

Let us consider the following integral:

$$I = \int_{-\infty}^{\infty} f(x)\, dx \qquad (4.26)$$

where the function $f(x)$ has a pole at $x = x_0$. One may write the above integral as

$$I = \lim_{R \to \infty} \int_{-R}^{R} f(x)\, dx \qquad (4.27)$$

There is a similarity of form between the integrals of (4.18) and (4.27). In both the cases, the integrand is singular. However, the integration in (4.18) is carried out over a contour in the complex plane, whereas it is carried out along the real axis in (4.27). In order to solve (4.27), we modify the integration along the real axis into a contour integral in the complex plane such that the real axis is included in the contour as in Figure 4.6(a). Also, the function $f(x)$ is analytically continued into the upper half plane $\mathrm{Im}(z) \geq 0$. Its analytic continuation is called $f(z)$ and is obtained by replacing the real variable x by the complex variable z.

The integral $\lim_{R \to \infty} \int_{-R}^{R} f(x)\, dx$ is therefore evaluated as $\oint_C f(z)\, dz$ and the value of I determined. The closed contour facilitates application of Cauchy integral

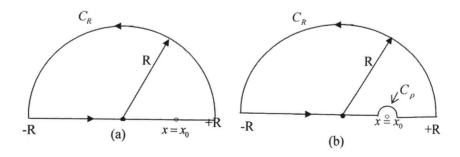

Figure 4.6 (a) An improper integral with integration along the real axis, analytically continued in the upper half z-plane for integration as a contour integral. (b) Contour of (a) deformed to exclude the pole.

formula and residue theorem. The curve C includes the range of actual integration, and a semicircle C_R of very large radius; that is,

$$\oint_C f(z)\, dz = \underset{R \to \infty}{lt} \left[\int_{-R}^{R} f(x)\, dx + \int_{C_R} f(z)\, dz \right] \tag{4.28}$$

The above approach is applicable if the function $f(z)$ satisfies the following conditions:

1. $f(z)$ should be analytic everywhere in the upper half plane defined by $\mathrm{Im}(z) \geq 0$, *except for a finite number of isolated singular points*.
2. $f(z)$ should vanish as strongly as $1/z^2$ for $|z| \to \infty$, $0 \leq \theta \leq \pi$. This condition implies that the integrand approaches zero over the semicircle C_R, and the contribution of arc C_R to the integral vanishes.

To apply the Cauchy integral formula, the integration along the closed contour C is now deformed to exclude the pole at $z = z_0 = x_0$, Figure 4.6(b). However, in the immediate vicinity of the isolated pole the integrand is analytic so that the deformation around the pole is in the form of a semicircle of vanishingly small radius. With this construction, (4.28) may be expressed as

$$\oint_C f(z)\, dz = \underset{R \to \infty}{lt} \left[\int_{C_R} f(z)\, dz + \int_{-R}^{z_0 - \rho} f(z)\, dz + \int_{C_\rho} f(z)\, dz + \int_{z_0 + \rho}^{+R} f(z)\, dz \right] \tag{4.29}$$

Of the various integrals we note that:

1. $\oint_C f(z)\, dz = 2\pi j \times$ residue of $f(z)$, for the pole included in the contour (by Cauchy integral formula). The value of this integral is zero here, because the pole is located outside the contour.

2. $\underset{R \to \infty}{lt} \int_{C_R} f(z)\, dz = 0$, in view of assumption 2 made above.

3. $\int_{C_\rho} f(z)\, dz = -j\pi \times$ residue of $f(z)$ at $z = z_0 = x_0$. The negative sign on the right-hand side appears because the semicircle C_ρ is traversed in the clockwise direction.

In view of 3 and the limit $\rho \to 0$, (4.29) becomes

$$\oint_C f(z)\ dz = \lim_{R \to \infty} \int_{-R}^{R} f(x)\ dx + \lim_{R \to \infty} \int_{C_R} f(z)\ dz - j\pi \times \text{residue of } f(z) \text{ at } z = x_0$$

$$(4.30)$$

Use of 1 and 2 gives

$$\int_{-\infty}^{\infty} f(x)\ dx = j\pi \times \text{residue of } f(z) \text{ at } z = x_0 \qquad (4.31)$$

Corr: In the above example, the contour may be deformed to include the pole as shown in Figure 4.7, and the integral may also be evaluated using the procedure described above. However, there are some modifications in the values of some of the integrals. These are:

1. $\oint_C f(z)\ dz = 2\pi j \times \text{residue of } f(z) \text{ at } z = x_0,$ because the pole is now included in the contour C.

2. $\int_{C_\rho} f(z)\ dz = j\pi \times \text{residue of } f(z) \text{ at } z = x_0.$

Substituting the values from 1 and 2 in (4.29), one obtains the same value for the given integral as (4.31).

Example 4.2.   Let us evaluate the following integral with a simple pole:

$$I = \int_{-\infty}^{\infty} \frac{1}{\alpha - k_0}\ d\alpha = \lim_{R \to \infty} \int_{-R}^{R} \frac{1}{\alpha - k_0}\ d\alpha \qquad (4.32)$$

The integrand has a simple pole at $\alpha = k_0$. Now, convert this integral into a contour integral in the complex plane through analytical continuation and call it I_C; that is,

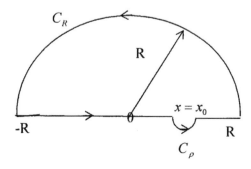

Figure 4.7   Contour of Figure 4.6(a) deformed to include the pole.

$$I_C = \oint_C \frac{1}{\alpha_c - k_0} \, d\alpha_c \tag{4.33}$$

Here, α_c is a complex variable with $\text{Re}(\alpha_c) = \alpha$. The contour C may be drawn to avoid the pole at $\alpha_c = k_0$, similar to that in Figure 4.6(b). The closed contour C is segmented as follows:

$$\oint_C \frac{1}{\alpha_c - k_0} \, d\alpha_c = \mathop{lt}_{R \to \infty} \int_{C_R} \frac{1}{\alpha_c - k_0} \, d\alpha_c + \mathop{lt}_{\substack{R \to \infty \\ \rho \to 0}} \int_{-R}^{k_0 - \rho} \frac{1}{\alpha_c - k_0} \, d\alpha_c \tag{4.34}$$

$$+ \mathop{lt}_{\rho \to 0} \int_{C_\rho} \frac{1}{\alpha_c - k_0} \, d\alpha_c + \mathop{lt}_{\substack{R \to \infty \\ \rho \to 0}} \int_{k_0 + \rho}^{R} \frac{1}{\alpha_c - k_0} \, d\alpha_c$$

Of the various integrals above, we note that:

$$(i) \ \oint_C \frac{1}{\alpha_c - k_0} \, d\alpha_c = 0, \ \text{by Cauchy integral formula} \tag{4.35a}$$

and

$$(ii) \ \mathop{lt}_{R \to \infty} \int_{C_R} \frac{1}{\alpha_c - k_0} \, d\alpha_c = 0, \ \text{since} \ \frac{1}{\alpha_c - k_0} \to 0 \ \text{as} \ \alpha_c \to \infty \tag{4.35b}$$

Substituting $\alpha_c = k_0 + \rho e^{j\phi}$ and integrating, we obtain

$$(iii) \ \mathop{lt}_{\rho \to 0} \int_{C_\rho} \frac{1}{\alpha_c - k_0} \, d\alpha_c = -j\pi \tag{4.35c}$$

Equation (4.34) therefore reduces to

$$\mathop{lt}_{\substack{R \to \infty \\ \rho \to 0}} \int_{-R}^{k_0 - \rho} \frac{1}{\alpha_c - k_0} \, d\alpha_c + \mathop{lt}_{\substack{R \to \infty \\ \rho \to 0}} \int_{k_0 + \rho}^{R} \frac{1}{\alpha_c - k_0} \, d\alpha_c = j\pi \tag{4.36}$$

Applying the limits gives

$$\int_{-\infty}^{\infty} \frac{1}{\alpha - k_0} \, d\alpha = j\pi \tag{4.37}$$

4.3.2   Fourier Transform Improper Integrals

Fourier transform integrals appear as a last step in the solution of PDE based on Fourier transform method (see Chapter 5). In general, these integrals have pole or branch-point singularities. We determine these integrals analytically next.

Consider the following Fourier transform integral:

$$I = \int_{-\infty}^{\infty} f(x) e^{j\alpha x} \, dx \tag{4.38}$$

with α real and positive. Following the approach described in the last section, the above integral can be formulated as

$$I = \lim_{R \to \infty} \int_{-R}^{+R} f(x) e^{j\alpha x} \, dx \tag{4.39}$$

This integral can be treated as a portion of the complex integral $\oint_C f(z) e^{j\alpha z} \, dz$ evaluated over the contour C as shown in Figure 4.6(a). Here, $f(z)$ is the analytic continuation of $f(x)$ in the upper-half complex plane. From the construction,

$$\oint_C f(z) e^{j\alpha z} \, dz = \int_{-R}^{R} f(x) e^{j\alpha x} \, dx + \int_{C_R} f(z) e^{j\alpha z} \, dz \tag{4.40}$$

In order to evaluate the second integral on the right side, the function $f(z)$ should satisfy certain conditions. These conditions are very similar to those described earlier for the determination of (4.28). Specifically,

1. $f(z)$ should be analytic in the upper-half plane except for a finite number of isolated singularities.
2. $f(z)$ should be uniformly convergent in the upper-half plane: that is,

$$|f(z)| = |f(\text{Re}^{j\theta})| < \epsilon \qquad \text{for } |z| \geq R$$

Under these conditions,

$$\lim_{R \to \infty} \int_{C_R} f(z) e^{j\alpha z} \, dz = 0 \tag{4.41}$$

because the factor $e^{j\alpha z}$ decays exponentially with z.

Using the contour shown in Figure 4.6(a), we have

$$\int_{-\infty}^{\infty} f(x)e^{jax}\,dx + \lim_{R\to\infty}\int_{C_R} f(z)e^{jaz}\,dz = 2\pi j \sum \begin{array}{l}\text{(residues of }f(z)e^{jaz}\\ \text{in the upper-half }z\text{-plane)}\end{array}$$

$$(4.42)$$

Since the second integral is zero, we have

$$\int_{-\infty}^{\infty} f(x)e^{jax}\,dx = 2\pi j \sum \text{(residues of }f(z)e^{jaz}\text{ in the upper-half }z\text{-plane)}$$

$$(4.43)$$

If there is a pole on the real axis, an additional term will arise due to the contribution of the integral over the semicircle $\lim_{\rho\to 0}\int_{C_\rho} f(z)e^{jaz}\,dz$.

Example 4.3.   Consider the following integral:

$$I = \int_{-\infty}^{\infty} \frac{e^{jar}}{\alpha - k}\,d\alpha \qquad k \text{ is a real number} \qquad (4.44a)$$

The integrand has a simple pole at $\alpha = k$, thus I is an improper integral. We can write the integral as

$$I = \lim_{R\to\infty} \int_{-R}^{R} \frac{e^{jar}}{\alpha - k}\,d\alpha \qquad (4.44b)$$

This integral can be treated as a portion of the complex integral $\oint_C \dfrac{e^{ja_c r}}{\alpha_c - k}\,d\alpha_c$ determined over the contour C as shown in Figure 4.6(a). To be able to apply the residue theorem we avoid the pole at $\alpha_c = k$ in some fashion. Let us do this by means of a semicircle C_ρ of small radius ρ in the upper-half z-plane as shown in Figure 4.6(b). Then we can write

$$\oint_C \frac{e^{ja_c r}}{\alpha_c - k}\,d\alpha_c = \left[\int_{-R}^{k-\rho} + \int_{C_\rho} + \int_{k+\rho}^{+R} + \int_{C_R}\right]\frac{e^{ja_c r}}{\alpha_c - k}\,d\alpha_c \qquad (4.45)$$

We can calculate the given integral in the limit $R \to \infty$, $\rho \to 0$ provided we can determine the other integrals. Of the various integrals on the right side of (4.45), we note that:

(i) $\underset{\rho \to 0}{lt} \int_{C_\rho} \dfrac{e^{j\alpha_c r}}{\alpha_c - k}\, d\alpha_c = -j\pi e^{jkr}$ (-ve sign because of clockwise movement on C_ρ)

$$(4.46)$$

(ii) $\int_{C_R} \dfrac{e^{j\alpha_c r}}{\alpha_c - k}\, d\alpha_c = 0$, because there is no pole on C_R

(iii) Similarly, $\oint_C \dfrac{e^{j\alpha_c r}}{\alpha_c - k}\, d\alpha_c = 0$

In the limit $R \to \infty$, we obtain from (4.45)

$$I = \underset{R \to \infty}{lt} \int_{-R}^{R} \frac{e^{j\alpha r}}{\alpha - k}\, d\alpha = j\pi e^{jkr} \qquad (4.47)$$

Example 4.4.   Let us consider the following integral with singularities:

$$I = \int_{-\infty}^{\infty} \frac{\alpha e^{j\alpha r}}{\alpha^2 - k^2}\, d\alpha, \qquad k > 0 \qquad (4.48)$$

This type of integral occurs in the solution of Helmholtz equation using Fourier transform method (Chapter 5). The term k may represent the resonant frequency of the resonator, for example. We use partial fractions to obtain separate integrals for the singularities at $\alpha = \pm k$; that is,

$$I = \frac{1}{2} \int_{-\infty}^{\infty} \frac{e^{j\alpha r}}{\alpha - k}\, d\alpha + \frac{1}{2} \int_{-\infty}^{\infty} \frac{e^{j\alpha r}}{\alpha + k}\, d\alpha \qquad (4.49)$$

$$= I_1 + I_2$$

Now, the integrals are converted into integrals in the complex plane and the method of contour integration can be used. There can be a number of possible contours for this problem. These are discussed next.

Case 1. The contour is drawn such that it excludes both the poles, as shown in Figure 4.8(a). From the contributions due to the analytical integrations about the semicircles, we have

$$I_1 = \pi j\, \frac{e^{jkr}}{2}, \qquad I_2 = \pi j\, \frac{e^{-jkr}}{2}$$

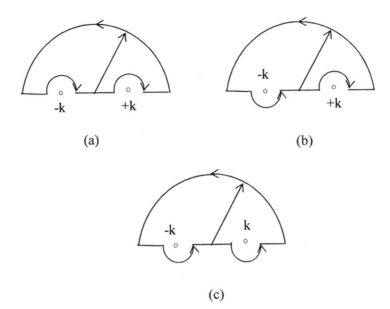

Figure 4.8 Three different possibilities for the contour for an improper integral with two poles on the real axis: (a) path going above the poles avoiding singularities; (b) the contour is drawn such that it excludes one pole, and includes the other pole; and (c) path going below both the poles to include both the singularities.

or

$$I = I_1 + I_2 = \pi j \cos(kr) \tag{4.50}$$

Case 2. The contour is drawn such that it excludes one pole, and includes the other pole as shown in Figure 4.8(b). The contributions of integrals I_1 and I_2 are found to be the same as earlier. Therefore, $I = \pi j \cos(kr)$.

Case 3. The new contour includes both the poles as shown in Figure 4.8(c). This situation can be analyzed similarly as above, and the value of the integral is found to be $I = \pi j \cos(kr)$.

Remarks. In the problem discussed above, the contour C cannot be closed in the lower-half α_c-plane because $e^{j\alpha r}$ ($\alpha = \alpha_r - j\alpha_i$ in the lower-half plane) will increase exponentially with r. However, the contour for the integral

$$I = \int\limits_{-\infty}^{\infty} \frac{\alpha e^{-j\alpha r}}{\alpha^2 - k^2} \, d\alpha \tag{4.51}$$

may be closed in the lower-half α_c-plane.

The Fourier transform integral of (4.48) produced a standing wave solution in the form of $\cos(kr)$. The propagating wave solutions for the same integral can also be obtained. For this, we attempt a slightly different approach.

Let us assume that the medium is lossy so that k is complex with $k \to k + j\epsilon$, where ϵ is positive but small and will eventually be made to approach zero; that is,

$$I(k) = \lim_{\epsilon \to 0} I(k + j\epsilon) \tag{4.52}$$

Also, the contour along the real axis need not be deformed in this case because the poles $\alpha = \pm(k + j\epsilon)$ will now shift away from the real axis due to losses. The contour for the integral

$$I = \int_{-\infty}^{\infty} \frac{\alpha e^{-j\alpha r}}{\alpha^2 - k^2} \, d\alpha \tag{4.53}$$

is shown in Figure 4.9. This integral may now be written as

$$I_c = \lim_{\epsilon \to 0} \int_{-\infty}^{\infty} \frac{\alpha_c e^{-j\alpha_c r}}{\alpha_c^2 - (k + j\epsilon)^2} \, d\alpha_c = -2\pi j \times (\text{residue at } \alpha_c = -k - j\epsilon) \tag{4.54}$$

The negative sign on the right side is introduced because of the clockwise movement on the contour C. The residue at $\alpha_c = \pm(k + j\epsilon)$ is obtained as

$$\lim_{\epsilon \to 0} \frac{e^{-j(-k-j\epsilon)r}}{2} = \frac{e^{jkr}}{2} \tag{4.55}$$

Therefore,

$$I = -\pi j e^{jkr} \tag{4.56}$$

The above solution represents an incoming wave. Substituting $k \to k - j\epsilon$ would have produced a solution in the form of an outgoing wave, e^{-jkr}. The standing wave solution $\cos(kr)$ is an average value of these two solutions and is called the Cauchy principal value of the integral; that is,

$$PV \int_{-\infty}^{\infty} \frac{\alpha e^{-j\alpha r}}{\alpha^2 - k^2} \, d\alpha = \frac{1}{2}\left[\lim_{\epsilon \to 0} \int_{-\infty}^{\infty} \frac{\alpha e^{-j\alpha r}}{\alpha^2 - (k + j\epsilon)^2} \, d\alpha + \lim_{\epsilon \to 0} \int_{-\infty}^{\infty} \frac{\alpha e^{-j\alpha r}}{\alpha^2 - (k - j\epsilon)^2} \, d\alpha \right] \tag{4.57}$$

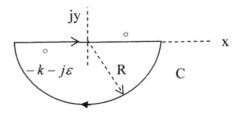

Figure 4.9 Contour for an improper integral along the real axis when the medium is lossy. The path need not be deformed near the poles.

It is now proved that three different solutions for the same integral are possible. This is true because the integral is an improper integral. The value of the integral is not uniquely defined until we define the particular limiting process, for example, averaging in this case.

Exercise.   Use the residue theorem to show that

$$\int_0^\infty \frac{dx}{1 + x^4} = \frac{\pi}{2\sqrt{2}}$$

4.3.3   Some Other Methods Useful for Solving Improper Integrals

The contour integration is perhaps the most commonly used method to solve improper integrals. Some of the other methods that may be employed are given next.

Singularity Subtraction Method
The details of this method are given in Appendix B at the end of the book. In summary, the numerator in this method is modified by subtracting a term so that the numerator and the denominator become zero simultaneously as the integration variable approaches the singular point as described below

$$\int_0^{2c} \frac{f(z)}{z - c}\, dz = \int_0^{2c} \frac{f(z) - f(c)}{z - c}\, dz + \int_0^{2c} \frac{f(c)}{z - c}\, dz \qquad (4.58)$$

The first integral is well behaved and can be evaluated numerically. The second integral is singular but simpler than the original integral; and is expected to be evaluated analytically.

Example 4.5.   Let

$$I = \int_0^2 \frac{\cos(z)}{z - 1}\, dz \qquad (4.59)$$

The integrand has a simple pole at $z = 1$. Extracting this singularity we can write

$$I = \int_0^2 \frac{\cos(z) - \cos(1)}{z - 1}\, dz + \int_0^2 \frac{\cos(1)}{z - 1}\, dz \qquad (4.60)$$

$$= \int_0^2 \frac{\cos(z) - \cos(1)}{z - 1}\, dz - j\pi \cos(1) \qquad \text{(using } z = 1 + e^{j\varphi})$$

Integration with Variable Substitution
In some types of improper integrals, substitution of the integration variable may be used to remove singularity. For example, consider

$$I = \int_{-a}^{a} \frac{1}{\sqrt{a^2 - x^2}} e^{jax} \, dx \qquad (4.61a)$$

The integrand has poles at $x = \pm a$, and can be removed by substituting $x = a \sin \theta$. With this substitution, the integration yields

$$I = \int_{-\pi/2}^{\pi/2} e^{j\alpha a \sin \theta} \, d\theta = J_0(\alpha a) \qquad (4.61b)$$

Other Method
A third method is as follows. Complete elliptic integrals of the first kind, and defined as

$$K(k) = \int_{0}^{\pi/2} \frac{d\theta}{\sqrt{1 - k^2 \sin^2 \theta}} \qquad (4.62)$$

may be expressed in a series form, and is discussed in Section 4.5.

4.4   Conformal Mapping of Complex Functions

The determination of distributed capacitance of a transmission line requires solution of the Laplace equation in two dimensions. For some geometrical shapes, it is difficult to solve Laplace equation in the natural coordinate system of the transmission line. Conformal mapping may be used in these cases to map the boundaries of the transmission line into parallel plate configuration for which the solution can be readily obtained. *The conformal mapping method is equivalent to a coordinate transformation* and its applications to planar transmission lines are described.

4.4.1   Mapping

Let us first consider mapping in the real variable domain. The equation $y = f(x)$ describes a correspondence between points x on the x-axis and points y on the y-axis; that is, points x are mapped onto points y. The graphical representation of $y = f(x)$ is a curve in the plane (x, y), and the shape of the curve is determined by f. For example, the expression $y = mx + c$, where m and c are real numbers, represents a straight line in the $x - y$ plane. In a similar fashion, we use a surface to exhibit graphically a real-valued function $y = f(x_1, x_2)$ of two real variables x_1 and x_2.

Now let us consider mapping between two complex variables $z = x + jy$ and $w = u + jv$, and denoted as $w = f(z)$. The function f performs mapping or transformation between the complex-domains (x, y) and (u, v) according to a rule described by the function f. Graphical representation of this mapping on a single graph is not useful because each of the real variables u and v are functions of x and y, and the mapped function in the complex-domain w may not represent a useful shape. For simplicity, we draw two graphs; one for the z-domain with x and y as the axis, and the other for the w-domain with u and v as the axis. Figure 4.10 illustrates the principle of mapping through the mapping of straight line $-\pi \le x \le \pi$, $y = c$ in the z-plane into a corresponding curve in the w-plane according to the transformation $w = \sin(z)$. For convenience in mapping we use the Cartesian coordinates

$$w = u + jv = \sin(x + jy) \tag{4.63}$$

or

$$u = \sin(x)\cosh(y) \quad \text{and} \quad v = \cos(x)\sinh(y) \tag{4.64}$$

The point-to-point mapping from A, B, C in the z-plane to the corresponding points A', B', C' according to (4.64) is illustrated in Figure 4.10. Thus, the given straight line is transformed into an ellipse by the transformation $w = \sin(z)$.

This example demonstrates that we can transform a straight line into a closed curve through mapping. Alternatively, we can use the mapping function $z = \sin^{-1}(w)$ to map the ellipse of Figure 4.10(b) into the straight line of Figure 4.10(a). In this case a complex geometry is transformed into a simple shape.

4.4.2 Properties of Conformal Mapping

A transformation $w = f(z)$ is said to be conformal at a point if the function $f(z)$ is analytic at that point. Conformal mapping or transformation preserves angles between every pair of curves; that is, if two curves in the z-plane intersect at a particular angle, the corresponding transformed curves will also intersect at the

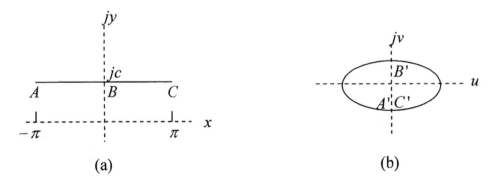

(a)   (b)

Figure 4.10 Mapping of the line segment $-\pi \le x \le \pi$, $y = c$ using the mapping function $w = \sin(z)$: (a) z-plane; and (b) w-plane.

same angle, although the transformed curves in w-plane may not have any resemblance to the original curves. Any conformal transformation would, therefore, transform parallel or orthogonal curves in the z-plane into parallel or orthogonal curves in the w-plane, respectively. As an illustration of conformal mapping, we consider the mapping $w = \ln(z)$. For convenience we use polar form $z = r \exp(j\theta)$, and obtain

$$u = \ln(r) \quad \text{and} \quad v = \theta \tag{4.65}$$

This equation indicates that the mapping $w = \ln(z)$ maps a circle in the z-plane into a straight line parallel to the v-axis in the w-plane. Also, an annular region $ABCDEF$ is transformed into a rectangular region $A'B'C'D'\,E'F'$ as shown in Figure 4.11. Further,

1. The r = constant arcs AE and BD map onto the u = constant segments $A'E'$ and $B'D'$, respectively;
2. The orthogonal curves AB and BD map onto the orthogonal curves $A'B'$ and $B'D'$, respectively;
3. The parallel arcs AE and BD map onto parallel segments $A'E'$ and $B'D'$, respectively.

It may be pointed out that the boundary conditions remain unchanged under conformal mapping; that is, Dirichlet boundary condition defined by, say, $H(x, y) = c$ in the z-plane transforms into the curve $H(u, v) = c$ in the w-plane [1]. This implies that instead of solving the boundary value problem directly in the z-plane, we may transform this problem into a simpler one in the w-plane, determine its solution, and transform back to the z-plane to obtain the solution of the original problem. We shall prove next that the capacitance per unit length of the transmission line remains unchanged under conformal transformation.

Consider the cross-section of a two-conductor transmission line shown in Figure 4.12. The medium between the conductors is assumed to be linear, homogeneous, isotropic, and lossless with permittivity ϵ and permeability μ_0. The energy per unit length stored in the electrostatic field of this transmission line is given by

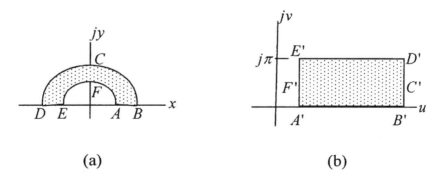

(a)   (b)

Figure 4.11 Mapping of an annular region into a rectangular region using the transformation $w = \ln(z)$ and describing the properties of conformal mapping. The dotted region in the z-plane is mapped into the dotted region in the w-plane. (a) z-plane; and (b) w-plane.

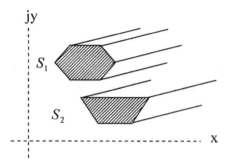

Figure 4.12 Cross-section of a two-conductor transmission line.

$$W_e = \frac{1}{2} \iint_S \epsilon |E|^2 \, ds \tag{4.66}$$

where E is the electric field and S is the cross-sectional area of the transmission line which includes the conductors S_1 and S_2. Since $E = -\nabla_t \phi$, with ϕ being the potential difference between the conductors, we can write

$$W_e = \frac{1}{2} \epsilon \iint_{y\ x} \left[\left(\frac{\partial \phi}{\partial x}\right)^2 + \left(\frac{\partial \phi}{\partial y}\right)^2 \right] dx \, dy \tag{4.67}$$

$$= \frac{1}{2} \epsilon \iint_{y\ x} \left[\left(\frac{\partial \phi}{\partial u}\right)^2 \left(\frac{\partial u}{\partial x}\right)^2 + \left(\frac{\partial \phi}{\partial v}\right)^2 \left(\frac{\partial v}{\partial y}\right)^2 \right] dx \, dy$$

Use of *Cauchy-Riemann* equation (4.7a) gives

$$W_e = \frac{1}{2} \epsilon \iint_{y\ x} \left[\left(\frac{\partial \phi}{\partial u}\right)^2 + \left(\frac{\partial \phi}{\partial v}\right)^2 \right] \frac{\partial u}{\partial x} \frac{\partial v}{\partial y} \, dx \, dy \tag{4.68}$$

$$= \frac{1}{2} \epsilon \iint_{v\ u} \left[\left(\frac{\partial \phi}{\partial u}\right)^2 + \left(\frac{\partial \phi}{\partial v}\right)^2 \right] du \, dv$$

This expression represents the electrostatic energy per unit length of the transformed line in the w-plane. Since the energy stored in the original and transformed planes is identical, the capacitance of the transmission line remains invariant under transformation. It means that transformation back to the original plane for the calculation of capacitance is not required; the calculation of capacitance in the transformed plane will suffice.

4.4.3   Applications of Conformal Mapping

Conformal transformation has been employed frequently to determine the characteristics of transmission lines, and discontinuities in them. The advantage of this approach is that the analytical solution based on conformal transformation leads to design equations. Also, an open geometry like that of planar lines is transformed into a closed geometry. The limitation of this method is that it provides solution for the static fields only, which may be used at best for the quasi-static cases. As a next step we show that the real and imaginary parts of an analytic function satisfy the Laplace equation.

Consider an analytic function $w = u(x, y) + jv(x, y)$ which satisfies (4.7). Now, differentiating (4.7a) with respect to x, (4.7b) with respect to y, and adding the resulting equations, gives

$$\frac{\partial^2 u}{\partial x^2} + \frac{\partial^2 u}{\partial y^2} = 0 \tag{4.69}$$

Similarly, differentiating (4.7a) with respect to y, (4.7b) with respect to x, and subtracting the resulting equations, we obtain

$$\frac{\partial^2 v}{\partial x^2} + \frac{\partial^2 v}{\partial y^2} = 0 \tag{4.70}$$

Thus, the real and imaginary parts of an analytic function separately satisfy the two-dimensional Laplace equation. Either of these functions may be used to represent the potential distribution and the other for the static electric field. It implies that we can solve for the potential and the corresponding static electric field in a transmission line structure by choosing an analytic function $w(z)$, determine its real part u and imaginary part v, and taking either u or v as the potential [6]. Boundary conditions dictate the selection of u or v for the potential. If u = constant curves coincide with the boundary of the structure, then u is selected as the potential function. Once the potential and electric field distributions are known in the transmission line, one can determine the capacitance per unit length of the line. Calculation of the capacitances of the transmission line with the dielectric present, and that of dielectric replaced by air can be used to determine the characteristic impedance and the effective dielectric constant as described in Section 1.13. In practice, several transformations in sequence are needed to yield a geometry whose solution is known or can be found easily.

4.5   Schwarz-Christoffel Transformation

The Schwarz-Christoffel transformations help solve an important class of boundary value problems that involve regions with polygonal boundaries. Our interest in this transformation is to use this technique for the design of planar transmission lines.

The Schwarz-Christoffel transformation maps the x-axis onto a polygonal contour, and upper-half of the z-plane $y > 0$ onto the interior of the polygon as shown in Figure 4.13. The polygon lies in the w-plane. Let the desired mapping be described by $w = f(z)$, and assume that its derivative is given as [7–10]

$$f'(z) = \frac{dw}{dz} = A(z - x_1)^{-k_1}(z - x_2)^{-k_2} \dots (z - x_N)^{-k_N} \qquad (4.71)$$

$$= A \prod_{n=1}^{N} (z - x_n)^{-k_n}$$

where A is a complex number, k_n ($n = 1, 2, 3, \dots, N$) are real numbers, and x_n ($n = 1, 2, \dots, N$) are points on the x-axis of domain D such that

$$x_1 < x_2 < \dots x_n < \dots x_{N-1} < x_N \qquad (4.72)$$

and are shown in Figure 4.13(a). One or two points out of these might be at infinity. The mapping $w = f(z)$ transforms domain D into a polygon G as shown there. The vertices of the polygon are $w_n = f(x_n)$.

From (4.71), the argument or the angle of $f'(z)$ can be obtained as

$$\arg f'(z) = \arg A - k_1 \arg(z - x_1) - k_2 \arg(z - x_2) - \dots - k_N \arg(z - x_N) \qquad (4.73)$$

Now consider a point z on the x-axis. Thus, $z = x$, and $z - x_n$ is positive if $z > x_n$, and negative if $z < x_n$. If the positive sense of the x-axis is to the right, then

$$\arg(z - x_n) = \begin{cases} 0 & \text{for } z > x_n \\ \pi & \text{for } z < x_n \end{cases} \qquad n = 1, 2, 3, \dots N - 1 \qquad (4.74)$$

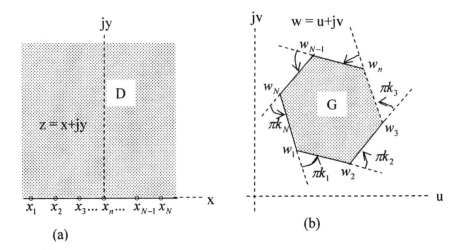

(a)

(b)

Figure 4.13 Schwarz-Christoffel transformation: (a) z-plane; and (b) w-plane.

It may be emphasized that $0 < \arg(z - x_n)$ for $y > 0$. Now assume that point z traverses to the right through different points x_n, and let $\theta_n = \arg f'(z)$ for $x_n < z < x_{n+1}$, we then have from (4.73) and (4.74)

$$\theta_n = \arg A - \pi(k_{n+1} + k_{n+2} + \ldots + k_N) \tag{4.75}$$

$$\theta_{n+1} = \arg A - \pi(k_{n+2} + k_{n+3} + \ldots + k_N) \tag{4.76}$$

or

$$\theta_{n+1} - \theta_n = \pi k_{n+1} \tag{4.77}$$

It should be apparent that as the point z moves along the x-axis passing through the point x_n, the corresponding point w traverses a polygon in the w-plane with the vertices defined by $w_n = f(x_n)$, and shown in Figure 4.13(b). Observe from (4.77) and Figure 4.13(b) that as the point z passes the point x_{n+1}, the direction of the path of the corresponding point w changes by an angle πk_{n+1}. This angle is the exterior angle of the polygon at the vertex w_{n+1}. The exterior angles are limited between $-\pi$ and π; therefore, $-1 < k_n < 1$. Also, since the sum of the exterior angles of a polygon is 2π, we obtain $k_1 + k_2 + \ldots k_N = 2$. It has been assumed in the mapping considered that the sides of the polygon do not cross each other, and the positive sense of traverse is counter-clockwise. If the interior angle of the polygon at the vertex w_{n+1} is defined as ϕ_{n+1}, then

$$\phi_{n+1} = \pi(1 - k_{n+1}) \tag{4.78}$$

Integrating (4.71) with respect to z gives the following expression for the Schwarz-Christoffel transformation:

$$w = f(z) = A \int_{z_0}^{z} (z' - x_1)^{-k_1}(z' - x_2)^{-k_2} \ldots (z' - x_N)^{-k_N} \, dz' + B$$

$$\tag{4.79a}$$

where $k_1 + k_2 + \ldots k_N = 2$, and B is an arbitrary constant that determines the position of the polygon. *The size and orientation of the polygon can be controlled through the magnitude and angle of A, respectively.* The integration in (4.79a) is carried out along any path in D that joins z_0 to z. The integrand consists of multivalued functions $(z' - x_n)^{-k_n}$ since $|k_n| < 1$, and the principal branches of these multivalued functions are selected such that these are a direct analytical continuation into the upper-half plane of the real valued functions $(x - x_n)^{-k_n}$, where $x > x_n$. Therefore, the integral (4.79a) is a single-valued analytic function in the upper-half plane $y > 0$, and the points x_n are the branch-point singularities of the integral. The inverse Schwarz-Christoffel transformation corresponding to (4.79a) is given by

$$z = F(w) = C \int_{w_0}^{w} (w' - w_1)^{-K_1}(w' - w_2)^{-K_2} \ldots (w' - w_N)^{-K_N} \, dw' + d$$

(4.79b)

where

$$K_1 + K_2 + \ldots K_N = 2$$
(4.80)

It may be noted that in Schwarz-Christoffel transformation we first create a complex number $w = u + jv$ by using u and v coordinates of a point in the geometry. The transformation is defined by a *line integral* in the complex plane, (4.79b). The transformed plane (z-plane) is described by the complex number $z = x + jy$, where x and y represent the coordinates in the z-plane.

Procedure for Schwarz-Christoffel Transformation

In practice, a polygon in the w-plane is given, and the transformation is determined such that the x-axis is mapped onto the given polygon. This requires the inverse mapping function giving z as a function of w. Generally, it is difficult to obtain this inverse transformation analytically, and we proceed by trial and error to set up the desired transformation. In practice, several transformations in sequence are needed to yield the desired geometry. The steps given below may be followed to arrive at the transformation:

First determine the exteriors angles of the polygon. This can be done easily from the given polygon.

Next, choose the points x_n. The selection of points often requires ingenuity. While choosing the points, the information about the polygon must be included. *The study of conformal mapping analyses of planar transmission lines indicates that the points x_n corresponding to the vertices of the polygon must be included as factors in* (4.79a). The points at infinity do not give rise to additional factors. The rest of the points in the two planes are linked by one-to-one correspondence of the mapping. The number of unknowns being greater than the number of equations the points x_n cannot be determined uniquely. A maximum of three points or three conditions on x_n can be chosen arbitrarily [1]. Hence, for polygons with more than three sides ($N > 3$), some of the points x_n must be determined so that the x-axis is transformed into the polygon.

Finally, evaluate the integral (4.79a). The integration involves singularities, and usually it is not possible to express it in a closed form. However, for $|k_n| = 0, 1/2, 1, 3/2, 2$ one may be able to express the integral in terms of standard functions like elliptic functions and a look-up table may be necessary. Otherwise, numerical integration is the best option. In such cases, the transformation may still be highly useful, but the value of w is obtained iteratively [11]. Integrals with singularities at the end points may be carried out using the function NIntegrate from Mathematica [12].

We next discuss a commonly used Schwarz-Christoffel transformation.

4.5.1 Elliptic Sine Function

The transformation

$$z = sn(w, k); \qquad z = x + jy \tag{4.81a}$$

is an important mapping function in the study of planar transmission lines [7]. The elliptic sine function $sn(w, k)$ transforms the boundary of a rectangle in the w-plane onto the real axis of the z-plane, and the interior of the rectangle is mapped onto the upper half of the z-plane. The inverse transformation is given by

$$w = sn^{-1}(z, k) \tag{4.81b}$$

and transforms the x-axis into a rectangle in the w-plane. The rectangle may represent a parallel plate capacitor.

The elliptic sine function is a generalization of the sine function to the complex domain and is defined as

$$\sin w = \sin(u + jv) = \sin u \cosh v + j \cos u \sinh v \tag{4.82}$$

The function $\sin w$ is periodic with period 2π along the u-axis and is plotted in Figure 4.14(a) for $-\pi/2 \le u \le \pi/2$, $0 \le v \le \infty$. The sine function goes from $-\infty$ to -1 and then from $+1$ to $+\infty$ along the contour ABCD, as shown in the figure. The mapping $z = \sin w$ transforms the contour ABCD into the strip A′B′C′D′ along the x-axis, as shown in Figure 4.14(b). It is described by

$$x = \sin u \cosh v \tag{4.83}$$

The points inside the rectangle are mapped onto the upper half of the z-plane.

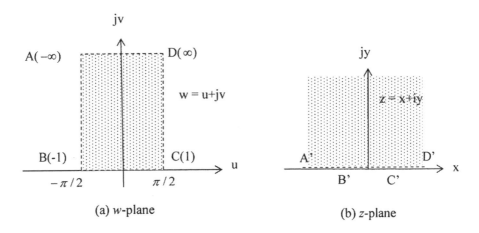

(a) w-plane

(b) z-plane

Figure 4.14 Mapping of an infinite strip onto the real axis using the function $z = \sin w$: (a) an infinite strip in w-plane described by $-\pi/2 \le u \le \pi/2$, $0 \le v \le \infty$; and (b) mapping of the rectangle ABCD using $z = \sin w$.

Unlike the sine function described above, the elliptic sine function (4.81) is periodic along both the axis. The periodicity of the elliptic sine function is $4K(k)$ along the u-axis and $2K'(k)$ along the v-axis. The parameter k is called the modulus. The inverse elliptic sine function is of a form similar to the inverse sine function

$$w = \sin^{-1} Z = \int_0^Z \frac{dz}{\sqrt{1 - z^2}}, \text{ and is given by [7]}$$

$$w(Z) = sn^{-1}(Z, k) = \int_0^Z \frac{dz}{\sqrt{(1 - z^2)(1 - k^2 z^2)}} \qquad (4.84)$$

For $k = 0$, the inverse elliptic function reduces to the inverse sine function.

The periods $K(k)$ and $K'(k)$ of the elliptic sine function determine the sides of the rectangle and are defined as

$$K(k) = \int_0^1 \frac{dz}{\sqrt{(1 - z^2)(1 - k^2 z^2)}} = sn^{-1}(1, k) \qquad (4.85)$$

and

$$K(k) + jK'(k) = \int_0^{1/k} \frac{dz}{\sqrt{(1 - z^2)(1 - k^2 z^2)}} \qquad (4.86a)$$

$$= \int_0^1 \frac{dz}{\sqrt{(1 - z^2)(1 - k^2 z^2)}} + \int_1^{1/k} \frac{dz}{\sqrt{(1 - z^2)(1 - k^2 z^2)}}$$

or

$$K'(k) = \int_1^{1/k} \frac{dz}{\sqrt{(z^2 - 1)(1 - k^2 z^2)}} \qquad \text{using (4.85)} \qquad (4.86b)$$

The integral in (4.85) may be expressed in a standard form by substituting $z = \sin\theta$,

$$K(k) = \int_0^{\pi/2} \frac{d\theta}{\sqrt{1 - k^2 \sin^2\theta}} = F\left(k, \frac{\pi}{2}\right) \qquad (4.87)$$

and is called the complete elliptic integral of the first kind. Similarly, substituting $k^2(z^2 - 1) = (1 - k^2)\cos^2\theta$ in (4.86b) gives

$$K'(k) = \int\limits_{0}^{\pi/2} \frac{d\theta}{\sqrt{1 - k'^2 \sin^2\theta}} = F\left(k', \frac{\pi}{2}\right) = K(k') \qquad (4.88)$$

where $k' = \sqrt{1 - k^2}$. For convenience, two useful series approximations for K and K' are given by [13, p. 905]

$$K(k) = K'(k') = \frac{\pi}{2}\left(1 + \left(\frac{1}{2}\right)^2 k^2 + \left(\frac{1 \times 3}{2 \times 4}\right)^2 k^4 + \left(\frac{1 \times 3 \times 5}{2 \times 4 \times 6}\right)^2 k^6 + \ldots\right)$$

(4.89a)

or

$$K'(k') = K(k) = \ell n \frac{4}{k'} + \left(\frac{1}{2}\right)^2 \left(\ell n \frac{4}{k'} - \frac{2}{1 \times 2}\right) k'^2$$

$$+ \left(\frac{1 \times 3}{2 \times 4}\right)^2 \left(\ell n \frac{4}{k'} - \frac{2}{1 \times 2} - \frac{2}{3 \times 4}\right) k'^4 \qquad (4.89b)$$

$$+ \left(\frac{1 \times 3 \times 5}{2 \times 4 \times 6}\right)^2 \left(\ell n \frac{4}{k'} - \frac{2}{1 \times 2} - \frac{2}{3 \times 4} - \frac{2}{5 \times 6}\right) k'^6 + \ldots$$

The convergence in (4.89) becomes very slow for values of k close to 1, in which case we can use the asymptotic formula

$$K(k) \approx \frac{1}{2} \ell n \frac{16}{1 - k^2}, \qquad \text{for } k \text{ nearly equal to 1} \qquad (4.90)$$

The ratio of complete elliptic integrals $K(k)/K(k')$ occurs very commonly in the design of planar lines. Fortunately, this ratio can be determined accurately to 1 part in 10^5 using the following simple expression [14]:

$$\frac{K(k)}{K(k')} = \frac{K(k)}{K'(k)} = \begin{cases} \dfrac{1}{\pi} \ell n \left(2 \dfrac{1 + \sqrt{k}}{1 - \sqrt{k}}\right), & \dfrac{1}{\sqrt{2}} \leq k \leq 1 \\[4ex] \dfrac{\pi}{\ell n \left(2 \dfrac{1 + \sqrt{k'}}{1 - \sqrt{k'}}\right)}, & 0 \leq k \leq \dfrac{1}{\sqrt{2}} \end{cases} \qquad (4.91)$$

4.5.2 Application to Coplanar Strips

In order to impress upon the utility of the elliptic sine function, let us consider the transformation of a pair of coplanar (metal) strips (CPS) into a parallel plate capacitor. The widths of the strips are $x_2 - x_1$ and $x_4 - x_3$, and they are separated by a gap (or slot) $x_1 - x_4$ as shown in Figure 4.15(a). Let us map the upper-half z-plane ($y > 0$) of this geometry onto the rectangle $A_1 A_2 A_3 A_4$ with vertices at $w = \pm a$, $\pm a + jb$, where $2a$ and b are the width and the height of the rectangle [Figure 4.15(b)]. We make use of (4.79a) to transform the pair of strips into the rectangle.

In the integrand of (4.79a), only three of the four points x_1, x_2, x_3, x_4 may be chosen arbitrarily. Since the rectangle is symmetric about the v-axis, we can choose the points along the x-axis symmetrically. Thus, for the right-half rectangle $OA_1 A_2 B$, let $w = 0$, a, jb correspond to $z = 0$, 1, ∞, respectively. Let point A_2 map to $z = 1/k$, $0 < k < 1$. Similarly, for the left-half rectangle $OBA_3 A_4$.

For a rectangle $k_n = 1/2$, therefore (4.79a) becomes

$$w = A \int_{z_0}^{z} (z' - x_1)^{-1/2} (z' - x_2)^{-1/2} (z' - x_3)^{-1/2} (z' - x_4)^{-1/2} dz' + B \quad (4.92)$$

Choosing x_1, x_2, x_3, x_4 as shown in Figure 4.15(a), we obtain

$$w = A \int_{z_0}^{z} \frac{dz'}{\sqrt{(z' - 1)(z' + 1)(z' - 1/k)(z' + 1/k)}} + B \quad (4.93)$$

For the point $z = 0$ to map onto $w = 0$, the constant B should be zero. Therefore,

$$w = A \int_{0}^{z} \frac{dz'}{\sqrt{(z' - 1)(z' + 1)(z' - 1/k)(z' + 1/k)}} \quad (4.94)$$

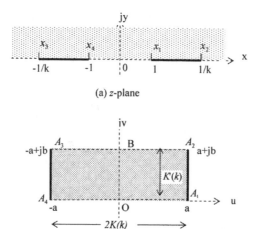

(a) z-plane

(b) w-plane

Figure 4.15 Mapping of coplanar strips in the z-plane into a parallel plate capacitor in the w-plane: (a) z-plane; and (b) w-plane.

The dimension $a \times b$ of the rectangle is determined from one-to-one correspondence between the points in the z-plane and w-plane, and is determined next.

Since $z = 1$ maps onto $w = a$, we obtain from (4.94)

$$a = A' \int_0^1 \frac{dz'}{\sqrt{(1 - z'^2)(1 - k^2 z'^2)}} \qquad kA' = A \tag{4.95}$$

where A' is an arbitrary complex constant, and can be determined if k is known and a is prescribed.

Also, since $z = 1/k$ maps onto $w = a + jb$, we obtain

$$
\begin{aligned}
a + jb &= A' \int_0^{1/k} \frac{dz'}{\sqrt{(1 - z'^2)(1 - k^2 z'^2)}} \\
&= A' \left[\int_0^1 \frac{dz'}{\sqrt{(1 - z'^2)(1 - k^2 z'^2)}} + \int_1^{1/k} \frac{dz'}{\sqrt{(1 - z'^2)(1 - k^2 z'^2)}} \right] \\
&= a + jA' \int_1^{1/k} \frac{dz'}{\sqrt{(z'^2 - 1)(1 - k^2 z'^2)}}, \text{ using } (4.95)
\end{aligned}
\tag{4.96}
$$

Thus, with $A' = 1$,

$$b = \int_1^{1/k} \frac{dz'}{\sqrt{(z'^2 - 1)(1 - k^2 z'^2)}} \tag{4.97}$$

Hence, we can determine A' and k from (4.95) and (4.96) if a and b are prescribed. On the other hand, if k is known and if we take $A' = 1$, then the values of a and b are determined from (4.95) and (4.96), respectively. The dimensions a and b need not be determined separately if we are interested only in the capacitance of the parallel plate capacitor represented by Figure 4.15(b). The ratio $b/2a$ determines the capacitance, and is given by [using (4.85) and (4.86)]

$$\frac{b}{2a} = \frac{\text{Plate width}}{\text{Plate separation}} = \frac{K'(k)}{2K(k)} \tag{4.98}$$

The above analysis may be applied to coplanar waveguide (CPW) geometry also. The CPW geometry is complimentary to the coplanar strips geometry. The continuous lines in Figure 4.15 now represent the slots and dotted lines the metallization for CPW line. The ratio $2a/b$ is now proportional to the line capacitance and is given by

$$\frac{2a}{b} = 2\,\frac{K(k)}{K'(k)} \tag{4.99}$$

The modulus k may be expressed in terms of the physical dimensions of the transmission line. For the CPS geometry, Figure 4.15(a), the modulus k is defined as

$$k = \frac{S/2}{W + S/2} \tag{4.100}$$

where S is the slot width $x_4 - x_1$ and W is the strip width $x_2 - x_1$.

The following points are noteworthy in the conformal mapping analysis:

1. An open geometry has been transformed into a closed geometry without the use of asymptotic boundary conditions or absorbing material at the open boundary.
2. The effect of charge singularity at the strip edges is included in the transformation through the factors $\sqrt{z' - 1}$, $\sqrt{z' + 1}$, $\sqrt{z' - 1/k}$, $\sqrt{z' + 1/k}$.
3. The conformal mapping leads to simple design expressions for the capacitance per unit length and therefore characteristics of the transmission line.

4.6 Quasi-Static Analysis of Planar Transmission Lines

Schwarz-Christoffel transformation is perhaps the most popular conformal analysis method for planar transmission lines like strip line, microstrip line, coplanar strips, coplanar waveguide, and microshield line. Its application to CPS and CPW lines was discussed in the last section.

The basic approach used in the conformal transformation of planar lines is to assume that all the dielectric interfaces in the structure, including slots, can be replaced by magnetic walls [6, 15]. This assumption is strictly valid for those geometries for which the electric field lies along the dielectric interfaces so that the magnetic field is normal to the interface. Under this assumption, the half-planes above and below the metallization plane of the planar transmission line can be analyzed separately for line capacitance. The total line capacitance is then the algebraic sum of the two capacitances. Further, if the dielectric substrate has finite thickness as in Figure 4.16 for CPW line, the contribution of the lower half-plane to the line capacitance can be determined as the sum of: (1) the free space capacitance, obtained by replacing dielectric by the air medium in the line, and (2) the capacitance of the dielectric layer alone assumed to have the permittivity $(\epsilon_r - 1)$ [15]. This approach yields exact results for infinitely thick substrate and for substrate thickness $h \to 0$. It has been found to give reasonable accuracy for most of the practical ranges of physical dimensions. We shall illustrate the application of this method to the strip transmission line. The mappings attempted for other planar lines will then be listed along with the results.

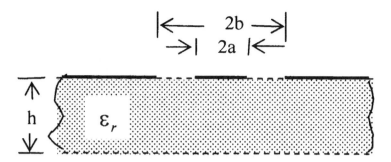

Figure 4.16 Cross-section of a coplanar waveguide with finite dielectric thickness.

4.6.1 Strip Line

The strip line consists of a thin metal strip inserted between two grounded dielectric sheets as shown in Figure 4.17. The dielectrics and conductors are assumed to be lossless, and the strip infinitely thin. The grounded dielectric sheets are assumed infinitely large. The strip may be assumed to be at potential V with respect to the ground planes. The electric and magnetic field distributions for the dominant TEM-mode in the cross-section is also shown in the figure. The magnetic field distribution in the strip line is such that it is normal to the symmetry planes at $x = 0$ and $y = 0$. Therefore, we can insert magnetic walls at these planes without affecting the field distribution. The utility of magnetic walls will be explained later.

We can now decompose the total capacitance (per unit-length) of the strip line into two independent capacitances: capacitance of the upper-half plane and that of the lower-half plane. Each of these capacitances is evaluated as though the other was not present. This is strictly possible due to the magnetic wall boundary condition at $y = 0$. Now we can use the magnetic wall boundary condition at $x = 0$ plane to reduce the geometry to be analyzed to one-fourth of the given geometry. The second quadrant of the strip line is drawn in Figure 4.18(a). The capacitance of the strip line is four times the capacitance of the structure in Figure 4.18(a). We visualize this geometry as a rectangle in the w-plane whose vertices are w_1, w_2, w_3, and w_4, and wish to transform it onto the upper-half of the z-plane as shown in Figure 4.18(b). Using (4.79a), mapping from the z-plane to the w-plane can be written as

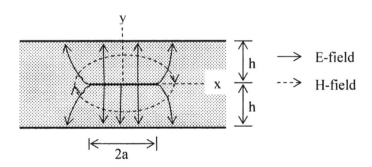

Figure 4.17 Cross-section of a stripline and the field distributions.

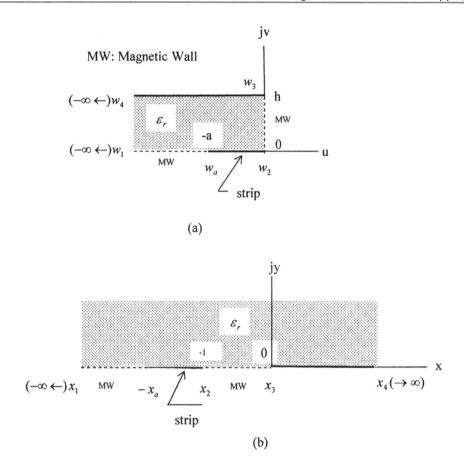

Figure 4.18 The Schwarz-Christoffel mapping of the second quadrant in the w-plane onto the z-plane: (a) w-plane; and (b) z-plane.

$$w = A \int_{z_0}^{z} (z' - x_1)^{-k_1} (z' - x_2)^{-k_2} (z' - x_3)^{-k_3} (z' - x_4)^{-k_4} \, dz' + B$$

$$(4.101)$$

The exterior angles of the given rectangle are $\pi/2$ each, and therefore $k_n = 1/2$. The vertices of the rectangle are: $w_1 \to -\infty$, $w_2 = 0$, $w_3 = jh$, and $w_4 \to -\infty + jh$. Since we can choose three points on the x-axis arbitrarily, we let $x_1 \to -\infty$ and $x_4 \to \infty$ in order to cover the entire x-axis. We also choose $x_2 = -1$ and $x_3 = 0$. The correspondence between the points on the x-axis and those on the w-plane is shown in Figure 4.18. The selected points are related to the vertices of the rectangle through the transformation. Assigning the values $x_2 = -1$ and $x_3 = 0$ is arbitrary. However, one of the points should be marked $x = 0$ because it helps in determining the unknown in (4.79a). Since B is as yet arbitrary, we may choose a lower limit for the integration, $z_0 = 0$. With these choices (4.101) reduces to

$$w = \frac{A}{\sqrt{x_1 x_4}} \int_0^z \left[(z' - x_2)(z' - x_3)\left(1 - \frac{z'}{x_1}\right)\left(1 - \frac{z'}{x_4}\right) \right]^{-1/2} dz' + B$$

$$(4.102)$$

For $x_1 \to -\infty$, $x_4 \to \infty$, $x_2 = -1$, and $x_3 = 0$, (4.102) reduces to

$$w = A' \int_0^z \frac{dz'}{\sqrt{z'(z' + 1)}} + B \qquad A' = \frac{A}{\sqrt{x_1 x_4}} \qquad (4.103)$$

If we multiply the numerator and the denominator by $\left(1/\sqrt{z'} + 1/\sqrt{z' + 1}\right)\left(\sqrt{z'} + \sqrt{z' + 1}\right)$, the integration can be carried out analytically:

$$w = A' \int_0^z \frac{\left(1/\sqrt{z'} + 1/\sqrt{z' + 1}\right)\left(\sqrt{z'} + \sqrt{z' + 1}\right)}{\left(\sqrt{z'} + \sqrt{z' + 1}\right)^2} dz' + B$$

$$= A' \int_0^z \frac{d\left[\sqrt{z'} + \sqrt{z' + 1}\right]^2}{\left[\sqrt{z'} + \sqrt{z' + 1}\right]^2} + B \qquad (4.104)$$

$$= 2A' \ell n \left[\sqrt{z} + \sqrt{z + 1}\right] + B$$

This transformation produces the following correspondence between the points in the z- and w-planes:

$$x_1 \to -\infty \leftrightarrow w_1 \to -\infty \qquad (4.105a)$$

$$x_2 \to -1 \leftrightarrow w_2 \to 0 \qquad (4.105b)$$

$$x_3 \to 0 \leftrightarrow w_3 \to jh \qquad (4.105c)$$

$$x_4 \to \infty \leftrightarrow w_1 \to -\infty + jh \qquad (4.105d)$$

By design we have satisfied $x_1 < x_2 < x_3 < x_4$ as required by (4.72). Applying the constraints (4.105b) and (4.105c) on (4.104), we can determine A' and B uniquely as

$$2A' \ln\sqrt{-1} = -B \qquad \text{or} \qquad A'j\pi = -B \qquad (4.106a)$$

$$B = jh \qquad (4.106b)$$

Therefore, $A' = -h/\pi$ and the transformation (4.104) is now given as

$$w = -\frac{2h}{\pi} \ell n \left[\sqrt{z} + \sqrt{z+1} \right] + jh \tag{4.107}$$

Using this mapping function we can now determine the unknown x_a, which corresponds to half strip width w_a. Substituting $z = -x_a$ and $w = w_a = -a$ in (4.107) gives

$$jx_a^{1/2} + j(x_a - 1)^{1/2} = e^{\pi a/2h} e^{j\pi/2} \tag{4.108}$$

or

$$x_a = \cosh^2 \left(\frac{\pi a}{2h} \right) \tag{4.109}$$

The geometry of Figure 4.18(b) cannot be analyzed readily for its capacitance. So, we transform it again to map it into a polygon in the $w' = u' + jv'$ plane. For convenience in calculation of the capacitance, we choose a rectangular polygon for the final transformed structure. The mapping between the z- and w'-planes is shown in Figure 4.19. It is to be noted that the strip and the ground plane, which are at different potentials and located on the same surface in the z-plane, have been transformed to different surfaces in the w'-plane. For the rectangular polygon in Figure 4.19(b), each of the exterior angles is $\pi/2$. For this transformation we choose the four points on the x-axis as the points defining the metal conductors; that is, $x_1 = -x_a$, $x_2 = -1$, $x_3 = 0$, and $x_4 = \infty$. These points are known from the previous mapping and have been selected because on transformation they define the vertices of the rectangle. The transformation for these points may be written as

$$w' = A_1 \int_{z_0}^{z} \frac{dz'}{\sqrt{z'(z'+1)(z'+x_a)}} + B_1 \tag{4.110}$$

where A_1 and B_1 are constants to be determined from the mapped points in the w'-plane. Since B_1 is as yet arbitrary, we may choose the lower limit for the integration arbitrarily, and set $z_0 = 0$ to obtain

$$w' = A_1 \int_{0}^{z} \frac{dz'}{\sqrt{z'(z'+1)(z'+x_a)}} + B_1 \tag{4.111}$$

The four points in the w'-plane corresponding to the selected four points in the z-plane are defined as

$$x_1 = -x_a = -\cosh^2(\pi a/2h) \leftrightarrow w_1' = 0 \tag{4.112a}$$

$$x_2 = -1 \leftrightarrow w_2' = 1 \tag{4.112b}$$

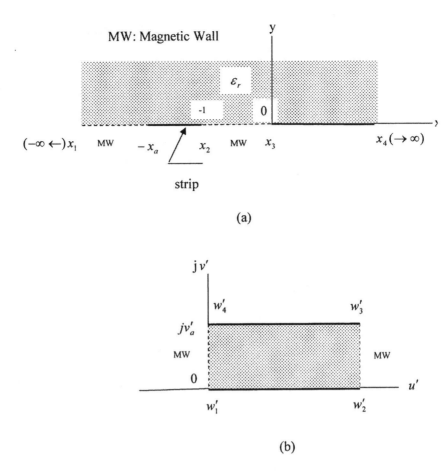

Figure 4.19 Conformal mapping from the upper-half of z-plane into the w'-plane: (a) z-plane; and (b) w-plane.

$$x_3 = 0 \leftrightarrow w_3' = 1 + jv_a' \tag{4.112c}$$

$$x_4 = \infty \leftrightarrow w_4' = jv_a' \tag{4.112d}$$

It may be noted that the four points in the z-plane define the metallization and are transformed to the parallel metal plates in the w'-plane.

To evaluate the integral in (4.111) we substitute $z' = -t^2$ and obtain

$$w' = A_1' \int_0^{\sqrt{-z}} \frac{dt}{\sqrt{(1 - t^2)(1 - t^2/x_a)}} + B_1, \qquad A_1' = \frac{2jA_1}{\sqrt{x_a}} \tag{4.113}$$

The integral can be expressed as inverse elliptic sine function defined in (4.84). Accordingly,

$$w' = A_1' \, sn^{-1}((-z)^{1/2}, x_a^{-1/2}) + B_1 \tag{4.114}$$

Now we use the conditions defined by (4.112a) through (4.112c) and obtain

$$A_1' \, sn^{-1}(x_a^{1/2}, x_a^{-1/2}) + B_1 = 0 \tag{4.115a}$$

$$A_1' \, sn^{-1}(1, x_a^{-1/2}) + B_1 = 1 \tag{4.115b}$$

$$B_1 = 1 + jv_a' \tag{4.115c}$$

Solving these equations algebraically gives

$$v_a' = \frac{-jsn^{-1}(1, k)}{sn^{-1}(k^{-1}, k) - sn^{-1}(1, k)}, \qquad k = \frac{1}{\sqrt{x_a}} \tag{4.116}$$

Use of (4.85) for the inverse elliptic sine function in terms of complete elliptic integral of the first kind $K(k)$ gives

$$v_a' = \frac{-jK(k)}{sn^{-1}(k^{-1}, k) - K(k)} \tag{4.117}$$

From the definition of $sn^{-1}(k^{-1}, k)$ we have

$$sn^{-1}(k^{-1}, k) = \int_0^{1/k} \frac{dt}{\sqrt{(1 - t^2)(1 - k^2 t^2)}}$$

$$= \int_0^1 \frac{dt}{\sqrt{(1 - t^2)(1 - k^2 t^2)}} + \int_1^{1/k} \frac{dt}{\sqrt{(1 - t^2)(1 - k^2 t^2)}} \tag{4.118}$$

$$= K(k) - jK(k')$$

Equation (4.117) therefore reduces to

$$v_a' = \frac{K(k)}{K(k')} \tag{4.119}$$

The geometry of Figure 4.19(b) is a parallel plate capacitor with the upper and lower conducting plates at different potentials. The fringing field in this capacitor is zero because of the assumed perfect magnetic walls at $u' = 0$ and $u' = 1$. The capacitance of the parallel plate capacitor is equal to one-fourth of the capacitance C of the strip line due to the two-fold symmetry. Hence, we have

$$C = \frac{4\epsilon_0 \epsilon_r}{v_a'} = 4\epsilon_0 \epsilon_r \frac{K(k')}{K(k)} \tag{4.120}$$

where

$$k = \text{sech}\,(\pi a/(2h)) \tag{4.121a}$$

$$k' = \sqrt{1 - k^2} = \tanh\,(\pi a/(2h)) \tag{4.121b}$$

The characteristic impedance Z_0 of the stripline is defined as [refer to (1.62a)]

$$Z_0 = \frac{1}{c\sqrt{CC_a}} \tag{4.122}$$

where c is the velocity of light in vacuum, and C_a is the capacitance per unit length of strip line with dielectric replaced by air. It is obtained from (4.120) as

$$C_a = 4\epsilon_0 \frac{K(k')}{K(k)} \tag{4.123}$$

Substituting the expressions for C and C_a in (4.122) gives

$$Z_0 = \frac{30\pi}{\sqrt{\epsilon_r}} \frac{K(k)}{K(k')} \tag{4.124}$$

4.6.2 Microstrip Line with a Cover Shield

The geometry of this planar transmission line is shown in Figure 4.20. This transmission line finds applications in microwave integrated circuits and its analysis is very similar to that of stripline. Comparing Figures 4.17 and 4.20 one finds that the difference between the two geometries is in the upper dielectric sheet, which has been replaced by air in microstrip line, and the thickness of this layer h_0 is different from that of the lower layer h. Due to this asymmetry along the y-axis, the plane $y = 0$ is no longer a perfect magnetic wall. The deviation from this assumption depends on the difference between the values of h and h_0, and ϵ_0 and $\epsilon_0\epsilon_r$. However, moderately accurate results for shielded microstrip lines can still be obtained using conformal mapping [6].

Again we decompose the total capacitance per unit length of the microstrip line into two independent capacitances: C_0 corresponding to the air layer, and C_r

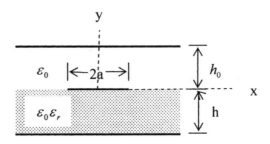

Figure 4.20 Cross-section of a microstrip line with a cover shield.

corresponding to the dielectric layer, *assuming a perfect magnetic wall at the plane* $y = 0$. Each of these capacitances is determined as though the other was not present. To determine these capacitances we proceed in the same way as for strip line. Divide each of these layers into two identical halves using the magnetic wall at the plane of symmetry $x = 0$. The one-fourth structures, with dielectric and with air, can be analyzed using the transformations detailed for the strip line. Using the results of earlier analysis we can write that the capacitance per unit length of the lower half of microstrip line, C_r, is given by [refer to (4.120)]

$$C_r = 2\epsilon_0 \epsilon_r \frac{K(k')}{K(k)} \tag{4.125}$$

where k and k' are given by (4.121). The capacitance per unit length of the upper half of the microstrip line, C_0, is similarly given by

$$C_0 = 2\epsilon_0 \frac{K(k_0')}{K(k_0)} \tag{4.126}$$

where

$$k_0 = \operatorname{sech}(\pi a/(2h_0)) \tag{4.127a}$$

$$k_0' = \sqrt{1 - k_0^2} = \tanh(\pi a/(2h_0)) \tag{4.127b}$$

The total capacitance per unit length C is the sum of C_r and C_0; that is,

$$C = 2\epsilon_0 \epsilon_r \frac{K(k')}{K(k)} + 2\epsilon_0 \frac{K(k_0')}{K(k_0)} \tag{4.128}$$

Due to the mixed dielectric nature of microstrip line, we determine the propagation constant of the wave in terms of effective relative dielectric constant, which is defined as

$$\epsilon_{re} = \frac{C}{C_a} \tag{4.129}$$

where C_a is the capacitance per unit length of the microstrip line with the lower dielectric sheet replaced by air; that is,

$$C_a = 2\epsilon_0 \frac{K(k')}{K(k)} + 2\epsilon_0 \frac{K(k_0')}{K(k_0)} \tag{4.130}$$

The effective dielectric constant is therefore given by

$$\epsilon_{re} = \frac{\epsilon_r \dfrac{K(k')}{K(k)} + \dfrac{K(k_0')}{K(k_0)}}{\dfrac{K(k')}{K(k)} + \dfrac{K(k_0')}{K(k_0)}} \tag{4.131}$$

It may be noted that the magnetic wall assumption along the plane $y = 0$ is true if $h_0 = h$. The results are exact in the static limit and the value of ϵ_{re} reduces to

$$\epsilon_{re} = \frac{\epsilon_r + 1}{2} \tag{4.132}$$

which is independent of the strip width and is equal to the average of two dielectric constants. The characteristic impedance of the shielded microstrip line can be computed from (4.122) to obtain

$$Z_0 = \frac{60\pi}{\sqrt{\epsilon_{re}} \left[\dfrac{K(k')}{K(k)} + \dfrac{K(k_0')}{K(k_0)} \right]} \tag{4.133}$$

We have described the conformal mapping analysis of a microstrip line with a cover shield. The analysis is valid for the static or quasi-static cases; that is, zero or very low frequency cases, and only if the assumption of magnetic wall at the air-dielectric interface is satisfied. For $h_0 \neq h$, *this assumption is not valid*. To determine the amount of error one needs to compute the results for Z_0 and ϵ_{re}, obtained from (4.131) and (4.133), versus $2a/h$ for different values of h_0/h with ϵ_r as parameter. These results are compared with those obtained from the accurate full-wave analysis. The comparison shows that (4.131) and (4.133) are accurate to within 6% for h_0/h up to 5, and the results for $h_0/h = 5$ can be used for an open microstrip line also, subject to the following conditions for k_0 and k_0' [6]:

$$k_0 = \begin{cases} \text{sech}(\pi a/(2h_0)), & 0 \le h_0/h < 5 \\ \text{sech}(\pi a/(10h)), & 5 \le h_0/h \end{cases} \tag{4.134a}$$

$$k_0' = \sqrt{1 - k_0^2} = \begin{cases} \tanh(\pi a/(2h_0)), & 0 \le h_0/h < 5 \\ \tanh(\pi a/(10h)), & 5 \le h_0/h \end{cases} \tag{4.134b}$$

When $\epsilon_r = 1$ and $h_0 = h$, the shielded microstrip line geometry reduces to the strip line geometry and the results are exact.

The popularity of the conformal mapping method to planar transmission lines arises from the fact that an open geometry is transformed into a closed polygonal geometry easily. In computational methods, on the other hand, an open region problem is solved by using either asymptotic boundary condition or by terminating the boundary by lossy material (see Chapter 9 on FDTD). These approaches increase the size of the problem. In the method of moments solution, the open region problem is handled analytically by means of the Green's function. Another advantage of conformal mapping is that metal strips are stretched and ground planes are compressed during transformation to parallel plate capacitor geometry. Due to the stretching of strips, the charge and current singularities on the strips vanish in the transformed plane. The capacitance of the transformed geometry may be determined using computational methods, and this does not require finer discretization near the strip edges [16].

4.7   Some Useful Mappings for Planar Transmission Lines

The conformal transformation analysis of some of the planar lines like strip line, microstrip line with a cover shield, coplanar strips, and CPW was described earlier. The change in metallization pattern on the substrate, lateral ground planes size, asymmetry in the structure, and so on, of these lines gives rise to a host of planar transmission lines useful for radio frequency (RF) and microwave circuit applications [6, 15]. Cross-sections of some of the CPW variants are shown in Figure 4.21. In addition, nonplanar transmission lines like microshield lines are being studied for use in microelectromechanical systems (MEMS) [11, 17]. All of these transmission lines have been analyzed for their quasi-static characteristics using conformal mapping methods. Instead of giving the details of conformal mapping for each of these transmission lines, we shall study the conformal mappings that are most common to them so as to bring out the useful features of these transformations.

From the analysis presented in Section 4.6 for strip line we see that the polygon of Figure 4.18(a) was transformed into parallel plate geometry in two steps. The first step involved transforming one quadrant of strip line into a symmetric structure with all the metallization on the same plane, and resting on an infinitely thick dielectric. The transformation of this geometry in the second step led to the parallel

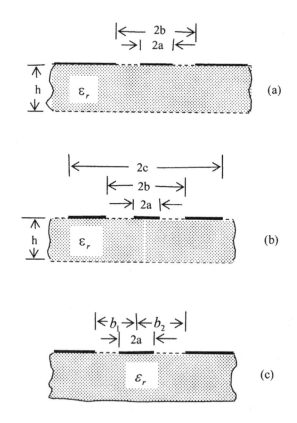

Figure 4.21   Cross-sections of a few coplanar waveguides (CPW): (a) CPW with finite dielectric thickness; (b) CPW with finite dielectric thickness and finite width lateral ground planes; and (c) asymmetric CPW.

plate capacitor. The two-step process was employed because the Schwarz-Christoffel mapping transforms the points on the real axis to a closed polygon and vice versa. It cannot be used to transform a polygon in one coordinate system to another polygon in the other coordinate system directly. The two-step or multistep mapping process has been found to be the most common feature for all the planar lines studied. Therefore, our endeavor will be to transform the given planar line into a symmetric structure with all the metalizations in the same plane, and resting on an infinitely thick dielectric. In the process we may transform finite dielectric thickness into infinite thickness, asymmetric geometry into a symmetric one, and metallization in different planes into one plane. Some of the basic and useful transformations are described next. Some of these are simply conformal transformations and the others are Schwarz-Christoffel transformations.

4.7.1   Transformation of Finite Dielectric Thickness to Infinite Thickness

Consider a coplanar waveguide with finite dielectric thickness h as shown in Figure 4.22(a). As before, we can divide this geometry into the upper-half region and the lower-half region along the assumed magnetic wall at the interface. Each of these halves can be further divided into two equal parts along the plane of symmetry, which coincides with the magnetic wall. The upper-half plane may be analyzed in the same way as discussed in Section 4.6. For the analysis of the lower-half plane let us consider the fourth quadrant of the CPW. The transformation from the finite dielectric thickness to infinite thickness is given by

$$t = \sinh(\pi z/(2h)) \tag{4.135a}$$

The mapping function is plotted in Figure 4.22(b), where

$$t_1 = \sinh(\pi a/(2h)) \quad \text{and} \quad t_2 = \sinh(\pi b/(2h)) \tag{4.135b}$$

To help visualize transformation of geometry, the points (1), (2), $\ldots$, are marked in Figure 4.22 before and after the transformation. The geometry of Figure 4.22(b) is the conventional CPW geometry (with infinite dielectric thickness). Its transformation to the parallel plate geometry is achieved using the transformation

$$w = \int_{t_0}^{t} \frac{dt}{\sqrt{(t - t_1)(t - t_2)}} \tag{4.136}$$

The upper half of CPW, Figure 4.22(a), can be mapped directly onto the parallel plate geometry using the mapping function similar to that of (4.136); that is,

$$w = \int_{z_0}^{z} \frac{dz}{\sqrt{(z - a)(z - b)}} \tag{4.137}$$

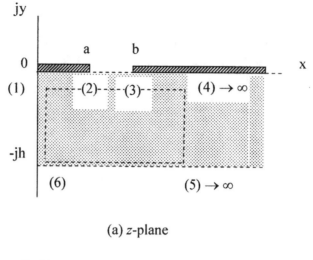

(a) z-plane

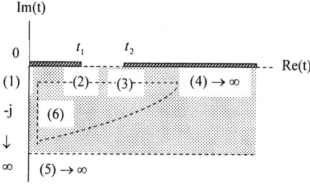

(b) t-plane

Figure 4.22 Transformation from finite substrate thickness h to infinite substrate thickness using the mapping $t = \sinh(\pi z/2h)$: (a) z-plane; and (b) t-plane.

4.7.2 Transformations for Finite Width Lateral Ground Planes and Finite Dielectric Thickness

Figure 4.21(b) shows a CPW with finite width lateral ground planes and finite dielectric thickness h. Let us first consider the upper half of the geometry. Using the symmetry considerations and the mapping function

$$t = z^2 \tag{4.138}$$

the first quadrant of the CPW can be transformed into a geometry covering the entire x-axis. This mapping is shown in Figure 4.23. The transformed points on the $\mathrm{Re}(t)$ axis are: $t_1 = a^2$, $t_2 = b^2$, $t_3 = c^2$. The next mapping function

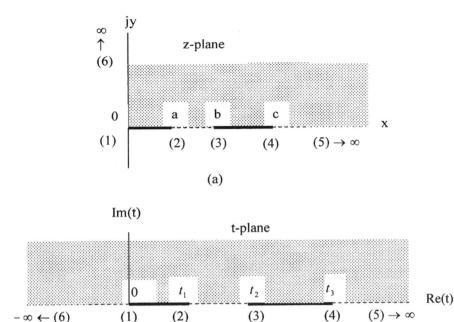

Figure 4.23 Mapping of the first quadrant of CPW with finite width lateral ground planes and finite dielectric thickness onto t-plane resulting in extension of given geometry to occupy $\text{Re}(t)$ axis from $-\infty$ to $+\infty$. The transformation used is $t = z^2$. (a) z-plane; and (b) t-plane.

$$w = \int_{t_0}^{t} \frac{dt}{\sqrt{t(t - t_1)(t - t_2)(t - t_3)}} \tag{4.139}$$

transforms Figure 4.23(b) into the desired parallel plate geometry.

The conformal mapping of the lower half of CPW of Figure 4.21(b) can be carried out using the following sequence of transformations:

$$t = \cosh^2(\pi z/(2h)) \tag{4.140}$$

$$w = \int_{t_0}^{t} \frac{dt}{\sqrt{(t - 1)(t - t_1)(t - t_2)(t - t_3)}} \tag{4.141}$$

The first mapping which transforms finite thickness h into infinite thickness is shown in Figure 4.24.

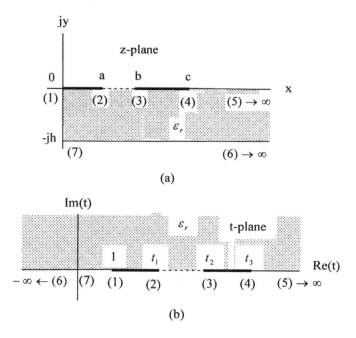

Figure 4.24 Transformation of a finite width lateral ground planes CPW with finite dielectric thickness into a geometry which extends to upper-half complex plane: (a) z-plane; and (b) t-plane.

4.7.3 Transformation from Asymmetric to Symmetric Metallization

Let us consider CPW with asymmetric metallization as shown in Figure 4.21(c). Assuming a magnetic wall along the air-dielectric interface, the structure can be divided into two halves. Consider now the upper-half plane, Figure 4.25(a). The first mapping used is $t = z/a$. This mapping function normalizes the strip width to unity, and the resulting geometry is shown in Figure 4.25(b). Here $k_1 = a/b_1$ and $k_2 = a/b_2$. Next we use the mapping

$$t_0^2 = \frac{t_3 + t_4}{2} \frac{t - t_3}{t - t_4} = \frac{1 + k_2}{2} \frac{t - 1}{k_2 t - 1} \tag{4.142}$$

to transform the asymmetric geometry into a symmetric one as shown in Figure 4.25(c). Here,

$$k^2 = \frac{2(k_1 + k_2)}{(1 + k_1)(1 + k_2)} \tag{4.143}$$

The transformation from this planar geometry to the parallel plate geometry is given by

$$w = \int \frac{dt_0}{\sqrt{(1 - t_0^2)(1 - k^2 t_0^2)}} \tag{4.144}$$

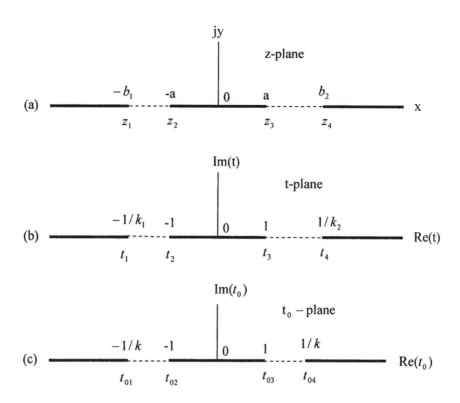

Figure 4.25 Transformations from an asymmetric CPW to a symmetric geometry: (a) z-plane; (b) t-plane; and (c) t_0-plane.

4.8   Summary

The functions of a complex variable $f(z)$ are useful for determining contour integrals, integrals involving singularities, and for conformal mapping of planar lines. The function $f(z)$ is assumed to be *analytic*, although it may not be so for a certain region about singularities. Cauchy-Riemann conditions are used to test the analytic nature of the function. The line integral of functions (with real variable) $f(z)$ with singularities (also called improper integrals) can be carried out analytically by converting the line integral into contour integral. The residue theorem helps in solving these integrals analytically. Other methods such as singularity subtraction, substitution of variable, and numerical integration are also available to determine the improper integrals. We come across these integrals in the Fourier transform–based, and in the Green's function–based solutions. An important application of $f(z)$ is based on conformal mapping, which is basically a coordinate transformation. This mapping preserves the capacitance of transmission line. Schwarz-Christoffel conformal mapping is found to be very useful to transform various planar and nonplanar transmission lines to parallel plate geometry so that the capacitance may be determined easily. In general, a number of transformations are needed to realize the final parallel plate configuration. The mapping procedure is illustrated through a number of examples which include strip line, coplanar strips, CPW, and microstrip with cover shield. Some useful mapping functions to transform finite dielectric thickness into infinite thickness, finite width lateral ground planes, and asymmetric metallization pattern to symmetric one are illustrated. The mapping functions properly capture the charge singularities at the edges of conductors. The conformal mapping of (open region) planar lines results in parallel plate geometry which is closed on all sides. Unlike the computational methods, the conformal mapping method does not require asymptotic boundary conditions or lossy material to truncate the open region. Also, the conformal mapping analysis of planar lines may result in closed-form design equations. Looking at the importance of singular integrals in analytical and computational methods, Appendix B is devoted to the evaluation of such integrals.

References

[1]   Churchill, R. V., and J. W. Brown, *Complex Variables and Applications*, 4th ed., New York: McGraw-Hill, 1984.

[2]   Davis, P. J., *Methods of Numerical Integration*, New York: Academic Press, 1975.

[3]   Arfken, G., *Mathematical Methods for Physicists*, 3rd ed., New York: Academic Press, 1985.

[4]   Kong, J. A., *Electromagnetic Wave Theory*, New York: John Wiley, 1986.

[5]   Butkov, E., *Mathematical Physics*, Reading, MA: Addison-Wesley, 1968.

[6]   Nguyen, C., *Analysis Methods for RF, Microwave, and Millimeter-Wave Planar Transmission Line Structures*, New York: John Wiley, 2000.

[7]   Collin, R. E., *Foundations for Microwave Engineering*, 2nd ed., New York: McGraw-Hill, 1992.

[8]   Kythe, P. K., *Computational Conformal Mapping*, Boston: Birkhauser, 1998.

[9]   Collin, R. E., *Field Theory of Guided Waves*, 2nd ed., New York: IEEE Press, 1991.

[10] Van Bladel, J., *Electromagnetic Fields*, Washington, D.C.: Hemisphere Publishing, 1985, Appendix 5.

[11] Cheng, K. K. M., and I. D. Robertson, "Quasi-TEM Analysis of V-Shaped Conductor Backed Coplanar Waveguide," *IEEE Trans. Microwave Theory Tech.*, Vol. MTT-43, 1995, pp. 1992–1994.

[12] http://documents.wolfram.co.jp/mathematica/book/section-3.5.

[13] Gradshteyn, I. S., and I. M. Ryzhik, *Tables of Integrals, Series, and Products*, New York: Academic Press, 1980.

[14] Hilberg, W., "From Approximations to Exact Relations for Characteristic Impedance," *IEEE Trans. Microwave Theory Tech.*, Vol. MTT-17, 1969, pp. 259–265.

[15] Gupta, K. C., et al., *Microstrip Lines and Slotlines*, 2nd ed., Norwood, MA: Artech House, 1996, Chapter 7.

[16] Shih, C., et al., "A Full-Wave Analysis of Microstrip Lines by Variational Conformal Mapping Technique," *IEEE Trans. Microwave Theory Tech.*, Vol. 36, 1988, pp. 576–581.

[17] Cheng, K. K. M., and I. D. Robertson, "Simple and Explicit Formulas for the Design and Analysis of Asymmetrical V-Shaped Microshield Line," *IEEE Trans. Microwave Theory Tech.*, Vol. MTT-43, 1995, pp. 2501–2504.

Problems

P4.1. Determine the singularities and corresponding residues of the following function:

$$f(z) = \frac{z}{z^2 + 1}$$

P4.2. Use contour integration to determine

$$I = \int_{-\infty}^{\infty} \frac{e^{-j\alpha r}}{\alpha - k} \, d\alpha$$

P4.3. Determine the integral given below

$$I = \int_{-\infty}^{\infty} \frac{\alpha e^{j\alpha r}}{k^2 - \alpha^2} \, d\alpha$$

by deforming the contour in the following ways:

1. Exclude pole at $\alpha_c = -k$ and include pole at $\alpha_c = k$;
2. Include both the poles;
3. Exclude both the poles.

P4.4. Show that [1, p. 172]

$$I = \int_{0}^{\infty} \frac{\sin x}{x} \, dx = \frac{\pi}{2}$$

using contour integration. Hint: determine $\oint \frac{e^{jz}}{z}$ to obtain I.

P4.5. Consider a metal strip of width W and zero thickness in free space. The strip is charged to a constant potential. Assume a magnetic wall boundary $\partial\varphi/\partial x = 0$ for $|x| > W/2$ coplanar with the strip as shown in Figure 4.26. Show using the mapping function $w = \sin^{-1} z$ that the charge density on the strip is given by

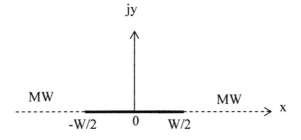

Figure 4.26 A metal strip of width w in free space.

$$\rho_s = \frac{Q}{\pi \sqrt{x - (W/2)^2}}$$

where Q is the total charge on the strip [7, p. 889].

P4.6. The coaxial transmission line of Figure 4.27 is filled with two different media with relative dielectric constants ϵ_{r1} and ϵ_{r2}. The media are assumed to be lossless, and the conductors are perfect. The potentials on the inner and outer conductors are V_0 and zero, respectively. Use conformal transformation to derive an expression for the characteristic impedance of coaxial line. Verify the derived expression with that of a uniformly filled coaxial line in the limit $\epsilon_{r1} = \epsilon_{r2} = \epsilon_r$ [6, p. 118].

P4.7 For this problem, the coaxial transmission line geometry of Figure 4.27 is filled uniformly with a medium with relative dielectric constant ϵ_r. Using the conformal mapping defined by $w = A\ell n(z) + B$ and choosing its real part u as the potential function, derive the potential distribution, electric and magnetic fields, capacitance per unit length, and the characteristic impedance of the coaxial line. Compare this expression with the well-known equation [6, p. 113]

$$Z_0 = \frac{60}{\sqrt{\epsilon_r}} \, \ell n(b/a)$$

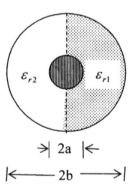

Figure 4.27   Cross-section of an inhomogeneously filled coaxial line.

CHAPTER 5
Fourier Transform Method

The mathematical objective of the Fourier transform or integral transform method is to reduce a difficult problem in the original domain to a simpler form in transform domain. To do this, an integral transformation $T\{f(x)\}$ is introduced. It changes the function $f(x)$ into another function of an auxiliary variable α, as follows:

$$T\{f(x)\} \equiv \int_a^b f(x)k(x, \alpha)\, dx = \tilde{F}(\alpha) \qquad (5.1)$$

The kernel $k(x, \alpha)$ and the limits of integration are the factors that distinguish different specific transforms such as Laplace transform, Fourier transform, Mellin transform, and Hankel transform from one another. The function $\tilde{F}(\alpha)$ is called a transform of function $f(x)$. We shall use $(\sim)$ to indicate a function in the transform domain. Similar to (5.1) one can define inverse transform also. In this operation, the function $\tilde{F}(\alpha)$ is integrated with a modified form of kernel over the range (a, b) to yield back $f(x)$. The functions $f(x)$ and $\tilde{F}(\alpha)$ are therefore called transform pair. It may be mentioned that the domain of the variable α is called the transform domain or more frequently the spectral domain. We shall restrict ourselves to the Fourier transform technique because of our interest in time harmonic solutions.

5.1 Introduction

The Fourier transform may be defined as

$$\tilde{F}(\alpha) = \int_{-\infty}^{+\infty} f(x)e^{j\alpha x}\, dx \qquad (5.2a)$$

The inverse Fourier transform corresponding to (5.2a) is defined as

$$f(x) = \frac{1}{2\pi} \int_{-\infty}^{+\infty} \tilde{F}(\alpha)e^{-j\alpha x}\, d\alpha \qquad (5.2b)$$

Sometimes the kernels of Fourier and inverse transforms are interchanged.

One point should be remembered that the Fourier or *direct transform* moves the problem in the transform domain, while the *inverse transformation* brings it back to the original domain. The solution of the problem in the original domain, if at all possible, may be difficult. When transformed into the new domain, the solution of the problem is expected to be easier [1]. However, the integral associated with the inverse transform, in general, involves singularity. The Fourier transform procedure is shown in Figure 5.1. It may be mentioned that the simplicity of the solution in the transform domain depends on the transformed differential equation and the boundary conditions.

The factor 2π in the denominator of (5.2b) is based on a certain convention. In some texts, this factor is used with (5.2a). In the modern convention, the factor $\sqrt{2\pi}$ is used in the denominators of both (5.2a) and (5.2b):

$$\tilde{F}(\alpha) = \frac{1}{\sqrt{2\pi}} \int\limits_{-\infty}^{+\infty} f(x) e^{j\alpha x} \, dx \tag{5.3a}$$

and

$$f(x) = \frac{1}{\sqrt{2\pi}} \int\limits_{-\infty}^{+\infty} \tilde{F}(\alpha) e^{-j\alpha x} \, d\alpha \tag{5.3b}$$

We shall mostly follow the modern convention while using the factor 2π. The Fourier transform of (5.3) can be easily generalized to more than one variable as discussed later.

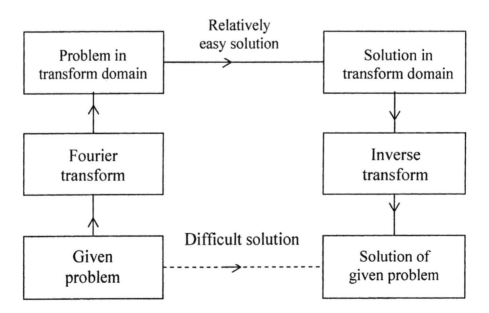

Figure 5.1 Typical steps in Fourier transform method.

Conditions for the Existence of Fourier Transform
The necessary, *but not sufficient*, condition for the existence of the Fourier transform of a function $f(x)$ is:

1. $f(x)$ should be differentiable.

2. $f(x)$ should be absolutely integrable (i.e., $\int\limits_{-\infty}^{+\infty} |f(x)|\, dx$ is finite).

3. Any discontinuities in $f(x)$ are finite or $f(x)$ is piecewise continuous.

A function $f(x)$ is said to be piecewise continuous in an interval (a, b) if the interval can be partitioned into a *finite* number of nonintersecting intervals

$$(a, a_1), (a_1, a_2), (a_2, a_3), \ldots, (a_{n-1}, b) \tag{5.4}$$

in each of which the function is continuous, and has finite limit as x approaches the end points of each of the subintervals, as shown in Figure 5.2.

Only in a few cases can one assign physical meaning to both the domains. For example, the spectrum of an electrical waveform $v(t)$ is given by

$$S(\omega) = \frac{1}{\sqrt{2\pi}} \int\limits_{-\infty}^{+\infty} v(t) e^{j\omega t}\, dt \tag{5.5}$$

Similarly, the waveform $v(t)$ can be synthesized from its spectral constituents as

$$v(t) = \frac{1}{\sqrt{2\pi}} \int\limits_{-\infty}^{+\infty} S(\omega) e^{-j\omega t}\, d\omega \tag{5.6}$$

The antenna current and the radiation pattern are also Fourier transform pairs. It is therefore advisable that one should avoid attaching physical meaning to the spectral domain.

Advantages of Fourier Transform Method
The Fourier transform method is almost indispensable for certain types of problems. The advantages associated with the Fourier transform method may be described as follows:

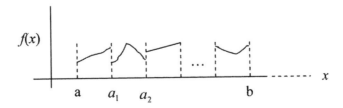

Figure 5.2 Illustration of a piecewise continuous function.

1. Reduces the PDE to ordinary differential equations or algebraic equations that can be solved more easily;
2. Solves PDE with complicated boundary conditions. For example, geometries with unbounded regions like laterally open structures, half-planes, and infinite/semi-infinite strips can be solved more easily when formulated in the Fourier domain.
3. The source function of the inhomogeneous PDE need not be a continuous function of variables. It can be a piecewise continuous function if it can be Fourier transformed.

These advantages have a penalty associated with them. The inverse Fourier transform, carried out to determine the solution in the original domain is, in general, an improper integral with a singular integrand, which is sometimes difficult to evaluate analytically. These types of integrals are discussed in Appendix B.

5.2   Reduction of PDE to Ordinary Differential Equation/Algebraic Equation Using Fourier Transform

In the solution of PDE using Fourier transform method, each term of the PDE is transformed by multiplying with $e^{j\alpha x}$ and integrated over the variable x. The boundary conditions are also likewise transformed. The resulting equations must be simpler than the original ones. Consequently, for this technique to be useful in the solutions of differential equations, the transform of the derivative $T\{df/dx\}$ (or some function of the derivative which appears in the differential equation) must be simply related to $\tilde{F}(\alpha)$.

Fourier Transform of Derivatives
When transforming the derivative of a function, the integration can be carried out by parts as follows:

$$T\left\{\frac{df}{dx}\right\} = \frac{1}{\sqrt{2\pi}} \int\limits_{-\infty}^{+\infty} \frac{df}{dx} e^{j\alpha x}\, dx$$

$$= \frac{1}{\sqrt{2\pi}} f(x) e^{j\alpha x}\Big|_{-\infty}^{+\infty} - j\alpha\tilde{F}(\alpha) \tag{5.7}$$

$$= -j\alpha\tilde{F}(\alpha) \tag{5.8}$$

The first term in (5.7) vanishes as $|x| \to \pm\infty$ because either $f(x)$ or $e^{j\alpha x}$ vanishes. The mathematical meaning of these assumptions is that one often includes a convergence factor when evaluating a Fourier transform, while the physical meaning is that finite sources do not produce a field at infinity in the presence of a physical medium with loss.

Expression (5.8) shows that the transform of the derivative is $(-j\alpha)$ times the transform of the original function. The operation of differentiation has been replaced by multiplication by the transform variable. *It is this property of the Fourier transform which makes it very powerful in solving PDE.* The higher order derivatives can be transformed similarly, giving rise to

$$T\left\{\frac{d^n f}{dx}\right\} = (-j\alpha)^n \tilde{F}(\alpha) \tag{5.9}$$

We shall apply Fourier transform method to solve PDE and ordinary differential equations.

5.3   Solution of Differential Equations with Unbounded Regions

Due to the limitation of computer resources, the unbounded region in computational methods is terminated by imposing either absorbing boundary conditions or radiation conditions. This methodology, although efficient, still requires extra computing resources. The use of Green's function does not require any such termination and provides the most efficient solution. The method of moments (Chapter 11) is based on the availability of Green's function. We shall employ the Fourier transform to determine Green's functions for homogeneous medium and layered dielectric medium. However, constructing Green's function for inhomogeneous dielectric configurations is a limitation.

5.3.1   Free-Space Green's Function in One Dimension

The free space being unbounded, its eigenvalue spectrum is continuous. Under this condition the summation in the series expansion of Green's function (Sections 3.2 and 3.3) becomes an integral. In order to associate physical meaning with the Green's function, we shall first determine Green's function for transmission lines match terminated at the two ends.

Consider an infinitely long transmission line of characteristic impedance Z_0 excited by a delta function voltage source at $z = z'$, as shown in Figure 5.3. The transmission line equations for this excitation are

$$\frac{dv}{dz} = -Zi + \delta(z - z') \tag{5.10}$$

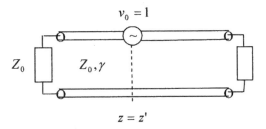

Figure 5.3   A transmission line terminated in matched loads at the ends and excited by a series voltage source.

$$\frac{di}{dz} = -Yv \qquad (5.11)$$

The line parameters Z and Y are the impedance and admittance per unit length, respectively. Eliminating v from (5.10) and (5.11), one obtains

$$\frac{d^2i}{dz^2} - YZi = -Y\delta(z - z')$$

or

$$\frac{d^2i}{dz^2} - \gamma^2 i = -Y\delta(z - z'), \qquad \text{since } \gamma^2 = YZ \qquad (5.12)$$

For a lossless transmission line $\gamma = j\beta$, and therefore

$$\frac{d^2i}{dz^2} + \beta^2 i = -Y\delta(z - z') \qquad (5.13)$$

Its solution can be obtained by employing the eigenfunction expansion method. For this, we solve the following eigenvalue problem:

$$\frac{d^2u}{dz^2} = -k^2 u \qquad (5.14)$$

The eigenfunctions of (5.14) can be assumed as $e^{\pm jkz}$, with the eigenvalue k no longer discrete. The reason for the continuous nature of k is that the structure is not bounded along z. It is clear from above that instead of series expansion for $i(z)$ in (5.13) we should seek its representation as a Fourier integral; that is,

$$i(z) = \frac{1}{\sqrt{2\pi}} \int_{-\infty}^{+\infty} \tilde{I}(k)e^{-jkz}\, dk \qquad (5.15)$$

The inverse Fourier transform is defined as

$$\tilde{I}(k) = \frac{1}{\sqrt{2\pi}} \int_{-\infty}^{+\infty} i(z)e^{jkz}\, dz \qquad (5.16)$$

In this spirit we subject the differential equation (5.12) to Fourier transformation. For this we multiply (5.12) by $e^{jkz}\, dz$ and integrate over z from $-\infty$ to $+\infty$ to obtain

$$\frac{1}{\sqrt{2\pi}} \int\limits_{-\infty}^{+\infty} \frac{d^2 i}{dz^2} e^{jkz}\, dz - \gamma^2 \frac{1}{\sqrt{2\pi}} \int\limits_{-\infty}^{+\infty} i(z) e^{jkz}\, dz = -Y \frac{1}{\sqrt{2\pi}} \int\limits_{-\infty}^{+\infty} \delta(z - z') e^{jkz}\, dz$$

$$(5.17)$$

Using the property of the Fourier transform of derivatives (5.9), and the definition of Fourier transform, (5.17) can be written as

$$-k^2 \tilde{I}(k) - \gamma^2 \tilde{I}(k) = \frac{-Y e^{jkz'}}{\sqrt{2\pi}} \tag{5.18}$$

or

$$\tilde{I}(k) = \frac{1}{\sqrt{2\pi}} \frac{Y e^{jkz'}}{k^2 + \gamma^2} \tag{5.19}$$

Now, substitute this expression in (5.15) to obtain the Green's function $i(z)$,

$$i(z) = \frac{Y}{2\pi} \int\limits_{-\infty}^{+\infty} \frac{e^{jk(z'-z)}}{\gamma^2 + k^2}\, dk \tag{5.20}$$

This expression is called the integral representation of Green's function. The integrand has singularities at $k = \pm j\gamma$, and may be determined using residue theorem (Chapter 4). We obtain for $z < z'$

$$i(z) = -2\pi j \frac{e^{\gamma(z'-z)}}{2j\gamma} \frac{Y}{2\pi} \tag{5.21}$$

or

$$i(z) = -\frac{Y}{2\gamma} e^{-\gamma(z'-z)} \quad \text{and} \quad v(z) = -\frac{1}{2} e^{-\gamma(z'-z)} \tag{5.22}$$

for $z > z'$

$$i(z) = \frac{Y}{2\gamma} e^{-\gamma(z-z')} \quad \text{and} \quad v(z) = -\frac{1}{2} e^{-\gamma(z-z')} \tag{5.23}$$

The above analysis may be employed to determine free space Green's function for one-dimensional problems.

The Green's function for the scalar wave equation is defined as

$$\frac{d^2 G}{dx^2} + k_0^2 G = -\delta(x - x') \tag{5.24}$$

subject to the radiation condition $G(\pm\infty, x') = 0$. The above differential equation may be obtained from (5.13) by substituting $Y = 1$ and $\beta = k_0$. The final solution may therefore be realized from (5.22) and (5.23). However, we shall now follow a purely mathematical approach to derive the Green's function.

Since $-\infty < x < \infty$, we may express the Green's function as a Fourier transform

$$G(x; x') = \sqrt{\frac{1}{2\pi}} \int\limits_{-\infty}^{\infty} \tilde{G}(\alpha; x') e^{j\alpha x}\, d\alpha \qquad (5.25)$$

and the delta function as (2.149a),

$$\delta(x - x') = \frac{1}{2\pi} \int\limits_{-\infty}^{\infty} e^{j\alpha(x - x')}\, d\alpha \qquad (5.26)$$

Substituting these expressions in (5.24) gives

$$\tilde{G}(\alpha; x') = \sqrt{\frac{1}{2\pi}} \frac{e^{-j\alpha x'}}{\alpha^2 - k_0^2}$$

or

$$G(x; x') = \frac{1}{2\pi} \int\limits_{-\infty}^{\infty} \frac{e^{j\alpha(x - x')}}{\alpha^2 - k_0^2}\, d\alpha \qquad (5.27)$$

Carrying out the integration using residue theorem gives the following expression for the free space Green's function:

$$G(x; x') = \frac{j}{2k_0} e^{jk_0|x - x'|} \qquad (5.28)$$

Exercise.   Derive the Green's function for the current $i(z)$ by taking the Fourier transform of (5.13).

5.3.2 Fourier Sine Transform and Half-Space Green's Function

The Fourier exponential transform (5.2a) reduces to the Fourier sine transform when the function $f(x)$ has odd symmetry. Similarly, it reduces to Fourier cosine transform when $f(x)$ is an even function. The symmetry property of a function is related to the boundary conditions satisfied by it. Therefore, Fourier sine and cosine transforms may be used in place of Fourier transform for solving PDE with Dirichlet or Neumann boundary conditions. We next consider an example to illustrate this point. The Fourier sine transform is defined as

$$\tilde{F}(\alpha) = \sqrt{\frac{2}{\pi}} \int_0^{+\infty} f(u) \sin(\alpha u) \, du \qquad (5.29a)$$

and the inverse transform as

$$f(u) = \sqrt{\frac{2}{\pi}} \int_0^{+\infty} \tilde{F}(\alpha) \sin(\alpha u) \, d\alpha \qquad (5.29b)$$

Fourier cosine transform-pair is defined similarly by replacing $\sin(\alpha u)$ by $\cos(\alpha u)$ in (5.29). If the range of variable u is (a, b), the limits of integration in (5.29a) are adjusted accordingly. However, (5.29b) does not hold in this case, and $f(u)$ is expressed as a series as described later.

Half-Space Green's Function in One Dimension
The Green's function is now defined as

$$\frac{d^2 G}{dx^2} + k_0^2 G = -\delta(x - x') \qquad (5.30)$$

subject to $G(0; x') = 0$ and the radiation condition $G(\infty; x') = 0$.

The geometry of the problem may be sketched as in Figure 5.4. Because of the boundary condition at $x = 0$ we now employ sine transform of the Green's function equation to convert it into ordinary differential equation. Using the sine transform (5.29b), we write

$$G(x; x') = \sqrt{\frac{2}{\pi}} \int_0^{\infty} \tilde{G}(\alpha) \sin(\alpha x) \, d\alpha \qquad (5.31a)$$

and the delta function as [2, p. 191]

$$\delta(x - x') = \frac{2}{\pi} \int_0^{\infty} \sin \alpha x' \sin(\alpha x) \, d\alpha \qquad (5.31b)$$

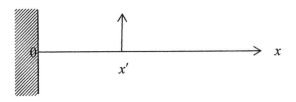

Figure 5.4 Half-space in one dimension with delta function source at x'.

Now substitute these expansions in (5.30) to obtain

$$\tilde{G}(\alpha) = \sqrt{\frac{2}{\pi}} \frac{\sin \alpha x'}{\alpha^2 - k_0^2} \tag{5.32}$$

Inverse Fourier sine transform according to (5.29b) gives

$$G(x; x') = \frac{2}{\pi} \int_0^\infty \frac{\sin \alpha x' \sin \alpha x}{\alpha^2 - k_0^2} \, d\alpha \tag{5.33}$$

Contour integration of the above yields the following expression for the half-space Green's function [2, p. 191]:

$$\tilde{G}(x; x') = \begin{cases} \dfrac{\sin k_0 x'}{k_0} e^{jk_0 x} & \text{for } x \geq x' \\[3mm] \dfrac{\sin k_0 x}{k_0} e^{jk_0 x'} & \text{for } 0 \leq x \leq x' \end{cases} \tag{5.34}$$

We next derive Green's function for two-dimensional cases. Any three-dimensional geometry with uniform cross-section, and excitation and boundary conditions also uniform normal to the cross-section, can be analyzed as a two-dimensional problem.

We shall assume uniformity along the z-direction.

5.3.3 Free-Space Green's Function in Two Dimensions

The Green's function wave equation in two dimensions is given by

$$\left(\frac{\partial^2}{\partial x^2} + \frac{\partial^2}{\partial y^2} + k_0^2 \right) G(x, y; x', y') = -\delta(x - x')\delta(y - y') \tag{5.35}$$

Following the mathematical approach described in the last section, we employ double Fourier transform representations of Green's function and delta function

$$G(x, y; x', y') = \frac{1}{2\pi} \int_{-\infty}^{\infty} \int_{-\infty}^{\infty} \tilde{G}(\alpha, \beta) e^{j\alpha(x-x')} e^{j\beta(y-y')} \, d\alpha \, d\beta \tag{5.36a}$$

$$\delta(x - x')\delta(y - y') = \frac{1}{(2\pi)^2} \int_{-\infty}^{\infty} \int_{-\infty}^{\infty} e^{j\alpha(x-x')} e^{j\beta(y-y')} \, d\alpha \, d\beta \tag{5.36b}$$

Substituting these expressions in (5.35), taking derivatives, and comparison gives

$$\tilde{G}(\alpha, \beta) = \frac{1}{2\pi} \frac{1}{\alpha^2 + \beta^2 - k_0^2}$$

or

$$G(x, y; x', y') = \frac{1}{(2\pi)^2} \int\limits_{-\infty}^{\infty} \int\limits_{-\infty}^{\infty} \frac{e^{j\alpha(x-x')} e^{j\beta(y-y')}}{\alpha^2 + \beta^2 - k_0^2} \, d\alpha \, d\beta \qquad (5.37)$$

The above expression is called the plane wave spectrum representation of Green's function. By employing plane-wave to cylindrical-wave transformation and residue calculus, the free space Green's function may be expressed as [2, p. 203]

$$G(x, y; x', y') = \frac{j}{4} H_0^{(1)}(k_0 |\rho - \rho'|) \qquad (5.38)$$

where $H_0^{(1)}(.)$ is the Hankel function of zero order and first kind, and

$$|\rho - \rho'| = \sqrt{(x - x')^2 + (y - y')^2} \qquad (5.39)$$

Alternative solution: An alternative method is to use Fourier transform along one of the directions and convert the two-dimensional wave equation into ordinary differential equation, which can be subsequently solved for one-dimensional free-space Green's function. Substituting

$$G(x, y; x', y') = \sqrt{\frac{1}{2\pi}} \int\limits_{-\infty}^{\infty} \tilde{G}(\alpha; y, y') e^{j\alpha(x-x')} \, d\alpha \qquad (5.40)$$

and (5.26) yields

$$\left(\frac{d^2}{dy^2} + k_0^2 - \alpha^2 \right) \tilde{G}(\alpha; y, y') = -\sqrt{\frac{1}{2\pi}} \delta(y - y') \qquad (5.41)$$

Solution of this ordinary differential equation is

$$\tilde{G}(\alpha; y, y') = \sqrt{\frac{1}{2\pi}} \frac{j}{2K} e^{jK|y-y'|}, \qquad K = \sqrt{k_0^2 - \alpha^2} \qquad (5.42a)$$

Inverse Fourier transform gives

$$G(x, x'; y, y') = \frac{1}{2\pi} \int\limits_{-\infty}^{\infty} \frac{j}{2K} e^{jK|y-y'|} e^{j\alpha(x-x')} \, d\alpha \qquad (5.42b)$$

and is another form of (5.38).

Example 5.1. Uniform line source in a grounded conducting trough.   Consider the cross-section of a grounded conducting trough as shown in Figure 5.5. The line source at (x', y') is uniform with respect to z. Due to the uniformity of the trough and the source the Green's function, corresponding to the Poisson equation, becomes

$$\nabla_t^2 G(x, y; x', y') = -\delta(x - x')\,\delta(y - y'), \qquad \nabla_t^2 \equiv \frac{\partial^2}{\partial x^2} + \frac{\partial^2}{\partial y^2} \qquad (5.43)$$

with $G = 0$ at $x = 0$, a; $y = 0$, $+\infty$. We next determine the Green's function.

Solution 1.   In view of the boundary condition $G = 0$ at $y = 0$, let us use the Fourier sine transform pair defined in (5.29). According to it, the function $f(u)$ is expressed as a superposition of an infinite number of sinusoidal functions, each of which is characterized by its period α and the amplitude $\tilde{F}(\alpha)$. The value of $\tilde{F}(\alpha)$ is obtained from (5.29a). We can use superposition of sine waves along the y-direction to describe the solution of (5.43) because each of these terms satisfies the boundary condition at $y = 0$. In order to implement this approach we take Fourier sine transform of (5.43). For this, we multiply both sides of (5.43) by $\sin(\alpha y)\,dy$ and integrate over y from 0 to $+\infty$, and obtain

$$\sqrt{\frac{2}{\pi}} \int_0^\infty \left[\frac{\partial^2}{\partial x^2} + \frac{\partial^2}{\partial y^2} \right] G(x, y; x', y')\, \sin(\alpha y)\, dy =$$

$$-\delta(x - x')\sqrt{\frac{2}{\pi}} \int_0^\infty \delta(y - y')\, \sin(\alpha y)\, dy$$

or

$$\frac{d^2}{dx^2}\, \tilde{G}(x, \alpha; x', y') + \sqrt{\frac{2}{\pi}} \int_0^\infty \frac{\partial^2 G}{\partial y^2}\, \sin(\alpha y)\, dy = -\sqrt{\frac{2}{\pi}}\, \delta(x - x')\, \sin(\alpha y')$$

$$(5.44)$$

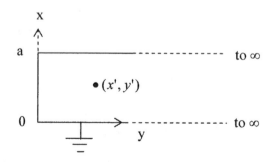

Figure 5.5 A uniform line source in a grounded trough.

Now, consider the second term on the left side of (5.44). Integrating it by parts and using the boundary conditions on G, one obtains

$$\sqrt{\frac{2}{\pi}} \int_0^\infty \frac{\partial^2 G}{\partial y^2} \sin(\alpha y)\, dy = -\alpha^2 \tilde{G}(x, \alpha; x', y') \qquad (5.45)$$

Equation (5.44) can therefore be written as

$$\left(\frac{d^2}{dx^2} - \alpha^2\right) \tilde{G}(x, \alpha; x', y') = -\sqrt{\frac{2}{\pi}}\, \delta(x - x') \sin(\alpha y') \qquad (5.46)$$

We notice that the Fourier transform has converted the PDE into an ordinary differential equation (in x) which can be solved using the direct construction approach of Chapter 3. The solution to (5.46) is

$$\tilde{G}_1(x, \alpha; x', y') = \sqrt{\frac{2}{\pi}} \frac{\sin(\alpha y') \sinh(\alpha x) \sinh[\alpha(a - x')]}{\alpha \sinh(\alpha a)} \qquad \text{for } x \le x' \quad (5.47a)$$

$$\tilde{G}_2(x, \alpha; x', y') = \tilde{G}_1(x', \alpha; x, y') \qquad \text{for } x \ge x' \quad (5.47b)$$

Equation (5.47b) follows from the symmetry property of the Green's function. It means that the function $\tilde{G}_2$ is obtained from $\tilde{G}_1$ by interchanging x and x'. It may be noted that solution (5.47) satisfies the boundary conditions at x.

Finally, the space domain Green's function is obtained by taking the inverse sine transform of (5.47). One obtains:

1. For $x \le x'$

$$G_1(x, y; x', y') = \frac{2}{\pi} \int_0^\infty \frac{\sin(\alpha y') \sinh(\alpha x) \sinh[\alpha(a - x')]}{\alpha \sinh(\alpha a)} \sin(\alpha y)\, d\alpha$$

$$(5.48a)$$

2. For $x \ge x'$,

$$G_2(x, y; x', y') = G_1(x', y'; x, y) \qquad (5.48b)$$

Observe that if we use the symmetry property of the Green's function, we do not need to carry out the integration for $x \ge x'$.

Solution 2.   An alternative solution in terms of Fourier sine transform along the x-direction [3]. We know that the factor $\sin(n\pi x/a)$ satisfies the boundary conditions at $x = 0, a$. In view of this, let us take Fourier sine transform of (5.43) along the x-direction. For this we multiply it by $\sin(n\pi x/a)\, dx$ and integrate over x from 0 to a,

$$\sqrt{\frac{2}{\pi}} \int_0^a \left[\frac{\partial^2}{\partial x^2} + \frac{\partial^2}{\partial y^2}\right] G(x, y; x', y') \sin(\alpha x)\, dx = \qquad (5.49)$$

$$-\delta(y - y')\sqrt{\frac{2}{\pi}} \int_0^a \delta(x - x') \sin(\alpha x)\, dx$$

where $\alpha = n\pi/a$. Integration by parts gives

$$\left[\frac{d^2}{dy^2} - \alpha^2\right] \tilde{G}(\alpha, y; x', y') = -\sqrt{\frac{2}{\pi}} \delta(y - y') \sin(\alpha x') \qquad (5.50)$$

where the Fourier sine transform of G is defined as

$$\tilde{G}(\alpha, y; x', y') = \sqrt{\frac{2}{\pi}} \int_0^a G(x, y; x', y') \sin(\alpha x)\, dx \qquad (5.51)$$

The boundary conditions to be satisfied by $\tilde{G}$ are

$$\tilde{G}(\alpha, 0; x', y') = 0 = \tilde{G}(\alpha, \infty; x', y') \qquad (5.52)$$

The ordinary differential equation of (5.50) may be solved by employing the methods described in Chapter 3. One obtains the following Green's function in spectral domain:

$$\tilde{G}(\alpha, y; x', y') = \begin{cases} \sqrt{\dfrac{2}{\pi}} \dfrac{1}{\alpha} \sin(\alpha x') \sinh(\alpha y) e^{-\alpha y'} & \text{for } y \leq y' \\[3mm] \sqrt{\dfrac{2}{\pi}} \dfrac{1}{\alpha} \sin(\alpha x') \sinh(\alpha y') e^{-\alpha y} & \text{for } y \geq y' \end{cases} \qquad (5.53)$$

To obtain $G(x, y; x', y')$ from $\tilde{G}(\alpha, y; x', y')$, let us proceed in the following manner:

$G(x, y; x', y')$ is periodic in x because of the boundary conditions along the x-direction, and therefore can be expanded in a Fourier series such as

$$G(x, y; x', y') = \sum_{n=1}^{\infty} g_n(y; x', y') \sin\left(\frac{n\pi x}{a}\right) \qquad (5.54)$$

Let us substitute this expansion in (5.51) to obtain

$$\tilde{G}(n, y; x', y') = \sqrt{\frac{2}{\pi}} \int_0^a \sum_{m=1}^{\infty} g_m(y; x', y') \sin\left(\frac{n\pi x}{a}\right) \sin\left(\frac{m\pi x}{a}\right) dx \qquad (5.55)$$

$$= \sqrt{\frac{2}{\pi}} \frac{a}{2} g_n(y; x', y')$$

Thus, (5.54) can be written as

$$G(x, y; x', y') = \frac{2}{a} \sqrt{\frac{\pi}{2}} \sum_{n=1}^{\infty} \tilde{G}(n, y; x', y') \sin\left(\frac{n\pi x}{a}\right) \qquad (5.56)$$

It may be clarified that $\tilde{G}(n, y; x', y')$ and $\tilde{G}(\alpha, y; x', y')$ are the same since $\alpha = n\pi/a$. Substituting (5.53) in (5.56) gives

$$G(x, y; x', y') = \begin{cases} \displaystyle\sum_{n=1}^{\infty} \frac{2}{n\pi} \sin(\alpha x') \sin(\alpha x) \sinh(\alpha y) e^{-\alpha y'} & \text{for } y \leq y' \\[2ex] \displaystyle\sum_{n=1}^{\infty} \frac{2}{n\pi} \sin(\alpha x') \sin(\alpha x) \sinh(\alpha y') e^{-\alpha y} & \text{for } y \geq y' \end{cases}$$

$$(5.57)$$

Example 5.2.   Probe excitation of TE-modes in a rectangular waveguide [3]. Consider a uniform unit line current extending across the rectangular waveguide, parallel with the y-axis and located at (x', z') as shown in Figure 5.6. The line current is an approximation of the excitation produced by the probe of coaxial-to-waveguide adapter. The current is assumed to be uniform along the y-coordinate. The field produced by the current will also not vary with y, meaning thereby $\partial/\partial y \equiv 0$; and therefore the problem is two-dimensional. The operator ∇^2 reduces to

$$\nabla_t^2 = \frac{\partial^2}{\partial x^2} + \frac{\partial^2}{\partial z^2} \qquad (5.58)$$

We shall use the vector potential approach to derive the field components. The relationship between the vector potential $\mathbf{A}$ and the excitation current $\mathbf{J}$ is [refer to (1.39)]

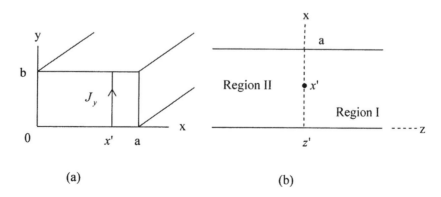

(a)   (b)

Figure 5.6   Probe current excitation of TE-modes in a rectangular waveguide: (a) isometric view; and (b) top view of the excitation of waveguide.

$$\left(\nabla_t^2 + k_0^2\right)\mathbf{A} = -\mu_0 \mathbf{J}(x, z) \tag{5.59}$$

Since $\mathbf{J} = \hat{y}J_y$, the nonzero component of $\mathbf{A}$ is A_y only, and (5.59) reduces to the following scalar equation:

$$\left(\nabla_t^2 + k_0^2\right)A_y = -\mu_0 J_y(x, z) \tag{5.60}$$

For the unit strength excitation current located at (x', z'), we write

$$J_y(x, z) = \delta(x - x')\,\delta(z - z') \qquad \text{amp/m} \tag{5.61}$$

Substituting for ∇_t^2 and J_y in (5.60), one obtains

$$\left(\frac{\partial^2}{\partial x^2} + \frac{\partial^2}{\partial z^2} + k_0^2\right)A_y = -\mu_0\,\delta(x - x')\,\delta(z - z') \tag{5.62}$$

The solution for (5.62) in terms of single series expansion was derived in Chapter 3. Here, we shall use Fourier transform method as an alternative approach.

Field Components Produced by A_y.   We know that $\mathbf{E} = -j\omega\mathbf{A}$; therefore,

$$E_y = -j\omega A_y \tag{5.63}$$

Also, from $\mu_0\mathbf{H} = \nabla \times \mathbf{A}$ we obtain

$$\mu_0 H_x = \frac{-\partial A_y}{\partial z}, \qquad \text{and} \qquad \mu_0 H_z = \frac{\partial A_y}{\partial x} \tag{5.64}$$

All other field components are zero. The modes described by E_y, H_x, and H_z can be classified as *TE-to-z* or *TM-to-y* modes.

Since $E_y = -j\omega A_y$, the wave equation (5.62) can be expressed in terms of E_y as

$$\left(\frac{\partial^2}{\partial x^2} + \frac{\partial^2}{\partial z^2} + k_0^2\right)E_y = j\omega\mu_0\,\delta(x - x')\,\delta(z - z') \tag{5.65}$$

with

$$E_y = 0 \text{ at } x = 0, a \tag{5.66}$$

Solution 1.   We take the Fourier sine transform of (5.65) along the x-direction; that is, multiply it by $\sin(n\pi x/a)\,dx$ and integrate over x giving the following (finite Fourier transform is taken here because $\sin(n\pi x/a)$ satisfies the boundary conditions on E_y at $x = 0, a$ and thus helps in evaluating the transform of derivatives of the function):

$$\left(k_0^2 - \alpha^2 + \frac{d^2}{dz^2}\right) \sqrt{\frac{2}{\pi}} \int_0^a E_y \sin(\alpha x)\, dx = j\omega\mu_0 \sin(\alpha x')\delta(z - z') \sqrt{\frac{2}{\pi}}$$

$$(5.67)$$

where $\alpha = n\pi/a$. Let us define the sine transform of the electric field Green's function as

$$\sqrt{\frac{2}{\pi}} \int_0^a E_y(x, z; x', z') \sin(\alpha x)\, dx = \tilde{E}_y(\alpha, z; x', z') \qquad (5.68)$$

Thus, $\tilde{E}_y(\alpha, z; x', z')$ must satisfy the following ordinary differential equation:

$$\left(\frac{d^2}{dz^2} - k_n^2\right)\tilde{E}_y(\alpha, z; x', z') = j\omega\mu_0 \sin(\alpha x')\delta(z - z')\sqrt{\frac{2}{\pi}} \qquad (5.69)$$

where $k_n^2 = \alpha^2 - k_0^2$. Solutions to (5.69) can be obtained in the usual manner by using *direct construction* procedure for the Green's function (Chapter 3). We obtain

$$\tilde{E}_y(\alpha, z; x', z') = \frac{-j\omega\mu_0}{2k_n} \sqrt{\frac{2}{\pi}} \sin(\alpha x') \begin{cases} e^{-k_n(z'-z)} & \text{for } z \leq z' \\ e^{k_n(z'-z)} & \text{for } z \geq z' \end{cases}$$

or

$$\tilde{E}_y(\alpha, z; x', z') = \frac{-j\omega\mu_0}{2k_n} \sqrt{\frac{2}{\pi}} \sin(\alpha x') e^{-k_n|z'-z|} \qquad (5.70)$$

The relationship between $E_y(x, z; x', z')$ and $\tilde{E}_y(\alpha, z; x', z')$ is in the form of a Fourier series and is given by [refer to (5.56)]

$$E_y(x, z; x', z') = \frac{2}{a} \sqrt{\frac{\pi}{2}} \sum_{m=1}^{\infty} \tilde{E}_y(m, z; x', z') \sin\left(\frac{m\pi x}{a}\right) \qquad (5.71)$$

Substituting for $\tilde{E}_y(m, z; x', z')$ from (5.70) gives

$$E_y(x, z; x', z') = \frac{2}{a}\left(\frac{-j\omega\mu_0}{2}\right) \sum_{n=1}^{\infty} \frac{1}{k_n} \sin(\alpha x) \sin(\alpha x') e^{-k_n|z-z'|} \quad (5.72)$$

$$= \left(\frac{-j\omega\mu_0}{a}\right) \sum_{n=1}^{\infty} \frac{1}{k_n} \sin(\alpha x) \sin(\alpha x') e^{-k_n|z-z'|}$$

where $k_n^2 = \alpha^2 - k_0^2$, and $\alpha = n\pi/a$.

The above problem can also be solved by taking Fourier exponential transform along the z-direction. This solution is presented next.

Solution 2.   In this solution we shall use Fourier exponential transform along the z-direction because the waveguide is infinitely long and therefore unbounded.

The wave equation (5.65) may be rewritten as

$$\left(\frac{\partial^2}{\partial x^2} + \frac{\partial^2}{\partial z^2} + k_0^2\right)\psi = \delta(x - x')\,\delta(z - z') \tag{5.73}$$

where $\psi \equiv E_y/(j\omega\mu_0)$. Now multiply both sides of (5.73) by $e^{j\beta z}$ and integrate over z from $-\infty$ to $+\infty$. We obtain after integration by parts

$$\frac{d^2\tilde{\psi}(x, \beta)}{dx^2} + \left(k_0^2 - \beta^2\right)\tilde{\psi}(x, \beta) = \frac{1}{\sqrt{2\pi}}\,e^{j\beta z'}\delta(x - x') \tag{5.74}$$

where $\tilde{\psi}(x, \beta) = \dfrac{1}{\sqrt{2\pi}}\displaystyle\int\limits_{-\infty}^{+\infty}\psi(x, z)\,e^{j\beta z}\,dz$ and the variables (x', z') have been sup-

pressed. The differential equation (5.74) is to be solved subject to the boundary conditions $\tilde{\psi}(x, \beta) = 0$ at $x = 0, a$. Using the *direct construction* method of Chapter 3 we obtain

$$\tilde{\psi}(x, \beta) = \frac{-1}{\sqrt{2\pi}}\frac{e^{j\beta z'}}{\alpha\sin(\alpha a)}\begin{cases}\sin(\alpha x)\sin(\alpha(a - x')) & \text{for } x \leq x' \\ \sin(\alpha x')\sin(\alpha(a - x)) & \text{for } x \geq x'\end{cases} \tag{5.75}$$

with $\alpha = \sqrt{k_0^2 - \beta^2}$. The inverse Fourier transform now gives the following solution for $\psi(x, z)$:

$$\psi(x, z; x', z') = \frac{-1}{2\pi}\int\limits_{-\infty}^{+\infty}\frac{e^{j\beta(z'-z)}}{\alpha\sin(\alpha a)}\,d\beta\begin{cases}\sin(\alpha x)\sin(\alpha(a - x')) & \text{for } x \leq x' \\ \sin(\alpha x')\sin(\alpha(a - x)) & \text{for } x \geq x'\end{cases}$$

$$\tag{5.76}$$

This expression emphasizes the fact that the behavior of $\psi(x, z; x', z')$ can be described in terms of plane wave variation along the z-direction, $e^{j\beta z}$. This is a consequence of the open boundary condition along the z-direction. If the integral in (5.76) is converted to contour integral, it may be evaluated in terms of the poles which occur at $\sin(\alpha a) = 0$ or $\alpha_n = n\pi/a$

$$k_0^2 - \beta^2 = \left(\frac{n\pi}{a}\right)^2 \Rightarrow \beta^2 = k_0^2 - \left(\frac{n\pi}{a}\right)^2 \tag{5.77}$$

Evaluation of the integral in terms of the residues at the poles leads to the eigenfunction expansion (5.72) derived earlier for this problem [2, p. 221].

Solution 3. Another alternative solution to (5.65) can be obtained if we combine the methodologies of solutions 1 and 2 by taking the Fourier sine transform along the x-direction and the Fourier exponential transform along the z-direction. Let

$$\tilde{\psi}(n, \beta) = \sqrt{\frac{2}{\pi}} \int\limits_{0}^{a} \tilde{\psi}(x, \beta) \sin\left(\frac{n\pi x}{a}\right) dx \qquad (5.78)$$

With this in mind let us multiply (5.74) by $\sin(n\pi x/a)\sqrt{2/\pi}$ and integrate over the x-variable. One obtains

$$\left[k_0^2 - \beta^2 - \left(\frac{n\pi}{a}\right)^2\right] \tilde{\psi}(n, \beta) = e^{j\beta z'} \sin\left(\frac{n\pi x'}{a}\right) \frac{1}{\pi}$$

or

$$\tilde{\psi}(n, \beta; x', z') = \frac{1}{\pi} \frac{e^{j\beta z'} \sin\left(\frac{n\pi x'}{a}\right)}{k_0^2 - \beta^2 - \left(\frac{n\pi}{a}\right)^2} \qquad (5.79)$$

Inverse Fourier transform of $\tilde{\psi}(n, \beta; x', z')$ can be written as

$$\psi(x, z; x', z') = \frac{1}{\sqrt{2\pi}} \frac{2}{a} \sqrt{\frac{\pi}{2}} \int\limits_{-\infty}^{\infty} \sum_{n=-\infty}^{\infty} \tilde{\psi}(n, \beta; x', z') \sin\left(\frac{n\pi x}{a}\right) e^{-j\beta z} d\beta$$

or

$$\psi(x, z; x', z') = \frac{1}{a\pi} \int\limits_{-\infty}^{\infty} \sum_{n=-\infty}^{\infty} \frac{\sin\left(\frac{n\pi x}{a}\right) \sin\left(\frac{n\pi x'}{a}\right)}{k_0^2 - \beta^2 - \left(\frac{n\pi}{a}\right)^2} e^{j\beta(z'-z)} d\beta \qquad (5.80)$$

In conclusion, the problem of excitation of TE-modes in a rectangular wave-guide by a line source of electric current can be solved in a number of ways. These are:

1. *Direct construction* of Green's function along the z-direction and Fourier expansion or eigenfunction expansion along the x-direction, (3.89);
2. Fourier sine transform along the x-direction and *direct construction* along the z-direction, (5.72);
3. Fourier exponential transform along the z-direction and *direct construction* along the x-direction, (5.76);
4. Fourier sine transform along the x-direction and Fourier exponential transform along the z-direction, (5.80).

Example 5.3.   If the boundary conditions of the PDE are such that the region is bounded in both directions, one may employ double Fourier transform to reduce the PDE to ordinary differential equation which can be solved easily. Let us solve the following inhomogeneous differential equation:

$$\frac{\partial^2 f}{\partial x^2} + \frac{\partial^2 f}{\partial y^2} = -g(x, y) \tag{5.81}$$

over the domain $0 \le x \le a$; $0 \le y \le b$, and subject to $f(x, y) = 0$ at $x = 0, a$; $y = 0, b$.

We shall use double Fourier sine transform here because of the stated boundary conditions. The double Fourier sine transforms may be defined as

$$\tilde{F}(m, n) = \frac{2}{\pi} \int\limits_0^b \int\limits_0^a f(x, y) \sin\left(\frac{m\pi x}{a}\right) \sin\left(\frac{n\pi y}{b}\right) dx\, dy \tag{5.82}$$

$$\tilde{G}(m, n) = \frac{2}{\pi} \int\limits_0^b \int\limits_0^a g(x, y) \sin\left(\frac{m\pi x}{a}\right) \sin\left(\frac{n\pi y}{b}\right) dx\, dy \tag{5.83}$$

Now, multiply both sides of (5.81) by $\sin\left(\dfrac{m\pi x}{a}\right) \sin\left(\dfrac{n\pi y}{b}\right) dx\, dy$ and integrate over the ranges of x and y to obtain

$$-\left[\left(\frac{m\pi}{a}\right)^2 + \left(\frac{n\pi}{b}\right)^2\right] \tilde{F}(m, n) = -\tilde{G}(m, n)$$

or

$$\tilde{F}(m, n) = \frac{\tilde{G}(m, n)}{\left[\left(\dfrac{m\pi}{a}\right)^2 + \left(\dfrac{n\pi}{b}\right)^2\right]} \tag{5.84}$$

Inverting $\tilde{F}(m, n)$ gives [refer to (5.56)]

$$f(x, y) = \frac{\pi}{2} \frac{2}{a} \frac{2}{b} \sum_{n=1}^{\infty} \sum_{m=1}^{\infty} \frac{\tilde{G}(m, n)}{\left(\dfrac{m\pi}{a}\right)^2 + \left(\dfrac{n\pi}{b}\right)^2} \sin\left(\frac{m\pi x}{a}\right) \sin\left(\frac{n\pi y}{b}\right) \tag{5.85}$$

A similar expression is obtained by using the eigenfunction expansion method, (3.59).

The coefficients $\tilde{G}(m, n)$ are obtained by integrating the product of the source density $g(x', y')$ and the factor $\sin\left(\dfrac{m\pi x'}{a}\right) \sin\left(\dfrac{m\pi y'}{b}\right)$ over x' and y'. Alternatively, $\dfrac{2}{a} \dfrac{2}{b} \tilde{G}(m, n)$ is the expansion coefficient of $g(x, y)$ in terms of eigenfunctions; that is,

$$g(x, y) = \sum_{n=1}^{\infty} \sum_{m=1}^{\infty} \tilde{G}(m, n) \sin\left(\frac{m\pi x}{a}\right) \sin\left(\frac{n\pi y}{b}\right) \tag{5.86}$$

5.3.4 Electric Line Source Above a Perfectly Conducting Ground Plane

This problem is a fundamental problem in the category of line source excitation of open space. It is similar to the excitation of grounded trough by a line source (Section 5.2.3) except that the parallel walls of the trough have receded to infinity. Therefore Fourier sine transform transverse to the walls of the trough cannot be used here. Instead, the corresponding eigenvalue spectrum will be continuous and Fourier exponential transform is necessary. The presentation closely follows that given in [4].

We consider that the source is time harmonic, uniform in the z-direction, and located at (x', y'). The ground plane is located at $y = 0$ as shown in Figure 5.7. The magnetic vector potential A_z associated with the current source $\mathbf{J} = \hat{z} J_z$ satisfies the two-dimensional Helmholtz equation in x and y; that is,

$$\left(\frac{\partial^2}{\partial x^2} + \frac{\partial^2}{\partial y^2} + k_0^2\right) A_z(x, y) = -\mu_0 I \delta(x - x') \delta(y - y') \tag{5.87}$$

where the current $I = \underset{A \to 0}{lt} \int_A J_z \, dA$, and A is the cross-sectional area of wire carrying the current. The boundary conditions to be satisfied are

$$A_z(x, y) = 0 \qquad \text{at} \qquad y = 0, \infty \tag{5.88}$$

The boundary condition at $y = 0$ arises due to the ground plane, and the boundary condition at $y = \infty$ is called the radiation condition.

Since the structure is infinite along the x-direction, the variation of field along this direction is expected to be of the form $e^{\pm jax}$. Consistent with it we multiply both sides of (5.87) by e^{jax} and integrate over x from $-\infty$ to $+\infty$; one obtains

$$\left(\frac{d^2}{dy^2} + \gamma^2\right) \tilde{A}_z(\alpha, y) = -B_1 \delta(y - y') \frac{1}{\sqrt{2\pi}} \tag{5.89}$$

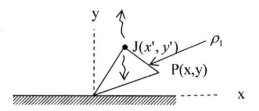

Figure 5.7 Current line source above a ground plane.

where

$$\tilde{A}_z(\alpha, y) = \frac{1}{\sqrt{2\pi}} \int\limits_{-\infty}^{+\infty} A_z(x, y) e^{j\alpha x} \, dx \tag{5.90}$$

$$\gamma^2 = k_0^2 - \alpha^2 \tag{5.91}$$

and

$$B_1 = I\mu_0 e^{j\alpha x'} \tag{5.92}$$

The ordinary differential equation (5.89) can be solved using the *direct construction procedure* described in Chapter 3. One obtains

$$\tilde{A}_z(\alpha, y) = \begin{cases} B_1 \dfrac{1}{\sqrt{2\pi}} \dfrac{\sin(\gamma y)}{\gamma} e^{-j\gamma y'} & \text{for } y \le y' \\[2ex] B_1 \dfrac{1}{\sqrt{2\pi}} \dfrac{\sin(\gamma y')}{\gamma} e^{-j\gamma y} & \text{for } y \ge y' \end{cases} \tag{5.93}$$

Taking its inverse transform, the solution to (5.89) then becomes

$$A_z(x, y; x', y') = \frac{I\mu_0}{2\pi} \int\limits_{-\infty}^{+\infty} \frac{1}{\gamma} \{\sin(\gamma y) e^{-j\gamma y'} u(y' - y) + \sin(\gamma y') e^{-j\gamma y} u(y - y')\} e^{-j\alpha(x-x')} \, d\alpha \tag{5.94}$$

where $u(.)$ is the unit step function. Converting the $\sin(.)$ function into exponentials reduces (5.94) to

$$A_z(x, y; x', y') = \frac{I\mu_0}{4\pi j} \int\limits_{-\infty}^{+\infty} \frac{e^{-j\alpha(x-x')}}{\gamma} \{e^{j\gamma(y-y')} u(y' - y) + e^{-j\gamma(y-y')} u(y - y')\} \, d\alpha \tag{5.95}$$

$$- \frac{I\mu_0}{4\pi j} \int\limits_{-\infty}^{+\infty} \frac{e^{-j\alpha(x-x')}}{\gamma} e^{-j\gamma(y+y')} \, d\alpha \qquad \text{for } y \ge 0$$

The factor $e^{j\gamma(y-y')} u(y' - y)$ in the first term represents a wave (from the source) going towards the ground plane, and the factor $e^{-j\gamma(y-y')} u(y - y')$ in the second term accounts for an upward going wave from the source (see Figure 5.7). The factor $e^{-j\gamma(y+y')}$ in the third term represents an upward going wave from the image of the source in the ground plane.

Equation (5.95) can be further simplified by employing the following identities [4]:

$$\frac{1}{\pi} \int\limits_{-\infty}^{\infty} \frac{e^{j\gamma(y-y')}e^{-j\alpha(x-x')}}{\gamma}\, d\alpha = H_0^{(1)}(\gamma_1 \rho_1) \qquad (5.96)$$

$$\frac{1}{\pi} \int\limits_{-\infty}^{\infty} \frac{e^{-j\gamma(y-y')}e^{-j\alpha(x-x')}}{\gamma}\, d\alpha = H_0^{(2)}(\gamma_1 \rho_1) \qquad (5.97)$$

where $\gamma_1^2 = \gamma^2 + \alpha^2$ and $\rho_1^2 = (x - x')^2 + (y - y')^2$. Potential Green's function therefore becomes

$$A_z(x, y; x', y') = \frac{I\mu_0}{4j} \left[H_0^{(1)}(\gamma_1 \rho_1) u(y' - y) + H_0^{(2)}(\gamma_1 \rho_1) u(y - y') \right] \qquad (5.98)$$

$$- \frac{I\mu_0}{4\pi j} \int\limits_{-\infty}^{\infty} \frac{e^{-j\gamma(y+y')}e^{-j\alpha(x-x')}}{\gamma}\, d\alpha$$

Exercise.   Solve the above problem by taking Fourier sine transform along the y-direction.

5.3.5   Free-Space Green's Function in Three Dimensions

The Green's function wave equation in three-dimensions is given by

$$\left(\frac{\partial^2}{\partial x^2} + \frac{\partial^2}{\partial y^2} + \frac{\partial^2}{\partial z^2} + k_0^2 \right) G(x, y, z; x', y', z') = -\delta(x - x')\delta(y - y')\delta(z - z')$$

$$(5.99)$$

Following the mathematical approach described for the two-dimensional case, we employ triple Fourier transform representations of Green's function and delta function

$$G(x, y, z; x', y', z') = \frac{1}{(2\pi)^{3/2}} \int\limits_{-\infty}^{\infty} \int\limits_{-\infty}^{\infty} \tilde{G}(\alpha, \beta, \gamma) e^{j\alpha(x-x')} e^{j\beta(y-y')} e^{j\gamma(z-z')}\, d\alpha\, d\beta\, d\gamma$$

$$(5.100)$$

$$\delta(x - x')\delta(y - y')\delta(z - z') = \frac{1}{(2\pi)^3} \int\limits_{-\infty}^{\infty} \int\limits_{-\infty}^{\infty} e^{j\alpha(x-x')} e^{j\beta(y-y')} e^{j\gamma(z-z')}\, d\alpha\, d\beta\, d\gamma$$

$$(5.101)$$

Substituting these expressions in (5.99), taking derivatives, and comparison gives

$$\tilde{G}(\alpha, \beta, \gamma) = \frac{1}{(2\pi)^{3/2}} \frac{1}{\alpha^2 + \beta^2 + \gamma^2 - k_0^2}$$

or

$$G(x, y, z; x', y', z') = \frac{1}{(2\pi)^3} \int_{-\infty}^{\infty} \int_{-\infty}^{\infty} \int_{-\infty}^{\infty} \frac{e^{j\alpha(x-x')} e^{j\beta(y-y')} e^{j\gamma(z-z')}}{\alpha^2 + \beta^2 + \gamma^2 - k_0^2} \, d\alpha \, d\beta \, d\gamma$$

$$(5.102)$$

The above expression is called the plane wave spectrum representation of Green's function. By transforming to spherical coordinate system and employing residue calculus for integration, the free space Green's function may be expressed as [2, p. 210]

$$G(x, y, z; x', y', z') = \frac{e^{jk_0|\mathbf{r}-\mathbf{r}'|}}{4\pi|\mathbf{r}-\mathbf{r}'|} \tag{5.103}$$

5.4 Radiation from Two-Dimensional Apertures

In this section we shall show that the radiation pattern of an aperture antenna is obtained from the Fourier transform of the electric field distribution in the aperture. The presentation here follows [5, 6].

Consider a two-dimensional rectangular aperture in an infinite conductor in the x-y plane as shown in Figure 5.8. The field in the region $z > 0$ can be determined from the knowledge of the electric field in the aperture. The electric field in free space is the solution of wave equation

$$(\nabla^2 + k_0^2)\mathbf{E}(x, y, z) = 0 \tag{5.104a}$$

with

$$\nabla \cdot \mathbf{E} = 0 \tag{5.104b}$$

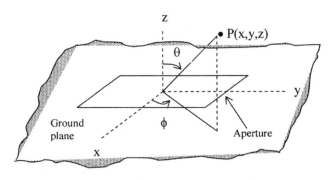

Figure 5.8 Aperture in a ground plane and the coordinate system.

Since the ground plane is infinite along the x and y-directions, the variation of the field is expected to be of the form $e^{\pm jk_x x}$ and $e^{\pm jk_y y}$, respectively. We may therefore express $E(x, y, z)$ as a superposition of these plane waves

$$E(x, y, z) = \frac{1}{2\pi} \int_{-\infty}^{\infty} \int_{-\infty}^{\infty} \tilde{E}(k_x, k_y, z) e^{-jk_x x - jk_y y} \, dk_x \, dk_y \qquad (5.105a)$$

where $\tilde{E}(k_x, k_y, z)$ is the Fourier transform of $E(x, y, z)$ with respect to x and y; that is,

$$\tilde{E}(k_x, k_y, z) = \frac{1}{2\pi} \int_{-\infty}^{\infty} \int_{-\infty}^{\infty} E(x, y, z) e^{jk_x x + jk_y y} \, dx \, dy \qquad (5.105b)$$

Using this decomposition, (5.104a) gets transformed to

$$\left[k_0^2 - k_x^2 - k_y^2 + \frac{d^2}{dz^2} \right] \tilde{E}(k_x, k_y, z) = 0$$

or

$$\left[k_z^2 + \frac{d^2}{dz^2} \right] \tilde{E}(k_x, k_y, z) = 0, \qquad k_z^2 = k_0^2 - k_x^2 - k_y^2 \qquad (5.106)$$

Solutions to (5.106) can be expressed as

$$\tilde{E}(k_x, k_y, z) = \tilde{f}(k_x, k_y) e^{-jk_z z} \qquad (5.107)$$

with

$$k_z = \begin{cases} +\sqrt{k_0^2 - k_x^2 - k_y^2} & \text{for } k_x^2 + k_y^2 < k_0^2 \\ -j\sqrt{k_x^2 + k_y^2 - k_0^2} & \text{for } k_x^2 + k_y^2 > k_0^2 \end{cases} \qquad (5.108)$$

The negative sign is chosen for $k_x^2 + k_y^2 > k_0^2$ in order to obtain a proper decay of fields for large values of z. The electric field in the region $z > 0$ can now be represented in the following form:

$$E(x, y, z) = \frac{1}{2\pi} \iint \tilde{f}(k_x, k_y) e^{-j(k_x x + k_y y + k_z z)} \, dk_x \, dk_y \qquad (5.109)$$

$$= \frac{1}{2\pi} \iint \tilde{f}(k_x, k_y) e^{-j\mathbf{k} \cdot \mathbf{r}} \, dk_x \, dk_y$$

where $\mathbf{k} = k_x\hat{\mathbf{x}} + k_y\hat{\mathbf{y}} + k_z\hat{\mathbf{z}}$. The expression (5.109) may be interpreted to mean that the field in space may be regarded as made up of a large number of plane waves of the form $\tilde{\mathbf{f}}e^{-j\mathbf{k}\cdot\mathbf{r}}$. The waves for which k_z is real denote propagating waves and contribute to the energy flow outward, while the waves for which k_z is a pure imaginary number are evanescent waves, and contribute to the energy stored in the fringing field near the aperture.

The function or the weight factor $\tilde{\mathbf{f}}(k_x, k_y)$ is determined by the excitation; that is, the electric field in the aperture. Before expressing $\tilde{\mathbf{f}}(k_x, k_y)$ in terms of aperture fields let us first discuss the stationary phase method of calculating radiation fields from (5.109).

5.5 Stationary Phase Method

The stationary phase method is an analytical technique employed for asymptotic evaluation of integrals with rapidly oscillating integrands. Such integrands occur, for example, in radiation fields and the asymptotic value of Bessel functions.

The electromagnetic field produced by an antenna at large distances is related to the current or the aperture field of the antenna through Fourier transforms of the form (5.109). An expression for the radiation pattern can be obtained by evaluating this field asymptotically and can be implemented using the method of stationary phase [7]. This method is based on the following observations:

1. *When r is very large*, the factor $e^{-j\mathbf{k}\cdot\mathbf{r}} = \cos(\mathbf{k}\cdot\mathbf{r}) - j\sin(\mathbf{k}\cdot\mathbf{r})$ oscillates very rapidly between equal positive and negative values except for certain range of values of k_z and k_y. This range is located at the stationary point of $\mathbf{k}\cdot\mathbf{r}$. The name of the method is derived from the stationary property of the phase factor $\mathbf{k}\cdot\mathbf{r}$. The other two observations mainly relate to simplifying the integrand.
2. *When a slowly varying function of k_x and k_y such as* $\mathbf{f}(k_x, k_y)$ *is multiplied by* $e^{-j\mathbf{k}\cdot\mathbf{r}}$ *and integrated over k_x and k_y*, the contribution to the integral arises mainly from those values of k_x and k_y for which the phase factor $e^{-j\mathbf{k}\cdot\mathbf{r}}$ remains stationary.
3. In the limit, as $r \to \infty$ the leading term in the expansion of the integral is given exactly by the contribution arising from the stationary points.

Determination of Stationary Phase Points
We shall first determine the stationary phase points; that is, the values of k_x, k_y, and k_z at which the phase factor $\mathbf{k}\cdot\mathbf{r} = k_x x + k_y y + k_z z$ becomes stationary. Substituting the expression for k_z into $\mathbf{k}\cdot\mathbf{r}$ gives

$$\mathbf{k}\cdot\mathbf{r} = k_x x + k_y y + z\sqrt{k_0^2 - k_x^2 - k_y^2} \qquad (5.110)$$

At the end, we shall express the radiation fields in the spherical coordinates. The transformation from the Cartesian-to-spherical coordinates is given as

$$x = r\sin(\theta)\cos(\phi) \qquad y = r\sin(\theta)\sin(\phi) \qquad z = r\cos(\theta) \qquad (5.111)$$

Equation (5.110) in spherical coordinates may be written as

$$\mathbf{k}.\mathbf{r} = r\left[k_x \sin(\theta)\cos(\phi) + k_y \sin(\theta)\sin(\phi) + \cos(\theta)\sqrt{k_0^2 - k_x^2 - k_y^2} \right]$$

$$(5.112)$$

The phase factor $\mathbf{k}.\mathbf{r}$ is stationary with respect to k_x and k_y (for r constant) if

$$\frac{\partial(\mathbf{k}.\mathbf{r})}{\partial k_x} = 0, \quad \text{and} \quad \frac{\partial(\mathbf{k}.\mathbf{r})}{\partial k_y} = 0 \qquad (5.113)$$

that is, if

$$r\left[\sin(\theta)\cos(\phi) - \frac{k_x \cos(\theta)}{k_z} \right] = 0, \Rightarrow k_x = k_z \frac{\sin(\theta)\cos(\phi)}{\cos(\theta)} \quad (5.114a)$$

and

$$r\left[\sin(\theta)\sin(\phi) - \frac{k_y \cos(\theta)}{k_z} \right] = 0, \Rightarrow k_y = k_z \frac{\sin(\theta)\sin(\phi)}{\cos(\theta)} \quad (5.114b)$$

Substituting these values of k_x and k_y in $k_y^2 = k_0^2 - k_x^2 - k_y^2$, we obtain the stationary phase point (k_{x0}, k_{y0}, k_{z0}) as

$$k_{z0} = k_0 \cos(\theta) \qquad (5.115a)$$

$$k_{x0} = k_x |_{k_{z0}} = k_0 \sin(\theta)\cos(\phi) \qquad (5.115b)$$

$$k_{y0} = k_y |_{k_{z0}} = k_0 \sin(\theta)\sin(\phi) \qquad (5.115c)$$

It may be verified that $\mathbf{k}.\mathbf{r}$ reduces to $k_0 r$ at the stationary phase point.
   In the vicinity of the stationary point (k_{x0}, k_{y0}, k_{z0}) the following approximations may be made:

1. $\tilde{f}(k_x, k_y)$ is slowly varying, and may be replaced by $\tilde{f}(k_{x0}, k_{y0})$. Taking this factor out of the integral sign, (5.109) reduces to

$$\mathbf{E}(\mathbf{r}) \approx \frac{1}{2\pi}\tilde{f}(k_{x0}, k_{y0}) \int_{-\infty}^{\infty} \int_{-\infty}^{\infty} e^{-j\mathbf{k}.\mathbf{r}}\, dk_x\, dk_y \qquad (5.116)$$

2. $\mathbf{k}.\mathbf{r}$ may be approximated by the first few terms in a Taylor series expansion as

$$\mathbf{k \cdot r} = \mathbf{k \cdot r}\big|_{k_{x0},k_{y0}} + \frac{\partial(\mathbf{k \cdot r})}{\partial k_x}\bigg|_{k_{x0},k_{y0}}(k_x - k_{x0}) + \frac{\partial(\mathbf{k \cdot r})}{\partial k_y}\bigg|_{k_{x0},k_{y0}}(k_y - k_{y0})$$

$$+ \frac{1}{2}\frac{\partial^2(\mathbf{k \cdot r})}{\partial k_x^2}\bigg|_{k_{x0},k_{y0}}(k_x - k_{x0})^2 + \frac{1}{2}\frac{\partial^2(\mathbf{k \cdot r})}{\partial k_y^2}\bigg|_{k_{x0},k_{y0}}(k_y - k_{y0})^2 \quad (5.117)$$

$$+ \frac{\partial^2(\mathbf{k \cdot r})}{\partial k_x \partial k_y}\bigg|_{k_{x0},k_{y0}}(k_x - k_{x0})(k_y - k_{y0})$$

The second and third terms on the right side are zero because of the stationary property of $\mathbf{k \cdot r}$. Also $\mathbf{k \cdot r} = k_0 r$, at the stationary point. Therefore, (5.117) may be expressed as

$$\mathbf{k \cdot r} = k_0 r - [Au^2 + Bv^2 + Cuv] \quad (5.118)$$

where $u = (k_x - k_{x0})$ and $v = (k_y - k_{y0})$ are the new variables, and A, B, and C are the constants obtained as

$$A = \frac{r}{2k_0}\left[1 + \frac{k_{x0}^2}{k_{z0}^2}\right], \ B = \frac{r}{2k_0}\left[1 + \frac{k_{y0}^2}{k_{z0}^2}\right], \ \text{and} \ C = \frac{r}{k_0}\frac{k_{x0}k_{y0}}{k_{z0}^2} \quad (5.119)$$

In the light of these approximations, (5.116) becomes

$$\mathbf{E(r)} \approx \tilde{\mathbf{f}}(k_{x0},k_{y0})\frac{e^{-jk_0 r}}{2\pi}\int_{-\infty}^{\infty}\int_{-\infty}^{\infty}e^{j(Au^2+Bv^2+Cuv)}\,du\,dv \quad (5.120)$$

The above integral can be evaluated as (see Appendix 5A)

$$\int_{-\infty}^{\infty}\int_{-\infty}^{\infty}e^{j(Au^2+Bv^2+Cuv)}\,du\,dv = \frac{1}{\sqrt{AB}}\sqrt{\pi}e^{j\pi/4}\sqrt{\frac{\pi}{1-C^2/(4AB)}}\,e^{j\pi/4} \quad (5.121)$$

$$= \frac{2\pi jk_0}{r}\cos(\theta)$$

Finally, (5.120) reduces to

$$\mathbf{E(r)} \approx j\frac{e^{-jk_0 r}}{r}k_0\cos(\theta)\tilde{\mathbf{f}}(k_{x0},k_{y0}) \quad (5.122)$$

5.5.1 Radiation Pattern

The radiation pattern is characterized by the variation of fields E_θ and E_ϕ as a function of angles θ and φ. For this, one needs to determine $\tilde{\mathbf{f}}(k_{x0},k_{y0})$ of (5.122) in terms of the known aperture field or antenna current [5, 7]. We shall determine $\tilde{\mathbf{f}}(k_{x0}, k_{y0})$ for the aperture antennas.

For an aperture in the x-y plane, the aperture field $\mathbf{E}_a$ is defined as

$$\mathbf{E}_a(x, y) = \mathbf{E}_t(x, y, 0) \tag{5.123}$$

where subscript t stands for the transverse part and $\mathbf{E}_t = \hat{\mathbf{x}}E_x + \hat{\mathbf{y}}E_y$. In terms of Fourier components, (5.123) may be expressed as

$$\mathbf{E}_a(x, y) = \frac{1}{2\pi} \int\limits_{-\infty}^{\infty} \int\limits_{-\infty}^{\infty} \tilde{\mathbf{f}}_t(k_x, k_y) e^{-j(k_x x + k_y y)} \, dk_x \, dk_y \tag{5.124}$$

Inversion gives

$$\tilde{\mathbf{f}}_t(k_x, k_y) = \frac{1}{2\pi} \int\limits_{-\infty}^{\infty} \int\limits_{-\infty}^{\infty} \mathbf{E}_a(x, y) e^{j(k_x x + k_y y)} \, dx \, dy \tag{5.125}$$

Therefore, $\tilde{\mathbf{f}}_t$ can be obtained once the aperture field is known. According to (5.122) the radiation field depends on $\tilde{\mathbf{f}}$, and $\tilde{\mathbf{f}} = \tilde{\mathbf{f}}_t + \hat{\mathbf{z}}\tilde{f}_z$. Therefore, the z-component of $\tilde{\mathbf{f}}$ is also needed for the evaluation of radiation fields in space. It is determined next.

To *determine* $\tilde{f}_z$ *in terms of* $\tilde{\mathbf{f}}_t$, we know that, in free space,

$$\nabla \cdot \mathbf{E} = 0 \tag{5.126}$$

Using (5.109) for $\mathbf{E}$ gives

$$\frac{1}{2\pi} \nabla \cdot \iint \tilde{\mathbf{f}} e^{-j\mathbf{k}\cdot\mathbf{r}} \, dk_x \, dk_y = 0 \tag{5.127}$$

Since the integration and differentiation are in different domains, their order can be interchanged as

$$\frac{1}{2\pi} \iint \tilde{\mathbf{f}} \cdot \nabla (e^{-j\mathbf{k}\cdot\mathbf{r}}) \, dk_x \, dk_y = 0$$

or

$$\frac{-1}{2\pi} \iint j\mathbf{k} \cdot \tilde{\mathbf{f}}(e^{-j\mathbf{k}\cdot\mathbf{r}}) \, dk_x \, dk_y = 0 \tag{5.128}$$

One of its possible solutions is

$$j\mathbf{k} \cdot \tilde{\mathbf{f}} e^{-j\mathbf{k}\cdot\mathbf{r}} = 0 \Rightarrow \mathbf{k} \cdot \tilde{\mathbf{f}} = 0$$

or

$$\tilde{f}_z = -\frac{k_x \tilde{f}_x + k_y \tilde{f}_y}{k_z} \tag{5.129}$$

Therefore,

$$\tilde{\mathbf{f}} = \tilde{\mathbf{f}}_t + \hat{z}\tilde{f}_z = \tilde{\mathbf{f}}_t - \hat{z}\frac{k_x \tilde{f}_x + k_y \tilde{f}_y}{k_z} = \tilde{\mathbf{f}}_t - \hat{z}\frac{\tilde{\mathbf{f}}_t \cdot \mathbf{k}_t}{k_z} \tag{5.130}$$

Since $\tilde{\mathbf{f}}_t$ is the Fourier transform of aperture field as per (5.125), expression (5.130) implies that only the aperture field need to be specified to determine the radiation pattern.

Assuming that the aperture field is known and $\tilde{\mathbf{f}}_t$ can be determined from it, we wish to determine the radiation field based on (5.122). At the stationary phase point (k_{x0}, k_{y0}, k_{z0}) the expression for $\tilde{\mathbf{f}}$ can be written as

$$\tilde{\mathbf{f}} = \tilde{\mathbf{f}}_t - \hat{z}\frac{k_x \tilde{f}_x + k_y \tilde{f}_y}{k_z} = \tilde{\mathbf{f}}_t(k_{x0}, k_{y0}) - \hat{z}\frac{1}{k_{z0}}(\tilde{f}_x k_{x0} + \tilde{f}_y k_{y0}) \tag{5.131}$$

Substituting for $\tilde{\mathbf{f}}$ in (5.122) produces

$$\mathbf{E}(\mathbf{r}) \approx j\frac{e^{-jk_0 r}}{r} k_0 \cos(\theta)\left[\tilde{\mathbf{f}}_t(k_{x0}, k_{y0}) - \hat{z}\frac{1}{k_{z0}}(\tilde{f}_x k_{x0} + \tilde{f}_y k_{y0})\right] \tag{5.132}$$

This result shows that the radiation pattern is related directly to the Fourier transform of the aperture field. To determine the field components E_θ and E_ϕ from (5.132), we use the following transformation between the unit vectors $\hat{x}$, $\hat{y}$ and $\hat{z}$; and $\hat{r}$, $\hat{\theta}$ and $\hat{\varphi}$

$$\begin{bmatrix} \hat{x} \\ \hat{y} \\ \hat{z} \end{bmatrix} = \begin{bmatrix} \sin(\theta)\cos(\phi) & \cos(\theta)\cos(\phi) & -\sin(\phi) \\ \sin(\theta)\sin(\phi) & \cos(\theta)\sin(\phi) & \cos(\phi) \\ \cos(\theta) & -\sin(\theta) & 0 \end{bmatrix} \begin{bmatrix} \hat{r} \\ \hat{\theta} \\ \hat{\varphi} \end{bmatrix} \tag{5.133}$$

This transformation and the evaluation at the stationary phase point gives

$$\mathbf{E}(\mathbf{r}) \approx jk_0\frac{e^{-jk_0 r}}{r}[\hat{\varphi}(\tilde{f}_y \cos\phi - \tilde{f}_x \sin\phi)\cos\theta + \hat{\theta}(\tilde{f}_x \cos\phi + \tilde{f}_y \sin\phi)] \tag{5.134}$$

that is,

$$E_\theta(\theta, \phi, r) \approx jk_0\frac{e^{-jk_0 r}}{r}(\tilde{f}_x \cos\phi + \tilde{f}_y \sin\phi) \tag{5.135a}$$

and

$$E_\phi(\theta,\,\phi,\,r) \approx jk_0 \frac{e^{-jk_0r}}{r}(\tilde{f}_y \cos\phi - \tilde{f}_x \sin\phi)\cos\theta \qquad (5.135b)$$

We may now use the following procedure for calculating the radiation pattern from the given aperture electric field:

1. Determine the Fourier transform of the aperture electric field as

$$\tilde{f}_t(k_x,\,k_y) = \frac{1}{2\pi} \iint\limits_{aperture} E_a(x,\,y,\,0)\,e^{\,j(k_x x + k_y y)}\,dx\,dy$$

2. Evaluate $\tilde{f}_t$ at the stationary phase point $(k_{x0},\,k_{y0})$ given by

$$k_{x0} = k_0 \sin(\theta)\cos(\phi), \qquad k_{y0} = k_0 \sin(\theta)\sin(\phi)$$

3. Determine the electric field components E_θ and E_ϕ from (5.135) in the principal planes defined by $\phi = 0$ and $\phi = \pi/2$.

We illustrate the above procedure by applying it to the problems given next.

Example 5.4.   Radiation pattern of a uniform aperture field. Let the electric field in the rectangular aperture of Figure 5.9 be defined as

$$\mathbf{E_a} = \begin{cases} \hat{y}E_0 & \text{for } |x| \le a/2,\, |y| \le b/2 \\ 0 & \text{otherwise} \end{cases} \qquad (5.136)$$

The uniform amplitude and phase distribution is an idealization of the distribution in very narrow apertures. The Fourier transform of the aperture field is given as

$$\tilde{f}_t = \hat{y}E_0' \int\limits_{-b/2}^{b/2} \int\limits_{-a/2}^{a/2} e^{\,j(k_x x + k_y y)}\,dx\,dy, \qquad E_0' = \frac{E_0}{2\pi}$$

or

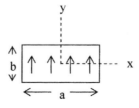

Figure 5.9   Electric field distribution in a narrow rectangular aperture in the ground plane.

$$\tilde{\mathbf{f}}_t = \hat{y} E_0' \, ab \, \sin c\left(\frac{k_x a}{2}\right) \sin c\left(\frac{k_y b}{2}\right), \qquad \sin c(x) = \frac{\sin(x)}{x} \qquad (5.137)$$

At the stationary phase point,

$$\tilde{\mathbf{f}}_t = \hat{y} E_0' \, ab \, \sin c\left(\frac{k_0 a \, \sin\theta \cos(\phi)}{2}\right) \sin c\left(\frac{k_0 b \, \sin\theta \sin(\phi)}{2}\right)$$

or

$$\tilde{f}_y = E_0' \, ab \, \sin c\left(\frac{k_0 a \, \sin\theta \cos(\phi)}{2}\right) \sin c\left(\frac{k_0 b \, \sin\theta \sin(\phi)}{2}\right) \qquad (5.138)$$

The radiation patterns in the principal planes are therefore given as follows:

1. $\phi = \pi/2$ *plane* *pattern:* from (5.135), $E_\phi = 0$ and $E_\theta(\theta, \phi) \propto \tilde{f}_y(\theta, \phi = \pi/2)$ or

$$E_\theta(\theta) \propto E_0' \, ab \, \sin c\left(\frac{k_0 b \, \sin(\theta)}{2}\right) \qquad (5.139a)$$

2. $\phi = 0$ *plane* *pattern:* from (5.135), $E_\theta = 0$ and $E_\phi(\theta, \phi) \propto \tilde{f}_y(\theta, \phi = 0) \cos\theta$ or

$$E_\phi(\theta) \propto E_0' \, ab \, \sin c\left(\frac{k_0 a \, \sin(\theta)}{2}\right) \cos\theta \qquad (5.139b)$$

The normalized pattern function $20 \log\left(E_\phi/\left(E_0' \, ab\right)\right)$ is plotted in Figure 5.10 for an aperture size of $a = 20\lambda_0$ and $b = 5\lambda_0$. Characteristics of this radiation pattern are:

1. The radiation pattern consists of a main lobe at $\theta = 0$ and a number of sidelobes on either side of the main lobe.
2. The first null occurs when $(k_0 b/2) \sin\theta = \pi$. For $k_0 b$ large, the corresponding value of θ is given to a good accuracy by

$$\frac{k_0 b}{2} \theta_{null} \approx \pi \Rightarrow \theta_{null} = \frac{\lambda_0}{b} \qquad (5.140)$$

that is, angular width of the main lobe is inversely proportional to the aperture dimension in wavelength.
3. The first sidelobe is 13.3 dB below the main beam peak. The sidelobes decrease in amplitude as the point of observation moves away from the direction of main beam.

Example 5.5. *Radiation pattern of a tapered aperture field.* The effect of tapering the aperture field to a smaller value at the edges of the aperture is equivalent to

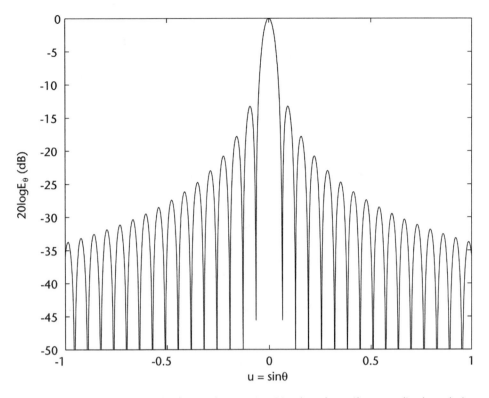

Figure 5.10 Radiation pattern in the *yz*-plane or $\phi = 90°$ plane for uniform amplitude and phase distribution in rectangular aperture ($a = 20\lambda_0$, $b = 5\lambda_0$), $u = \sin\theta$.

an effective reduction in the aperture area, and will result in a broader main lobe and lower directivity.

Let the aperture electric field produced by the TE_{10} mode in an open ended waveguide terminated by a flange (Figure 5.11) be modeled as

$$\mathbf{E_a} = \hat{\mathbf{y}} E_0 \cos\left(\frac{\pi x}{a}\right) \qquad \text{for } |x| \leq \frac{a}{2}, |y| \leq \frac{b}{2} \qquad (5.141)$$

Implementing the various steps as in Example 5.4, one obtains

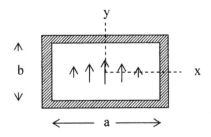

Figure 5.11 A rectangular aperture with tapered aperture electric field.

$$\tilde{f}_y = \frac{1}{2\pi} E_0 b \, \sin c\left(\frac{k_{yo}b}{2}\right) \frac{\pi a}{2} \frac{\cos\left(\frac{k_{x0}a}{2}\right)}{\left(\frac{\pi}{2}\right)^2 - \left(\frac{k_{x0}a}{2}\right)^2} \tag{5.142}$$

where $k_{x0} = k_0 \sin(\theta) \cos(\phi)$, and $k_{y0} = k_0 \sin(\theta) \sin(\phi)$. The radiation patterns, as before, are given as follows:

1. $\phi = 0$ *plane pattern*: $E_\theta = 0$ and $E_\phi(\theta, \phi) \propto \tilde{f}_y(\theta, \phi = 0) \cos \theta$ or

$$E_\phi(\theta) \propto E_0 \frac{ab}{4} \frac{\cos\left(\frac{k_0 a \, \sin(\theta)}{2}\right)}{\left(\frac{\pi}{2}\right)^2 - \left(\frac{k_0 a \, \sin(\theta)}{2}\right)^2} \cos \theta \tag{5.143a}$$

2. $\phi = \pi/2$ *plane pattern*: $E_\phi = 0$ and $E_\theta(\theta, \phi) \propto \tilde{f}_y(\theta, \phi = \pi/2)$ or

$$E_\theta(\theta) \propto E_0 \frac{ab}{4} \left(\frac{2}{\pi}\right)^2 \sin c\left(\frac{k_0 b \, \sin(\theta)}{2}\right) \tag{5.143b}$$

The pattern $E_\phi(\theta)$ is plotted in Figure 5.12 for $a/\lambda_0 = 3$, and its characteristics are summarized as follows:

1. The angular width of the mainlobe is $2\theta \approx 2 \sin \theta = 3\lambda_0/a$.
2. The half-power beamwidth is $1.2\lambda_0/a$ (compared with λ_0/a for the uniform field).
3. The first sidelobe has amplitude 23 dB below the main lobe maximum (compared with 13.3 dB for the uniform aperture field). It is because of the factor $(\pi/2)^2 - (k_0 a \sin(\theta)/2)^2$ in the denominator of the pattern function, (5.143a).

Exercise.   Determine the radiation pattern of an aperture field with linear phase variation modeled as

$$\mathbf{E_a} = \hat{y} E_0 e^{-j\frac{2\pi}{\lambda}z} e^{-\alpha z} \qquad \text{for } 0 \leq z \leq L, |y| \leq \frac{b_0}{2}$$

Here L is the slot length, b_0 is the slot width, and α is the attenuation constant due to the leakage of power. This type of aperture field is obtained in a leaky wave antenna, for example, by cutting a narrow longitudinal slot in a rectangular waveguide propagating TE_{10} mode.

5.5.2   Asymptotic Value of Bessel Functions

The stationary phase method was used in the last section to determine the radiation integral. The integrand was oscillating rapidly away from the stationary phase

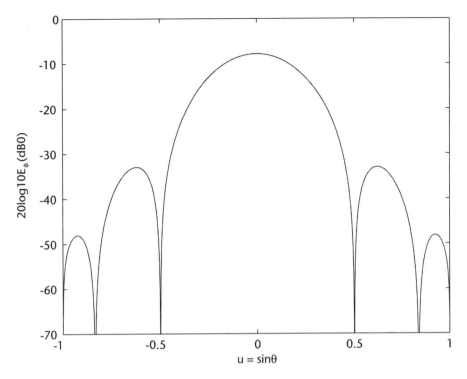

Figure 5.12 Radiation pattern in the xz-plane or $\phi = 0°$ plane for cosine tapered amplitude distribution in rectangular aperture ($a = 3\lambda_0$), $u = \sin \theta$.

point. In this section, we apply this method to determine the asymptotic value of Bessel functions.

The Bessel function of the nth order may be defined as [5, p. 344]

$$J_n(\rho) \triangleq \mathrm{Re}\left\{ \frac{1}{\pi} \int_0^\pi e^{j\rho \sin(x)} e^{-jnx}\, dx \right\} \qquad (5.144)$$

As the distance $\rho \to \infty$, the term $e^{j\rho \sin(x)}$ oscillates very rapidly. The contribution to the integral comes from a small region near the stationary phase point, which is obtained as

$$\frac{d}{dx}(\rho \sin x) = 0 \Rightarrow \rho \cos x = 0 \text{ or } x = \frac{\pi}{2} \qquad (5.145)$$

The asymptotic value of the Bessel function may therefore be calculated about the stationary phase point as

$$J_n(\rho) \sim \mathrm{Re}\left\{ \frac{1}{\pi} \int_{\pi/2-\epsilon}^{\pi/2+\epsilon} e^{j\rho \sin(x)} e^{-jnx}\, dx \right\} \qquad (5.146)$$

where ϵ is a very small positive number. It may be noted that the limits of integration represent a small region about the stationary point. The factor e^{-jnx} is slowly varying compared to $e^{j\rho \sin(x)}$ (for large values of ρ). Therefore, it can be approximated by $e^{-jn\pi/2}$, and (5.146) reduces to

$$J_n(\rho) \sim \text{Re}\left\{ \frac{e^{-jn\pi/2}}{\pi} \int_{\pi/2 - \epsilon}^{\pi/2 + \epsilon} e^{j\rho \sin(x)}\, dx \right\} \tag{5.147}$$

Now expand $\sin(x)$ in a Taylor series about the point $x = \pi/2$ to obtain

$$\sin x \approx \sin x \Big|_{x = \pi/2} + \frac{1}{2}\left(x - \frac{\pi}{2}\right)^2 (-\sin x)\Big|_{x = \pi/2} \tag{5.148}$$

$$\approx 1 - \frac{1}{2}\left(x - \frac{\pi}{2}\right)^2$$

Expression for the Bessel function therefore reduces to

$$J_n(\rho) \cong \text{Re}\left\{ \frac{e^{-jn\pi/2}}{\pi} \int_{\pi/2 - \epsilon}^{\pi/2 + \epsilon} e^{j\rho\{1 - (x - \pi/2)^2/2\}}\, dx \right\} \tag{5.149}$$

$$\cong \text{Re}\left\{ \frac{e^{j(\rho - n\pi/2)}}{\pi} \int_{\pi/2 - \epsilon}^{\pi/2 + \epsilon} e^{-j\rho(x - \pi/2)^2/2}\, dx \right\}$$

Since

$$\int_{\pi/2 - \epsilon}^{\pi/2 + \epsilon} e^{-j\rho(x - \pi/2)^2/2}\, dx \equiv \int_{-\infty}^{\infty} e^{-j\rho(x - \pi/2)^2/2}\, dx \tag{5.150}$$

$$= \sqrt{\frac{\pi}{\rho/2}}\, e^{-j\pi/4}$$

$$J_n(\rho) \cong \text{Re}\left\{ \frac{e^{j(\rho - n\pi/2 - \pi/4)}}{\pi} \sqrt{\frac{2\pi}{\rho}} \right\} \tag{5.151}$$

$$\cong \sqrt{\frac{2}{\pi\rho}} \cos\left(\rho - \frac{n\pi}{2} - \frac{\pi}{4}\right)$$

that is, the Bessel function behaves like a $\cos(.)$ function for large arguments.

5.6   Green's Function for the Quasi-Static Analysis of Microstrip Line

This topic is considered here because of the importance of microstrip line in microwave integrated circuits, and the Green's function in closed form for this geometry can be obtained only in the spectral domain. This is due to the mixed dielectric nature which gives rise to the inhomogeneous boundary conditions.

Consider a microstrip line of strip width w on a grounded substrate of dielectric constant ϵ_r and dielectric thickness h, as shown in Figure 5.13. A quasi-static analysis of the problem is carried out by solving the Poisson equation for the electric potential in a plane transverse to the direction of wave propagation,

$$\frac{\partial^2 \varphi(x, y)}{\partial x^2} + \frac{\partial^2 \varphi(x, y)}{\partial y^2} = -\frac{\rho_s(x)}{\epsilon} \tag{5.152}$$

where $\varphi(x, y)$ is the unknown potential and ρ_s is the charge density on the strip. The Green's function, charge density, and the potential at the plane $y = h$ are related as

$$\varphi(x, h) = \frac{1}{\epsilon_0} \int_{-w/2}^{w/2} \rho_s(x') G(x, h; x', h) \, dx' \tag{5.153}$$

However, it is difficult to determine $\varphi(x, h)$ than the spectral domain Green's function $\tilde{\varphi}(\alpha, h)$ because of the inhomogeneous boundary conditions at $y = h$. Also, Green's function in spectral domain can be easily obtained if $\tilde{\varphi}(\alpha, h)$ is known because [6]

$$\tilde{\varphi}(\alpha, h) = \frac{1}{\epsilon_0} \tilde{\rho}_s(\alpha) \tilde{G}(\alpha, h) \tag{5.154}$$

The expression for $\tilde{\varphi}(\alpha, h)$ is obtained by solving the Laplace equation for $\tilde{\varphi}$ subject to the boundary conditions at the strip. The Laplace equation is given by

$$\left(-\alpha^2 + \frac{d^2}{dy^2} \right) \tilde{\varphi}_i(\alpha, y) = 0 \qquad y \neq h, \, i = 1, 2 \tag{5.155}$$

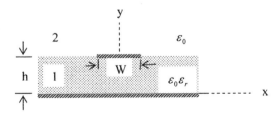

Figure 5.13 Cross-section of a microstrip line.

The subscript i identifies the regions marked in Figure 5.13. The interface conditions to be satisfied by $\tilde{\phi}_i(\alpha, y)$ are the Fourier transform of the interface conditions for $\phi_i(x, y)$ and may be listed as

$$\tilde{\phi}_1(\alpha, 0) = 0 \tag{5.156a}$$

$$\tilde{\phi}_1(\alpha, h) = \tilde{\phi}_2(\alpha, h) \tag{5.156b}$$

$$\tilde{\phi}_2(\alpha, \infty) = 0 \tag{5.156c}$$

The effect of charge on the strip is included through the condition

$$\tilde{D}_{n2}(\alpha, h) - \tilde{D}_{n1}(\alpha, h) = \tilde{\rho}_s(\alpha) \tag{5.156d}$$

where $\tilde{D}_n$ is the component of displacement vector normal to the interface. Equation (5.156d) may be expressed in terms of potential function as ($\mathbf{E} = -\nabla\varphi$)

$$\epsilon_r \frac{d}{dy} \tilde{\varphi}_1(\alpha, h) - \frac{d}{dy} \tilde{\varphi}_2(\alpha, h) = \frac{\tilde{\rho}_s(\alpha)}{\epsilon_0} \tag{5.157}$$

The Fourier transform is defined as in (5.3a).

The solution of (5.155) in the two regions is well known and may be written as

$$\tilde{\varphi}_1(\alpha, y) = A \sinh(\alpha y) \tag{5.158a}$$

$$\tilde{\varphi}_2(\alpha, y) = B e^{-|\alpha|y} \tag{5.158b}$$

The choice of potential functions in (5.158) takes into account the boundary conditions (5.156a) and (5.156c). The constants A and B can be determined by using the remaining boundary conditions. One obtains finally

$$\tilde{\varphi}(\alpha, h) = \frac{\tilde{\rho}_s(\alpha)}{\epsilon_0 |\alpha| \left[1 + \epsilon_r \coth(|\alpha|h) \right]} \tag{5.159}$$

Next, we use (5.154), the relationship between $\tilde{\varphi}(\alpha, h)$ and $\tilde{G}(\alpha, h)$, to determine the Green's function as

$$\tilde{G}(\alpha, h) = \frac{1}{|\alpha| \left[1 + \epsilon_r \coth(|\alpha|h) \right]} \tag{5.160}$$

5.7 Summary

The Fourier transform method is an elegant approach to solve differential equations with unbounded regions, the solution of which using computational techniques is inefficient. The transform method converts the given differential equation into ordinary differential equation or algebraic equation which is easier to solve. The simplification is based on the property that Fourier transform of the derivative of a function is proportional to the Fourier transform of the function itself. The bounded as well as semi-bounded region problems can also be solved by employing Fourier transform method. Some examples of Fourier transform method include: current excitation of unbounded transmission line, line excitation of grounded trough, probe excitation of modes in a rectangular waveguide, electric line source above a ground plane, and radiation from apertures. The layered dielectric problems such as planar transmission lines can be worked out in the Fourier/spectral domain. Alternative forms of the solution can be obtained by carrying out part of the analysis in physical domain. The final solution of the problem requires taking inverse transform, and this sometimes involves solving improper integrals. Residue calculus helps in solving these integral analytically. The radiation field is related to the Fourier transform of current on the antenna. The transform is called radiation integral and its properties are such that it can be evaluated in closed-form by employing stationary phase method.

References

[1]   Sneddon, I. H., *The Use of Integral Transforms*, New York: McGraw-Hill, 1972.

[2]   Eom, H. J., *Electromagnetic Wave Theory for Boundary Value Problems*, Berlin: Springer, 2004.

[3]   Collins, R. E., *Field Theory of Guided Waves*, 2nd ed., New York: IEEE Press, 1991.

[4]   Stinson, D. C., *Intermediate Mathematics of Electromagnetics*, Englewood Cliffs, NJ: Prentice-Hall, 1976.

[5]   Kong, J. A., *Electromagnetic Wave Theory*, New York: John Wiley, 1986, Chapter 4.

[6]   Yamashita, E., "Variational Method for the Analysis of Microstrip-Like Transmission Lines," *IEEE Trans., Microwave Theory Tech.*, Vol. MTT-16, 1968, pp. 529–535.

[7]   Collins, R. E., and Zucker, *Antenna Theory—Part 1*, New York: McGraw-Hill, 1969.

[8]   Van Bladel, J., *Electromagnetic Fields*, Washington, D.C.: Hemisphere Publishing Corp., 1985.

Appendix 5A: Evaluation of the Integral in (5.120)

The integrand in (5.120) can be expressed in a standard form by using the following substitution:

$$x = u\sqrt{A}, \qquad y = v\sqrt{B} \tag{5A.1}$$

so that

$$\int\limits_{-\infty}^{\infty} \int\limits_{-\infty}^{\infty} e^{j(Au^2 + Bv^2 + Cuv)} \, du\, dv = \frac{1}{\sqrt{AB}} \int\limits_{-\infty}^{\infty} \int\limits_{-\infty}^{\infty} e^{j(x^2 + y^2 + Cxy/\sqrt{AB})} \, dx\, dy \tag{5A.2}$$

Now, complete the square in x to obtain

$$\int_{-\infty}^{\infty}\int_{-\infty}^{\infty} e^{j(Au^2 + Bv^2 + Cuv)}\, du\, dv = \frac{1}{\sqrt{AB}}\int_{-\infty}^{\infty}\int_{-\infty}^{\infty} e^{j(x + Cy/(2\sqrt{AB}))^2}\, e^{jy^2(1 - C^2/(4AB))}\, dx\, dy$$

$$(5A.3)$$

Use of the following standard integral

$$\int_{-\infty}^{\infty} e^{\pm j\alpha(\xi - \xi_0)^2}\, d\xi = e^{\pm j\pi/4}\sqrt{\frac{\pi}{\alpha}} \qquad (5A.4)$$

in (5A.3) gives

$$\int_{-\infty}^{\infty}\int_{-\infty}^{\infty} e^{j(Au^2 + Bv^2 + Cuv)}\, du\, dv = \frac{1}{\sqrt{AB}}\sqrt{\pi}e^{j\pi/4}\sqrt{\frac{\pi}{1 - C^2/(4AB)}}e^{j\pi/4} \quad (5A.5)$$

$$= \frac{2\pi jk_0}{r}\cos(\theta)$$

Problems

P5.1. The following functions are used as basis functions in MoM for solving differential and integral equations. Determine their Fourier transforms.

$$\text{Triangular: } J(z) = \begin{cases} 1 - |z|/L & -L \le z \le L \\ 0 & \text{elsewhere} \end{cases}$$

$$\text{Piecewise sinusoidal: } J(z) = \begin{cases} \dfrac{\sin\left[k_0(L - |z|)\right]}{\sin(k_0 L)} & -L \le z \le L \\ 0 & \text{elsewhere} \end{cases}$$

$$\text{Constant: } J(z) = 1 \qquad -\infty \le z \le \infty$$

P5.2. Calculate the Fourier transforms of the following functions:

1. (i) $f(x) = |x|$ $-W/2 \le x \le W/2$
2. $f(x) = I_0 e^{-jkx}$ $-a \le x \le a$
3. $f(x) = \delta(x)$ $-W \le x \le W$
4. $f(x) = \begin{cases} 1 & -W \le x \le W \\ 0 & \text{elsewhere} \end{cases}$

5. $J_x(x) = \dfrac{1}{\sqrt{1 - (x/W)^2}}$ $-W \le x \le W$

P5.3. Finite Fourier transform of a function $f(x, y)$ is defined as

$$\tilde{f}(\alpha_n, y) = \int_0^a f(x, y) e^{j\alpha_n x}\, dx \qquad \alpha_n = \frac{n\pi}{a}$$

Write down the inverse Fourier transform.

P5.4. Solution to the following differential equation:

$$\left(\nabla^2 + k_0^2\right) G(x, y, z; x', y', z') = -\delta(x')\,\delta(y')\,\delta(z')$$

subject to the boundary condition $G \to \infty$ as $r \to \infty$ is a spherical wave described by [5, p. 346]

$$G(x, y, z; x', y', z') = \frac{e^{-jk_0 r}}{4\pi r}$$

In order to determine its spectral representation, solve the above differential equation by taking its Fourier transform defined as

$$\tilde{G}(k_x, k_y, k_z; x', y', z') = \iiint G(x, y, z; x', y', z') e^{j(k_x x + k_y y + k_z z)}\, dx\, dy\, dz$$

and show that

$$\tilde{G}(k_x, k_y, k_z; x', y', z') = \frac{-1}{k_0^2 - k_x^2 - k_y^2 - k_z^2}$$

The above procedure implies that

$$\frac{e^{-jk_0 r}}{4\pi r} = \frac{1}{(2\pi)^3} \iiint \frac{-1}{k_0^2 - k_x^2 - k_y^2 - k_z^2}\, e^{-j(k_x x + k_y y + k_z z)}\, dk_x\, dk_y\, dk_z$$

Now use contour integration to show that the above integration can be simplified to yield (assuming that $k_0 = k_{0r} + j k_{0i}$)

$$\frac{e^{-jk_0 r}}{r} = \frac{1}{2\pi} \iint \frac{e^{-j(k_x x + k_y y + k_z |z|)}}{k_z}\, dk_x\, dk_y$$

where $k_z = \sqrt{k_0^2 - k_x^2 - k_y^2}$. Show the contour also.

P5.5. Consider the problem of diffraction by a slit of width 2ℓ in the x-direction and infinite length in the y-direction, as shown in Figure 5.14. A plane wave polarized in the y-direction is normally incident upon the slit from $z \leq 0$. At the slit aperture, assume the field to be [5, p. 463]

$$\mathbf{E}(x, z = 0) = \hat{y}E_0\, U(\ell - |x|)$$

where $U(\ell - |x|)$ is the unit step function. The field at any distance z can be viewed as a superposition of plane waves with the wave vector component k_x spanning from $-\infty$ to ∞,

$$\mathbf{E}(x, z) = \hat{y} \int\limits_{-\infty}^{\infty} \tilde{E}(k_x) e^{jk_x x + jk_z z}\, dk_x$$

where $k_z = \sqrt{k^2 - k_x^2}$. Show that at the aperture where $z = 0$

$$\tilde{E}(k_x) = \frac{E_0\ell}{\pi} \frac{\sin(k_x\ell)}{k_x\ell}$$

The major contribution to the integral comes from the interval $[-\pi/\ell, \pi/\ell]$ on the k_x-axis, where the peak of $\tilde{E}(k_x)$ is located. Assume $k \gg \pi/\ell$, then $k_x \ll k$ in the region where the integrand is significantly large. Approximate

$$k_z = \sqrt{k^2 - k_x^2} \approx k - k_x^2/2k + \ldots$$

Using the above approximation, the expression for the field becomes

$$\mathbf{E}(x, z) = \hat{y}e^{jkz}e^{j(kx^2/2z)} \int\limits_{-\infty}^{\infty} \tilde{E}(k_x) e^{-j(z/2k)[k_x - (kx/z)]^2}\, dk_x$$

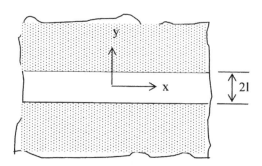

Figure 5.14 A slit of width 2ℓ in the ground plane.

Determine the stationary-phase point. Show that the far field as evaluated by the stationary phase method yields

$$E(x, z) \approx \hat{y} e^{jkz} e^{j(kx^2/2z)} \sqrt{\frac{2\pi k}{jz}} \; \tilde{E}\left(\frac{kx}{z}\right)$$

Thus the far-field pattern is proportional to the Fourier transform of the aperture field. Show that for uniform illumination at the aperture,

$$E(x, z) = \hat{y} E_0 \ell e^{jkz} e^{j(kx^2/2z)} \sqrt{\frac{2k}{j\pi z}} \; \frac{\sin\left(\frac{k\ell x}{z}\right)}{\frac{k\ell x}{z}}$$

P5.6. *Excitation of surface waves by a current source above a grounded dielectric sheet.* Figure 5.15 illustrates a grounded, lossless dielectric sheet of thickness t, and relative dielectric constant ϵ_r. An electric current source is located at $x = d$, $z = 0$ and is parallel to the y-axis. The current source may be represented as $\mathbf{J} = \hat{y} \, \delta(z) \, \delta(x - d)$. Using *direct construction method* (Chapter 3) along the x-direction and double Laplace-transform along the z-direction, show that [3, p. 726]

$$E_y(x, z) = \frac{j\omega\mu_0}{4\pi} \int\limits_{-\infty}^{+\infty} \left(\frac{e^{-jl|x-d|}}{l} + \frac{Re^{-jl|x+d|}}{l} \right) e^{-\gamma z} \, d\gamma$$

where

$$R = \frac{jl - h\cot(ht)}{jl + h\cot(ht)}, \qquad h^2 = \gamma^2 + \epsilon_r k_0^2, \; l^2 = \gamma^2 + k_0^2$$

P5.7. Consider an infinitely long transmission line excited by a current source $I_0 e^{j\omega t}$ at z' and shown in Figure 5.16. Solve

$$\frac{d^2 V}{dz^2} + k_0^2 V = -j\omega L \delta(z - z'), \qquad k_0^2 = \omega^2 LC(1 - jG/(\omega C))$$

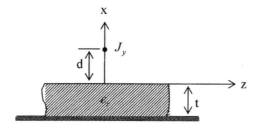

Figure 5.15 Line current source excitation of a grounded dielectric sheet.

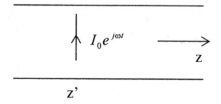

Figure 5.16   Excitation of transmission line by a current source.

by means of Fourier transform [3]. Hint: The presence of shunt conductance will displace the poles away from the real axis. For $z > z'$ the inversion contour can be closed in the lower-half plane, and for $z < z'$ it can be closed in the upper-half plane. Thus the inverse transform can be evaluated in terms of residues.

P5.8. Consider a lossy transmission line of length l short-circuited at the ends $z = 0$ and $z = l$. The voltage Green's function on this line satisfies the following differential equation [8, p. 533]:

$$\frac{\partial^2 v}{\partial z^2} - RC\frac{\partial v}{\partial t} - LC\frac{\partial^2 v}{\partial t^2} = \delta(z - z')\,\delta(t - t')$$

where R, L, and C are line constants. Take the Fourier transform of this equation with t and show that the Fourier transform $\tilde{V}(z, \omega)$ satisfies the following ordinary differential equation:

$$\frac{\partial^2 \tilde{V}}{\partial z^2} - -j\omega RC\tilde{V} + \omega^2 LC\tilde{V} = \frac{1}{\sqrt{2\pi}}\delta(z - z')e^{j\omega t'}$$

You may use the *direct construction method* of Chapter 3 to solve this differential equation, and show that

$$\tilde{V}(z, \omega) = \frac{1}{\sqrt{2\pi}}
\begin{cases}
\dfrac{1}{\gamma}\dfrac{\sin(\gamma z)\,\sin(\gamma(z' - l))}{\sin(\gamma l)}e^{-j\omega t'} & \text{for } z \le z' \\[4mm]
\dfrac{1}{\gamma}\dfrac{\sin(\gamma z')\,\sin(\gamma(z - l))}{\sin(\gamma l)}e^{-j\omega t'} & \text{for } z \ge z'
\end{cases}$$

where $\gamma^2 = \omega^2 LC - j\omega RC$. The inverse Fourier transform gives the following expression for the voltage on the transmission line:

$$v(z, t) = \frac{1}{2\pi}\int\limits_{-\infty}^{+\infty}\frac{\sin(\gamma z)\,\sin(\gamma(z' - l))}{\gamma\,\sin(\gamma l)}e^{j\omega(t - t')}\,d\omega \qquad \text{for } z < z'$$

and a similar formula for $z > z'$. Show that the poles of the above integrands are given by $\sin(\gamma l) = 0$ and not at $\gamma = 0$; that is, at

$$\omega = j\frac{R}{2L} \pm \sqrt{\left(\frac{n\pi}{l}\right)^2 \frac{1}{LC} - \frac{R^2}{4L^2}} \qquad n = 1, 2, 3, \ldots$$

Using the Cauchy residue theorem, show that $v(z, t)$ is zero for $t < t'$. Determine the voltage on the line for $t > t'$.

P5.9. Consider a microstrip line as shown in Figure 5.13. The conducting strip and the ground plane are assumed to be perfect conductors. The strip thickness is assumed to be negligible, and the dielectric substrate is assumed to be lossless. Derive an expression for the capacitance per unit length of the microstrip line assuming the following charge distribution on the strip:

$$\rho_s(x) = \begin{cases} 1 + \left|\dfrac{2x}{W}\right|^3 & -\dfrac{W}{2} \leq x \leq \dfrac{W}{2} \\ 0 & \text{otherwise} \end{cases}$$

P5.10. Repeat Problem P5.9 using the following charge distribution [6, p. 145]:

$$\rho_s(x) = \begin{cases} \dfrac{1}{\sqrt{1 - \left(\dfrac{|x|}{W/2}\right)^2}} & \dfrac{W}{2} \leq |x| \\ 0 & \text{otherwise} \end{cases}$$

P5.11. An alternative approach for solving (5.43) for the half-space Green's function is to use direct construction approach. Assume the following [2, p. 190]:

$$G(x; x') = \begin{cases} A e^{jk_0 x} & \text{for } x \geq x' \\ B \sin k_0 x & \text{for } 0 \leq x \leq x' \end{cases}$$

and show that the Green's function is given by (5.34).

CHAPTER 6

Introduction to Computational Methods

The boundary value problems in electromagnetics may be represented by the operator equation $Lf = g$. The computer-based solution of this equation is obtained by discretizing it; that is, by expressing it in the form of matrix equation $[A][x] = [b]$, where $[x]$ is the vector corresponding to the unknown function f, $[A]$ is the matrix corresponding to operator L, and the vector $[b]$ represents excitation function g. The vector and matrix representations are obtained in an appropriate function space. Here we emphasize the use of function space consisting of subdomain functions in order that the solution method can be applied to arbitrary shapes. The accuracy of the solution depends on the size of vectors and the matrix. An optimum size is realized as a compromise between the accuracy of solution and computer resources required. This gives rise to the issues such as convergence, accuracy, and stability of the solution. These aspects of computational methods are discussed in general in this chapter. The specific computational methods are discussed in later chapters.

In general, we can express the unknown function in the form of a series

$$f(\mathbf{r}) = \sum_n \alpha_n f_n(\mathbf{r}) \tag{6.1}$$

where $f_n(\mathbf{r})$ are known expansion or basis functions and α_n are the complex coefficients to be determined. The unknown distribution depends on factors like geometrical shape, dielectric filling, frequency, excitation, and boundary conditions of the electromagnetic device. The various methods of analysis differ in the choice of expansion/basis functions and in the determination of coefficients.

6.1 Elements of Computational Methods

The analysis of a problem using computational methods involves three major steps. This is shown as block schematics in Figure 6.1.

Step 1, Governing Equation: The different computational methods differ in preprocessing of Maxwell's equations to arrive at the governing equation. In the finite-difference time-domain (FDTD) method, the Maxwell's equations are expressed in finite difference form in space and time. In the finite-difference method (FDM), Helmholtz or wave equations such as the scalar inhomogeneous wave equation

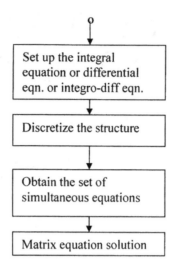

Figure 6.1 Major steps in the computational methods.

$$\nabla^2 \psi + k^2 \psi = g \tag{6.2}$$

is employed and expressed in finite-difference form to arrive at the algebraic equation. The Maxwell's equations or Helmholtz equations are processed to obtain either functional form or weak wave equation form for finite element method (FEM). The functionals are derived in Chapter 9. For the method of moments (MoM), the integral equations are developed from the inhomogeneous wave equation with delta function as the source term.

Step 2, Device Domain Discretization: This step distinguishes computational methods from analytical methods. In order to solve the governing equations for an arbitrary shaped device with dielectric inhomogeneity, the device geometry is discretized in the form of nonoverlapping cells. For this, the device geometry is first marked by a mesh or grid. For illustration, the rectangular grid on an L-shaped geometry in two dimensions is shown in Figure 6.2. Nodes are placed at the intersection of grid lines. Some of the nodes are marked there. Next, the nodes are joined by the straight lines as shown to form rectangular or triangular or quadrilateral cells or elements or subdomains. The whole device geometry is divided into one type of nonoverlapping cells or combination of different types. The discretiza-

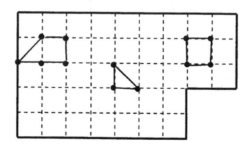

Figure 6.2 Discretization of device geometry into a number of cells or elements.

tion is carried out such that the dielectric is homogeneous in each cell. This completes device discretization.

The device discretization has not only separated out the dielectric inhomogeneity, it has discretized the function also into subdomains. The size of subdomain is equal to the cell size. The basic cells or elements used in discretization are shown in Figure 6.3.

The cell size is selected such that the function may be assumed to be uniform or linearly varying over the cell. Quadratic and cubic variations over the cell are used to improve efficiency at the cost of increased complexity in formulation. In commercially available software, discretization is carried out using an automatic mesh generator. Discretization of the device geometry results in a system of linear simultaneous equations and can be described by the matrix equation

$$[A][x] = [b] \tag{6.3}$$

The size of the matrix is equal to the number of unknowns, sometimes more than a few hundred.

Step 3, Matrix Equation Solver: The simultaneous equations of (6.3) can be solved for the unknowns using the matrix equation solvers. Since the matrix size may be large, it is important to employ efficient algorithms. These may include; Gauss elimination, *L-U* factorization, and conjugate gradient method. Some of these solution methods are discussed in Appendix A at the end of the book.

Step 4, Postprocessing: The raw data obtained in step 3 may be processed to obtain characteristics of interest for the device. These may include capacitance and characteristic impedance of waveguiding structures, resonant frequency, S-parameters of circuits or antennas, radiation patterns, and RCS of scatterers.

The discretization of device geometry into a number of cells/subdomains is equivalent to discretization of the unknown function $f(\mathbf{r})$ into a number of subdomain expansion/basis functions $f_n(\mathbf{r})$ as described by (6.1). We now discuss the various types of basis functions.

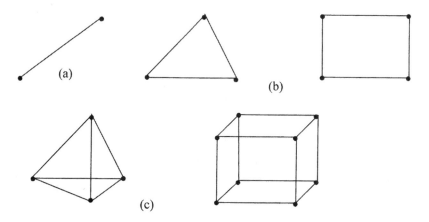

Figure 6.3 Basic cells or elements used in discretization of geometry: (a) one-dimensional; (b) two-dimensional cells: triangle and rectangle; and (c) three-dimensional cells: tetrahedron and rectangular brick.

6.2 Basis Functions

There are two principle classes of basis functions. These are entire-domain, and subdomain basis functions. The subdomain basis functions are discussed next.

6.2.1 Subdomain Basis Functions

The subdomain basis functions are versatile and take the guesswork out of the unknown distribution. For this, the domain of the problem is first divided into a number of subdomains of same size or different sizes. Each subdomain is spanned preferably by a single function called subdomain function. The subdomain basis functions are nonzero over the subdomain and zero outside, and can therefore model the unknown distribution efficiently.

6.2.1.1 One-Dimensional Subdomain Basis Functions

Pulse or Window Basis Functions
Let us consider the range of interest or domain size as $0 \leq x \leq 1$, and divide this range into $N + 1$ equal subintervals/subdomains of width

$$h_x = \frac{1}{N + 1} \tag{6.4}$$

centered about the points x_m defined as

$$x_m = \frac{m}{N + 1}, \qquad m = 1, 2, 3, \ldots \tag{6.5}$$

The subintervals and points x_m are shown in Figure 6.4(a) for $N = 4$. A pulse or window function which is nonzero over only one subinterval and is centered about x_m is defined as [1]

$$P(x - x_m) = \begin{cases} 1 & \text{for } x_m - \dfrac{h_x}{2} \leq x \leq x_m + \dfrac{h_x}{2} \\ 0 & \text{elsewhere} \end{cases} \tag{6.6}$$

The function $P(x - x_2)$ is shown in Figure 6.4(a). A linear combination of pulse functions according to $f(x) = \sum_{n=1}^{N} \alpha_n P(x - x_n)$ gives a step or staircase approximation as shown in Figure 6.4(b). It may be noted that pulse functions are orthogonal in nature because they do not overlap with others. A pulse function can be differentiated only once and therefore cannot be employed to approximate a function which is to be differentiated twice.

Triangular Basis Functions
A triangular function or piecewise linear function (PWL) can be differentiated twice and is defined as

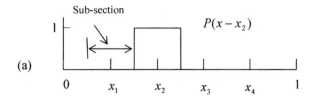

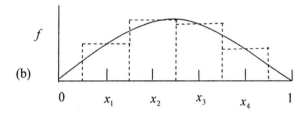

Figure 6.4 Pulse or window basis functions and approximation using them: (a) a pulse function; and (b) step approximation of a function using pulse functions.

$$T(x - x_m) = \begin{cases} \dfrac{x - x_{m-1}}{x_m - x_{m-1}} & \text{for } x_{m-1} \leq x \leq x_m \\[2ex] \dfrac{x_{m+1} - x}{x_{m+1} - x_m} & \text{for } x_m \leq x \leq x_{m+1} \\[2ex] 0 & \text{elsewhere} \end{cases} \qquad (6.7)$$

For $N = 4$, the function $T(x - x_2)$ is plotted in Figure 6.5(a), and a linear combination of these functions gives a piecewise linear approximation as shown in Figure 6.5(b). The triangular functions are deceptively simple because of linear variation and resulting integrability. It may be noted that the triangular functions exist over two connected subdomains, and a subdomain is spanned by two basis functions. The *partial* overlap between them makes triangular functions nonorthogonal.

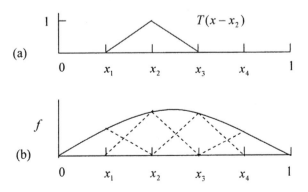

Figure 6.5 Triangular basis functions and approximation using them: (a) a triangular function; and (b) piecewise linear approximation of a function using triangular functions.

Some of the commonly occurring mathematical operations with triangular functions are as follows:

$$\frac{d^2}{dx^2} T(x - x_n) = \frac{1}{\Delta x} [\delta(x - x_{n-1}) - 2\delta(x - x_n) + \delta(x - x_{n+1})] \quad (6.8a)$$

$$\int_{x_{m-1}}^{x_{m+1}} T(x - x_m) T(x - x_n) \, dx = \begin{cases} 2\Delta x/3, & \text{for } n = m \\ \Delta x/6, & \text{for } |n - m| = 1 \\ 0 & \text{for } |n - m| > 1 \end{cases} \quad (6.8b)$$

$$\int_{x_m - \Delta x/2}^{x_m + \Delta x/2} P(x - x_m) T(x - x_n) \, dx = \begin{cases} 3\Delta x/4, & \text{for } n = m \\ \Delta x/8, & \text{for } |n - m| = 1 \\ 0 & \text{for } |n - m| > 1 \end{cases} \quad (6.8c)$$

The element size Δx is defined as $\Delta x = x_{m+1} - x_m$.

Spline Basis Functions

A spline of order n has polynomial degree $n - 1$ with order zero denoting a Dirac delta function, order 1 denoted by pulse functions, and order 2 by triangle functions. The order 3 spline function is called quadratic spline. It is continuous, has a continuous first derivative, and is therefore smoother than triangular function. The quadratic splines are defined as [2, p. 192]

$$\Omega(x - x_m) = \begin{cases} 0 & \text{for } x \leq x_{m-3/2} \\ \frac{9}{8} + \frac{3}{2}\left(\frac{x - x_m}{\Delta}\right) + \frac{1}{2}\left(\frac{x - x_m}{\Delta}\right)^2 & \text{for } x_{m-3/2} \leq x \leq x_{m-1/2} \\ \frac{3}{4} - \left(\frac{x - x_m}{\Delta}\right)^2 & \text{for } x_{m-1/2} \leq x \leq x_{m+1/2} \\ \frac{9}{8} - \frac{3}{2}\left(\frac{x - x_m}{\Delta}\right) + \frac{1}{2}\left(\frac{x - x_m}{\Delta}\right)^2 & \text{for } x_{m+1/2} \leq x \leq x_{m+3/2} \\ 0 & \text{for } x \geq x_{m+3/2} \end{cases}$$

$$(6.9)$$

The function $\Omega(x - x_2)$ is plotted in Figure 6.6. It may be noted that each spline function spans three subintervals and each subinterval is spanned by three basis functions. The subinterval Δ is defined as $3\Delta = x_{m+3/2} - x_{m-3/2}$.

Various other subdomain basis functions including Lagrangian functions are described in [2].

Exercise. Determine and plot $\dfrac{d\Omega(x - x_m)}{dx}$ and $\dfrac{d^2\Omega(x - x_m)}{dx^2}$ over its domain.

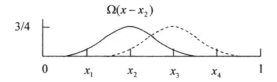

Figure 6.6 Quadratic spline functions. Each spline spans three subintervals.

6.2.1.2 Two-Dimensional Subdomain Basis Functions

If the expected distribution $f(x, y)$ can be described as separable in the two dimensions, it may be expressed as

$$f(x, y) = f_1(x) f_2(y) \tag{6.10}$$

The functions $f_1(x)$ and $f_2(y)$ may now be expressed in terms of one-dimensional subdomain basis functions; that is,

$$f_1(x) = \sum_{m=1}^{M} a_m f_{1m}(x) \quad \text{and} \quad f_2(y) = \sum_{n=1}^{N} b_n f_{2n}(y) \tag{6.11}$$

We now describe some of the popular subdomain basis functions in two dimensions.

Pulse-Pulse Basis Functions
The pulse basis functions along the x- and y-directions are obtained from (6.6) and are given by

$$f_{1m}(x) = \begin{cases} 1 & \text{for } x_m - \dfrac{h_x}{2} \le x \le x_m + \dfrac{h_x}{2} \\ 0 & \text{elsewhere} \end{cases} \tag{6.12a}$$

and

$$f_{2n}(y) = \begin{cases} 1 & \text{for } y_n - \dfrac{h_y}{2} \le y \le y_n + \dfrac{h_y}{2} \\ 0 & \text{elsewhere} \end{cases} \tag{6.12b}$$

where $h_x = x_{m+1} - x_m$ and $h_y = y_{n+1} - y_n$ are the subdomain lengths along the x- and y-directions, respectively. The pulse-pulse subdomain basis functions for $J_x(x, y)$ defined over a pair of cells are shown in Figure 6.7(a).

Roof-Top Basis Functions
These basis functions are one of the most popular two-dimensional basis functions. The roof-top basis function is a combination of piecewise linear (PWL) function along one direction and pulse function along the other direction. For PWL variation along the x-direction, the function is written as

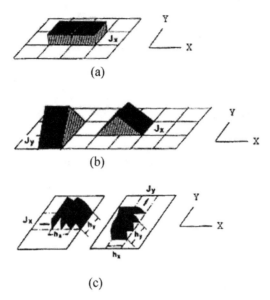

Figure 6.7 Two-dimensional subdomain basis functions for current distribution: (a) pulse-pulse basis functions for J_x over a pair of cells; (b) roof-top basis functions for J_x and J_y; and (c) PWS-pulse basis functions for J_x and J_y.

$$f_{1m}(x) = \begin{cases} (x - x_{m-1})/h_x & \text{for } x_{m-1} \leq x \leq x_m \\ (x_{m+1} - x)/h_x & \text{for } x_m \leq x \leq x_{m+1} \\ 0 & \text{otherwise} \end{cases} \qquad (6.13)$$

where $x_m - x_{m-1} = h_x = x_{m+1} - x_m$. The pulse variation along the y-direction is described by (6.12b). The roof-top basis functions for $J_x(x, y)$ and $J_y(x, y)$ is drawn in Figure 6.7(b).

PWS-Pulse Basis Functions
These basis functions are a combination of piecewise sinusoidal (PWS) and pulse basis functions. For PWS variation along the x-direction, the function is defined as

$$f_{1m}(x) = \begin{cases} \sin[k(x - x_{m-1})]/\sin(kh_x) & \text{for } x_{m-1} \leq x \leq x_m \\ \sin[k(x_{m+1} - x)]/\sin(kh_x) & \text{for } x_m \leq x \leq x_{m+1} \\ 0 & \text{otherwise} \end{cases} \qquad (6.14)$$

where k is a constant. The pulse variation along the y-direction is described by (6.12b). The PWS-pulse basis functions for $J_x(x, y)$ and $J_y(x, y)$ are sketched in Figure 6.7(c).

6.2.2 Entire Domain Basis Functions

As the name suggests, the entire domain basis functions are defined over the entire domain, that is, the range of space variables over which the problem is defined.

The expansion/basis functions employed in the analytical methods are of the entire domain type and orthogonal in nature, as discussed in Chapter 2. One may also use general orthogonal functions such as Bessel functions, Legendre functions, Chebyshev polynomials, and power series. Use of entire domain functions is supposed to make the analysis efficient because only the first few functions may be needed to approximate $f(\mathbf{r})$ to the desired accuracy. The efficiency improves if the expansion functions match the eigenfunctions of the problem. The entire domain basis functions are useful in describing smooth variations such as those on *unloaded* regular shaped geometries such as rectangular and circular patches. Also, the distributions that are fast varying may be determined using subdomain basis functions and modeled for later use as a single-term entire domain function (e.g., charge and current distributions on a strip or slot).

6.2.2.1   One-Dimensional Entire Domain Basis Functions

Series Expansion
Series expansion as defined in (6.1) is the most general form of expressing an unknown distribution. One may write the power series expansion of one-dimensional function $f(x)$ as

$$f(x) = \sum_n \alpha_n f_n(x) \text{ with } f_n(x) = \left(\frac{x}{w}\right)^n, \, n = 0, \, 1, \, 2, \, \ldots \qquad (6.15)$$

where w is the domain size. For the basis functions to represent even symmetric distribution in x, only the even values of n in (6.15) should be selected. The above basis functions can be easily Fourier transformed for their usage in Fourier transform method.

Next we describe some efficient basis functions used in the study of metal strips and slots.

Basis Functions with Edge Singularity
Consider an infinitely long metal strip of width w as shown in Figure 6.8. The metal strip is common to microstrip line, strip line, coplanar strips, coplanar waveguide, and printed dipole antenna. The charge density distribution on the strip can be described by a single basis function as

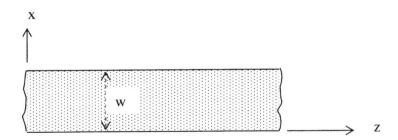

Figure 6.8   Geometry of a metal strip of infinite length and width w.

$$f(x) = \begin{cases} \left(1 - \left(\dfrac{2x}{w}\right)^2\right)^{-1/2} & \text{for } -\dfrac{w}{2} \le x \le \dfrac{w}{2} \\ 0 & \text{otherwise} \end{cases} \tag{6.16}$$

It may be noted that the function $f(x)$ models a symmetric distribution with the charge density suddenly increasing at the edges, $x = \pm w/2$. This behavior is called *edge singularity*, and is due to the Coulomb forces of repulsion between the charges. A similar behavior for the current distribution is observed along the length of a wire (Section 11.2).

Generalized Current Basis Functions

The conduction current $J(x)$ on a metal strip (Figure 6.8) may be decomposed into the longitudinal part $\mathbf{J}_z(x)$ and the transverse part $\mathbf{J}_x(x)$ as

$$\mathbf{J}(x) = \hat{x} J_x(x) + \hat{z} J_z(x) \tag{6.17}$$

The current components may be discretized as

$$J_x(x) = \sum_{m=1}^{M} a_m J_{xm}(x) \qquad \text{and} \qquad J_z(x) = \sum_{n=1}^{N} b_n J_{zn}(x) \tag{6.18}$$

with

$$J_{xm}(x) = \sin\left(\frac{2m\pi x}{w}\right), \qquad m = 1, 2, \ldots M \qquad \text{for } -\frac{w}{2} \le x \le \frac{w}{2} \tag{6.19a}$$

and

$$J_{zn}(x) = \cos\left[\frac{2(n-1)\pi x}{w}\right], \qquad n = 1, 2, \ldots N \qquad \text{for } -\frac{w}{2} \le x \le \frac{w}{2} \tag{6.19b}$$

as the basis functions. For better numerical efficiency of the solution, the current basis functions with edge singularity are employed to model the longitudinal current distribution on metal strips or wires, and are given by

$$J_{zn}(x) = \frac{\cos\left[\dfrac{2(n-1)\pi x}{w}\right]}{\sqrt{1 - \left(\dfrac{2x}{w}\right)^2}}, \qquad n = 1, 2, \ldots N, \qquad \text{for } -\frac{w}{2} \le x \le \frac{w}{2} \tag{6.20}$$

The transverse current J_x is not singular.

The basis functions described above are orthogonal in nature, and J_{zn} are plotted in Figure 6.9 for $n = 1, 2, 3$. The singularity in current is related to the singularity in charge by the equation $\nabla . \mathbf{J} = -j\omega\rho$.

The current basis functions can be Fourier transformed for their usage in analysis in the spectral domain. The Fourier transform is defined as

$$\tilde{J}(\alpha) = \frac{1}{\sqrt{2\pi}} \int_{-w/2}^{w/2} J(x) e^{j\alpha x} \, dx \tag{6.21}$$

The Fourier transform of the z-directed current is given by [3, p. 342]

$$\tilde{J}_{zn}(\alpha) = \frac{1}{\sqrt{2\pi}} \frac{\pi w}{4} \left[J_0\left(\left| \frac{w\alpha}{2} + (n-1)\pi \right| \right) + J_0\left(\left| \frac{w\alpha}{2} - (n-1)\pi \right| \right) \right] \tag{6.22}$$

where $J_0(.)$ is the Bessel function of order zero. The electric current basis functions on the strips may be used to model the equivalent magnetic current distribution in the slots because of the complementary nature of strips and slots.

6.2.2.2   Two-Dimensional Entire Domain Basis Functions

The two-dimensional basis functions are needed to describe the distribution on surfaces when both the dimensions are comparable to the wavelength. These include

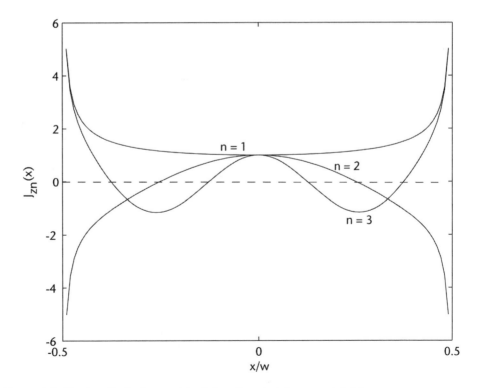

Figure 6.9   The longitudinal current basis functions $J_{zn}(x)$ on a metal strip.

metal plates, wide slots in conductors, microstrip antennas, resonators, and micro-strip circuits.

The complete set of entire domain basis functions for regular shaped geometries like rectangles, disks, circular rings, a few triangles, and circular sectors may be obtained from the eigenfunctions of the geometry. The electric or magnetic wall cavity eigenfunctions are the most suitable and can be determined from the modal analysis of the cavity. The eigenfunctions suitable for use in microstrip patch antennas are available in [4]. For a rectangular patch in the xy-plane, the electric current may be expressed as

$$\mathbf{J}(x, y) = \hat{\mathbf{x}} J_x(x, y) + \hat{\mathbf{y}} J_y(x, y) \tag{6.23}$$

It is easier to describe $\mathbf{J}(x, y)$ mathematically if $J_x(x, y)$ and $J_y(x, y)$ are separable in x and y; that is, variations of these functions with respect to x is independent of the variation with respect to y. We can therefore utilize the strip current basis functions to develop the basis functions for a rectangular patch. Consider a rectangular patch of dimensions $l \times w$ oriented as shown in Figure 6.10. The current components may be expanded as

$$J_x(x, y) = \sum_m \sum_n a_{mn} J_{mn}^x(x, y) \tag{6.24a}$$

and

$$J_y(x, y) = \sum_p \sum_q b_{pq} J_{pq}^y(x, y) \tag{6.24b}$$

For the x-directed current, one may write

$$J_{mn}^x(x, y) = J_m^x(x) J_n^y(y) = \sin\left(\frac{m\pi}{l}\left[x + \frac{l}{2}\right]\right) \cos\left(\frac{n\pi}{w}\left[y + \frac{w}{2}\right]\right) \tag{6.25a}$$

and the basis functions for the y-directed current are

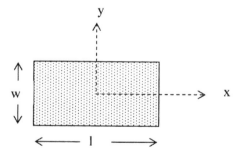

Figure 6.10 Geometry of a rectangular patch with the coordinate system.

$$J_{pq}^y(x, y) = J_p^y(x)J_q^y(y) = \cos\left(\frac{p\pi}{l}\left[x + \frac{l}{2}\right]\right)\sin\left(\frac{q\pi}{w}\left[y + \frac{w}{2}\right]\right) \quad (6.25b)$$

The above basis functions are orthogonal. It may be noted that unlike the basis functions for the current density on a strip or wire, the above basis functions do not include the edge singularity factor. If required for faster convergence of the solution, (6.25a) may be multiplied by the factor $1/\sqrt{1 - (2y/w)^2}$, and (6.25b) by the factor $1/\sqrt{1 - (2x/l)^2}$. Figure 6.11 illustrates the behavior of several types of basis functions for a rectangular patch. The basis functions are of the type $(1, 0)$ and $(3, 0)$ for the x-directed current, and type $(2, 1)$ for the y-directed current.

The advantage of using entire domain basis functions lies in their efficiency; that is, we may need fewer number of basis functions to model the expected distribution. However, the major disadvantages with these functions are as follows: (1) it is necessary to guess the function we want to simulate; (2) the entire domain functions are not versatile (i.e., they may not be able to model or may require a large number of basis functions to model an arbitrary distribution, especially if there is a notch or dip in the distribution); (3) the use of entire domain basis may make the evaluation of matrix elements difficult; and (4) the stability of the matrix solution techniques depends on the condition number of the matrix, and the condition number of the matrix increases with the order of matrix. The use of entire domain basis functions is likely to increase the condition number very rapidly with

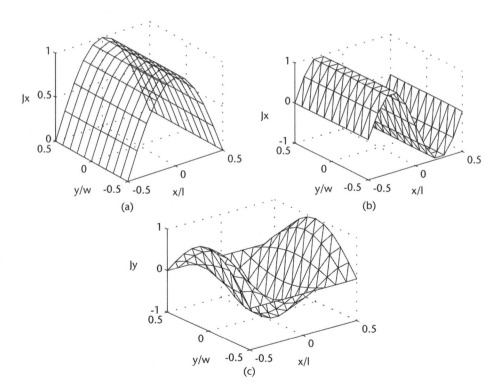

Figure 6.11 Some entire domain basis functions for the surface current on a rectangular patch: (a) $J_x(1, 0)$; (b) $J_x(3, 0)$; and (c) $J_y(2, 1)$.

the order of matrix. Therefore, only the first few basis functions may be useful before the instability sets in.

There are infinitely many possible sets of basis functions for a given problem. Some sets of functions may give faster convergence than others, or give matrix elements which are easier to evaluate. Thus, the choice of basis functions determines the accuracy and efficiency of the solution. A poor choice may give a divergent solution. For any particular problem our task is to choose basis sets which are well suited to the problem.

The computer-based solutions suffer from the computer related problems like finite word length, computer memory size, and processor speed. The finite word length affects the precision of stored values and gives rise to round-off error. The memory and processor time requirements increase with the decrease in cell size. These limitations show up in the form of approximate, and/or false solutions. Next, we discuss the effect of discretization or cell size on the convergence, discretization error, and numerical stability.

6.3   Convergence and Discretization Error

The discretization of the unknown function in terms of basis functions, described by (6.1), gives rise to error in the solution, called *discretization error*, if a finite number of terms are used. Also, since we do not know the distribution $f(\mathbf{r})$ a priori, we rely on the *convergence* behavior of approximate $f(\mathbf{r})$ as the number of terms in the expansion is increased. These characteristics of the solution are similar to that of numerical integration of a function. The numerical integration is carried out by dividing the range of integration into a number of subintervals and by fitting the unknown function in each subinterval by a pulse or linear or parabolic or polynomial approximation. The quadrature algorithms are mostly function independent.

We shall illustrate the typical characteristics of computer method based solutions by evaluating the following integral numerically:

$$I = \int \frac{dx}{\sqrt{a^2 - x^2}} \qquad (6.26)$$

Since numerical integration is carried out by discretizing the integrand, the effect of step size on the accuracy of the integral is very similar to that observed in computational methods due to discretization of unknown distribution. Also, the integrand is singular at $x = a$, and the singularity is typical of the charge and current distributions at the end points of a finite length conductor. The diagonal matrix elements for MoM are also singular. The above integral can be determined analytically for comparison with numerical results and is obtained as $I = \sin^{-1}(x/a)$.

For numerical integration we divide the range of integration assumed $(0, a/2)$ into N equal steps of size $h = a/(2N)$, sample the integrand at the mid-point of each step, and sum the contributions as

$$I = h \sum_{i=1}^{N} f_i, \quad f_i = \frac{1}{\sqrt{a^2 - \left(ih - \dfrac{h}{2}\right)^2}} \tag{6.27}$$

This process is shown in Figure 6.12 for $a = 1$, $N = 5$. The value of I is given in Table 6.1 for various values of h. It is observed that the value of integral approaches $\sin^{-1}(1/2) = \pi/6 = 0.523599$ as h is decreased. This behavior is called convergence. Also, the error reduces by a factor of 4 as the value of h is made half. The last column in Table 6.1 corresponds to the evaluation of (6.27) according to Simpson's rule, which employs parabolic interpolation over the step. It is observed that in this case $N = 5$ only results in the exact value of the integral. This is because the

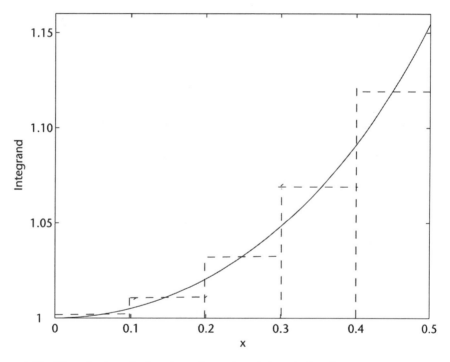

Figure 6.12   Discretization of a function using pulse subdomain functions.

Table 6.1   Convergence of $I = \displaystyle\int_{0}^{0.5} \dfrac{dx}{\sqrt{1 - x^2}}$ with Step Size h

N	h	I (Mid-point)	% Error	I (Simpson)
5	0.1000	0.523279	0.062	0.523599
7	0.0714	0.523435	0.031	0.523599
10	0.0500	0.523519	0.015	0.523599
15	0.0333	0.523563	0.007	0.523599
20	0.0250	0.523579	0.004	0.523599
30	0.0166	0.523590	0.002	0.523599

$$I \text{ (exact)} = 0.523599$$

shape of the curve over the step resembles a parabola and the Simpson's rule is parabolic interpolation.

We have described convergence as a phenomenon in which the difference between the computed and analytical values decreases monotonically as the step size is reduced. In most of the situations, the analytical value or the expected value is not known; how to determine the convergence of a function is described next.

The rate of convergence describes how fast I approaches the correct value $\sin^{-1}(1/2)$. A measure of this is called the *order of convergence* and is obtained by fitting the numerical integration data to the expression

$$I = \sin^{-1}\left(\frac{x}{a}\right) + ch^p \qquad (6.28)$$

where c is a constant and p is called the order of convergence. Obviously, I approaches $\sin^{-1}(x/a)$ when h approaches zero. Alternatively, one may carry out the convergence test on the data to determine if the solution is approaching the true value, and this is discussed next.

6.3.1 Convergence Test

The convergence test is based on the work of Richardson and is very well described in [5]. Consider a function $f(x)$. Let its value be f_1, f_2, and f_3 corresponding to the number of divisions (per unit x) n_1, n_2, and n_3, respectively, with $n_1 > n_2 > n_3$. The monotonic behavior of the solution can be tested by performing the *ratio* test defined as [6]

$$R = \frac{b_2 f_2 - b_1 f_1}{b_3 f_3} \qquad (6.29)$$

where

$$b_1 = n_1^2\left(n_3^2 - n_2^2\right), \qquad b_2 = n_2^2\left(n_3^2 - n_1^2\right), \qquad b_3 = n_3^2\left(n_2^2 - n_1^2\right)$$

For true monotonic convergence $R = 1$, and closeness of R to unity gives a measure of convergence for practical cases. For the data given in Table 6.1, for the mid-point rule, we find that $R = 1.000001$, indicating that the solution is converging.

6.3.2 Order of Convergence

The rate of convergence describes how fast the computed value $f(h)$ for the step size h approaches the asymptotic value f_0. A measure of this is called the *order of convergence*. The function $f(h)$ may be expanded about f_0 as

$$f(h) = f_0 + a_1 h^p + a_2 (h^p)^2 + \ldots \qquad (6.30)$$

where p is called the order of convergence. In the vicinity of convergence, the step size h is very small, and (6.30) may be approximated as

$$f(h) \approx f_0 + a_1 h^p \tag{6.31}$$

To determine the order of convergence we compute $f(h)$ for a number of step sizes h_i chosen in geometric progression; that is,

$$\frac{h_i}{h_{i+1}} = \frac{h_{i+1}}{h_{i+2}} \tag{6.32}$$

If $f(h_i)$ denotes the value of f corresponding to h_i, the order of convergence as obtained from (6.31) and (6.32) is

$$p = \frac{\ln\left[\dfrac{f(h_i) - f(h_{i+1})}{f(h_{i+1}) - f(h_{i+2})}\right]}{\ln\left[\dfrac{h_i}{h_{i+1}}\right]} \tag{6.33}$$

Fitting the data of Table 6.1 for the mid-point integration rule to (6.33) we find that $p = 2$. This is confirmed from the observation that the error reduces by a factor of 4 each time step size h is halved. A large value of p denotes faster convergence.

6.3.3   Disctretization Error and Extrapolation

From the analysis of the example described above we have observed that the step size h determines the accuracy of the numerical solution and the associated discretization error can be reduced by reducing h. There is, however, a limit to this decrease, which is set by the round-off error generated by the computer word length. Also, finer discretization requires more of computer memory and processor time.

Extrapolation
It is possible to predict the asymptotic value f_0 very closely without attempting a very fine discretization. For this, one may carry out computations for a number of discretization sizes and then extrapolate the curve obtained to predict the solution for very fine discretization or the continuous case (zero step size) provided the solution converges. Extrapolation requires polynomial fitting of the curve generated if the number of points are three or more. Quadratic fitting of three data points is described by Culver [7]. Let the value of the function $f(x)$, to be extrapolated, be f_1, f_2, and f_3, corresponding to the number of divisions n_1, n_2, and n_3, respectively with $n_1 > n_2 > n_3$. Provided the function converges as described by (6.29), the extrapolated or asymptotic value of $f(x)$, called f_0, is given as [7]

$$f_0 = f_1 a_1 + f_2 a_2 + f_3 a_3 \tag{6.34}$$

where

$$a_1 = \frac{n_1^4\left(n_3^2 - n_2^2\right)}{D}, \qquad a_2 = \frac{n_2^4\left(n_1^2 - n_3^2\right)}{D}, \qquad a_3 = \frac{n_3^4\left(n_2^2 - n_1^2\right)}{D}$$

$$D = n_1^2 n_2^2\left(n_2^2 - n_1^2\right) + n_2^2 n_3^2\left(n_3^2 - n_2^2\right) + n_3^2 n_1^2\left(n_1^2 - n_3^2\right)$$

The extrapolated value of I based on the last three data points in Table 6.1 for the mid-point rule is found to be $I_0 = 0.523598$ which is very close to the exact value 0.523599.

Effect of Using Higher-Order Numerical Integration Schemes

The pulse-fitting integration scheme, also called Euler's rule or midpoint rule, is the most primitive type. The rate of convergence of integration can be improved if one uses higher order fitting of the integrand, say, for example, linear or quadratic fitting. Linear approximation of the function is described well by the trapezoidal rule, and Simpson's rule implements parabolic fitting of the function. These integration schemes are defined as

Trapezoidal rule

$$I = h \sum_{i=1}^{N-1} f_i + \frac{h}{2}\left(f_0 + f_n\right) \tag{6.35a}$$

Simpson's rule

$$I = \frac{h}{3}\left(f_0 + 4f_1 + 2f_2 + 4f_3 + \ldots + 2f_{N-2} + 4f_{N-1} + f_N\right) \tag{6.35b}$$

The results of Simpson's rule are included in Table 6.1. The integration is found to converge to the analytical value for $N = 5$. However, the performance of trapezoidal rule is poorer compared to the mid-point rule. This is associated with the shape of the curve being fitted. In general, the rate of convergence improves with the use of higher-order fitting polynomials.

Effect of Singularity of the Function on Convergence

We have determined the rate of convergence when the integrand was well behaved, that is, it did not have any singularity. Next we determine the effect of singularity on the rate of convergence for (6.26). For this, we choose the range of x as $(0, a)$ with singularity at $x = a$. Assuming $a = 1$, the analytical value of the integral is given by $I(exact) = \sin^{-1}(1) = \pi/2 = 1.570796$. The value of the integral for different step sizes is given in Table 6.2. It is seen that I approaches the true value. However, comparison of Tables 6.1 and 6.2 shows that the percentage error, for the same value of h, is more now compared to the earlier case, suggesting that the rate of convergence must be slower now. Trapezoidal and Simpson's rules cannot be used in this case because the function needs to be computed very close to the singularity also. When fitted with (6.33), the value of p is found to be -0.4973 compared to $p = 2$ for the nonsingular integral discussed earlier. This example shows that the rate of convergence depends very much on the nature of the function. The data in

Table 6.2   Convergence of $I = \int\limits_{0}^{1} \dfrac{dx}{\sqrt{1-x^2}}$ with Step Size h

N	h	I (Mid-point)	% Error
5	0.2000	1.380492	12.11
7	0.1428	1.409720	10.25
10	0.1000	1.435880	8.59
15	0.0667	1.460544	7.02
20	0.0500	1.475274	6.08
30	0.0333	1.492770	4.97
40	0.0250	1.503209	4.30
50	0.0200	1.510337	3.85
60	0.0167	1.515600	3.51

I (exact) = 1.570796

Table 6.2 was plotted as a function of $\sqrt{h}$ based on the convergence estimate obtained earlier. Figure 6.13 shows this plot. Linear fitting of the data gives $I = 1.5706 - 0.42635\sqrt{h}$.

In electromagnetics we frequently come across singular functions (e.g., diagonal elements of MoM matrix, charge and current distributions at the end points of a conductor, and so forth).

The integrands with end-point singularity can be determined numerically using special integration schemes. One such scheme is available from Mathematica in the form of NIntegrate [8]. Some of the special integration schemes are described in Appendix B at the end of the book.

6.3.4   Discretization of Operators

The computational methods provide the solution of operator equation $Lf = g$ when expressed in the form of matrix equation. This requires discretizations of operator L and functions f and g. We can discretize the functions in basis functions and

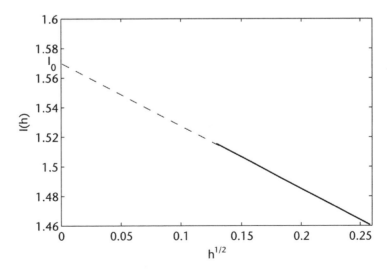

Figure 6.13   Plot of the data of Table 6.2 as a function of $\sqrt{h}$.

express them in the form of vectors of expansion coefficients as discussed in Chapter 2. We need to discretize the operator L also using the same basis functions so that the operator equation translates into a matrix equation that can be solved using computers. The discretization of operators was discussed in Chapter 2, and was implemented using eigenfunctions as basis functions to utilize the benefits of orthogonality of functions. This severely limits the range of applications of computational methods. However, the concept of discretization of operators is general and can be applied with reference to many types of basis functions, as illustrated next.

The discretization of a function is based on the concept of projecting a function on the basis functions to determine the coefficients of expansion. It is implemented mathematically as inner product between the functions (Section 2.5). The same approach is applicable to the discretization of operators also. Let us illustrate discretization of operators by considering the differential operator $L = d^2/dx^2$ and determine its matrix representation using various types of basis functions: subdomain or entire domain type. In the process of discretization, the operator operates on the basis function and the resultant function is projected (through inner product) on another function called the test function, in the language of MoM. The test functions may be the same as the basis functions or another type.

Linear Basis Representation of Operator
The linear bases in one dimension are defined in (6.7) and can be differentiated twice (for $L = d^2/dx^2$) to obtain

$$LT_n = [\delta(x - x_{n-1}) - 2\delta(x - x_n) + \delta(x - x_{n+1})]/\Delta x \qquad (6.36)$$

We take its inner product with another linear function T_m and obtain the following expression for the matrix element:

$$l_{mn} = \langle T_m, LT_n \rangle = \frac{1}{(\Delta x)^2} \begin{cases} 1 & \text{for } |m - n| = 1 \\ -2 & \text{for } m = n \end{cases} \qquad (6.37)$$

The operator matrix defined by (6.37) is tridiagonal and therefore sparse. The combination of linear basis for expansion and pulse function for testing also produces the same operator matrix.

Entire Domain Basis Representation of Operator
Let us choose the entire domain basis as

$$f_n = x - x^{n+1} \qquad (6.38)$$

These functions are also called power basis, and when operated upon by L give

$$Lf_n = -n(n + 1)x^{n-1} \qquad (6.39)$$

and

$$l_{mn} = <f_m, Lf_n> = \frac{mn}{m+n+1} \qquad \text{assuming } 0 \le x \le 1 \qquad (6.40)$$

This operator matrix is full with all the elements nonzero.

Eigenfunction Basis Representation of Operator
The eigenfunctions for the given operator subject to $f(0) = 0 = f(1)$ are given by $f_n(x) = \sqrt{2}\sin(n\pi x)$ with eigenvalues $\lambda_n = -(n\pi)^2$. Using these functions as basis, one obtains

$$Lf_n(x) = -\sqrt{2}(n\pi)^2 \sin(n\pi x) \qquad (6.41)$$

and

$$l_{mn} = <f_m, Lf_n> = -(n\pi)^2 \delta_{mn} \qquad (6.42)$$

The operator matrix in this case is diagonal with $l_{nn} = \lambda_n$. This is due to the orthogonality of eigenfunctions.

We shall have occasions to determine the operator matrix using FDM, FEM, and MoM in the respective chapters. The structure of the operator matrix influences the matrix solution technique and the stability of the matrix solution. The effect of condition number of matrix on the stability of solution is discussed in Section 6.4.

6.3.5 Discretization Error in FDM, FDTD, and FEM

Now we discuss the effect of discretization on the accuracy of solution with reference to the computational methods discussed in the book. For this, we consider sinusoidal wave propagation in an infinite one-dimensional space. The analytical solution is $e^{-jk_0 z}$ and $e^{jk_0 z}$ for the waves propagating along $+z$ and $-z$ directions, respectively. Here, k_0 is the free space wave number. For computational methods, we discretize the media in steps of size h. The numerical solution denoted as wavenumber $\bar{k}$ is found to be the function of h. Also, the value of $\bar{k}$ varies with the computational method and the order of basis functions employed. The results in the form of dispersion relations are summarized next, and the details are included in the respective chapters.

Finite difference method, (7.66b) for pulse basis functions:

$$\bar{k}h = \cos^{-1}\left(1 - \frac{(k_0 h)^2}{2}\right) \qquad (6.43a)$$

or

$$(\bar{k}h)^2 \approx 6\left(-1 + \sqrt{1 + \frac{1}{3}(k_0 h)^2}\right) \qquad \text{for } k_0 h \ll 1 \qquad (6.43b)$$

Finite-difference time-domain method, (8.22) for pulse basis functions:

$$\overline{k}h = 2\sin^{-1}\left(\frac{1}{\alpha}\sin\left[\frac{\alpha k_0 h}{2}\right]\right), \qquad \alpha = \frac{c\Delta t}{h} \qquad (6.44\text{a})$$

or

$$\overline{k}h \approx k_0 h\left(1 - \frac{\alpha^2}{24}(k_0 h)^2\right) \qquad \text{for } k_0 h \ll 1 \qquad (6.44\text{b})$$

Finite element method, (10.51) for triangular elements:

1. Linear basis functions

$$\overline{k}h = \cos^{-1}\left(\frac{6 - 2(k_0 h)^2}{6 + (k_0 h)^2}\right) \qquad (6.45\text{a})$$

or

$$(\overline{k}h)^2 \approx 6\left(-1 + \sqrt{1 - \frac{2}{3}\left(\frac{6 - 2(k_0 h)^2}{6 + (k_0 h)^2}\right)}\right) \qquad \text{for } k_0 h \ll 1 \quad (6.45\text{b})$$

2. Quadratic basis functions [9, p. 83]

$$\overline{k}h = \frac{1}{2}\cos^{-1}\left(\frac{15 - 26(k_0 h)^2 + 3(k_0 h)^4}{15 + 4(k_0 h)^2 + (k_0 h)^4}\right) \qquad (6.46)$$

3. Cubic basis functions [9, p. 84]

$$\overline{k}h = \frac{1}{3}\cos^{-1}\left(\frac{2800 - 11520(k_0 h)^2 + 4860(k_0 h)^4 - 324(k_0 h)^6}{2800 + 1080(k_0 h)^2 + 270(k_0 h)^4 + 81(k_0 h)^6}\right) \quad (6.47)$$

The phase velocity defined as $v_{ph} = \omega/\overline{k}$ is computed from the dispersion relation, and is found to vary with discretization size h/λ. Next we compute the associated phase error per wavelength defined as

$$\delta\varphi = 360\left|\frac{\overline{k} - k_0}{k_0}\right| \qquad \text{(degrees)} \qquad (6.48)$$

The plot of phase error with h/λ is given in Figure 6.14 for the FDM, FDTD, and FEM with linear basis. The value of α is assumed to be 0.5 for FDTD. It may be noted that in the figure h/λ is plotted decreasing (equivalent to λ/h in increasing order) along the axis. The convergence rate p, determined by the slope of the curves, is same for all the three computational methods, and $p = 2$.

The use of higher order basis functions in computational methods is known to reduce the discretization error, and the speed of convergence therefore increases.

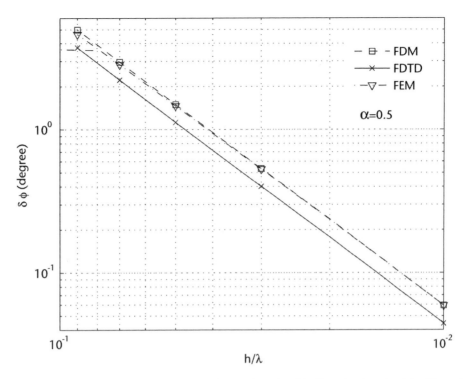

Figure 6.14 Variation of phase error per unit wavelength with discretization size h/λ. The value of h/λ is decreasing along the x-axis.

The order of a basis function is defined by the highest degree of polynomial used to define the function. The computed phase error for the linear, quadratic and third-order basis functions in FEM, (6.45) through (6.47), is plotted in Figure 6.15. The order of convergence is $p = 2$ for linear basis functions, $p = 4$ for quadratic, and $p = 6$ for cubic basis functions, indicating the higher efficiency of higher-order basis functions. The main disadvantage with the higher order basis functions is the complexity of formulation.

To determine the value of h/λ for the desired accuracy, let us consider a linear resonator. The phase error determines the corresponding error in resonant frequency. For 1% accuracy in phase and therefore frequency, the maximum phase error over one wavelength should be 3.6 degrees. The data from Figure 6.14 indicates that about 13 cells per wavelength are needed when FDM or FDTD is used for analysis. The corresponding number of cells per wavelength or resolution required for FEM (with linear basis) is about 12 cells per wavelength.

The phase error accumulates with the increase in distance traveled by the wave. The importance of reducing phase error for large distances or large device sizes therefore increases, and one needs to use higher order basis functions to limit the phase error. For example, the phase error is about 6 degrees per λ at $h/\lambda = 0.1$ for FEM with linear basis functions. Over a distance of 10 wavelengths the phase error becomes 60 degrees and might not be desirable. If the acceptable phase error is 10 degrees, then either the resolution should be increased or higher order basis functions should be employed. For FEM with linear basis, the resolution required

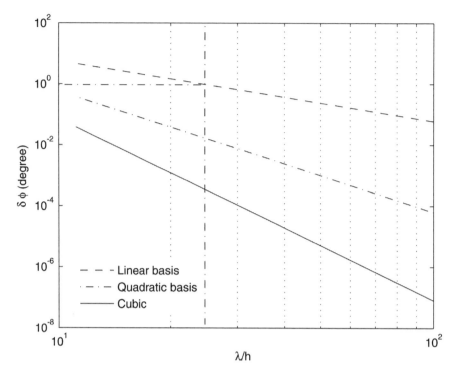

Figure 6.15 Effect of higher order basis functions on phase error in FEM analysis.

is about 25 cells (Figure 6.15), whereas the use of quadratic basis would require a resolution of less than 10 cells. This amounts to a considerable savings in computational costs for analyzing three-dimensional devices. However, the MoM solution using an exact Green's function spanning the whole domain will not accumulate error [10, p. 200] because the medium of propagation is not discretized in MoM. The effect of phase error on the device performance could be very important for applications like design of low sidelobe antennas, radar cross-section calculations, and the design of highly selective filters.

Convergence Rate for Practical Cases
It may be mentioned here that the convergence rate discussed above is the maximum one can expect for a given step size h/λ, order of basis function, and the computational method used. It does not apply to nonuniform discretization. The convergence rate, in actuality, will be lower because of singularities of the current/charge distribution at the edges and corners of conductors and singularity associated with the diagonal elements of MoM. In addition, the device geometry may be discretized nonuniformly for better computational efficiency. The convergence in such cases may be oscillatory, and it is often possible to extract a major portion of the order of convergence using the procedure described above [10, p. 16]. An improvement over this procedure is to use linear fit of the data to h^p, where p is the estimated order of convergence.

6.3.6   Vector and Matrix Norms

Once the computer solution for f in the form of a data vector has been obtained, one would like to describe it in the form of simple parameters so that it may be compared against the expected value and the error determined. Norm of a vector is a real number that provides a measure of the *length* or *size* of the vector, and the Euclidean norm is defined as

$$\|x\| = \sqrt{\sum_{i=1}^{N} (x_i)^2} = x^t.x \qquad \text{for real valued vector } x \qquad (6.49a)$$

$$\|x\| = \sqrt{\sum_{i=1}^{N} (x_i)^2} = x^t.x^* \qquad \text{for complex vector } x \qquad (6.49b)$$

The superscript t stands for transpose and * denotes complex conjugate. The norm of a matrix is defined similarly and is given by

$$\|A\| = \sqrt{\sum_{i=1}^{N} \sum_{j=1}^{N} |A_{ij}|^2} \qquad (6.50)$$

The norm may be used to test the robustness or stability of the numerical system; that is, a small change Δb in b should produce a corresponding small change Δx in x for a linear system described by $[A][x] = [b]$.

Although norm of x may be used to test the stability of the matrix solution, the principle contributor to this behavior is the matrix A. We now discuss the stability of matrix $[A]$.

6.4   Stability of Numerical Solutions

While analyzing an electromagnetic device using computational methods, we may find that the numerical solution is diverging instead of converging to the expected value as the discretization is made finer. This behavior is called numerical instability, and may be defined as an increase in error as we reduce the step size. The increase in error may be modeled as a growth factor g. If ϵ^n is the error in the solution associated with the nth step size, and ϵ^{n+1} is associated with the smaller step size $n + 1$, then

$$\epsilon^{n+1} = g\epsilon^n \qquad (6.51)$$

The growth factor is greater than 1 for unstable solution process. The error therefore goes on increasing with finer discretization. There are varied reasons for the instability of the solution process. Some of these are discussed next.

6.4.1 Stability of FDTD Solution

One of the reasons for the numerical instability in FDTD is the violation of causality; that is, information cannot be transmitted from one node to the next with a speed exceeding the speed of light, c. The causality puts an upper limit on the time step Δt for the stability of the numerical solution, and is given by $\Delta t \leq \Delta x/c$.

The effect of increasing time step on the pulse amplitude is plotted in Figure 6.16. Here α is defined as $\alpha = c\Delta t/\Delta x$. The upper figure compares the propagation of a Gaussian pulse in the discretized one-dimensional space at $n = 185$ for $\alpha = 0.5$ and $\alpha = 1.0006$, i denotes the node number, and n is the number of time step. For $\alpha = 1.0006$, the pulse has traveled far away compared to the case for $\alpha = 0.5$ because of the larger value of $n\Delta t$ and velocity associated with $\alpha = 1.0006$. Also, the tail of this pulse has nonzero amplitude. The reason for this behavior is that a Gaussian pulse consists of a large number of spectral components, and the periodic arrangement of nodes in a grid acts like a low pass filter. Therefore, higher spectral components are attenuated more. Also, these components travel slowly compared to others and are left behind. These spectral components violate causality and show growing amplitude or instability with increasing n and α. The lower figure at $n = 200$ shows the increase in amplitude of the tail while the peak remains unaffected. The instability can be avoided by using $\alpha < 1$.

The instability in the matrix equation based solution (e.g., FDM, FEM, and MoM) is linked to the instability of the matrix and is discussed next.

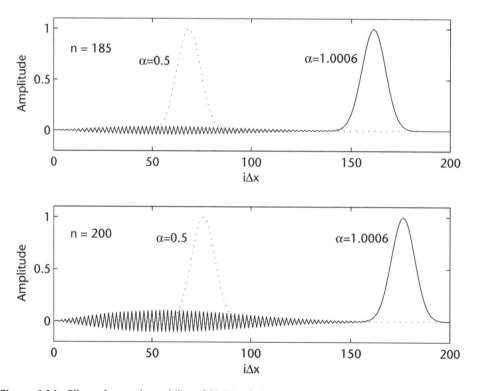

Figure 6.16 Effect of α on the stability of FDTD solution.

6.4.2 Stability of Matrix Solution

The condition number of matrix A in the equation $[A][x] = [b]$ often provides a good indication of numerical stability of the solution. If the condition number is very large, it indicates a nearly singular matrix and therefore the matrix is ill conditioned. Apart from the condition number, there are other checks, described next, which may be performed to determine the *condition* of a matrix [11].

Determinant of the matrix: Unusually small values of the determinant may give an indication that the matrix is ill conditioned. The condition number of such a matrix is found to be large.

Sensitivity of the solution: If the solution $[x]$ is found to be excessively sensitive to small variations (1% to 2%) in some of the elements of the matrix or the excitation vector, it indicates that the behavior is nonphysical and therefore the matrix is ill conditioned. The condition number of such a matrix is found to be large.

Condition number of a matrix: Condition number of matrix $[A]$ is defined as $k(A) = \|A\| \|A^{-1}\|$, where $\|A\|$ denotes the norm of matrix $[A]$ and is defined in (6.50). $\mathbf{A}^{-1}$ is the inverse of $[A]$. A *good* condition number is small, near to 1; and $k = 1$ for the identity matrix. The condition number of a matrix increases with the size of the matrix.

The effect of changes in $[A]$ and $[b]$ on the solution $[x]$ has been analyzed in [2]. The change in $[A]$ may represent the approximation and round-off errors, while $[b]$ may be changed intentionally to determine the sensitivity of the solution. The solution will therefore differ from $[x]$ by an amount $[\Delta x]$. If the increment in $[A]$ is denoted as $[\Delta A]$, and that in $[b]$ as $[\Delta b]$, the new matrix equation is

$$[A + \Delta A][x + \Delta x] = [b + \Delta b] \tag{6.52}$$

Considering all the increments as perturbations, the use of the Schwarz inequality gives [2, p. 147]

$$\frac{\|\Delta x\|}{\|x\|} \leq k(A) \frac{\|\Delta A\|}{\|A\|}, \qquad \text{for } [\Delta b] = 0 \tag{6.53a}$$

$$\frac{\|\Delta x\|}{\|x\|} \leq k(A) \frac{\|\Delta b\|}{\|b\|}, \qquad \text{for } [\Delta A] = 0 \tag{6.53b}$$

The condition number of the matrix will therefore amplify the changes in A or b. It may be noted that for an ill-conditioned matrix, minor changes in computational algorithm also may lead to large differences in final results. If we approximate the fractional change in $[A]$ as

$$\frac{\|\Delta A\|}{\|A\|} = 10^{-t} \tag{6.54}$$

the corresponding change in $[x]$ is given by

$$\frac{\|\Delta x\|}{\|x\|} = 10^{-t}k(A) \tag{6.55}$$

Specifically, a change in the seventh decimal place for the norm of A translates to a change in the third decimal place for the norm of x if $k(A) = 10^4$.

The discretization of operator $L = d^2/dx^2$ for subdomain and entire domain bases was studied in Section 6.3. The condition number of the corresponding matrix is compared in Table 6.3 for linear bases, power bases, and eigenfunction bases.

It can be seen from Table 6.3 that the condition number of matrix increases very rapidly with the size of matrix for entire domain basis whereas it increases slowly for linear basis. The matrix solution is therefore more stable for subdomain basis. This is another reason for the preference of subdomain basis over entire domain basis functions.

A study of charge distribution on a perfectly conducting wire is carried out in Chapter 11. The variation of $\|\Delta f\|/\|f\|$ and $k(A)$ was studied as a function of order of matrix N. The data is compared in Table 6.4 for pulse and entire domain basis functions. The vector f is known analytically and is given by $f(x) = 1/\sqrt{1 - x^2}$. $k(A)$ and $\|\Delta f\|$ are obtained from MoM using point matching. It may be noted that the matrix condition number k increases with N although slowly for pulse basis and rapidly for entire domain basis. However, the error in the unknown function is found to be similar for the basis functions considered.

The stability of the matrix solution is linked to the eigenvalues of the matrix. The matrix is a discretized version of the operator. If the eigenvalue of the operator vanishes, the eigenvalue of the corresponding matrix may not be zero because of discretization error but is expected to be very small. When this occurs, the matrix is nearly singular and cannot be inverted or factored into LU form [2, p. 235]. The numerical solution under these conditions becomes less stable. Therefore, large condition number alone may not give rise to instability of the solution; *the corresponding eigenvalue of the matrix must be nearly zero*. The study of charge distribution on wire using 20 entire domain basis functions produces a stable

Table 6.3 Condition Number of Matrix Operator
$L = d^2/dx^2$ for Various Types of Basis Functions and Matrix Size

Matrix Size N	Linear Bases	Power Bases	Eigenfunction Bases
6	10.8	9.96e07	36
8	20.3	1.067e11	64
10	33.1	1.15e14	100

Table 6.4 Effect of Increase of Matrix Size N on the Solution Accuracy and the Matrix Condition Number $k(A)$

N	Pulse Basis $\|\Delta f\|/\|f\|$	$k(A)$	Entire Domain Basis $\|\Delta f\|/\|f\|$	$k(A)$
5	0.2436	1.72	0.247	63.84
10	0.3265	2.27	0.3308	1.743×10^4
15	0.3653	2.78	0.3715	5.427×10^6
20	0.3899	3.29	0.3969	2.274×10^8

solution even when the condition number of the matrix was of the order of 10^8. However, the use of entire domain basis functions in Example 11.1 produced erroneous solution for $N > 9$.

6.5 Accuracy of Numerical Solutions

The question of accuracy of the solution is very important if it is to be reliable and useful. The accuracy of the approximate solution determines its closeness to the exact solution, assuming it exists. There are three principal sources of errors in computational methods. These are [12]:

1. Modeling error;
2. Truncation error;
3. Round-off error.

Each of these error types affects the accuracy and therefore degrades the solution.

6.5.1 Modeling Errors

The modeling errors arise due to the simplifying assumptions made in arriving at the mathematical model of the problem. For example, an irregular boundary may be represented by a stair-step approximation. This type of modeling error may be reduced by using a finer mesh.

6.5.2 Truncation Error

The truncation errors may arise due to the fact that an infinite series is implemented numerically by a finite number of terms. For example, in series expansion for a function we take a finite number of terms due to the finite computer memory. This introduces truncation error. Truncation errors may be reduced by using a large number of terms.

6.5.3 Round-Off Error

The round-off error symbolizes the fact that the computations can be carried out on the computer with a finite precision only. The source of error in this case is the limited size of the registers in the arithmetic unit of the computer. This error is also termed *finite word length error*. The round-off error can be reduced by using double-precision arithmetic. An alternative is to code all mathematical operations using integer arithmetic. This is hardly possible in most practical situations. The round-off error in MATLAB (double precision) is 10^{-16}.

It may appear from the above description that an increased accuracy of the solution may be achieved by using a finer mesh to reduce discretization and modeling errors. However, there is a lower limit to the mesh size. This limit is set by the round-off error. As we reduce the mesh size, the number of mathematical operations increase and therefore an increase in the round-off error. A point is reached where the total error (sum of truncation error and round-off error) becomes minimum. This point is illustrated qualitatively in Figure 6.17. The lower limit on the total error and the optimum step size is set by the algorithm used and the word length.

An important feature of the matrix methods is that the condition number of the matrix $k(A)$ increases with the order of matrix as the solution attains better accuracy, so much so that finally the matrix may become singular. Therefore, although the condition number of the matrix $k(A)$ is an indicator of the sensitivity of the solution to precision and other errors, it is also a good indicator of the accuracy of the solution. Therefore, a large condition number indicates a better accuracy provided the solution is not nonphysical. This argument follows from the expression

$$k(A) \geq \frac{\|\Delta x\|}{\|x\|} \cdot \frac{\|A\|}{\|\Delta A\|} \tag{6.56}$$

6.5.4 Validation

Once the stability and accuracy issues are addressed, the following tests may be carried out before accepting the results of computational methods:

- *Validation of the software:* For validation, a problem with the known solution should be solved on the software developed. The accuracy of the computed solution can also be estimated by comparing with the known solution.
- *Validation of the solution:* The exact solution for the problem may not be known for comparison and validation. However, the parameters of the given problem may be modified so that it reduces to a problem whose solution is known. The solution of this problem is carried out and compared with the known solution. If the results agree, then it is expected that the results for the given class of problems produced by the software will also be correct. This type of validation may not always be possible.

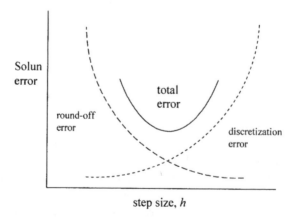

Figure 6.17 Variation of round-off error, discretization error, and the total error in the numerical solution with the step size.

6.6 Spurious Solutions

Sometimes we find that the numerical solution we have obtained after going through the rigorous checks is not the proper solution; that is, it is not consistent with physical reasoning. This type of solution is called the spurious/false solution. There are many reasons for this situation. The principal reason is the deviation of the computer implementation of the solution from the rigorous theoretical formulation. Some of the issues are discussed next.

When the proper boundary conditions are imposed using analytical functions, the unique solutions are generated. However, in computational methods the boundary conditions are enforced in an average sense. The approximate satisfaction of the boundary conditions is not sufficient to guarantee correct solutions [13].

In FDM and FDTD methods, the partial differential equations are replaced by the difference equations. Improper replacement may lead to spurious solutions. For example, in our endeavor to improve accuracy we may use difference equations which are of higher order than the differential equations they represent.

The spurious solutions in FEM were observed while solving the eigenvalue problem for material discontinuities using node-based elements. These solutions were found to be related to the property that node-based elements impose continuity of tangential and normal components of field across the interface; whereas normal component of field is not continuous at the interface, thus violating interface condition, (1.13c) and (1.13d). Edge-based elements, on the other hand, are found to be free from spurious modes [9, p. 198].

For a given excitation, a number of modes may get excited; the contribution of each of these modes to the solution depends on the coefficient of excitation and coupling between them. If any of the excited modes is a spurious mode, the overall solution may become erroneous depending on its contribution. It becomes difficult to identify the spurious modes from the solution. Therefore, the eigenvalue analysis of the system may be carried out to identify the spurious modes among them.

6.7 Formulations for the Computational Methods

The computational methods provide us the approximate numerical solution of the given boundary value problem. Two methods of formulating such an approximate solution are the Ritz or Raleigh-Ritz method, and the method of weighted residuals. The method of weighted residuals is discussed in Chapters 9 and 11 and is employed in the implementations of MoM and FEM.

In the Ritz method, functions f are sought which make a functional stationary. This is equivalent to solving the operator equation $L(f) = g$. The MoM approach can also be fitted into this formalism. The Ritz method is discussed in Chapter 9.

Some of the useful texts on computational methods include [1–3, 9, 10, 13–20].

6.8 Summary

The major steps in the analysis of boundary value problems based on computational methods include development of governing equations, discretization of device geometry into subdomains, application of governing equation to each of the sub-domains, and the solution of resulting simultaneous equations. This process is equivalent to expressing the unknown function as a sum of known functions with unknown coefficients of expansion. The various computational methods differ in the choice of expansion functions and in the determination of coefficients. Subdomain expansion functions are employed to make the solution technique applicable to arbitrary shaped geometry and dielectric configuration. The various types of subdomain and entire domain expansion functions are described. Single-term basis functions may be developed by fitting the expected distribution. The basic characteristics of the solution such as convergence and discretization error are discussed with numerical integration as an example. The phase error as a function of discretization size in FDM, FDTD, and FEM is compared. The cumulative phase error might not be permissible with certain phase sensitive devices such as low sidelobe antennas. The phase error may be reduced by employing higher order basis functions. The stability of the solution is described in terms of causality for FDTD, and matrix condition number for matrix based solutions such as FDM, FEM, and MoM. The matrix representation of operator with respect to various types of basis functions is discussed. The condition number of the operator matrix increases slowly if the subdomain basis functions are employed. The origin of spurious solutions is briefly discussed.

References

[1]   Harrington, R. F., *Field Computation by Momemt Methods*, Malabar, FL: R.E. Kreiger Publishing Co., 1982.

[2]   Peterson, A. F., S. L. Ray, and R. Mittra, *Computational Methods for Electromagnetics*, New York: IEEE Press, 2001.

[3]   Itoh, T., (ed.), *Numerical Techniques for Microwave and Millimeter-Wave Passive Structures*, New York: Wiley, 1989, Chapter 5.

[4]   Garg, R., et al., *Microstrip Antenna Design Handbook*, Norwood, MA: Artech House, 2000, Chapter 2.

[5]   Atkinson, K. E., *An Introduction to Numerical Analysis,* New York: Wiley, 1989.

[6]   Green, H. E., "The Numerical Solution of Some Important Transmission Line Problems," *IEEE Trans. Microwave Theory Tech.*, Vol. MTT-13, 1965, pp. 676–692.

[7]   Culver, R. C., "The Use of Extrapolation Techniques with Electrical Network Analogue Solutions," *Brit. J. Appl. Phys.*, Vol. 3, 1952, pp. 376–378.

[8]   http://documents.wolfram.co.jp/mathematica/book/section-3.5.

[9]   Jin, J., *The Finite Element Method in Electromagnetics*, 2nd ed., New York: Wiley, 2002.

[10]   Bondeson, A., T. Rylander, and P. Ingelstrom, *Computational Electromagnetics*, Berlin: Springer, 2005.

[11]   Iskander, M. F., et al., "A New Course on Computational Methods in Electromagnetics," *IEEE Trans. Education*, Vol. 31, 1988, pp. 101–114.

[12]   Stutzman, W. L., and G. A. Thiele, *Antenna Theory and Design*, New York: John Wiley, 1981.

[13]   Collin, R. E., *Field Theory of Guided Waves*, 2nd ed., New York: IEEE Press, 1991.

[14]   Volakis, J. L., A. Chatterjee, and L. C. Kempel, *Finite Element Method for Electromagnetics*, Hyderabad: Universities Press (India), 2001.

[15]   Booton, R. C., Jr., *Computational Methods for Electromagnetics and Microwaves*, New York: John Wiley, 1992.

[16]   Chari, M. V. K., and S. J. Salon, *Numerical Methods in Electromagnetism*, New York: Academic Press, 2000.

[17]   Sadiku, M. N. O., *Numerical Techniques in Electromagnetics*, 2nd ed., Boca Raton, FL: CRC Press, 2001.

[18]   Davidson, D. B., *Computational Electromagnetics for RF and Microwave Engineering*, New York: Cambridge University Press, 2005.

[19]   Taflove, A., and S. C. Hagness, *Computational Electrodynamics: The Finite-Difference Time-Domain Method*, 3rd ed., Norwood, MA: Artech House, 2005.

[20]   Sullivan, D. M., *Electromagnetic Simulation Using the FDTD Method*, New York: IEEE Press, 2000.

Problems

P6.1. Wire dipole antenna is analyzed in Section 11.2 using pulse expansion and point matching. Assume the following parameters for the dipole antenna: $\lambda = 1$m, $a = 0.005\lambda$, $l = 0.47\lambda$. The value of input resistance R_{in} as a function of the number of segments N is given in the table below. Perform the ratio test (6.29) on the last three values and show that $R = 1.0071$. Implement (6.34) and show that the extrapolated value from the last three data is found to be $R_{in} = 82.43$ ohms. The exact value for the dipole is near 80.0 ohms.

Number of Divisions N	R_{in}
21	47.11
41	67.92
61	76.17
81	78.94

P6.2 The sine integral is defined as

$$Si(x) = \int_0^x \frac{\sin t}{t}\, dt$$

The value of $Si(2)$ for various divisions of the range of integration is given in the table below for mid-point and Simpson's rules of integration. Determine the rate of convergence p and the asymptotic value of $Si(2)$. Comment on the poor performance of the mid-point rule of integration.

N	h	I (Mid-point)	I (Simpson)
5	0.40	1.3046	1.5996
10	0.20	1.3021	1.6040
20	0.10	1.3015	1.6051
40	0.05	1.3014	1.6053

I (exact) = 1.60541

P6.3. The MoM analysis of strip line is carried out in Section 11.2. Pulse expansion and point matching is employed to solve the integral equation. The expression for the matrix element is given in (11.85) and coded in the software *stripline.m*. Determine the condition number of the matrix for the integral operator and plot it as a function of N for $N = 8:8:32$. Comment on the stability of the matrix solution.

P6.4. Consider discretizing the operator $L = d^2/dx^2$ by employing quadratic spline functions of (6.9) as basis functions. Show that the following expression for the matrix element is obtained in each case:

1. Point testing

$$l_{mn} = \begin{cases} -2/\Delta^2 & \text{for } m = n \\ 1/\Delta^2 & \text{for } |m - n| = 1 \\ 0 & \text{for } |m - n| > 1 \end{cases}$$

2. Pulse testing

$$l_{mn} = \begin{cases} -2/\Delta & \text{for } m = n \\ 1/\Delta & \text{for } |m - n| = 1 \\ 0 & \text{for } |m - n| > 1 \end{cases}$$

3. Triangle testing

$$I_{mn} = \begin{cases} -5/(4\Delta) & \text{for } m = n \\ 1/(2\Delta) & \text{for } |m - n| = 1 \\ 1/(8\Delta) & \text{for } |m - n| = 2 \\ 0 & \text{for } |m - n| > 2 \end{cases}$$

P6.5. The potential distribution in a rectangular trough is analyzed in Section 7.1 using finite difference method. The computed data and the exact theoretical solution, (7.20), is given there for 6×6 discretization of the trough. Use this data to determine $\|\varphi_{exact}\|$ and the error $\|\varphi_{fdm} - \varphi_{exact}\|$.

CHAPTER 7

Method of Finite Differences

Many engineering problems in electromagnetics may be formulated as partial differential equations (PDE). This may include electrostatic problems formulated as Laplace or Poisson equations; or dynamic problems such as wave equations. Only a few of the useful geometries can be solved using analytical methods. Computers may be used to directly solve the boundary value problems to include the effects of dielectric inhomogeneities and irregular boundary. For the direct solution of PDE, the differential equations should be expressed in a form suitable for computers. The difference form of the derivatives is most suitable and can be programmed in a computer. The method of finite differences is based on the solution of PDE when expressed in finite difference form. This method is least analytical and therefore easy to follow. It is a versatile method, but its efficiency is poor. This method is taught to the undergraduate students as a part of numerical solution of PDE. With this preparation, the students can be easily introduced to the computational methods. A very good exposition of finite difference method (FDM) with its applications in electromagnetics is available in [1–4].

7.1 Finite Difference Approximations

The method of finite differences involves the following major steps:

1. The difference form of the derivatives is used to express the differential equation in the finite difference form, and the governing equation obtained.
2. The device is discretized in the form of cells, creating a number of nodes.
3. The finite difference form of the differential equation is applied at each and every node creating a number of linear simultaneous equations.
4. The set of simultaneous equations is solved to determine the unknown nodal potential.
5. The potential distribution may be used to derive engineering-important characteristics like capacitance, characteristic impedance, cutoff wavelength, resonant frequency, and so on.

7.1.1 Difference Form of the First Derivative

Consider the function $f(x)$ in Figure 7.1. The derivative of the function at x_0 is defined as

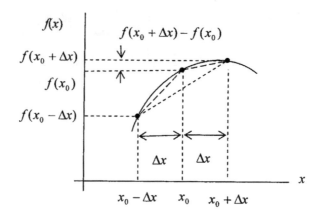

Figure 7.1   Illustration for the difference form of the first derivative.

$$f'(x_0) = \lim_{x \to x_0} \frac{f(x) - f(x_0)}{x - x_0} \tag{7.1}$$

Although this definition is exact, it is not computer friendly. The value of the function might not be available for all values of x, but only at discrete points. For this reason, we may *approximate* the derivative as follows:

$$f'(x_0) = \lim_{\Delta x \to 0} \frac{f(x_0 + \Delta x) - f(x_0)}{\Delta x} \approx \frac{f(x_0 + \Delta x) - f(x_0)}{\Delta x} \tag{7.2a}$$

where Δx is the discretization length along x. Expression (7.2a) is called the *forward difference approximation* of the derivative. This form of derivative is approximate when Δx is nonzero. Another possibility is

$$f'(x_0) = \lim_{\Delta x \to 0} \frac{f(x_0) - f(x_0 - \Delta x)}{\Delta x} \approx \frac{f(x_0) - f(x_0 - \Delta x)}{\Delta x} \tag{7.2b}$$

This expression is called the *backward difference approximation* of the derivative. Still another possibility is

$$f'(x_0) = \lim_{\Delta x \to 0} \frac{f(x_0 + \Delta x) - f(x_0 - \Delta x)}{2\Delta x} \approx \frac{f(x_0 + \Delta x) - f(x_0 - \Delta x)}{2\Delta x}$$
$$\tag{7.2c}$$

This equation is called the *central difference approximation* of the derivative.

The three different forms of the first derivative are found to give varying amount of approximation to the derivative for nonzero value of Δx. Let us first determine the approximation involved. Consider the Taylor series expansion of $f(x)$ about the point x_0,

$$f(x_0 \pm \Delta x) = f(x_0) \pm \Delta x \left.\frac{df}{dx}\right|_{x_0} + \frac{(\Delta x)^2}{2}\left.\frac{d^2 f}{dx^2}\right|_{x_0} \pm \frac{(\Delta x)^3}{3!}\left.\frac{d^3 f}{dx^3}\right|_{x_0} + \cdots \quad (7.3)$$

Using this expansion in the forward difference approximation (7.2a) gives

$$\frac{f(x_0 + \Delta x) - f(x_0)}{\Delta x} = \left.\frac{df}{dx}\right|_{x_0} + \frac{\Delta x}{2}\left.\frac{d^2 f}{dx^2}\right|_{x_0} + \frac{(\Delta x)^2}{3!}\left.\frac{d^3 f}{dx^3}\right|_{x_0} + \cdots \quad (7.4a)$$

The second, third, and subsequent terms on the right side represent the error terms for $\Delta x \neq 0$. Similarly, the backward difference approximation gives

$$\frac{f(x_0) - f(x_0 - \Delta x)}{\Delta x} = \left.\frac{df}{dx}\right|_{x_0} - \frac{\Delta x}{2}\left.\frac{d^2 f}{dx^2}\right|_{x_0} + \frac{(\Delta x)^2}{3!}\left.\frac{d^3 f}{dx^3}\right|_{x_0} + \cdots \quad (7.4b)$$

Adding (7.4a) and (7.4b) gives the following expression for the central difference approximation [1–7]:

$$\frac{f(x_0 + \Delta x) - f(x_0 - \Delta x)}{2\Delta x} = \left.\frac{df}{dx}\right|_{x_0} + \frac{(\Delta x)^2}{3!}\left.\frac{d^3 f}{dx^3}\right|_{x_0} + \cdots \quad (7.4c)$$

It may be observed that the error in finite difference approximation is of the order of Δx in (7.4a) and (7.4b), and it is of the order of $(\Delta x)^2$ for the central difference approximation, (7.4c). The error is lower in case of (7.4c), because $\Delta x < 1$. The error is called discretization error and may be reduced for the same step size by using more nodal values about the central node [4]. Care must be exercised in applying the higher-order approximations in finite difference methods because the resulting expression may represent higher order derivative than the intended second order derivative and may lead to spurious solution.

7.1.2 Difference Form of the Second Derivative

Solutions of Laplace, Poisson, or wave equations, using the method of finite differences, require the knowledge of the difference form of the second derivative and the associated errors. In order to find the difference approximation for $d^2 f/dx^2$, let us consider three nodes in the grid at x_0, $x_0 + \Delta x$, and $x_0 - \Delta x$, as shown in Figure 7.1. Using the Taylor expansion (7.3) gives

$$f(x_0 + \Delta x) + f(x_0 - \Delta x) = 2f(x_0) + (\Delta x)^2 \left.\frac{d^2 f}{dx^2}\right|_{x_0} + 2\frac{(\Delta x)^4}{4!}\left.\frac{d^4 f}{dx^4}\right|_{x_0} + \cdots$$

or

$$\left.\frac{d^2f}{dx^2}\right|_{x_0} = \frac{f(x_0 + \Delta x) - 2f(x_0) + f(x_0 - \Delta x)}{(\Delta x)^2} - \frac{(\Delta x)^2}{12}\left.\frac{d^4f}{dx^4}\right|_{x_0} + \ldots \quad (7.5)$$

It is obvious from above that if d^2f/dx^2 is approximated as

$$\left.\frac{d^2f}{dx^2}\right|_{x_0} \approx \frac{f(x_0 + \Delta x) - 2f(x_0) + f(x_0 - \Delta x)}{(\Delta x)^2} \quad (7.6a)$$

the error in (7.6a) is of the order of Δx^2; that is, $O(\Delta x^2)$. Similarly, for the function $f(x, y)$ of two independent variables x and y, the derivative $\partial^2 f/\partial y^2$ may be approximated as

$$\frac{\partial^2 f}{\partial y^2} \approx \frac{f(x_0, y_0 + \Delta y) - 2f(x_0, y_0) + f(x_0, y_0 - \Delta y)}{(\Delta y)^2} \quad (7.6b)$$

7.1.3   Difference Form of Laplace and Poisson Equations

Let us consider the solution of electrostatic problems governed by the Laplace equation,

$$\nabla^2 \varphi = 0 \quad (7.7)$$

or

$$\frac{\partial^2 \varphi}{\partial x^2} + \frac{\partial^2 \varphi}{\partial y^2} = 0 \quad \text{in two-dimensions} \quad (7.8)$$

To derive the difference form of (7.8), let us consider four nodes surrounding the central node in a square grid with node spacing h as shown in Figure 7.2. The value of the function $\varphi(x, y)$ at the surrounding nodes are φ_1, φ_2, φ_3, and φ_4. Using (7.5) for the second order derivative along the x-direction, we have

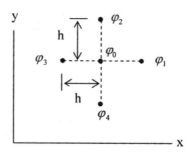

Figure 7.2   Nodes in a square grid.

$$\frac{\partial^2 \varphi}{\partial x^2} = \frac{\varphi_1 - 2\varphi_0 + \varphi_3}{h^2} - \frac{h^2}{12}\frac{\partial^4 \varphi}{\partial x^4} + \dots \qquad (7.9a)$$

Similarly, the second order derivative along the y-direction is given by

$$\frac{\partial^2 \varphi}{\partial y^2} = \frac{\varphi_2 - 2\varphi_0 + \varphi_4}{h^2} - \frac{h^2}{12}\frac{\partial^4 \varphi}{\partial y^4} + \dots \qquad (7.9b)$$

Add (7.9a) and (7.9b), use Laplace equation (7.8), and multiply throughout by h^2 to obtain

$$\varphi_0 = \frac{1}{4}(\varphi_1 + \varphi_2 + \varphi_3 + \varphi_4) - \frac{h^4}{48}\left[\frac{\partial^4 \varphi}{\partial x^4} + \frac{\partial^4 \varphi}{\partial y^4}\right] + \dots \qquad (7.10)$$

If the value at the central node is approximated by the first term on the right side, the last term contributes to the error due to the nonzero value of h. The error is of the order of h^4. We may approximate (7.10) as

$$\varphi_0 \approx \frac{1}{4}(\varphi_1 + \varphi_2 + \varphi_3 + \varphi_4) \qquad (7.11)$$

Equation (7.11) may be generalized for any node (i, j) as

$$\varphi(i, j) \approx \frac{1}{4}[\varphi(i + 1, j) + \varphi(i, j + 1) + \varphi(i - 1, j) + \varphi(i, j - 1)] \qquad (7.12)$$

Equation (7.12) is the governing equation for the finite difference analysis of Laplace equation. It is also called the five-point finite difference approximation of Laplace equation. It implies that the potential at the central node is the average of the potentials at the four neighboring nodes, and *may be used to determine the potential distribution for a problem satisfying Laplace equation.* For this, (7.12) is written for each and every node in the problem space. Arranging these equations gives a set of simultaneous equations which can be solved for the unknowns. Some of these equations may be eliminated by the boundary conditions. The set of equations is singular unless the potential of at least one node is specified.

The Poisson equation defined as

$$\frac{\partial^2 \varphi}{\partial x^2} + \frac{\partial^2 \varphi}{\partial y^2} = \rho(x, y) \qquad (7.13)$$

may also be discretized in the same way as the Laplace equation to obtain the following approximate expression for the potential at the central node:

$$\varphi_0 \approx \frac{1}{4}(\varphi_1 + \varphi_2 + \varphi_3 + \varphi_4 - \rho(x_0, y_0)h^2) \qquad (7.14)$$

For a *rectangular* grid with node spacing Δx along the x-direction and Δy along the y-direction, the discretized form of the Poisson equation for the potential at the central node at (i, j) may be written as

$$\varphi_{i,j}\left(\frac{\Delta y}{\Delta x} + \frac{\Delta x}{\Delta y}\right) \approx \frac{1}{2}\left[\frac{\Delta y}{\Delta x}(\varphi_{i+1,j} + \varphi_{i-1,j}) + \frac{\Delta x}{\Delta y}(\varphi_{i,j+1} + \varphi_{i,j-1}) - \rho(x_i, y_j)\Delta x\, \Delta y\right]$$

$$(7.15)$$

This expression reduces to (7.14) for $\Delta x = \Delta y = h$.

Quiz 7.1. The discretization error associated with the approximation $f'\left(x_0 + \dfrac{h}{2}\right) \approx \dfrac{f(x_0 + h) - f(x_0)}{h}$ is of the order of

(a) h
(b) h^2
(c) None of the above

Answer. (b). The given expression is central difference approximation about the point $(x_0 + h/2)$. Or expand $f(x_0 + h)$ and $f(x_0)$ about $f(x_0 + h/2)$.

Example 7.1. The geometry in Figure 7.3 consists of an infinitely long square metal pipe with insulation at the corners so that the different faces of the pipe can be raised to different potentials. We would like to determine the potential distribution, V inside this pipe.

Since the geometry is invariant along the length, we expect the potential distribution also to be invariant along the length. Therefore, the problem reduces to a two-dimensional problem in the cross-sectional plane, xy.

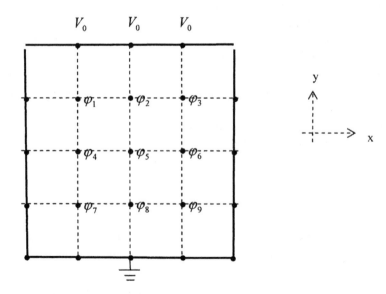

Figure 7.3 Geometry of a square pipe showing grid lines, nodes, and boundary conditions.

The inside of the pipe is discretized in square cells of size $h \times h$. The nodes are introduced at the crossings of grid lines. There are 23 nodes in Figure 7.3, with 9 nodes distributed inside the pipe and 14 on the surfaces. The nodes on the surfaces are called fixed nodes because the potential on these nodes is fixed by the boundary conditions. The internal nodes are called free nodes because the potential at these nodes is free to adjust to satisfy the boundary conditions. The governing Laplace equation should be satisfied at each and every point of the geometry. Let us apply the discretized version of Laplace equation, (7.12), at the various nodes of Figure 7.3. Next the boundary conditions due the voltage at the four conducting strips constituting the pipe are applied. Let these potentials be: 0V, 0V, 0V, and V_0. Since the potentials on the nodes residing at the four surfaces is specified, we need not write (7.12) for these nodes. This way we are left with only nine nodes, distributed inside the pipe, at which the potential is to be determined. The symmetry of the pipe and the boundary conditions about the mid-plane parallel to the y-axis indicate that $\varphi_1 = \varphi_3$, $\varphi_4 = \varphi_6$, and $\varphi_7 = \varphi_9$, leaving six unknowns, φ_1, φ_2, φ_4, φ_5, φ_7, and φ_8. Implementing (7.12) at the six nodes gives the following equations:

$$4\varphi_1 = 0 + V_0 + \varphi_2 + \varphi_4 \qquad (7.16a)$$

$$4\varphi_2 = \varphi_1 + V_0 + \varphi_1 + \varphi_5, \qquad \text{because } \varphi_3 = \varphi_1 \qquad (7.16b)$$

$$4\varphi_4 = 0 + \varphi_1 + \varphi_5 + \varphi_7 \qquad (7.16c)$$

$$4\varphi_5 = \varphi_4 + \varphi_2 + \varphi_4 + \varphi_8, \qquad \text{because } \varphi_6 = \varphi_4 \qquad (7.16d)$$

$$4\varphi_7 = 0 + \varphi_4 + \varphi_8 + 0 \qquad (7.16e)$$

$$4\varphi_8 = \varphi_7 + \varphi_5 + \varphi_7 + 0, \qquad \text{because } \varphi_9 = \varphi_7 \qquad (7.16f)$$

The above set of equations may be expressed in the form of a matrix equation with known quantities transferred to the right side. One obtains

$$\begin{bmatrix} 4 & -1 & -1 & 0 & 0 & 0 \\ -2 & 4 & 0 & -1 & 0 & 0 \\ -1 & 0 & 4 & -1 & -1 & 0 \\ 0 & -1 & -2 & 4 & 0 & -1 \\ 0 & 0 & -1 & 0 & 4 & -1 \\ 0 & 0 & 0 & -1 & -2 & 4 \end{bmatrix} \begin{bmatrix} \varphi_1 \\ \varphi_2 \\ \varphi_4 \\ \varphi_5 \\ \varphi_7 \\ \varphi_8 \end{bmatrix} = \begin{bmatrix} V_0 \\ V_0 \\ 0 \\ 0 \\ 0 \\ 0 \end{bmatrix} \qquad (7.17)$$

Solution Methods for the Set of Simultaneous Equations
The application of the finite difference method to the Laplace equation has led to a set of simultaneous equations as described above. Their efficient solution is a major problem in itself. For matrices of small sizes, one may use a direct approach in the form of matrix inversion or by using the Gauss elimination method (see Appendix A). The indirect or iterative methods are more efficient for matrices which are diagonally dominant, sparse, and large. The above matrix is diagonally dominant and sparse.

Iterative Methods

An iterative method is one in which a first approximation is used to calculate a second approximation, which in turn is used to calculate the third approximation, and so on until a specified tolerance is attained. The three common iterative methods are: Jacobi, Gauss-Seidel, and successive over-relaxation (SOR). We discuss the implementation of SOR next.

The value at the free nodes may be set equal to zero to start with in the absence of any educated guess. Now consider node φ_1 and apply (7.12) as in (7.16a). We raster scan, from top to bottom and left to right, through all the nodes at which the potential is to be determined. This procedure is iterated a number of times such that the potential at a node does not change much in the next iteration and is within a specified tolerance. For example, if the specified tolerance is 1%, then the iteration is continued until

$$\left| \varphi_{i,j}^{n+1} - \varphi_{i,j}^{n} \right| < \left| 0.01 \varphi_{i,j}^{n} \right| \tag{7.18}$$

where the superscript n denotes the iteration number. The problem of Example 7.1 was solved using this method. It was found that after six iterations

$$\varphi_1 = 0.4274 V_0, \; \varphi_2 = 0.5256 V_0, \; \varphi_4 = 0.1863 V_0 \tag{7.19}$$

$$\varphi_5 = 0.2488 V_0, \; \varphi_7 = 0.0708 V_0, \; \varphi_8 = 0.0976 V_0$$

The accuracy of this solution can be determined by comparing it with the values given by the exact solution [8],

$$\varphi(x, y) = \frac{2 V_0}{\pi} \sum_{n=1}^{\infty} [1 - (-1)^n] \frac{\sinh\left(\dfrac{n \pi y}{a}\right)}{n \sinh\left(\dfrac{n \pi b}{a}\right)} \sin\left(\dfrac{n \pi x}{a}\right) \tag{7.20}$$

where a is the width and b is the height of the pipe. The values obtained from (7.20) are as follows:

$$\varphi_1 = 0.4320 V_0, \; \varphi_2 = 0.5405 V_0, \; \varphi_4 = 0.1820 V_0$$

$$\varphi_5 = 0.2500 V_0, \; \varphi_7 = 0.0680 V_0, \; \varphi_8 = 0.0954 V_0$$

The difference between the analytical solution and the finite difference solution is found to be about 4%. The error can be reduced by reducing the cell size h, that is, by increasing the number of nodes inside the pipe.

Successive Over-Relaxation (SOR)

The simplistic structure of the FDM needs very fine discretization for accurate results. The number of nodes, therefore, becomes very large, resulting in large computer storage and the number of iterations. The iterations can be reduced by using SOR. The relaxation formula for the new value $\varphi_{i,j}^{n+1}$ in terms of the old value $\varphi_{i,j}^{n}$ is given by

$$\varphi_{i,j}^{n+1} = \varphi_{i,j}^{n} + \Omega_0 R_{i,j}^{n+1} \tag{7.21a}$$

where Ω_0 is the relaxation factor which determines the rate of convergence and $R_{i,j}^{n+1}$ is the residual at each node and is defined as

$$R_{i,j}^{n+1} = \frac{1}{4}\left(\varphi_{i+1,j}^{n} + \varphi_{i-1,j}^{n+1} + \varphi_{i,j-1}^{n+1} + \varphi_{i,j+1}^{n}\right) - \varphi_{i,j}^{n} \tag{7.21b}$$

It may be noted that updated nodal values are included for the nodes surrounding node (i, j). The residual $R_{i,j}^{n+1}$ may be regarded as a correction, which must be added to $\varphi_{i,j}^{n}$ to make it nearer to the correct value. As convergence to the correct value is approached, $R_{i,j}^{n+1}$ tends to zero. The process of successive correction of the nodal value for $\Omega_0 = 1$ is called relaxation. An optimum value of Ω_0 may reduce the number of iterations considerably. The optimum value of Ω_0 is determined by trial and error and is found to lie in general between 1 and 2 [3]. For the rectangular cells the optimum relaxation factor is given by the root of the quadratic equation [9]

$$\Omega_0^2 r^2 - 16\Omega_0 + 16 = 0 \tag{7.22a}$$

where $r = \cos(\pi/N_x) + \cos(\pi/N_y)$, and N_x and N_y are the number of cells per unit length along the x- and y-directions, respectively. Therefore,

$$\Omega_0 = \frac{8 - \sqrt{64 - 16r^2}}{r^2} \tag{7.22b}$$

It has been found that the optimum value Ω_0 holds for sufficiently large node density.

The trough problem of Example 7.1 is now analyzed for a large number of nodes. For this, a software *trough.m* is developed. This software employs an iterative method for the solution of the matrix equation. SOR is used to improve efficiency. The number of iterations as a function of Ω for 1% tolerance is plotted in Figure 7.4. The node density $N = N_x = N_y = 30$.

The software determines the optimum value of Ω from the number of iterations and compares this value with the value given by (7.22b). The optimum value in this case is found to be 1.8170, whereas the analytical value is 1.8107. The number of iterations required is 14. The equipotential contours are plotted in Figure 7.5. As expected, the potential decreases away from the top plate and is symmetric about the mid-plane because of the symmetry of the boundary conditions.

Exercise.   Determine the optimum value of Ω as a function of node density for square cells using software *trough.m*. Assume $N > 20$ and tolerances of 1% and 0.1%. Compare the values with those obtained from (7.22b).

Quiz 7.2.   In Example 7.1, the potential distribution is symmetric about the y-axis. It is due to:

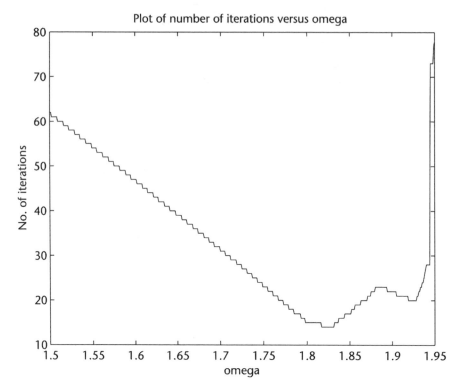

Figure 7.4 Plot of number of iterations as a function of relaxation factor, Ω. The node density N: 30 and tolerance: 1%.

(a) The symmetry of the trough about the y-axis
(b) The symmetry of the boundary conditions about the y-axis
(c) Both (a) and (b)
(d) None of the above

Prove your point through simulations after making suitable changes in the problem.

Answer.   (c).

Quiz 7.3.   In Example 7.1, the potential distribution depends on the discretization size h. In other words, the potential distribution changes with change in the value of h. Also, the boundary condition is satisfied at different number of points for different values of h. The change in potential value at a common node is:

(a) Due to the change in discretization error
(b) Due to the modified boundary conditions
(c) Due to both (a) and (b)
(d) None of the above

Please choose the correct entries and explain why.

Answer.   (c).

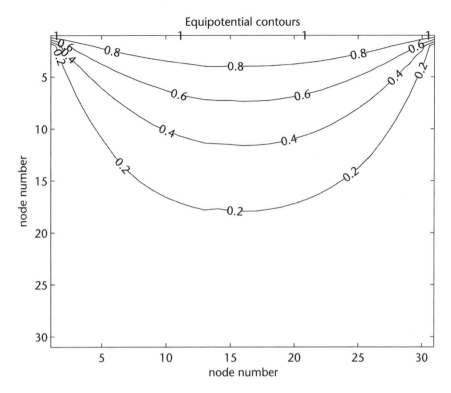

Figure 7.5   Equipotential contours for the trough. Node density N: 30, tolerance: 1%.

7.2   Treatment of Interface and Boundary Conditions

One of the reasons why we use computational methods is the ease with which the material inhomogeneity and the complicated geometries can be handled. These may give rise to nodes at the dielectric interface, nodes at the corner, and nodes on the edge [1]. The finite difference equations for these nodes are different from the equation for the node in the interior. Let us first consider the nodes on an interface between two dielectrics.

7.2.1   Nodes on the Interface

The dielectric interface occurs in geometries with dielectric inhomogeneities such as planar transmission lines and partially filled waveguides. The analysis of these geometries invariably involves placing nodes on the interface. Consider Figure 7.6 with nodes at the interface of two dielectrics ϵ_1 and ϵ_2. Let us determine the finite difference equation for the central node with assumed potential φ_0. If there is no charge on the interface, then the application of Gauss's law will give rise to an equation for φ_0 in terms of the potentials at the nodes in the neighborhood. Using the dotted line cell as the Gaussian surface, we determine the flux leaving this surface. The Gauss law for the electric field states that the electric flux leaving a Gaussian surface is equal to the free charge enclosed by the surface [2]; that is,

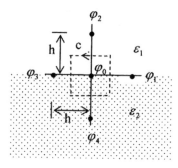

Figure 7.6 Nodes at the interface of two dielectrics.

$$\oiint_{s} \epsilon E . ds = q = 0 \tag{7.23}$$

Since there is no *free* charge enclosed by the Gaussian surface, q is set equal to zero. Substituting $E = -\nabla \varphi$ gives

$$-\oint_{c} \epsilon \nabla \varphi . dc = 0 \tag{7.24}$$

where the two-dimensional closed surface has been replaced by the closed contour c in Figure 7.6. Denoting the derivative of φ on the contour by $\partial \varphi / \partial n$, we obtain

$$-\oint_{c} \epsilon \frac{\partial \varphi}{\partial n} dc = 0 \tag{7.25}$$

where n is the unit *outward* normal to the contour. The flux leaving the right side of the contour is

$$\psi_R = \epsilon_1 \frac{\varphi_1 - \varphi_0}{h} \frac{h}{2} + \epsilon_2 \frac{\varphi_1 - \varphi_0}{h} \frac{h}{2} \tag{7.26a}$$

Similarly, the flux leaving the left side is

$$\psi_L = \epsilon_1 \frac{\varphi_3 - \varphi_0}{h} \frac{h}{2} + \epsilon_2 \frac{\varphi_3 - \varphi_0}{h} \frac{h}{2} \tag{7.26b}$$

The flux leaving the top side is

$$\psi_T = \epsilon_1 \frac{\varphi_2 - \varphi_0}{h} h \tag{7.26c}$$

and the flux leaving the bottom side is

$$\psi_B = \epsilon_2 \frac{\varphi_4 - \varphi_0}{h} h \qquad (7.26d)$$

Setting the total flux leaving the Gaussian surface to zero, we obtain

$$\epsilon_1 \frac{\varphi_1 - \varphi_0}{h} \frac{h}{2} + \epsilon_2 \frac{\varphi_1 - \varphi_0}{h} \frac{h}{2} + \epsilon_1 \frac{\varphi_3 - \varphi_0}{h} \frac{h}{2} + \epsilon_2 \frac{\varphi_3 - \varphi_0}{h} \frac{h}{2} \qquad (7.27)$$

$$+ \epsilon_1 \frac{\varphi_2 - \varphi_0}{h} h + \epsilon_2 \frac{\varphi_4 - \varphi_0}{h} h = 0$$

Rearranging the terms gives

$$\frac{\epsilon_1 + \epsilon_2}{2} \varphi_0 = \frac{1}{4} \left(\frac{\epsilon_1 + \epsilon_2}{2} \varphi_1 + \epsilon_1 \varphi_2 + \frac{\epsilon_1 + \epsilon_2}{2} \varphi_3 + \epsilon_2 \varphi_4 \right) \qquad (7.28)$$

It may be noted that (7.28) reduces to (7.11) for $\epsilon_1 = \epsilon_2$. Also, (7.28) can be written down by inspection from (7.11) if the average of the permittivity surrounding the node is used as the permittivity multiplier; that is, $(\epsilon_1 + \epsilon_2)/2$ for the nodes on the interface and the appropriate ϵ value for the nodes in the medium. If the dielectric interface occurs along the line joining the nodes 2 and 4, the expression for φ_0 in this case may be obtained from the simple permutation of (7.28) [1].

7.2.2   Dielectric Inhomogeneity in One Quadrant About a Node

The geometry of dielectric inhomogeneity in one quadrant about a node is shown in Figure 7.7. Here, the dielectric ϵ_2 occupies the fourth quadrant of the Gaussian surface while the other three quadrants are occupied by ϵ_1. Using the procedure described in the last subsection, the total flux leaving the Gaussian surface is found to be

$$\epsilon_1 \frac{\varphi_1 - \varphi_0}{h} \frac{h}{2} + \epsilon_2 \frac{\varphi_1 - \varphi_0}{h} \frac{h}{2} + \epsilon_1 \frac{\varphi_2 - \varphi_0}{h} h + \epsilon_1 \frac{\varphi_3 - \varphi_0}{h} h \qquad (7.29)$$

$$+ \epsilon_1 \frac{\varphi_4 - \varphi_0}{h} \frac{h}{2} + \epsilon_2 \frac{\varphi_4 - \varphi_0}{h} \frac{h}{2} = 0$$

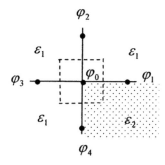

Figure 7.7   Node at the corner of dielectric inhomogeneity.

Rearranging (7.29) gives for the potential at the central node

$$\frac{3\epsilon_1 + \epsilon_2}{4} \varphi_0 = \frac{1}{4}\left(\frac{\epsilon_1 + \epsilon_2}{2}\varphi_1 + \epsilon_1\varphi_2 + \epsilon_1\varphi_3 + \frac{\epsilon_1 + \epsilon_2}{2}\varphi_4\right) \qquad (7.30)$$

Again, (7.30) can be written down by inspection if the average permittivity surrounding the node is used in (7.11). If the dielectric inhomogeneity occurs in a quadrant other than the fourth quadrant, then the simple permutation of the terms in (7.30) can be used to obtain the expression for φ_0 [1].

7.2.3   Neumann Boundary Condition and the Nodes on the Edge

Sometimes the normal derivative of the potential at the boundary is specified to determine the unique solution of the Laplace or Poisson equation. This condition is called the Neumann boundary condition. In this case, we may obtain the finite difference expression as follows.

Consider the geometry of Figure 7.8. In this case the potential at the central node is to be determined in terms of the potential at nodes 1, 2, 3, and the boundary condition about the line joining nodes 1 and 3. We know that if node 0 is the interior node, (7.11) can be used to determine φ_0 in terms of φ_1, φ_2, φ_3, and φ_4. However, node 4 condition is outside the solution domain and its value φ_4 is not known. This situation may occur if the nodes 0, 1, and 3 lie on an edge of the problem space. The outward normal derivative of potential at node 0, about the line joining the nodes 1 and 3, is expressed as

$$\frac{\partial\varphi}{\partial n} = \frac{\partial\varphi}{\partial y} = \frac{\varphi_2 - \varphi_4}{2\Delta y} \qquad (7.31)$$

The value of φ_4 is thus obtained as

$$\varphi_4 = -2\Delta y\left.\frac{\partial\varphi}{\partial y}\right|_{\text{at the edge 31}} + \varphi_2 \qquad (7.32)$$

Substituting for φ_4 in (7.11) gives

$$\varphi_0 \approx \frac{1}{4}\left(\varphi_1 + 2\varphi_2 + \varphi_3 - 2\Delta y\left.\frac{\partial\varphi}{\partial y}\right|_{\text{at the edge 31}}\right) \qquad (7.33)$$

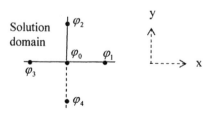

Figure 7.8   Node at an edge with the Neumann boundary condition specified normal to the edge 1–3.

A special but very common case is the *homogeneous* Neumann boundary condition $\partial\varphi/\partial n = 0$. It means $\varphi_2 = \varphi_4$ and the equipotential lines are perpendicular to the edge. For the Neumann boundary condition, (7.33) reduces to

$$\varphi_0 \approx \frac{1}{4}(\varphi_1 + 2\varphi_2 + \varphi_3) \tag{7.34}$$

Expression (7.34) applies to the nodes on the bottom edge of the problem space, with normal derivative zero at the edge. For the top edge, left-hand edge, and the right-hand edge, the corresponding expressions for φ_0 may be obtained by the simple permutation of (7.34) [1].

The homogeneous Neumann boundary condition may be used to reduce the size of the problems with magnetic wall or electric wall at the symmetry plane, for example, Example 7.1, shielded centered strip with two planes of symmetry (Problem 7.3), square coaxial line (Problem 7.4), double strip (Problem 7.5), and coaxial line with circular inner conductor (Problem 7.6).

Quiz 7.4.   Consider the arrangement of nodes as shown in Figure 7.9. The expression for φ_5 subject to $\partial\varphi/\partial n = 0$ is given by ($\Delta x = \Delta y$):

(a) $\epsilon_0 \varphi_5 = \dfrac{1}{4}(\epsilon_0 \varphi_1 + (\epsilon_0 + \epsilon)\varphi_6 + \epsilon_0 \varphi_9)$

(b) $\epsilon_0 \varphi_5 = \dfrac{1}{4}(\epsilon_0 \varphi_1 + 2\epsilon_0 \varphi_6 + \epsilon_0 \varphi_9)$

(c) $\epsilon_0 \varphi_5 = \dfrac{1}{4}(\epsilon_0 \varphi_1 + 2\epsilon\varphi_6 + \epsilon_0 \varphi_9)$

Choose the correct expression and justify.

Answer.   The expression (b) is correct, because similar to Figure 7.6 the contour is drawn enclosing node 5 and mid-way between node 5 and the surrounding nodes.

Quiz 7.5.   For the arrangement of nodes shown in Figure 7.9 the expression for φ_6 is given by ($\Delta x = \Delta y$):

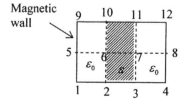

Figure 7.9   An arrangement of nodes for Quiz 7.4.

(a) $\dfrac{\epsilon_0 + \epsilon}{2} \varphi_6 = \dfrac{1}{4}\left(\epsilon_0 \varphi_5 + \dfrac{\epsilon_0 + \epsilon}{2}(\varphi_2 + \varphi_7 + \varphi_{10})\right)$

(b) $\dfrac{\epsilon_0 + \epsilon}{2} \varphi_6 = \dfrac{1}{4}\left(\epsilon_0 \varphi_5 + \dfrac{\epsilon_0 + \epsilon}{2}(\varphi_2 + \varphi_{10}) + \epsilon\varphi_7\right)$

(c) $\dfrac{\epsilon_0 + \epsilon}{2} \varphi_6 = \dfrac{1}{4}\left(2\epsilon_0 \varphi_5 + \dfrac{\epsilon_0 + \epsilon}{2}(\varphi_2 + \varphi_7 + \varphi_{10})\right)$

Choose the correct expression and justify.

Answer.   The expression (b) is correct, because the contour is drawn enclosing node 6 and mid-way between node 6 and the surrounding nodes.

7.2.4   Node at a Corner

Consider the left-hand side bottom corner of the geometry shown in Figure 7.10. It consists of two intersecting edges and three nodes, 0, 1, and 2. The value φ_0 is to be determined in terms of φ_1, φ_2, and the normal derivatives $\partial\varphi/\partial x$ and $\partial\varphi/\partial y$ at the two edges of the corner. Proceeding in the same manner as in the last section, we can write

$$\varphi_4 = -2\Delta y \left.\frac{\partial\varphi}{\partial y}\right|_{\text{at the edge 01}} + \varphi_2 \qquad (7.35a)$$

and

$$\varphi_3 = -2\Delta x \left.\frac{\partial\varphi}{\partial x}\right|_{\text{at the edge 02}} + \varphi_1 \qquad (7.35b)$$

Use of (7.35) in (7.11) gives the following expression for φ_0:

$$\varphi_0 \approx \frac{1}{2}\left(\varphi_1 + \varphi_2 - \Delta y \left.\frac{\partial\varphi}{\partial y}\right|_{\text{at the edge 01}} - \Delta x \left.\frac{\partial\varphi}{\partial x}\right|_{\text{at the edge 02}}\right) \qquad (7.36)$$

For the special case of homogeneous Neumann conditions $\partial\varphi/\partial x = \partial\varphi/\partial y = 0$, the above expression becomes

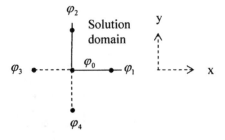

Figure 7.10   Node at a corner with the Neumann boundary conditions specified along two edges crossing at the corner.

$$\varphi_0 \approx \frac{1}{2}(\varphi_1 + \varphi_2) \tag{7.37}$$

The expressions for the potential at the nodes of right-hand bottom corner, left-hand top corner, and right-hand top corner can be written down by inspection of (7.37).

7.2.5 Node at an Edge with Dielectric Inhomogeneity About the Node

Consider the node at an edge with dielectric inhomogeneity about it as shown in Figure 7.11(a). The dielectric constant of the medium is ϵ in the second quadrant and ϵ_0 in the first quadrant. The edge condition is converted into symmetry condition about the x-axis by change in dielectric in the third quadrant as shown in Figure 7.11(b). The new geometry is similar to Figure 7.6. From (7.28) we can write

$$\varphi_0 \frac{\epsilon + \epsilon_0}{2} \approx \frac{1}{4}\left(\epsilon_0 \varphi_1 + \varphi_2 \frac{\epsilon + \epsilon_0}{2} + \epsilon \varphi_3 + \varphi_4 \frac{\epsilon + \epsilon_0}{2}\right) \tag{7.38}$$

Since $\partial/\partial y = 0$ means $\varphi_4 = \varphi_2$, the above expression becomes

$$\varphi_0(\epsilon + \epsilon_0) \approx \frac{1}{2}(\epsilon_0 \varphi_1 + \varphi_2(\epsilon + \epsilon_0) + \epsilon \varphi_3) \tag{7.39}$$

7.2.6 Treatment of Curved Boundaries

The finite difference method relies on a regular grid, which means only those geometries can be analyzed which can be fitted into regular grid patterns such as rectangular grid. In order to extend the validity of finite difference method to geometries with irregular shapes, some approximation in the contour of the given geometry or the modification of (7.11) for the nodes near the boundary may be

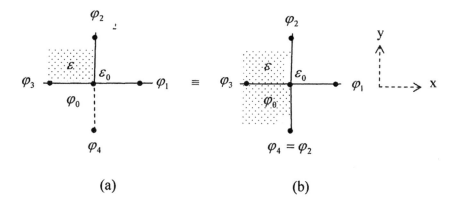

(a) (b)

Figure 7.11 Neumann boundary condition with dielectric inhomogeneity about the central node: (a) original problem; and (b) equivalent problem.

attempted. The *stair-step* approximations to the curved boundary shown in Figure 7.12 may be attempted. The dotted curve a is the inside fit to the curved boundary, whereas curve b is the outside fit to the boundary. For better approximation, one can solve the problem twice for the dotted boundaries and take the average. Another approximation is the zigzag approximation through the curved boundary as shown in curve c of Figure 7.12. Modification of (7.11) for the nodes on the boundary is discussed next.

The nodes on an irregular boundary may be expressed as unequal separation between the nodes. Consider the irregular boundary and the finite difference nodes in Figure 7.13. The nodes 1 and 2 are on the boundary. Let the distance between the nodes 0 and 1 be $c_1 h$, and that between nodes 0 and 2 be $c_2 h$. Let us assume $c_3 = c_4 = 1$ to start with. The constants c_1 and c_2 lie between zero and one, and h is the normal spacing of the nodes. In this case, the potential φ_0 will be expressed in terms of φ_1, φ_2, φ_3, and φ_4.

Let us now assume that $\varphi(x)$ varies as a quadratic function of x in the vicinity of the boundary with [5]

$$\varphi(x) = \alpha_0 + \alpha_1 x + \alpha_2 x^2 \tag{7.40}$$

The constants α_0, α_1, and α_2 can be determined from the potential on three nodes along the x-direction. Using φ_0 at $x = 0$, φ_1 at $x = c_1 h$, and φ_3 at $x = -h$, we obtain

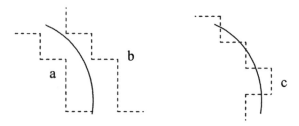

Figure 7.12 Stair-step approximations to the curved boundary in rectangular grid.

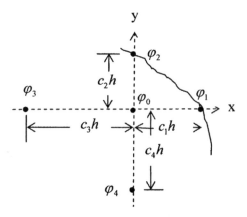

Figure 7.13 Curved boundary with nodes on it.

$$\varphi_0 = \alpha_0 \tag{7.41a}$$

$$\varphi_1 = \alpha_0 + \alpha_1 c_1 h + \alpha_2 c_1^2 h^2 \tag{7.41b}$$

$$\varphi_3 = \alpha_0 - \alpha_1 h + \alpha_2 h^2 \tag{7.41c}$$

Solution of these equations for α_2 yields

$$\alpha_2 = \frac{\varphi_1 - (1 + c_1)\varphi_0 + c_1\varphi_3}{c_1 h^2 (1 + c_1)} \tag{7.42}$$

Taking the double derivative of (7.40) gives

$$\frac{\partial^2 \varphi}{\partial x^2} = 2\alpha_2 \tag{7.43}$$

Therefore,

$$\frac{\partial^2 \varphi}{\partial x^2} = 2 \frac{\varphi_1 - (1 + c_1)\varphi_0 + c_1\varphi_3}{c_1 h^2 + c_1^2 h^2} \tag{7.44}$$

This expression reduces to the standard finite difference form for $c_1 = 1$. Similarly,

$$\frac{\partial^2 \varphi}{\partial y^2} = 2 \frac{\varphi_2 - (1 + c_2)\varphi_0 + c_2\varphi_4}{c_2 h^2 + c_2^2 h^2} \tag{7.45}$$

Adding (7.44) and (7.45) and applying the Laplace equation gives

$$\varphi_0 \left[\frac{1}{c_1} + \frac{1}{c_2} \right] = h^2 \frac{\varphi_1 + c_1\varphi_3}{c_1 h^2 (1 + c_1)} + \frac{\varphi_2 + c_2\varphi_4}{c_2 h^2 (1 + c_2)} \tag{7.46}$$

Expression (7.46) should be used in place of (7.11) for nodes near the irregular boundary. It reduces to (7.11) for $c_1 = c_2 = 1$.

The above procedure may be generalized for an arbitrary separation between the nodes as shown in Figure 7.13

$$\varphi_0 \left[\frac{1}{c_1 c_3} + \frac{1}{c_2 c_4} \right] = \left[\frac{c_3\varphi_1 + c_1\varphi_3}{c_1 c_3 (c_1 + c_3)} + \frac{c_4\varphi_2 + c_2\varphi_4}{c_2 c_4 (c_2 + c_4)} \right] \tag{7.47}$$

One of the attractive features of the finite difference method is the conversion of the differential form of the equation to the algebraic form like (7.12). If the parameters of the problem are chosen judiciously, coarse discretization and symmetry properties may be used to reduce the number of simultaneous equations. Analytical solutions of these equations may be attempted to obtain design guidelines. This step may be followed by accurate analysis with finer discretization. An example is attempted next and in Section 7.3.4 to describe the procedure.

7.2.7   Finite Difference Analysis of an Inhomogeneously Filled Parallel Plate Capacitor

As an application of interface and Neumann boundary conditions, let us analyze the parallel plate capacitor of Figure 7.14 using FDM. The upper plate is charged to 1 volt and the lower plate is grounded. The side walls are terminated in magnetic wall boundary condition to limit the size of the problem in FDM. This amounts to neglecting the fringing fields of the capacitor. Analytical solution for this geometry is available [8, p. 124] with which we can compare the FDM solution.

For FDM analysis, the capacitor geometry is discretized into square cells with $\Delta x = \Delta y = a$. The boundary conditions at the top and bottom plates lead to $\varphi_9 = \varphi_{10} = \varphi_{11} = \varphi_{12} = 1V$; $\varphi_1 = \varphi_2 = \varphi_3 = \varphi_4 = 0$. The unknown potentials are then φ_5, φ_6, φ_7, and φ_8. From the symmetry of the structure about the y-axis we can write $\varphi_5 = \varphi_8$ and $\varphi_6 = \varphi_7$.

We now use the interface condition of Section 7.2.1 and Neumann boundary condition of Section 7.2.3 to write the following expressions for the potentials at nodes 6 and 5:

$$\frac{\epsilon_0 + \epsilon}{2} \varphi_6 = \frac{1}{4}\left(\frac{\epsilon_0 + \epsilon}{2} \varphi_5 + \frac{\epsilon_0 + \epsilon}{2} \varphi_7 + \epsilon_0 \varphi_{10} + \epsilon \varphi_2 \right) \tag{7.48}$$

$$\frac{\epsilon_0 + \epsilon}{2} \varphi_5 = \frac{1}{4}\left(\epsilon \varphi_1 + 2\frac{\epsilon_0 + \epsilon}{2} \varphi_6 + \epsilon_0 \varphi_9 \right) \tag{7.49}$$

Applying the boundary and symmetry conditions leads to the following solution:

$$\varphi_5 = \varphi_6 = \frac{\epsilon_0}{\epsilon_0 + \epsilon} = \varphi_7 = \varphi_8 \tag{7.50}$$

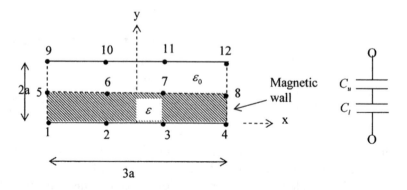

Figure 7.14   An inhomogeneously filled parallel plate capacitor and its equivalent circuit.

In order to determine the capacitance per unit length C_0, we calculate the energy stored in the electric field as

$$W_e = \frac{1}{2} \iint \epsilon |\nabla \varphi|^2 \, ds = \frac{1}{2} \iint \epsilon |\mathbf{E}|^2 \, ds \qquad (7.51)$$

where the integration is carried out over the cross-section of the capacitor. The electric field intensity is given by

$$E_y(y) = -\left(\frac{\varphi(y + \Delta y) - \varphi(y)}{\Delta y} \right) \text{V/m} \qquad (7.52)$$

For the region between the top plate and the interface, the electric field is constant and is given by

$$E_y = -\left(\frac{\varphi_9 - \varphi_5}{a} \right) \text{V/m} = -\left(\frac{\epsilon}{a(\epsilon_0 + \epsilon)} \right) \text{V/m} \qquad (7.53)$$

The energy stored in this region is therefore obtained as

$$W_e^{\epsilon_0} = \frac{1}{2} \epsilon_0 \int\limits_{a}^{2a} \int\limits_{-3a/2}^{3a/2} \left(\frac{\epsilon}{a(\epsilon + \epsilon_0)} \right)^2 dx \, dy = \frac{3}{2} \epsilon_0 \left(\frac{\epsilon}{\epsilon + \epsilon_0} \right)^2 \qquad (7.54a)$$

The energy stored in the lower half of the capacitor is similarly given by

$$W_e^{\epsilon} = \frac{1}{2} \epsilon \int\limits_{0}^{a} \int\limits_{-3a/2}^{3a/2} \left(\frac{\epsilon_0}{a(\epsilon + \epsilon_0)} \right)^2 dx \, dy = \frac{3}{2} \epsilon \left(\frac{\epsilon_0}{\epsilon + \epsilon_0} \right)^2 \qquad (7.54b)$$

The total energy stored in the capacitor is therefore

$$W_e = W_e^{\epsilon_0} + W_e^{\epsilon} = \frac{3}{2} \frac{\epsilon \epsilon_0}{\epsilon + \epsilon_0} \qquad (7.55)$$

Equating W_e to $\frac{1}{2} C_0 V^2$, where V is the potential difference between the plates ($V = 1$ volt), yields the following expression for capacitance:

$$C_0 = \frac{3 \epsilon \epsilon_0}{\epsilon + \epsilon_0} \qquad (7.56)$$

The analytical expression for the capacitance may be obtained by modeling the parallel plate geometry as a series combination of capacitors due to the upper half and lower half portions, as shown in Figure 7.14 [8]. The capacitance of the upper half portion is given by $C_u = 3\epsilon_0$, whereas $C_l = 3\epsilon$ for the lower half geometry. Series combination of C_u and C_l results in C_0, the FDM expression derived above.

It may be observed from the above analysis that in spite of crude discretization used here, the FDM result matches exactly with the analytical expression; that is, the discretization error is zero in this case. The reason for this accuracy lies in the potential distribution we are modeling using FDM. The expected potential distribution is linearly varying along the y-direction and is constant along the x-direction.

7.3 Finite Difference Analysis of Guiding Structures

The method of finite differences has been used successfully to solve many diverse problems [7, 10, 11] including transmission lines [12–15], waveguides [16–19], and microwave circuits [20–22]. Finite difference analysis of microstrip line, rectangular waveguide, and ridge waveguide is discussed next. The effect of discretization error on the propagation constant leading to numerical dispersion is also included.

7.3.1 Analysis of Enclosed Microstrip Line

Consider the cross-section of a microstrip line as shown in Figure 7.15. It consists of a metal strip of width W on a grounded dielectric substrate of thickness d and dielectric constant ϵ_1. The free space above the strip and the dielectric substrate are infinite in size. The open region geometry gives rise to an infinite number of nodes, which is impossible to solve on computers due to limited computer resources. The metal enclosure around the microstrip line has been placed here to limit the number of nodes. The characteristics of the line remain unaffected if the enclosure size is sufficiently large.

The characteristic impedance and the phase velocity of the transmission line can be determined from the capacitance per unit length of the line as described in Section 1.14.

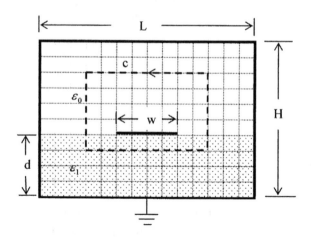

Figure 7.15 Cross-section of an enclosed microstrip line.

The potential distribution in the microstrip line is determined by discretizing the geometry and satisfying the Laplace equation at the free nodes, as described in Example 7.1. The only difference is that the nodes at the interface are subjected to (7.28) and not (7.12). Once the potential distribution has been obtained, the charge on the strip may be obtained using Gauss law, described as

$$\oiint_s \epsilon \mathbf{E}.\mathbf{ds} = q \tag{7.57a}$$

Since the cross-section of the microstrip line does not change along its length, the charge distribution on the strip also remains same with length. Integrating over unit length gives

$$\oint_c \epsilon \mathbf{E}.\mathbf{dc} = q_l$$

where q_l is the charge per unit length, and c is the contour enclosing the strip. The above expression may be written as

$$-\oint_c \epsilon \nabla \varphi.\mathbf{dc} = q_l, \qquad \text{since } \mathbf{E} = -\nabla \varphi$$

or

$$\oint_c \epsilon \frac{\partial \varphi}{\partial n} dc = -q_l \qquad \text{C/m} \tag{7.57b}$$

The charge density q_l can be determined by utilizing the nodal potential values about the contour. The contour c for computing the charge enclosed is shown separately in Figure 7.16. The contour integration is carried out here in the form of summation by dividing the contour in smaller segments. Applying the definition of $\partial \varphi / \partial n$ along this contour, we obtain [2]

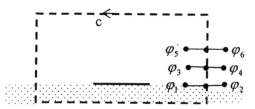

Figure 7.16   Contour c and the nodes employed in calculating the charge.

$$\ldots + \epsilon_0 \left(\frac{\varphi_6 - \varphi_5}{2h} \right) h + \epsilon_0 \left(\frac{\varphi_4 - \varphi_3}{2h} \right) h + \epsilon_0 \left(\frac{\varphi_2 - \varphi_1}{2h} \right) \frac{h}{2} + \epsilon_1 \left(\frac{\varphi_2 - \varphi_1}{2h} \right) \frac{h}{2} + \ldots$$

+ contribution from other sides of the contour = q_l (7.58)

Here h is the separation between the nodes in either direction.

The line capacitance is given by

$$C_0 = \frac{q_l}{V} \qquad\qquad (7.59)$$

where V is the assumed potential difference between the strip and the ground plane, normally 1 volt.

The microstrip line is next completely filled with air and the above procedure is repeated to calculate C_0^a, the capacitance per unit length with air as dielectric. These values of capacitances are used to determine the characteristic impedance Z_0 and the effective dielectric constant ϵ_{re} according to (1.62). A Fortran source code for the finite difference analysis of microstrip line is included in [4].

In order to compare the results with the data reported in [2], we select the following parameters (in arbitrary units) for the microstrip line of Figure 7.15:

$$H = 2, \; L = 7.0, \; d = 1.0, \; W = 1.0, \; \epsilon_1 = 9.6\epsilon_0$$

The dimensions are normalized to d. The actual dimensions, whether expressed in meters or microns or any other unit, is not important here.

There are a number of issues to be discussed before we present the final results. These are related to the convergence aspect of the FDM, the number of cells per unit length N, the number of iterations I used for solving the set of simultaneous equations iteratively, and accuracy of results.

Convergence

All the computations reported here are carried out using the source code *mstrip.m*. Figure 7.17 shows the effect of number of nodes or node density on the characteristic impedance of microstrip line. It may be noted that the impedance value converges for $N \geq 24$. The number of iterations used for each point of calculation is 1,000 because SOR is not used here. A large number of nodes on the strip are required for accurate calculation of Z_0. This is related to the physical charge density distribution on the strip. Due to the mutual repulsion between free charges on the strip, the charge density should be large at the ends and flat in the middle for a zero thickness strip. The charge density is plotted in Figure 7.18 for two different values of N. It is seen that $N = 24$ gives rise to a larger peak/minimum value of charge density compared to $N = 20$ case, and therefore more accurate value of Z_0.

We next discuss the optimum number of iterations, I. The value of Z_0 is expected to be sensitive to I because the calculation of charge on the strip involves derivative of potential, (7.58). *A sufficiently large number of iterations will produce an accurate potential distribution about the strip from which we can determine an*

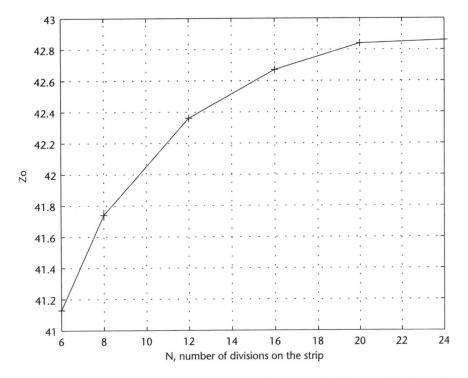

Figure 7.17 Computed characteristic impedance of the microstrip line as a function of node density N. $W/d = 1$, $\epsilon_r = 9.6$, $I = 1,000$.

accurate value of $\partial\varphi/\partial n$. Variation of Z_0 as a function of number of iterations I is plotted in Figure 7.19 for $N = 20$. The converged value of I is 2,000. The equipotential contours about the strip are plotted in Figure 7.20. For this we have used the MATLAB command *contour*. This plot may be used for the diagnostics of the data generated. It may be noted that the equipotential contours are similar to the magnetic field lines encircling the strip. Finally, the electric field distribution about the strip is plotted in Figure 7.21. The electric field is not tangential to the interface because of refraction there.

The computed characteristic impedance and the effective dielectric constant for the microstrip line are given in Table 7.1 for various values of W/d. The value of I used is 1,800 and $N = 24$. For comparison, the impedance reported in [2] for $W/d = 1$ is 42.76 ohms. It may be noted from Table 7.1 that the effective dielectric constant ϵ_{re} for the given geometry is nearly equal to $(\epsilon_r + 1)/2$. This is because the microstrip line behaves like a strip line for $H = 2d$. The characteristic impedance Z_0 of strip line is given by (11.115) with ϵ_r replaced by $(\epsilon_r + 1)/2$. The exact value for strip line for $W/d = 1$, $\epsilon_r = 9.6$, $b = 2d$ is 43.65 ohms, which is very close to the finite difference value of 42.98 ohms. The error in the computed value based on FDM is due to the discretization of microstrip line.

This code *mstrip.m* may also be employed to analyze a shielded strip line by homogeneous filling of the enclosed microstrip configuration. Use of SOR reduced the number of iterations from 2,000 to 250. Square grid with the value of omega, given by (7.22b), was used for the purpose.

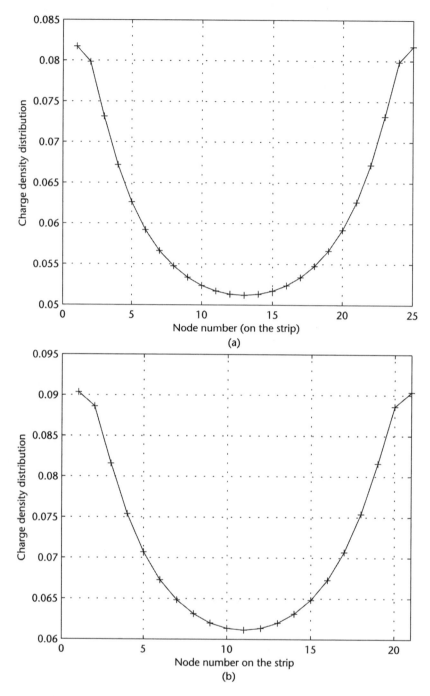

Figure 7.18 Charge density distribution on the strip for $W/d = 1$, $\epsilon_r = 9.6$: (a) $N = 24$, $I = 2,000$; and (b) $N = 20$, $I = 2,000$.

Effect of Strip Thickness

It has been assumed in the analysis described above that the metal strip thickness is zero. The effect of strip thickness on the line impedance can be included in the FDM analysis easily. For this, the discretization of the line geometry is carried out

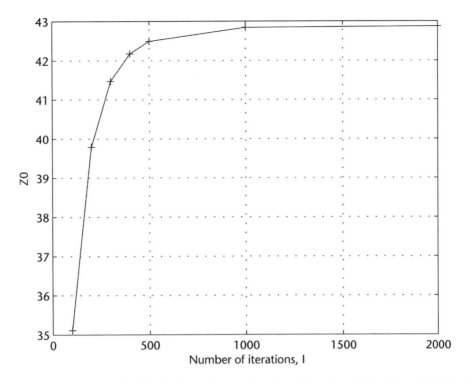

Figure 7.19   Variation of W/d of microstrip as a function of number of iterations I. $W/d = 1$, $\epsilon_r = 9.6$, $N = 20$.

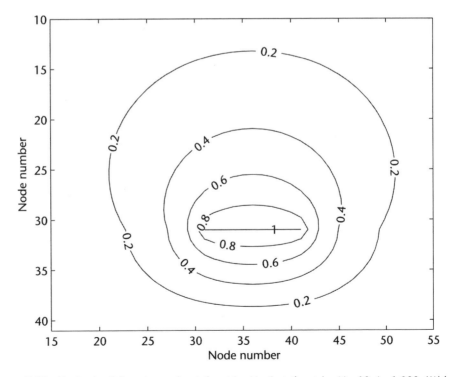

Figure 7.20   Equipotential contours about the strip, $V = 1$ at the strip. $N = 10$, $I = 1,000$, $W/d = 1$, $H/d = 5$, $L/d = 7$, $\epsilon_r = 9.6$.

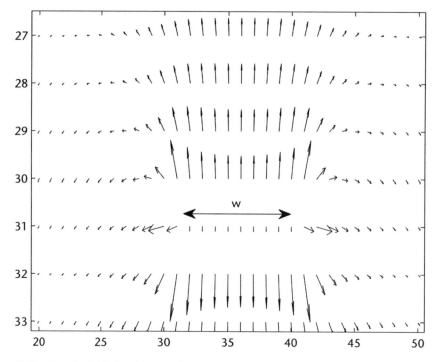

Figure 7.21 Electric field distribution about the strip, $V = 1$ at the strip. $N = 10$, $I = 1,000$, $W/d = 1$, $H/d = 5$, $L/d = 7$, $\epsilon_r = 9.6$.

Table 7.1 Computed FDM Results for a Microstrip Line with $H = 2$, $L = 7$, $\epsilon_r = 9.6$

Characteristic → W/d ↓	Z_0	ϵ_{re}
1	42.98	5.297
1.5	33.97	5.299
2	28.11	5.299
2.5	23.98	5.299

as before and all the nodes on the strip are assigned the same potential, 1 volt in the present case. The discretized strip with strip thickness $t = \Delta y$ is shown in Figure 7.22. The Gaussian contour c, for the determination of q_l, may have to be redrawn so that it encloses the strip.

Nonuniform Discretization

It may be observed from Figure 7.18 that the charge density on a metal strip is maximum at the edges and minimum at the center. This distribution can be modeled

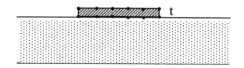

Figure 7.22 Discretization of metal strip for nonzero value of strip thickness t, $\Delta x = 2\Delta y$.

more efficiently if we discretize the strip nonuniformly; that is, the step size h should be smaller near the edges and larger in the middle. One may use the following expression to describe the coordinates of nodes on a strip of width W:

$$x_i = -\frac{W}{2}\cos\frac{i\pi}{N}, \qquad i = 1, 2, 3, \ldots N \qquad \text{for } -\frac{W}{2} < x < \frac{W}{2} \qquad (7.60)$$

7.3.2 Analysis of Geometries with Open Boundaries

In the finite difference analysis in the last section, we enclosed the microstrip line in a metal box to limit the number of nodes for the open region problem. But, we do not know the appropriate size of the box. For this one may carry out a number of simulations for increasing the size of the box. The resulting data for Z_0 may be used to plot a curve as a function of box size. This curve can be extrapolated to determine the value of Z_0 for an open region problem. The accuracy of the solution depends on the size of the box used to enclose the device. An alternative approach for the open boundary problems is to use a fictitious boundary [23] outside the device, or the use of asymptotic boundary conditions [24]. The first method is an iterative method in which the approximate potential distribution inside the fictitious boundary is used to calculate the total charge enclosed. This charge distribution is next used to determine the potential on the boundary. This procedure is repeated iteratively to produce the correct potential distribution on the fictitious boundary. For a microstrip line geometry, the results at the interior nodes are found to be independent of the location of the boundary for a grid size larger than 18×12 [23].

The asymptotic boundary condition (ABC) is a (local) radiation boundary condition expressed as a partial differential equation. The enclosure of Figure 7.15 at which $\varphi = 0$ has been assumed, may be called the 0th order ABC. The first order ABC at the $x = $ constant boundary is defined as

$$\left.\frac{\partial\varphi}{\partial x}\right|_{boundary} = -\frac{1}{x}\left(\varphi + y\left.\frac{\partial\varphi}{\partial y}\right|_{boundary}\right) \qquad (7.61)$$

Similarly, the ABC at the other boundaries can be written down. Second order ABC is also described in [24]. The partial derivatives in (7.61) can be described in terms of the nodal values near the boundary, which are already known. The use of first-order and second-order ABC may be used to bring the boundary close to the device without affecting the device characteristics. This reduces the size of the matrix to be solved. However, the number of iterations required (if iterative method is used) to determine the nodal values increase considerably compared to the 0th order ABC, because the nodal values at the boundaries are not fixed and depend on the value at neighboring nodes and change with iteration.

Another approach for analyzing open boundary problems is to combine conformal mapping and FDM for problems in electrostatics. Conformal mapping may be used to reduce the open region problems like that of planar lines to closed geometry (Chapter 4). FDM may now be used to determine the capacitance of this geometry [25].

Effect of the discretization error on dispersion is discussed next.

7.3.3 Wave Propagation and Numerical Dispersion

We have discussed the error associated with the discretization of function and by implication the discretization of domain in Section 6.3. The discretization is necessary so that computers may be used to solve problems with arbitrary shape and dielectric inhomogeneity. When FDM is used to solve problems based on wave propagation, we come across a phenomenon called numerical dispersion. In this case the wave suffers dispersion, and the propagation constant becomes a function of frequency, even though the medium may be dispersionless. To describe this behavior mathematically we consider the solution of one-dimensional scalar Helmholtz equation [26],

$$\frac{d^2 E_y}{dz^2} + k_0^2 E_y(z) = 0 \tag{7.62}$$

The traveling wave solution of this wave equation is of the form

$$E_y(z) = E_0 e^{\pm jk_0 z} \tag{7.63}$$

Let us now discretize (7.62), determine the propagation constant, and compare it with k_0 to determine the numerical dispersion.

We assume a large free space with no boundaries and discretize it uniformly with cell size Δz, as shown in Figure 7.23. The central difference approximation of the double derivative in (7.62) gives rise to the following form of discretized Helmholtz equation about node i:

$$2E_y(i) - E_y(i-1) - E_y(i+1) - (k_0 \Delta z)^2 E_y(i) = 0 \tag{7.64}$$

We now solve this equation using a discretized version of (7.63); that is,

$$E_y(i) = E_0 e^{\pm j\beta(i\Delta z)} \tag{7.65}$$

where β is the propagation constant in the discretized medium. Substituting the proposed solution in (7.64) gives

$$(2 - (k_0 \Delta z)^2) E_0 - E_0 e^{j\beta \Delta z} - E_0 e^{-j\beta \Delta z} = 0$$

or

$$2 - (k_0 \Delta z)^2 = 2 \cos(\beta \Delta z) \tag{7.66a}$$

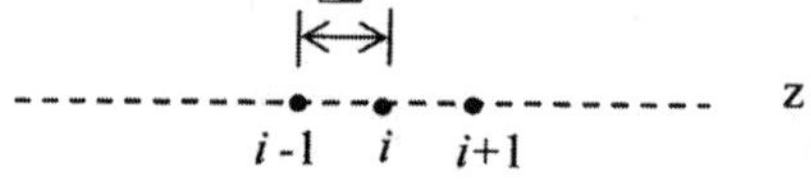

Figure 7.23 Discretized one-dimensional space for wave propagation.

$(\varphi = 0)$ is applicable for the *TM* modes, and Neumann condition $(\partial\varphi/\partial n = 0)$ for the *TE* modes. The Neumann condition is specified by (7.34). By applying (7.72) to all the nodes in the cross-section of the waveguide, we obtain a set of simultaneous equations which can be cast into the matrix equation

$$([A] - \lambda[I])[\varphi] = [0] \quad \text{or} \quad [A][\varphi] = \lambda[\varphi] \tag{7.73}$$

where A is the coefficient matrix of known integer elements, and

$$\lambda = 4 - h^2 k_c^2 \tag{7.74}$$

is the unknown eigenvalue, I is an identity matrix, and $\varphi = (\varphi_1, \varphi_2, \varphi_3, \dots)$ is the eigenvector. The set of simultaneous equations represented by (7.73) will have nonzero solution only if the determinant of the matrix vanishes; that is,

$$\det[A - \lambda I] = 0 \tag{7.75}$$

There are several ways of determining λs and the corresponding eigenfunctions φs. We describe two approaches here [2, 4].

Direct Solution Method
In this method, $\det[A - \lambda I]$ is set equal to zero, and results in a polynomial in λ which can be solved for the various eigenvalues λs. For each of these eigenvalues the corresponding eigenfunction φ is obtained from the matrix equation $A\varphi = \lambda\varphi$. This method requires storing the matrix elements and is therefore used for small sized matrices.

Example 7.2.   Let us determine the cutoff frequency of modes in a rectangular waveguide. The dimensions of the waveguide and discretization are selected such that it produces a small sized matrix which can be solved almost analytically. In the process we will clarify some of the computational aspects. The rectangular waveguide is analyzed for the *TE* mode cut off.

As shown in Figure 7.25(a), the waveguide with aspect ratio 2:1 is discretized in square cells of size $h \times h$. The total number of nodes is 15 and all of these are free nodes for the *TE* modes. In order to keep the matrix size small, we shall assume TE_{n0} modes for analysis. For constant field distribution E_y along the height of the waveguide, the potential φ is also constant along the height; that is,

$$\varphi_{11} = \varphi_6 = \varphi_1 \qquad \varphi_{12} = \varphi_7 = \varphi_2 \qquad \varphi_{13} = \varphi_8 = \varphi_3 \tag{7.76}$$

and so on. Therefore, we need to determine the potential at nodes 1, 2, 3, 4, and 5 only. The problem therefore becomes one-dimensional. The size of the problem can be reduced further if we restrict the solutions for $n =$ odd or even modes only. For odd modes, the potential function $\varphi = H_z$ is odd symmetric about the plane passing through 13-3-8 and maximum at the side walls. Therefore, the device geometry reduces to that shown in Figure 7.25(b) with $\varphi_3 = 0$. Applying (7.72) for the potential at the nodes 1 and 2 gives

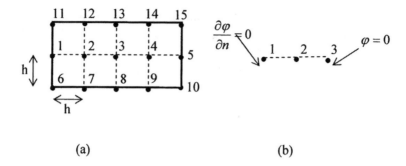

(a)   (b)

Figure 7.25 Simplification of the analysis of a rectangular waveguide using symmetry. (a) Discretized geometry before symmetry consideration. (b) Waveguide geometry for the analysis of TE_{n0} (n, odd) modes. Odd symmetry about 13-3-8 plane results in $\varphi = 0$ at this plane.

$$\lambda \varphi_1 = \varphi_{11} + 2\varphi_2 + \varphi_6 \qquad \text{using (7.34) for the edge condition} \qquad (7.77)$$

$$\lambda \varphi_2 = \varphi_1 + \varphi_{12} + \varphi_3 + \varphi_7 \qquad\qquad (7.78)$$

where λ has been defined in (7.74). Use of (7.76) and $\varphi_3 = 0$ gives

$$(\lambda - 2)\varphi_1 = 2\varphi_2 \qquad \text{and} \qquad (\lambda - 2)\varphi_2 = \varphi_1 \qquad (7.79)$$

or

$$\begin{bmatrix} \lambda - 2 & -2 \\ -1 & \lambda - 2 \end{bmatrix} \begin{bmatrix} \varphi_1 \\ \varphi_2 \end{bmatrix} = 0 \qquad\qquad (7.80)$$

The eigenvalue equation is obtained by setting the determinant to zero. We obtain $(\lambda - 2)^2 = 2$, and the eigenvalues are $\lambda_1 = 2 + \sqrt{2}$ and $\lambda_2 = 2 - \sqrt{2}$. The corresponding values of k_c obtained from (7.74) are $h^2 k_{c1}^2 = 2 - \sqrt{2}$ and $h^2 k_{c2}^2 = 2 + \sqrt{2}$. The first eigenvalue gives

$$\left(\frac{2\pi h}{\lambda_{c1}} \right)^2 = 2 - \sqrt{2}, \qquad k_c = \frac{2\pi}{\lambda_c}$$

or

$$\lambda_{c1} = \frac{2\pi h}{\sqrt{0.586}} \qquad\qquad (7.81)$$

For $h = a/4$, where a is the width of the waveguide, one obtains

$$\lambda_{c1} = \frac{\pi a}{2\sqrt{0.586}} = 2.05a \qquad\qquad (7.82)$$

This value is very close to the exact value of $2a$ for the TE_{10} mode. The eigenfunction for the mode is obtained by substituting $\lambda = \lambda_1 = 2 + \sqrt{2}$ in (7.80) and solving for φ_1 and φ_2. It yields, $\varphi_2 = \varphi_1/\sqrt{2}$, which is exactly the value obtained from $\cos(\pi x/a)$.

The second eigenvalue $\lambda_2 = 2 - \sqrt{2}$ leads to

$$\lambda_{c2} = \frac{\pi a}{2\sqrt{3.414}} = 0.849a \tag{7.83}$$

The exact cutoff wavelength for the TE_{30} mode is $0.667a$.

Iterative Solution Method

In this method, we begin with the assumed values of φ_1, φ_2, φ_3, ... and the eigenvalue λ. For the solution of wave equation, (7.21a) is modified as

$$\varphi_{i,j}^{n+1} = \varphi_{i,j}^{n} + \frac{\Omega_0 R_{i,j}^{n+1}}{\left(4 - h^2 k_c^2\right)} \tag{7.84}$$

Since we start with an assumed value of k_c, it needs to be updated from one iteration to the next. When we have the exact value of $\varphi_{i,j}^{n}$, the value of k_c can be determined from (7.70) yielding

$$k_c^2 = -\frac{\nabla_t^2 \varphi}{\varphi}, \qquad \nabla_t^2 = \frac{\partial^2}{\partial x^2} + \frac{\partial^2}{\partial y^2} \tag{7.85}$$

This expression is applied at each node (i, j). Since $\varphi_{i,j}^{n}$ is known only approximately, the value of k_c may vary from node to node. One may take the average of k_c over all the free nodes and use this value for the next iteration. A better choice than (7.85) is its weighted average with respect to φ [26]. Expression for this average is obtained by multiplying (7.70) with φ and integrating over the cross-section; one obtains

$$k_c^2 = \frac{-\displaystyle\iint_S \varphi \nabla_t^2 \varphi\, ds}{\displaystyle\iint_S \varphi^2\, ds} \tag{7.86}$$

This expression is called the Rayleigh formula, and it is found to be variational in nature [6, 14].

Now we replace $\nabla_t^2 \varphi$ by its finite difference equivalent based on (7.9) and carry out the integration using the discrete values of $\varphi_{i,j}^{n}$, resulting in

$$k_c^2 h^2 = \frac{-\displaystyle\sum_{i=1}^{N}\sum_{j=1}^{M} \varphi(i, j)[\varphi(i, j+1) + \varphi(i, j-1) + \varphi(i+1, j) + \varphi(i-1, j) - 4\varphi(i, j)]}{\displaystyle\sum_{i=1}^{N}\sum_{j=1}^{M} \varphi^2(i, j)} \tag{7.87}$$

where the summation is carried out over all the free nodes in the device geometry. We use this value of k_c to update (7.84) and continue the iterations until convergence is reached. The cutoff wavelength for the mode is obtained from the final value of $(k_c h)^2$ as

$$(k_c h)^2 = \left(\frac{2\pi h}{\lambda_c}\right)^2 \tag{7.88}$$

We now apply this method to determine the cutoff wavelength of the ridge waveguide of Figure 7.26. The waveguide dimensions are $a_1 = 2$ cm, $b_1/a_1 = 0.5$, and the discretization used is $\Delta x = \Delta y = 0.01$ cm. The ridge parameters were varied as follows: $a_2/a_1 = 0.1$ to 0.9 in steps of 0.1, and $b_2/b_1 = 0.80$ to 0.95 in steps of 0.05. The computed cutoff wavelength for the dominant mode of the ridge waveguide λ_c' was normalized to the cutoff wavelength in the waveguide without ridge, $\lambda_c = 2a_1$. The software used was *waveguide.m*, and the results are plotted in Figure 7.27. It is seen that the λ_c' increases with increase in the ridge height b_2. It also increases with a_2 until it becomes maximum at $a_2/a_1 = 0.5$. The cutoff frequency for the dominant mode of the waveguide may therefore be decreased using a single ridge by a factor of about 3.8.

Once the cutoff wavenumber k_c is obtained from the analysis, it may be used to determine the propagation constant β for the mode by using $k_c^2 = k^2 - \beta^2$, where $k^2 = \omega^2 \mu \epsilon$. The constant H_z lines for the dominant mode are plotted in Figure 7.28, and are similar to the electric field lines for the mode. This plot may be employed for diagnostics of the data.

The FDM is a very simple, versatile, but inefficient method. It requires large node density for accurate results. The asymptotic value corresponding to the zero discretization size may be obtained by extrapolating the data obtained for decreasing discretization size [6]. The extrapolation technique is discussed in Chapter 6. The basic concepts of finite difference method are utilized in the time-domain finite-difference method discussed in the next chapter.

7.4 Summary

The finite difference method is a simple and versatile differential equation solver. However, its efficiency is poor because the node density required for accurate

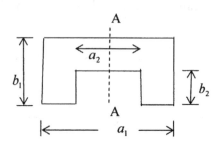

Figure 7.26 Cross-section of a ridged waveguide.

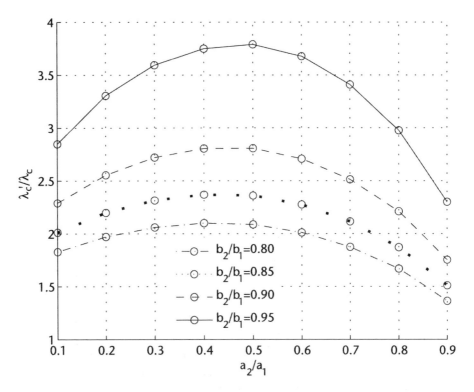

Figure 7.27 Ratio of the cutoff wavelength of the ridge waveguide λ'_c and the corresponding rectangular waveguide without ridge $\lambda_c(a_1 = 2$ cm, $b_1/a_1 = 0.5)$, $\Delta x = \Delta y = 0.01$ cm.

solution is more than 20. For the application of this method, the derivatives in the differential equation are expressed in the finite difference form resulting in an algebraic equation. This equation is applied at each node after the discretization of the device geometry. Some of the nodes are called fixed nodes because the value at these nodes is determined by the boundary conditions or excitation. The other nodes are called free nodes at which the potential values are obtained from the solution of simultaneous equations. The matrix equation is solved using iterative methods because the size of the matrix obtained for accurate solution is very large. Successive over relaxation may be used to reduce the number of iterations considerably. A simple expression is given to determine the optimum value of relaxation factor. Various types of boundary conditions one may come across while attempting FDM solution, like node on the interface, magnetic wall boundary condition, and node on a corner, are analyzed and expressed in finite difference form. The FDM is illustrated by its application to problems like inhomogeneously filled parallel plate capacitor, microstrip line, and rectangular and ridge waveguide. The effect of discretization on the accuracy of the solution is addressed through numerical dispersion determination. Examples are so designed that FDM analysis is carried out in a tutorial fashion without the use of software. For this the size of the problem is kept small through crude discretization and the use of symmetry conditions. The same problem is next worked out using the software and fine discretization to obtain accurate results.

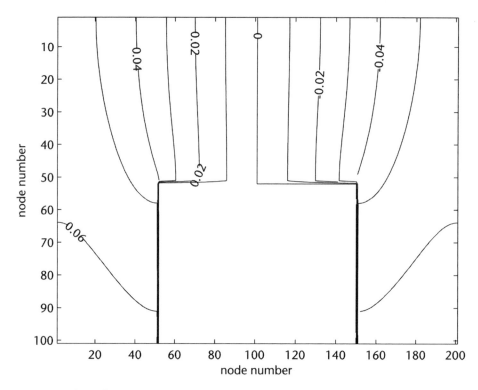

Figure 7.28 Plot of constant H_z contours for the dominant mode. The parameters are: $a_2/a_1 = 0.5$, $b_2/b_1 = 0.5$, $a_1 = 0.02$m, $b_1/a_1 = 0.5$.

References

[1]   Green, H. E., "The Numerical Solution of Some Important Transmission-Line Problems," *IEEE Trans. Microwave Theory Tech.*, Vol. MTT-13, 1965, pp. 676–692.

[2]   Iskander, M. F., et al., "A New Course on Computational Methods in Electromagnetics," *IEEE Trans. Educ.*, Vol. 31, 1988, pp. 101–115.

[3]   Schneider, M. V., "Computation of Impedance and Attenuation of TEM-Lines by Finite Difference Methods," *IEEE Trans. Microwave Theory Tech.*, Vol. MTT-13, 1965, pp. 793–800.

[4]   Sadiku, M. N. O., *Numerical Techniques in Electromagnetics*, Boca Raton, FL: CRC Press, 1992.

[5]   Chari, M. V. K., and S. J. Salon, *Numerical Methods in Electromagnetism*, New York: Academic Press, 2000.

[6]   Booton, R. C., *Computational Methods for Electromagnetics and Microwaves*, New York: John Wiley, 1992.

[7]   Wexler, A., "Computation of Electromagnetic Fields," *IEEE Trans. on Microwave Theory Tech.*, Vol. MTT-17, 1969, pp. 416–439.

[8]   Neff, H. P., Jr., *Basic Electromagnetic Fields*, 2nd ed., New York: Harper & Row, 1987.

[9]   de Vahl Davis, G., *Numerical Methods in Engineering and Science*, New York: Van Nostrand Reinhold, 1986.

[10] "Computer-Oriented Microwave Practices," special issue of *IEEE Trans. on Microwave Theory and Techniques*, Vol. MTT-17, No. 8, August 1969.

[11] Vemuri, V., and W. J. Karplus, *Digital Computer Treatment of Partial Differential Equations*, Englewood Cliffs, NJ: Prentice-Hall, 1981.

[12] Metcalf, W. S., "Characteristic Impedance of Rectangular Transmission Lines," *Proc. IEE*, Vol. 112, 1965, pp. 2033–2039.

[13] Gupta, R. R., "Accurate Impedance Determination of Coupled TEM Conductors," *IEEE Trans. Microwave Theory Tech.*, Vol. MTT-17, 1969, pp. 479–489.

[14] Yamashita, E., et al., "Characterization Method and Simple Design Formulas of MDS Lines Proposed for MMIC's," *IEEE Trans. on Microwave Theory Tech.*, Vol. MTT-35, 1987, pp. 1355–1362.

[15] Molberg, J. R., and D. K. Reynolds, "Iterative Solutions of the Scalar Helmholtz Equations in Lossy Regions," *IEEE Trans. on Microwave Theory Tech.*, Vol. MTT-17, 1969, pp. 460–477.

[16] Davies, J. B., and C. A. Muilwyk, "Numerical Solution of Uniform Hollow Waveguides with Boundaries of Arbitrary Shape," *Proc. IEEE*, Vol. 113, 1966, pp. 277–284.

[17] Hornsby, J. S., and A. Gopinath, "Numerical Analysis of a Dielectric-Loaded Waveguide with a Microstrip Line-Finite-Difference Methods," *IEEE Trans. on Microwave Theory Tech.*, Vol. MTT-17, 1969, pp. 684–690.

[18] Beubien, M. J., and A. Wexler, "An Accurate Finite-Difference Method for Higher-Order Waveguide Modes," *IEEE Trans. on Microwave Theory Tech.*, Vol. MTT-16, 1968, pp. 1007–1017.

[19] Collins, J. H., and P. Daly, "Calculations for Guided Electromagnetic Waves Using Finite-Difference Methods," *J. Electronics & Control*, Vol. 14, 1963, pp. 361–380.

[20] Muilwyk, C. A., and J. B. Davies, "The Numerical Solution of Rectangular Waveguide Junctions and Discontinuities of Arbitrary Cross Section," *IEEE Trans. on Microwave Theory Tech.*, Vol. MTT-15, 1967, pp. 450–455.

[21] Sinnott, D. H., et al., "The Finite Difference Solution of Microwave Circuit Problems," *IEEE Trans. on Microwave Theory Tech.*, Vol. MTT-17, 1969, pp. 464–478.

[22] Corr, D. G., and J. B. Davies, "Computer Analysis of the Fundamental and Higher Order Modes in Single and Coupled Microstrip," *IEEE Trans. Microwave Theory Tech.*, Vol. MTT-20, 1972, pp. 669–678.

[23] Sandy, F., and J. Sage, "Use of Finite Difference Approximations to Partial Differential Equations for Problems Having Boundaries at Infinity," *IEEE Trans. on Microwave Theory Tech.*, Vol. MTT-29, 1981, pp. 484–486.

[24] Jordon, R. K., and S. H. Hook, "A Finite Difference Approach that Employs an Asymptotic Boundary Condition on a Rectangular Outer Boundary for Modeling Two-Dimensional Transmission Line Structures," *IEEE Trans. on Microwave Theory Tech.*, Vol. 41, 1993, pp. 1280–1286.

[25] Chang, C. N., Y. C. Wong, and C. H. Chen, "Hybrid Quasistatic Analysis for Multilayer Coplanar Lines," *IEE Proc.-H*, Vol. 138, 1991, pp. 307–312.

[26] Peterson, A. F., S. L. Ray, and R. Mittra, *Computational Methods for Electromagnetics*, New York: IEEE Press, 2001.

[27] Naiheng, Y., and R. F. Harrington, "Characteristic Impedance of Transmission Lines with Arbitrary Dielectric Under the TEM Approximation," *IEEE Trans. on Microwave Theory Tech.*, Vol. MTT-34, 1986, pp. 472–475.

[28] Kim, W., M. F. Iskander, and C. M. Krowne, "Modified Green's Function," *IEEE Trans. on Microwave Theory Tech.*, Vol. 55, 2007, pp. 402–408.

Problems

P7.1. Consider the trough problem discussed in Example 7.1. Use a 6 × 6 set of square cells to determine the potentials at the nodes for a given tolerance. Next, increase the number of cells to 12 × 12 and again determine the potential distribution for the same tolerance. Compare your computed values with those given by (7.20) and comment on the accuracy of the solution with the decrease in cell size.

P7.2. Consider the trough shown in Figure 7.29. The lower half of the trough is filled with a dielectric having $\epsilon_r = 3$ and the upper half with $\epsilon_r = 1$. Show that the solution based on finite-difference method (FDM) is [7]

$$\varphi_1 = 0.3929V_0 \qquad \varphi_2 = 0.4777V_0 \qquad \varphi_4 = 0.0938V_0$$

$$\varphi_5 = 0.1250V_0 \qquad \varphi_7 = 0.0357V_0 \qquad \varphi_8 = 0.0491V_0$$

P7.3. The geometry of a shielded strip line is shown in Figure 7.30. Determine the potential distribution and the characteristic impedance of the line Z_0 with $w/b = 0.8$, $t/b = 0, 0.1, 0.2,$ and 0.3, $s/b = 0.1$ to 0.7 in steps of 0.1 [3]. Plot the potential

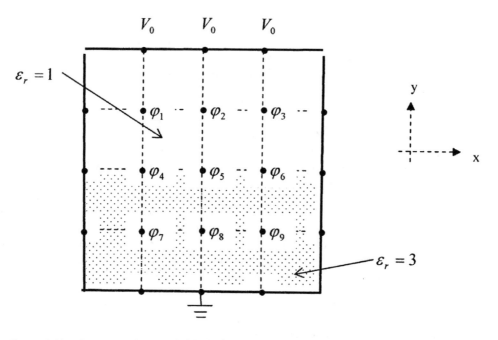

Figure 7.29 Geometry of a trough filled inhomogeneously with dielectric.

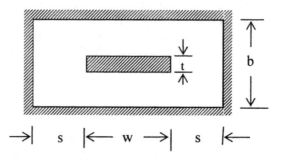

Figure 7.30 Cross-section of shielded strip line.

distribution in the cross-section. You may use the symmetry of the structure about the width and height to reduce the problem domain to its quarter and use the Neumann boundary conditions at the symmetry planes. Compare the results for $w/b = 1.5$, $(2s + w)/b = 4$ with the laterally open strip line case solved using MoM (Section 11.2.2). MoM result is 47.92 ohms for $w/b = 1.5$, $\epsilon_r = 1$. The FDM result is 47.12 ohms for $L/W = 3$ (using *mstrip.m*).

P7.4. Square coaxial line is a configuration similar to that given in Figure 7.30. Consider the coaxial line shown in Figure 7.31. Using the symmetry considerations, the problem can be reduced to one-fourth of its size. Show that

$$\varphi_1 = 140/322\,V_0 \qquad \varphi_2 = 56/322\,V_0 \qquad \varphi_3 = 91/322\,V_0$$
$$\varphi_4 = 42/322\,V_0 \qquad \varphi_5 = 21/322\,V_0$$

Also, show that the capacitance per unit length of the line is given by

$$C_0 = 4\epsilon\,\frac{V_0 - \varphi_1}{V_0} = 2.26\epsilon$$

where ϵ is the dielectric constant of the medium filling the space between two conductors [7].

P7.5. The geometry of a suspended coupled strip line is shown in Figure 7.32. Determine the characteristic impedance of the line for $a = 5.0$ cm, $b = 5.0$ cm, $h = 1.0$ cm, $w = 2.0$ cm, $t = 0.001$ cm, $\epsilon_1 = \epsilon_0$, and $\epsilon_2 = 2.35\epsilon_0$. The reported impedance value is 65.02. The geometry of Figure 7.32 can propagate two different modes depending on the potential at the strips. If the strips are at the same potential, the mode of propagation is called the even mode. The symmetry plane parallel to

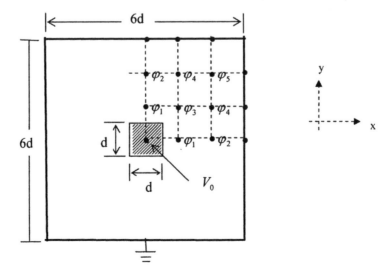

Figure 7.31   Cross-section of square coaxial line.

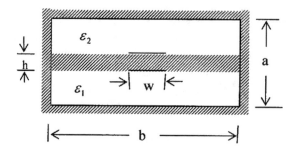

Figure 7.32 Cross-section of suspended coupled strip line.

the strips and passing through the substrate is the magnetic wall. When the strips are at different potentials, the mode of propagation is called the odd mode and the symmetry plane is the electric wall or ground plane. Placing this symmetry plane will convert the given geometry into a microstrip line geometry. Another symmetry consideration will reduce the problem to one-quarter size [27].

P7.6. The geometry of a coaxial line with circular inner conductor and rectangular outer conductor is shown in Figure 7.33. Determine the characteristic impedance of the line for $a = 1.0$ cm, $b = 1.25$ cm, $d = 0.51$ cm, and $\epsilon = \epsilon_0$. The reported impedance value is 50.43 ohms. You may use symmetry considerations to reduce the size of the problem [27].

P7.7 Solve the following wave equation:

$$\frac{\partial^2 \varphi}{\partial x^2} - \frac{1}{c^2} \frac{\partial^2 \varphi}{\partial t^2} = 0, \qquad 0 < x < 1, \qquad t \geq 0$$

subject to the boundary conditions

$$\varphi(0, t) = 0 = \varphi(1, t), \qquad t \geq 0$$

and the initial conditions

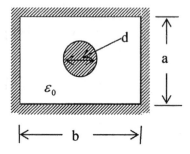

Figure 7.33 Cross-section of a coaxial line with circular inner conductor and rectangular outer conductor.

$$\varphi(x, 0) = \sin(\pi x), \qquad 0 < x < 1$$

$$\frac{\partial \varphi(x, 0)}{\partial t} = 0, \qquad 0 < x < 1$$

Compare your solution with the analytical value obtained from

$$\varphi(x, t) = \sin(\pi x) \cos(\pi c t)$$

Fortran code for a similar wave equation problem is available in [4].

P7.8. Solve the following Poisson equation:

$$\frac{\partial^2 \varphi}{\partial x^2} + \frac{\partial^2 \varphi}{\partial y^2} = -\frac{\rho_s}{\epsilon_0}, \qquad 0 \le x, y \le 1$$

subject to the boundary conditions

$$\varphi(x, 0) = 0V; \qquad \varphi(0, y) = -10V; \qquad \varphi(1, y) = 10V; \qquad \varphi(x, 1) = 20V$$

Assume $\rho_s = x(y - 1) \times 10^{-9}$ C/m^2, divide the region into 4×4 cells, and plot the potential at the free nodes. You may use successive over-relaxation. The analytical solution to this problem may be obtained by using (7.21) and superposition. If φ_1 is the solution to the Laplace equation $\nabla^2 \varphi_1 = 0$ subject to the boundary conditions given above, and φ_2 is the solution to the Poisson equation $\nabla^2 \varphi_2 = -\rho_s/\epsilon_0$ subject to $\varphi_2 = 0$ at $x = 0, 1; y = 0, 1$, then

$$\varphi = \varphi_1 + \varphi_2$$

where φ_2 is obtained by the eigenfunction expansion method and is given by

$$\varphi_2 = \sum_{m=1}^{\infty} \sum_{n=1}^{\infty} A_{mn} \sin\left(\frac{m\pi x}{a}\right) \sin\left(\frac{n\pi y}{b}\right)$$

with

$$A_{mn} = -\int_0^a \int_0^b \frac{\rho_s}{\epsilon_0} \sin\left(\frac{m\pi x}{a}\right) \sin\left(\frac{n\pi y}{b}\right) dx\, dy$$

For the given problem $a = 1$, $b = 1$ and $\rho_s = x(y - 1) \times 10^{-9}$ C/m^2, one obtains

$$A_{mn} = \frac{1 - (1 - (-1)^n)}{(m\pi)^2 + (n\pi)^2} \frac{144(-1)^{m+n}}{mn\pi} = \frac{-1}{(m\pi)^2 + (n\pi)^2} \frac{144(-1)^{m+n}}{mn\pi}, \text{ for } n \text{ odd}$$

The Fortran source code based on SOR and the solution is available in [4].

P7.9. Use FDM to solve the following one-dimensional scalar Helmholtz equation:

$$\frac{d^2 E_y(x)}{dx^2} + Ck^2 E_y(x) = 0$$

where $k = 2\pi/\lambda$ defines the wavelength in the medium. Show that the value of C such that the propagation constant β in the discretized medium is equal to k, the propagation constant in the continuous medium, is given by the expression

$$C = \left(\frac{\sin\left(\frac{k\Delta x}{2}\right)}{\frac{k\Delta x}{2}} \right)^2$$

Calculate C for $\Delta x/\lambda = 1/8$, $1/12$, $1/20$, and $1/40$ and compare with the values obtained in Problem 7.10.

P7.10. The field distribution in a transmission line resonator is described by the following scalar wave equation:

$$\frac{d^2 E_y(x)}{dx^2} + \pi^2 E_y(x) = g(x), \qquad 0 < x < 1$$

with the boundary conditions

$$E_y = 0 \text{ at } x = 0 \text{ and } x = 1$$

Assume the resonator length to be 1m and therefore the resonant frequency is 150 MHz. Now discretize the wave equation and use finite difference approximation to obtain the following expression for the ith cell:

$$2E_y(i) - E_y(i-1) - E_y(i+1) - (\pi\Delta z)^2 E_y(i) = g(i)$$

To fine tune the resonator and to determine the phase velocity in the discretized domain, we introduce a correction factor C in the above expression to give

$$2E_y(i) - E_y(i-1) - E_y(i+1) - C(\pi\Delta z)^2 E_y(i) = g(i)$$

Assume $g(i) = 1$ at the central cell and 0 for others. Tune the resonator to 150 MHz by varying C for three different cell sizes $\Delta x = 0.25$m, 0.2m, and 0.1m. Determine the amplitude distribution over the resonator as a function of C. Interpret the results in terms of phase velocity of the wave (see also Section 10.2.2).

P7.11 (a) Derive the finite difference implementation of the boundary condition $D_{in} = D_{2n}$ for V_0 located on the dielectric boundary as shown in Figure 7.6.

(b) Derive the finite difference implementation of the symmetry condition $\partial V/\partial n = 0$ along the y-axis as shown in Figure 7.34.

P7.12 Consider the rectangular waveguide of Figure 7.24 with $b = a/2$. The waveguide is half filled with a dielectric of $\epsilon = \epsilon_0 \epsilon_r$ so that the nodes 1, 2, 3, ... are on the interface as shown in Figure 7.35. We are interested in finding the approximate closed-form expression for the cutoff wavelength for the dominant TE_{10} mode. For this we use symmetry condition about $x = a/2$ plane and coarse discretization to model the problem as shown. The potential function φ represents H_z field for the TE mode. Therefore, the boundary conditions are: $\varphi = 0$ at 13, 3, and 8. Also assume that $\varphi_6 = \varphi_{11} = \varphi_1$ and $\varphi_{12} = \varphi_7 = \varphi_2$ for the mode. Show that the cutoff wavelength λ_c for $b = a/4$ is given by

$$\lambda_c = \frac{\pi a}{2} \frac{\sqrt{\epsilon_r + 1}}{\sqrt{\left(4 - \sqrt{2}\right)(\epsilon_r + 1) - 4}}$$

where a is the width of the waveguide.

P7.13. Consider a square waveguide with dimension $b = a$ as shown in Figure 7.36(a). Use FDM to determine the approximate solution for the cutoff wavelength for the dominant TM_{11} mode. For this we use symmetry conditions about $x = a/2$

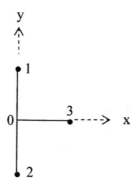

Figure 7.34 Geometry of nodes with Neumann boundary condition.

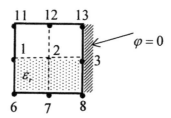

Figure 7.35 Geometry of a rectangular waveguide half filled with dielectric and symmetry condition employed.

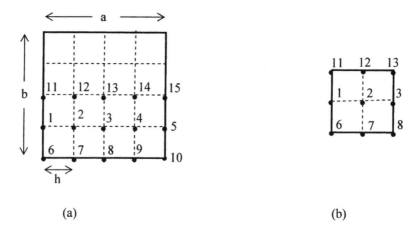

Figure 7.36 Geometry of a square waveguide: (a) discretized waveguide geometry; and (b) waveguide with two-fold symmetry.

and $y = b/2$ planes and coarse discretization with square grid to model the problem as shown in Figure 7.36(b). The potential function φ represents E_z field for the TM mode. Therefore, the boundary conditions are: $\varphi = 0$ at 11, 1, 6, 7, and 8; and $\partial\varphi/\partial n = 0$ at 12, 13, and 3. Show that the eigenvalues are given by $\lambda = \pm 2\sqrt{2}$, and for $h = a/4$ the cutoff wavelength λ_c for the mode is obtained as $\lambda_c = 1.4517a$ compared with the exact value of $1.414a$, an error of less than 3%.

P7.14. Consider a rectangular waveguide of dimension $a \times b$, and $b = a/2$. Determine the cutoff wavenumber k_c for the TM_{11} mode using FDM. Use $\Delta x = a/2$, $\Delta y = b/2$ for the discretization of the geometry. Compare with the analytical solution $k_c = \pi\sqrt{\left(\dfrac{1}{a^2} + \dfrac{1}{b^2}\right)}$. You may use $a = 2$ cm for comparison.

P7.15. Consider a parallel plate capacitor inhomogeneously filled with the dielectric as shown in Figure 7.37. The upper plate is charged to 1 volt and the lower plate is grounded. The boundary conditions therefore are: $\varphi_i = 0$ for $i = 1$ to 7, and $\varphi_i = 1$ for $i = 15$ to 21. The side walls of the capacitor are terminated in magnetic

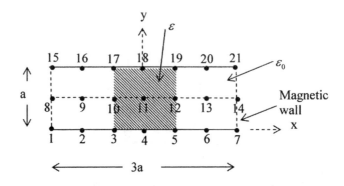

Figure 7.37 Geometry of a parallel plate capacitor inhomogeneously filled with dielectric.

walls to limit the size of FDM domain. The geometry is discretized as shown with $\Delta x = \Delta y = a/2$. Determine the capacitance per unit length. Compare the computed capacitance with the analytical value $C = (2\epsilon_0 + \epsilon)$.

P7.16. The singularity of charge density distribution on the metal strips in planar lines is principally determined by the dielectric configuration below the strip. In CPW line, the charge distribution is similar to the one that shown in Figure 7.18 if the substrate is single layered. Let us now introduce two thin dielectric layers between the main substrate and the metal strip as shown in Figure 7.38. Determine the charge density distribution on the central strip and show that the increase in density towards the edges is almost linear for $w = 30$ μm, $s = 15$ μm, $\epsilon_{r1} = 3.8$, $\epsilon_{r2} = 440(1 - j0.01)$, and $\epsilon_{r3} = 23.5$ [28].

P7.17. Use FDM to solve the following differential equation for the box shown in Figure 7.39:

$$\frac{\partial^2 V}{\partial x^2} + \frac{\partial^2 V}{\partial y^2} + 50 = 0, \qquad 0 \le x, y \le 1$$

subject to the boundary conditions $V = 10$ at $x = 0, 1$; $\partial V/\partial y = 40$ at $y = 0$, and $\partial V/\partial y = -20$ at $y = 1$. Assume $\Delta x = \Delta y = h = 1/3$ and

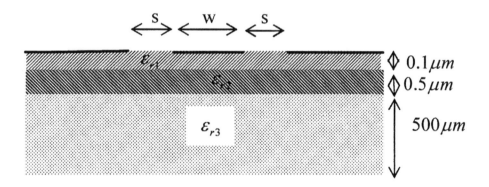

Figure 7.38 A multilayered CPW geometry.

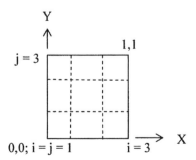

Figure 7.39 Discretized rectangular box geometry.

1. Obtain an expression for $V(i, j)$ in terms of the values at the four nodes surrounding it.
2. To implement the boundary condition at $y = 0$, obtain an expression for $V(i, 1)$ in terms of values at the nodes surrounding it.
3. To implement the boundary condition at $y = 1$, obtain an expression for $V(i, 4)$ in terms of values at the nodes surrounding it.
4. How can you reduce the number of unknowns? Write down the final matrix equation to be solved.
5. Develop a computer code to solve the same problem using $h = 0.05$, 0.1, 0.2, 0.25, and 0.5.

CHAPTER 8
Finite-Difference Time-Domain Analysis

Most of the full wave computational methods produce the wave analysis of the device when subjected to a time harmonic excitation of the type $\exp(\omega t)$. The final result is the steady state behavior of the field. Between the excitation and steady state, the wave undergoes reflection in to-and-fro manner and the time history of the process is not available. The measurements at RF and microwave frequencies are generally carried out with time harmonic excitations and steady state behavior is measured. The measurements are therefore compatible with time harmonic analysis of the wave equation. The time domain measurements are not carried out due to the nonavailability of fast oscilloscopes at these frequencies. However, things are changing now. Not only is the analysis carried out to determine the time domain or transient behavior, but fast oscilloscopes are available now to record events at the RF and microwave frequencies. FDTD method is compatible with changing scenario in electromagnetics in this respect. The impetus for the progress in FDTD method is provided by the unique contributions of Yee [1].

To simulate time varying electromagnetic fields in any linear, isotropic medium, Maxwell's curl equations are sufficient, and the solution of these equations generates time history of the event. This is called *animation*, which separates FDTD from other computational methods. This feature of FDTD is very useful for diagnostics of the device, for broadband information, and as a teaching aid.

FDTD analysis simulates propagation behavior in time domain. The steady state behavior of a device is the cumulative effect of the time domain behavior and therefore these behaviors are different from each other. The excitation is not assumed to be time harmonic to bring out the usefulness of FDTD. Since we are mainly trained for steady state analysis, it takes time to learn FDTD analysis. Some of the basic concepts of FDTD, like reflection at an impedance mismatch, standing wave and resonance formation, can be understood from the wave propagation in transmission lines. This is discussed next.

8.1 Pulse Propagation in a Transmission Line

Consider a transmission line characterized by impedance Z_0 and the phase velocity v_p. We assume the transmission line to be lossless and the phase velocity to be constant, independent of frequency. Let us launch a rectangular pulse of width Δt as shown in Figure 8.1. The source impedance is R_g and the load impedance is R_L.

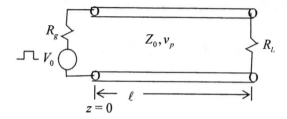

Figure 8.1 A transmission line terminated with load resistance R_L and driven by a pulsed source of internal resistance R_g.

We assume that no signal exists on the line at $t < 0$, and the pulse is launched at $t = 0$. The pulse originates at $z = 0$ and travels towards the load. Let the voltage of the pulse be V^+ and the current be I^+ at $z = 0$. These values can be determined from the lumped circuit of Figure 8.2(a). It may be noted that the load resistance R_L is not shown in the circuit at $t = 0$ because the wave has not reached the other end of the line where the load is located. This is different from the lumped circuit behavior where the effect of transient in any part of the circuit is assumed felt instantaneously in other parts of the circuit. *It is not so in the transmission line or distributed circuits because travel delay is comparable to 1/frequency.* Applying Kirchhoff's voltage law to the circuit in Figure 8.2(a) gives

$$V^+ = V_0 - I^+ R_g \qquad (8.1)$$

Also, $I^+ = V^+/Z_0$. Therefore,

$$V^+ = V_0 \frac{Z_0}{R_g + Z_0}, \qquad I^+ = \frac{V^+}{Z_0} = \frac{V_0}{R_g + Z_0} \qquad (8.2)$$

The voltage wave V^+ travels towards the load with velocity v_p and reaches the end of the line at time $t = \ell/v_p$. The wave gets reflected at the load because of the mismatch between the line impedance Z_0 and the load R_L. Let the reflected wave be described by voltage amplitude V^- and current amplitude I^-. The total voltage across R_L is therefore $V^+ + V^-$ and the total current through the load is $I^+ + I^-$ as shown in Figure 8.2(b). Applying Kirchhoff's voltage law at the load gives

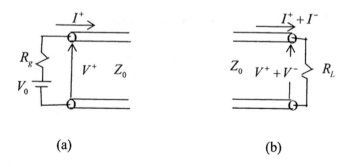

(a)   (b)

Figure 8.2 Equivalent circuits for the wave: (a) at $z = 0$, $t = 0$; and (b) at $z = l$.

$$V^+ + V^- = R_L(I^+ + I^-) \tag{8.3}$$

We know that $I^+ = V^+/Z_0$ and $I^- = -V^-/Z_0$ (by convention the current is assumed negative when directed backward); therefore,

$$V^- = V^+ \frac{R_L - Z_0}{R_L + Z_0} \tag{8.4}$$

With the voltage reflection coefficient of the load defined as

$$\Gamma_L = \frac{V^-}{V^+} = \frac{R_L - Z_0}{R_L + Z_0} \tag{8.5}$$

the reflected voltage is given by

$$V^- = V^+ \Gamma_L \tag{8.6}$$

This wave travels towards the generator and gets reflected there with voltage reflection coefficient of the generator defined as

$$\Gamma_g = \frac{R_g - Z_0}{R_g + Z_0} \tag{8.7}$$

The process of rereflection continues in time until the energy of the pulse is absorbed in the resistors.

In order to illustrate the effect of reflections on pulse amplitude in time, let us assume that the velocity of the wave and length l of the line are such that the travel time on the line $\ell/v_p = 1$ ns. The pulse width Δt is assumed to be 100 ps, that is, one-tenth of travel time. The velocity of the wave is assumed to be independent of frequency so that there is no dispersion. The other parameters are assumed as: $Z_0 = 50\Omega$, $R_L = 30\Omega$, $R_g = 20\Omega$, and $V_0 = 2$V. Therefore, $\Gamma_L = -0.250$, $\Gamma_g = -0.429$, and $V^+ = 1.429$V. Let us plot the pulse amplitude at the mid-point $z = l/2$ as a function of time. The (incident) pulse of amplitude 1.429V reaches this point at $t = 0.5$ ns. After reflection from the load, the pulse of amplitude $\Gamma_L V^+ = -0.357$V reaches $z = l/2$ at $t = 1.5$ ns. This pulse will get reflected from the source side and reach this point again at $t = 2.5$ ns with amplitude $\Gamma_g \Gamma_L V^+ = 0.429 \times 0.357 = 0.153$V. The progress of the pulse with time is plotted in Figure 8.3. It may be observed that the amplitude of the pulse decreases steadily with time due to absorption in the resistors.

If the wave velocity v_p is a function of frequency, the various frequency components of the pulse will take different times to reach the load, and the pulse will get broadened.

Review Question 8.1.   Consider the transmission line circuit of Figure 8.1. For $V_0 = 2$V, $R_g = Z_0 = 50\Omega$, $R_L = \infty$, the voltage $V(z)$ for $t > 2\ell/v_p$ is given by:

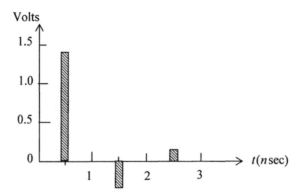

Figure 8.3 Progress of voltage pulse in a terminated transmission line at $z = l/2$. The parameters are: $Z_0 = 50\Omega$, $R_L = 30\Omega$, $R_g = 20\Omega$, $V_0 = 2V$, pulse width = 100 ps, $\ell/v_p = 1$ ns.

1. 0V
2. 0.5V
3. 2V
4. 1V

Choose the correct entry and explain why?

Answer 1. The pulse reflected from R_L travels towards the source and gets absorbed there, since $R_g = Z_0$.

8.2 FDTD Analysis in One Dimension

Let us now analyze the transmission line problem of Section 8.1 in the language of FDTD. To simplify the implementation of FDTD we shall discuss plane wave propagation in one dimension in free space. We assume TEM-wave propagation along the x-direction although the z-direction would have been consistent with the analysis in Section 8.1. The Maxwell's curl equations (1.1) for TEM wave propagation along the x-direction reduce to the following (for $\partial/\partial y = 0$, $\partial/\partial z = 0$, $J_S = 0$)

$$\epsilon_0 \frac{\partial E_y}{\partial t} = -\frac{\partial H_z}{\partial x} \tag{8.8a}$$

$$\mu_0 \frac{\partial H_z}{\partial t} = -\frac{\partial E_y}{\partial x} \tag{8.8b}$$

For the FDTD analysis, we express the derivatives in terms of finite difference approximations as in Chapter 7. The central difference approximation is used for higher accuracy and is defined as

$$\frac{\partial f}{\partial u}\bigg|_{u_0} = \frac{f\left(u_0 + \dfrac{\Delta u}{2}\right) - f\left(u_0 - \dfrac{\Delta u}{2}\right)}{\Delta u}\Bigg|_{\Delta u \to 0} + O(\Delta u)^2 \qquad (8.9)$$

where $O(.)$ stands for *the order of*. Expression (8.9), when applied to (8.8), implies that the E and H fields should be known in space and time at discrete points (x_i, t_n) only, where $x_i = i\Delta x$ and $t_n = n\Delta t$ with Δx and Δt representing the step size.

Staggered Grids

First we discretize the space domain in segments of size Δx and the time domain in steps of Δt. Nodes are introduced at the end points and at the mid-point of the segments. Next, we *arbitrarily* locate the electric field nodes E_y at integer values of i and the magnetic field nodes H_z at half-integer values of i such that electric fields nodes are separated by magnetic field nodes and vice versa. This is called *interleaving* in space and is shown in Figure 8.4. For (8.8a), we apply the central finite-difference approximation at $x_i = i\Delta x$ and $t_n = (n + 1/2)\Delta t$, giving rise to

$$\frac{E_y^{n+1}(i) - E_y^n(i)}{\Delta t} = \frac{-1}{\epsilon_0} \frac{H_z^{n+1/2}\left(i + \dfrac{1}{2}\right) - H_z^{n+1/2}\left(i - \dfrac{1}{2}\right)}{\Delta x} \qquad (8.10a)$$

This approximation is also called staggered grid approximation. Similarly, (8.8b) is subjected to derivatives at $((i + 1/2)\Delta x, n\Delta t)$ for consistency with (8.10a) and grid arrangement,

$$\frac{H_z^{n+1/2}\left(i + \dfrac{1}{2}\right) - H_z^{n-1/2}\left(i + \dfrac{1}{2}\right)}{\Delta t} = \frac{-1}{\mu_0} \frac{E_y^n(i + 1) - E_y^n(i)}{\Delta x} \qquad (8.10b)$$

The electric field values are computed at $n\Delta t$ instants, whereas magnetic field values are computed at $(n + 1/2)\Delta t$ instants; that is, field calculations are interleaved in time also. The alternate updating of electric and magnetic fields in time reduces the processor time by half. Averaging may be used to determine the fields at intermediate nodes or time instants if needed.

Equations (8.10) may be arranged in an algorithm as

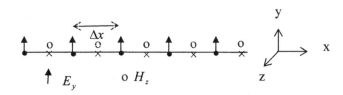

Figure 8.4   Positions of E and H nodes for TEM wave propagation along the x-direction.

$$H_z^{n+1/2}\left(i+\frac{1}{2}\right) = H_z^{n-1/2}\left(i+\frac{1}{2}\right) - \frac{\Delta t}{\mu_0 \Delta x}\left[E_y^n(i+1) - E_y^n(i)\right] \qquad i = 1, 2, \ldots$$
$$(8.11a)$$

$$E_y^{n+1}(i) = E_y^n(i) - \frac{\Delta t}{\epsilon_0 \Delta x}\left[H_z^{n+1/2}\left(i+\frac{1}{2}\right) - H_z^{n+1/2}\left(i-\frac{1}{2}\right)\right] \qquad i = 1, 2, \ldots$$
$$(8.11b)$$

An alternative to the first order coupled differential equations of (8.8) is the second order differential equation in E_y or H_z, called wave equation. Eliminating H_z from (8.8) yields the following wave equation in E_y:

$$\frac{\partial^2 E_y}{\partial t^2} = \frac{1}{\mu_0 \epsilon_0}\frac{\partial^2 E_y}{\partial x^2} \qquad\qquad (8.12a)$$

The discretization of this equation on integer grid points in t and x leads to

$$\frac{E_y^{n+1}(i) - 2E_y^n(i) + E_y^{n-1}(i)}{(\Delta t)^2} = \frac{1}{\epsilon_0 \mu_0}\frac{E_y^n(i+1) - 2E_y^n(i) + E_y^n(i-1)}{(\Delta x)^2}$$
$$(8.12b)$$

The solution of this equation and that based on the staggered grid approach of (8.11) are found to be equally robust in dispersion and stability [2, p. 65].

Space, Time Marching of Fields

The expression (8.11a) shows that the new value of H_z is calculated from its previous value and the most recent value of E_y. The computation of E_y is carried out similarly. This process is depicted graphically in Figure 8.5 and is called leapfrog marching [3]. The rectangular grid in space-time domain is also shown there. Time marching of fields is similar to the marching of a two-legged creature with one leg representing the electric field and the other representing the magnetic field. Expressions (8.11) may be coded as follows in MATLAB:

```
% nmax is the number of time steps for which the simulations are
carried out.
% imax is the number of nodes.

    for n = 1 to nmax
        for i = 1 to imax
    hz(i) = hz(i) - b*(ey(i+1)-ey(i))        %b=Δt/(μ₀Δx)
    ey(i) = ey(i)-a*(hz(i) -hz (i-1))        %a=Δt/(ε₀Δx)
        end
    end
```

The time index n is implicit in these expressions. The values on the left side are the new values (at a later instant) being calculated, whereas similar entries on the right side are the old values of the available field components. The space index

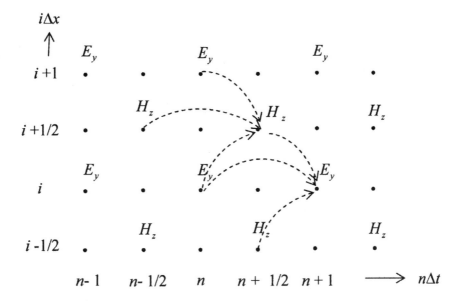

Figure 8.5 Discretized space-time domain and graphical representation of time marching of fields according to (8.11). (*After:* [4].)

i is explicit. However, indices $i + 1/2$ and $i - 1/2$ for the array $hz(.)$ have been rounded off to i and $i - 1$, respectively, for the ease of book keeping [5].

Let us try to understand the physical process and the mathematical procedure involved in space-time marching of fields represented by (8.11). Consider the discretized space along the x-direction and shown in Figure 8.6. The E-nodes are numbered ... 4, 5, 6, 7 ... in this figure.

Let us assume for simplicity: $\Delta x = 1$ mm, and $c\Delta t = \Delta x$. With this choice, (8.11) reduces to

$$H_z^{n+1/2}\left(i + \frac{1}{2}\right) = H_z^{n-1/2}\left(i + \frac{1}{2}\right) - \frac{1}{120\pi}\left[E_y^n(i + 1) - E_y^n(i)\right] \qquad i = 1, 2, \ldots$$

$$(8.13a)$$

$$E_y^{n+1}(i) = E_y^n(i) - 120\pi\left[H_z^{n+1/2}\left(i + \frac{1}{2}\right) - H_z^{n+1/2}\left(i - \frac{1}{2}\right)\right] \qquad i = 1, 2, \ldots$$

$$(8.13b)$$

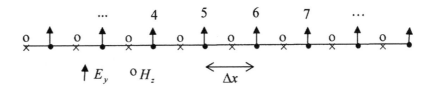

Figure 8.6 Discretized space along the x-direction showing positions of E- and H-nodes.

The following starting conditions are assumed: $E_y^0(i) = 0$, $H_z^0(i) = 0$. Let us now introduce excitation in the grid through initial conditions by defining $E_y^0(5) = 1$, $E_y^0(6) = 1$. The update of the fields at the various nodes results in the following values ($\eta = 120\pi$):

nodes →	3	3½	4	4½	5	5½	6	6½	7	7½	8
$n = 0$	0	0	0	0	1	0	1	0	0	0	0
$n = \frac{1}{2}(H_z^{n+1/2})$	—	0	—	$-1/\eta$	—	0	—	$1/\eta$	—	0	—
$n = 1(E_y^{n+1})$	0	—	1	—	0	—	0	—	1	—	0
$n = 1\frac{1}{2}$	—	$-1/\eta$	—	0	—	0	—	0	—	$1/\eta$	—
$n = 2$	1	—	0	—	0	—	0	—	0	—	1

$$(8.14)$$

The excitation is located at nodes 5 and 6 as indicated by initial conditions. The nodal values for various time instants show that the excitation marches outward as the time progresses. This is the essence of propagation phenomenon. The phase velocity can be determined from the displacement of excitation with time. Since the excitation moves one node in Δt, the velocity of propagation is c, the velocity of light. If Δt is reduced to half its present value, the number of time steps required will double for the same distance traveled, resulting in the same wave velocity.

Exercise 8.1.   Show that if Δt is reduced to half (i.e., $c\Delta t = \Delta x/2$) in the example discussed above) the new table is obtained as follows:

nodes →	3	3½	4	4½	5	5½	6	6½	7	7½	8
$n = 0$	0	0	0	0	1	0	1	0	0	0	0
$n = \frac{1}{2}$	—	0	—	$-1/2\eta$	—	0	—	$1/2\eta$	—	0	—
$n = 1$	0	—	1/4	—	3/4	—	3/4	—	1/4	—	0
$n = 1\frac{1}{2}$	—	$-1/8\eta$	—	$-3/4\eta$	—	0	—	$3/4\eta$	—	$1/8\eta$	—
$m = 2$	1/16	—	9/16	—	3/8	—	3/8	—	9/16	—	1/16

We next clarify some important issues such as step sizes Δx, Δt, source excitation, absorbing boundary condition and stability of the solution before we work out a one-dimensional problem using FDTD.

8.2.1   Spatial Step Δx and Numerical Dispersion

In Chapter 7, we discussed the effect of step size Δx on the accuracy of solution. Associated with nonzero step size Δx is the phenomenon of *numerical* dispersion. However, in FDTD the numerical dispersion depends on time step Δt also.

The finite difference implementation of the derivatives makes the solution approximate. Consequently, the phase constant varies with frequency in an otherwise dispersionless medium. This dispersion is called *numerical dispersion* [3, 4]. The amount of dispersion depends on the wavelength, and the discretization size Δx, and Δt. The effect of numerical dispersion is equivalent to filling the medium with a dielectric constant different from the actual dielectric constant of the medium.

To illustrate the phenomenon of numerical dispersion let us consider the wave equation of (8.12a), repeated here for convenience,

$$\frac{1}{c^2} \frac{\partial^2 E_y}{\partial t^2} = \frac{\partial^2 E_y}{\partial x^2} \tag{8.15}$$

Consider the following monochromatic, sinusoidal, traveling wave solution,

$$E_y(x, t) = e^{j(\omega t - kx)} \tag{8.16}$$

with propagation constant,

$$k = \pm \frac{\omega}{c} \tag{8.17}$$

The phase velocity, group velocity, and guide wavelength or grid wavelength for the wave are defined as

$$\text{phase velocity, } v_p = \frac{\omega}{k} = \pm c \tag{8.18a}$$

$$\text{group velocity, } v_g = \frac{d\omega}{dk} = \pm c \tag{8.18b}$$

$$\text{grid wavelength, } \lambda_g = \frac{2\pi}{k} \tag{8.18c}$$

The phase velocity being independent of frequency implies *dispersionless propagation*. Now, we shall determine the finite-difference solution of (8.15) and the associated numerical dispersion.

Use of the central difference approximation for the double derivatives in (8.15) gives

$$\frac{E_y^{n+1}(i) - 2E_y^n(i) + E_y^{n-1}(i)}{(\Delta t)^2} + O[(\Delta t)^2] = c^2 \left\{ \frac{E_y^n(i+1) - 2E_y^n(i) + E_y^n(i-1)}{(\Delta x)^2} + O[(\Delta x)^2] \right\} \tag{8.19}$$

Let us assume a discretized version of solution (8.16). This is obtained by sampling the continuous solution at the discrete space-time point (x_i, t_n),

$$E_y^n(i) = E_0 e^{j(\omega n \Delta t - \bar{k} i \Delta x)} \tag{8.20}$$

Here, $\overline{k}$ is the propagation constant of the *numerical* traveling wave in the finite-difference grid. In general, $\overline{k}$ differs from k and this difference causes numerical errors resulting in numerical dispersion artifacts.

We now substitute (8.20) in (8.19) and neglecting the error term on both sides obtain the following expression:

$$e^{j(\omega(n+1)\Delta t - \overline{k}i\Delta x)} = \left(\frac{c\Delta t}{\Delta x}\right)^2 \left[e^{j(\omega n\Delta t - \overline{k}(i+1)\Delta x)} - 2e^{j(\omega n\Delta t - \overline{k}i\Delta x)} + e^{j(\omega n\Delta t - \overline{k}(i-1)\Delta x)}\right]$$

$$+ \left[2e^{j(\omega n\Delta t - \overline{k}i\Delta x)} - e^{j(\omega(n-1)\Delta t - \overline{k}i\Delta x)}\right] \tag{8.21}$$

After deleting the common factors on both sides, we obtain

$$e^{j\omega\Delta t} = \left(\frac{c\Delta t}{\Delta x}\right)^2 \left(e^{-j\overline{k}\Delta x} - 2 + e^{j\overline{k}\Delta x}\right) + (2 - e^{-j\omega\Delta t})$$

or

$$\cos(\omega\Delta t) = \left(\frac{c\Delta t}{\Delta x}\right)^2 [\cos(\overline{k}\Delta x) - 1] + 1$$

or

$$\sin\left(\frac{\omega\Delta t}{2}\right) = \pm\left(\frac{c\Delta t}{\Delta x}\right)\sin\left(\frac{\overline{k}\Delta x}{2}\right) \tag{8.22}$$

Equation (8.22), relating ω and $\overline{k}$, is called the *dispersion relation*. It can be verified that this expression reduces to the exact solution (8.17) in the limit $\Delta x \to 0$ and $\Delta t \to 0$. It is interesting to note that (8.22) again reduces to (8.17) for the time step $c\Delta t = \Delta x$, independent of the choice of space and time steps (coarse or fine). The wave propagation is therefore dispersionless for this combination of Δt and Δx. Because of the unexpected behavior of the numerical wave, the step size $\Delta t = \Delta x/c$ is called *magic time step*. The reason for zero numerical dispersion is that (8.19) is no longer an approximate form of (8.15); it becomes exact with error term on both sides of (8.19) canceling each other. The magic time step is a useful feature of one-dimensional FDTD analysis. It allows (ideal) FDTD analysis of the problem without incurring the errors associated with discretization.

Defining $\overline{v}_p = \omega/\overline{k}$, (8.22) may also be expressed as

$$\frac{\overline{v}_p}{c} = \frac{\pi\Delta x}{\lambda_0} \frac{1}{\sin^{-1}\left(\frac{\sin\left(\frac{\pi\alpha\Delta x}{\lambda_0}\right)}{\alpha}\right)} \tag{8.23}$$

where $\alpha = c\Delta t/\Delta x$ is called the stability ratio, and λ_0 is the free space wavelength. The phase velocity $\bar{v}_p/c$ is plotted in Figure 8.7 as a function of the cell size $\Delta x/\lambda_0$ with α as a parameter. These curves are compared with the exact solution $\alpha = 1$. We observe that the numerical wave defined by (8.20) shows error in phase velocity with discretization. For small values of Δx, the errors are small. In other words, the numerical dispersion can be reduced by using more accurate finite difference approximations of second order derivatives. For a given value of Δx, the use of small relative time steps ($\alpha < 1$) also increases phase error, consistent with the magic time step.

Another measure of numerical dispersion is the grid wavelength defined as $\bar{\lambda}_g = 2\pi/\bar{k}$. We obtain the following relationship:

$$\frac{\bar{v}_p}{c} = \frac{\omega}{\bar{k}}\frac{1}{c} = \frac{\bar{\lambda}_g}{2\pi}\frac{\omega}{c} = \frac{\bar{\lambda}_g}{\lambda_0} \tag{8.24}$$

For $\bar{v}_p < c$, the grid wavelength $\bar{\lambda}_g < \lambda_0$. The discretized free space may therefore be modeled as a medium with equivalent dielectric constant $\epsilon_e > 1$ which varies with frequency.

To get an idea about the phase error introduced by dispersion, consider an example with $\Delta x = \lambda_0/5$ and $\alpha = 1/2$. Inserting the assumed values in (8.23), we obtain $\bar{v}_p/c = 0.943$. The numerical phase velocity for this wave is seen to be in error of -5.7% compared to the free space phase velocity c. It implies that for a physical wave propagating over a distance of $10\lambda_0$ (50 space cells), the numerical

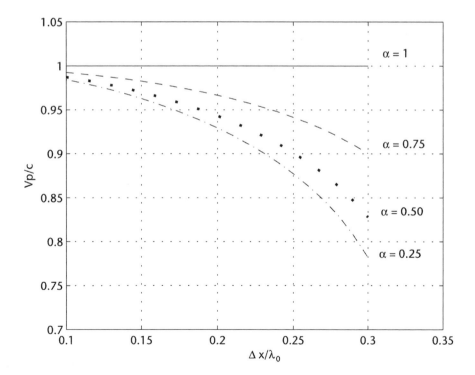

Figure 8.7 Illustration of numerical dispersion in one dimension, which is variation of normalized phase velocity with normalized cell size with α as a parameter.

wave would propagate only $50 \times 0.943 = 47.15$ cells. It amounts to a phase error of $(50 - 47.15) \times 360°/5$, or $205.2°$, and is considerable. The phase error due to numerical dispersion can be reduced by reducing the cell size. For example, in the above example if the cell size is reduced by half, the phase error will reduce by a factor of 4 (i.e., $51.3°$). The factor of four reduction in phase error is due to the second order accuracy of central difference approximation.

We know that the origin of pulse dispersion is the difference in phase velocity of the different frequency components of the pulse. To investigate this phenomenon further, one may carry out the following simulations. First, the propagation of the time pulse is observed in the FDTD grid as described above. Next the discrete spectral components of the pulse are obtained using the fast Fourier transform (FFT). Each spectral component is multiplied by the appropriate phase factor, added and inverse FFT taken to reconstruct the pulse after n time steps. The reconstructed pulse is found to compare well with the FDTD solution [6]. On the other hand, if the phase velocities are replaced by the true speed c (of propagation in continuous medium), no distortion in waveform is observed.

Rule of Thumb for the Step Size Δx

For FDTD analysis, the medium in the electromagnetic device is discretized in cells such that the material properties and the field distribution in the cells are assumed to be uniform. Smaller cell size means more number of cells for a given device size and therefore better accuracy at the cost of increased computational requirement. The cell size is a function of frequency, material parameters, and the expected field distribution where the cell is located. As a compromise between the accuracy and computational resources, the rule of thumb is that in the worst case situation, the cell size should not exceed $\lambda/10$ at the highest frequency and the highest value of material loading. For an accuracy of about 1%, one may reduce the cell size to $\lambda/20$. For initial estimates or for crude approximation one may even choose $\lambda/5$ for the cell size.

8.2.2   Time Step Δt and Stability of the Solution

Once the cell size Δx has been selected, we next choose the time step Δt since the two step sizes are related. The minimum time taken for the wave to travel a distance Δx between two consecutive nodes is $\Delta x/v_{max}$, where v_{max} is the maximum phase velocity in the device. The maximum time step permitted is therefore $\Delta t_{max} = \Delta x/v_{max}$. If we choose $\Delta t > \Delta t_{max}$, the distance traveled by the wave over the time interval Δt will be more than Δx and the wave will miss the next node implying that the FDTD grid is not causally connected, and leads to instability in the solution. The condition

$$\Delta t < \frac{\Delta x}{v_{max}} \tag{8.25}$$

is called Courant-Friedrich-Levy (CFL) *stability condition* [3], and $\alpha = c\Delta t/\Delta x$ is called the stability factor. Due to the inhomogeneity of the medium in the device, the phase velocity may vary from cell to cell and from one frequency to the other. We may therefore introduce a safety margin and choose $\Delta t = (1/2)\Delta x/v_{max}$

uniformly to simplify coding. It implies that $\alpha = 1/2$ and the wave takes $2\Delta t$ to travel to the next node.

Example 8.1.   In this example we study the propagation of TEM wave in free space. Our aim is to quantify the effect of dispersion on the pulse shape. Let the wave consist of E_y and H_z components with propagation along the x-direction. The propagation of the discretized version of the wave is described by (8.11). For the simulation we use a rectangular pulse. In order to limit the size of space domain for simulations, we use the first order Mur analytical boundary condition (to be discussed later), at the ends. This condition is defined as

$$E_y^{n+1}(N + 1) = E_y^n(N) + \left(\frac{\alpha - 1}{\alpha + 1}\right)\left(E_y^{n+1}(N) - E_y^n(N + 1)\right) \qquad (8.26)$$

Figure 8.8 shows the propagation of a rectangular pulse in the discretized space. The excitation is implemented as a rectangular pulse of width $T = 500$ ps on the node $i = 0$. The values of α used are 1, 0.99, and 0.5. As seen in Figure 8.8(a) for $n = 250$, the pulse shape in space is a rectangular pulse with spatial pulse width given by $cT = 3 \times 10^8 \times 500 \times 10^{-12} = 0.15$m. Since Δx used in the simulations is 1.5 mm, the spatial width of the pulse is 100 space steps. The distance traveled by the leading edge of the pulse over 250 time steps is $250\Delta x$ because $\alpha = 1$. The trailing edge of the pulse is located at $i = 150$ as shown in the figure. Figure 8.8(b) depicts the same pulse at $n = 250$ for $\alpha = 0.99$. The significance of this value of alpha is that although the time step $\Delta t = \alpha\Delta x/c$ is very close to the magic time step, the various frequency constituents of the pulse travel at different phase velocities giving rise to distortion in pulse shape. In Figure 8.8(c), we have selected $\alpha = 0.5$. The value of Δt therefore becomes half and it takes a larger number of time steps for the pulse to travel the same distance. It took 480 time steps now for the leading edge of the pulse to reach $i = 250$, as shown in Figure 8.8(c). Also, the received pulse is broader due to dispersion. The calculations are carried out using the source code *dispersion.m*. There is a provision in the code to select the Gaussian pulse.

The numerical stability of the solution is illustrated through Figure 8.9. Here we plot the propagation of Gaussian pulse in discretized space. The value of $\alpha = 1.00004$ has been purposely chosen to violate the CFL stability condition, (8.25). Figure 8.9(a) shows that the instability has set in at $n = 480$ and is described by the nonzero amplitude at the trailing edge of the pulse. The instability is mainly contributed by the high frequency components of the pulse traveling at lower speeds. As time progresses, the amplitude at the site of instability increases as shown in Figure 8.9(b) for $n = 550$.

It may be noted from Figure 8.8 that a rectangular pulse (in time) launches a rectangular pulse in space. Similarly, a Gaussian pulse in time will launch a Gaussian pulse in space domain. The transformation from time domain to space domain is independent of the shape of the waveform because of the linear relation $x = vt$. For example, a window pulse in time described by

$$E(t) = E_0, \qquad \text{for } t_1 \leq t \leq t_2 \qquad (8.27a)$$

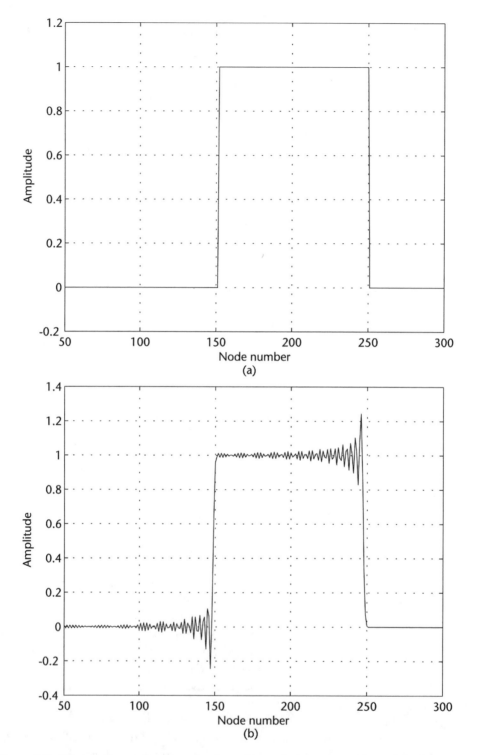

Figure 8.8 Effect of dispersion on the pulse shape: (a) magic time step $\alpha = 1$, $n = 250$, no distortion; (b) $\alpha = 0.99$, $n = 250$, distortion at the leading and trailing edges of the pulse; and (c) $\alpha = 0.5$, $n = 480$, broader pulse due to dispersion.

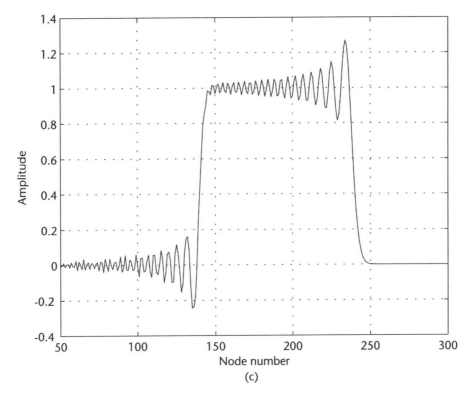

Figure 8.8 (continued).

will produce a window function in space with the spatial width given by

$$E(x) = E_0, \qquad vt_1 \le x \le vt_2 \tag{8.27b}$$

Hence, we may specify the excitation or source in FDTD in time domain only. However, if we are exciting a particular mode like TE_{mn}-mode in a waveguide, the field distribution in space matching that of TE_{mn}-mode will have to be specified as a part of source function.

8.2.3 Source or Excitation of the Grid

In the FDTD analysis, we launch a particular time waveform to carry out narrowband or wideband analysis of the device. This waveform is called excitation or source. The source may be introduced in two different ways. These are based on the initial conditions, and a point source specification. The excitation of the grid in (8.14) is through the initial conditions specified as $E_y^0(5) = 1$, $E_y^0(6) = 1$. This is a rectangular pulse excitation with pulse width equal to $2\Delta x$. One may choose Gaussian or other forms of excitation by specifying the source distribution in space as initial condition. A Gaussian pulse is defined as

$$E_y(x, t) = \exp\left[-\left(\frac{x - x_c - ct}{w}\right)^2\right] \tag{8.28}$$

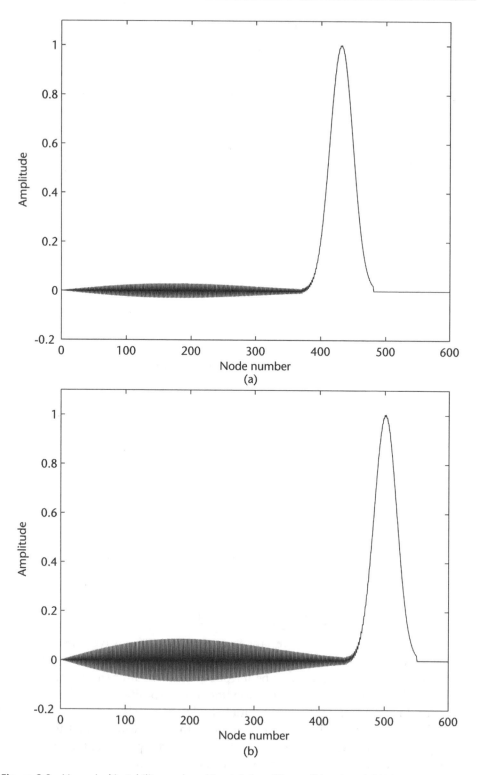

Figure 8.9 Numerical instability produced by violating CFL condition ($\alpha = 1.00004$, $\Delta x = 1.5$ mm, pulse width = 500 ps): (a) $n = 480$, sign of instability; and (b) $n = 550$, growth of instability with time.

where w is the width of the pulse in space. The discretized version of (8.28) at $t = 0$ for all nodes in the grid is given by

$$E_y^0(i) = \exp\left[-\left(\frac{i - i_s}{i_w}\right)^2\right] \qquad (8.29)$$

Here i_s stands for the peak position of the pulse and i_w is its width, both expressed in index i. The initial condition excitation is not a compact representation because it requires specifying the source amplitude at a number of nodes.

In the *point source excitation,* the excitation is introduced at the source node while other nodes are specified to be zero at $t = 0$. The value at the source node is updated with time according to the amplitude variation of the source. For the Gaussian pulse, the value at the source node i_s is varied as

$$E_y^n(i_s) = \exp\left(-\left[\frac{n - n_0}{n_w}\right]^2\right) \qquad (8.30)$$

where n_0 is the center of the pulse and n_w is the width of the pulse, both defined in time step Δt. As the value of n is increased from 0 to n_0, the source amplitude increases in the Gaussian fashion and then decreases. The algorithm described earlier may be modified in the following manner to include the point source:

```
for n = 1 to nmax%  time loop
pulse = Gaussian/ sinusoidal/window function in time

    for i = 1 to imax    % node loop
hz(i) = hz(i) - b*(ey(i + 1) - ey(i))     %b=Δt/(μ₀Δx)
ey(i) = ey(i) - a*(hz(i) - hz(i - 1))     %a=Δt/(ε₀Δx)
ey (ic) = pulse    % ic is the source node
end
    end
```
$$(8.31)$$

In the above algorithm, the statement on the source node *ic* overrides what was previously calculated. The normal FDTD update now propagates the source pulse away from the source node.

Hard Source and Soft Source.   The expression (8.30) implies that the source function should be *on* during the entire simulation. However, this may give rise to what is called *hard source* conditions. As time stepping is continued for sufficiently large values of n, the reflected numerical wave generated by the scatterer eventually returns to the source node. Because the source condition, $E_y^n(i_s)$, fixes the total electric field at the node i_s, regardless of any impinging wave there, it is equivalent to a spurious, retro-reflection of these waves at i_s back toward the scatterer. This type of source, called *hard source*, is nontransparent to reflections from the scatterer because it prevents the reflected signal to travel through the source node. The hard source condition can be easily understood. For example, the electric field given by (8.30) decays to zero for $(n - n_0) \gg n_w$. The source point then simulates a perfectly conducting barrier having zero tangential electric field.

It is possible to make a hard source transparent to impinging fields by specifying the field at the source node, (8.31), as

$$ey(ic) = pulse + ey(ic), \qquad \% \text{ ic is the source node} \tag{8.32}$$

The Choice of Time Varying Function for the Source
The time variation of the source is chosen depending on the bandwidth of the device to be analyzed. For applications where frequency dependent data is to be generated, a pulse is used such that its frequency content covers the desired frequency range. A Gaussian pulse is frequently used in FDTD analysis because of its smooth amplitude variation and large frequency content. It may be defined as

$$E_y(t) = \exp\left[-\frac{(t - t_0)^2}{T^2}\right] \tag{8.33a}$$

where t_0 defines the center of the pulse and T governs the width of the pulse. The values of t_0 and T are selected so that the truncation of the pulse does not introduce unwanted high frequencies in the spectrum, and yet does not waste computation time on determining values of the fields that are essentially zero. Normally, t_0 is selected so that when the source is turned on at $t = 0$, $E_y(t) = e^{-9}$; that is, $t_0 = 3T$. The pulse turns off at $t = 2t_0$. Therefore, pulse width is $2t_0 = 6T$. The Gaussian pulse of (8.33a) is plotted in Figure 8.10.

A modulated Gaussian pulse may be used to generate passband about f_c. This excitation is defined by

$$E_y(t) = \exp\left[-\frac{(t - t_0)^2}{T^2}\right] \cos\{2\pi f_c(t - t_0)\} \tag{8.33b}$$

The spectrum of this pulse is similar to that of (8.33a) except for a shift in frequency by f_c. The bandpass Gaussian pulse is plotted in Figure 8.11. Its frequency spectrum is obtained by taking the Fourier transform of (8.33b) and is shown in Figure 8.12 for two different values of t_0 and $f_c = 500$ MHz. It is observed that the narrow Gaussian pulse with $t_0 = 5$ ns is relatively broadband in nature. The frequency spectrum of modulated Gaussian pulse is compared with the modulated rectangular pulse in Figure 8.13. The pulse width is assumed to be 60 ns for the rectangular pulse which corresponds to $t_0 = 30$ ns for the Gaussian pulse. The center frequency is 500 MHz. The bandwidth over which the spectral components are relatively large defines the useful spectrum of the pulse. The useful bandwidth of the Gaussian pulse is found to be much wider.

To generate a continuous sinusoidal wave of frequency f_c, the time varying function may be selected as

$$E_y(t) = \sin(2\pi f_c t) \qquad \text{or} \qquad E_y(t) = \cos(2\pi f_c t) \tag{8.33c}$$

This source function is suitable for narrowband studies.

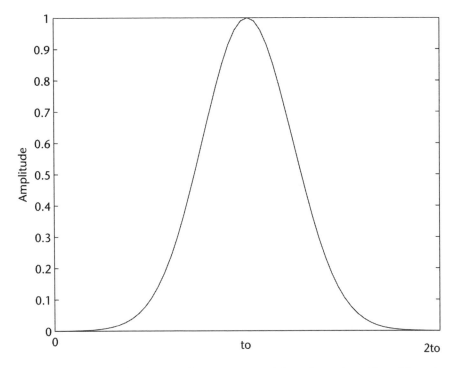

Figure 8.10 A Gaussian pulse in time. The parameter t_0 defines the peak position of the pulse and $2t_0$ is the width of the pulse.

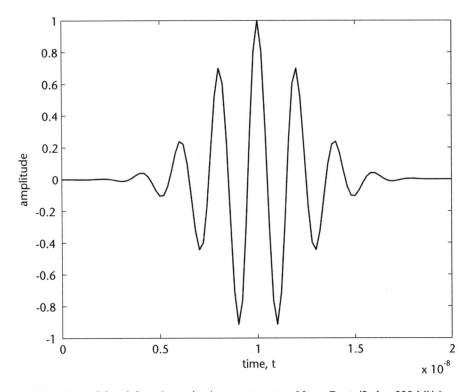

Figure 8.11 A modulated Gaussian pulse (parameters: $t_0 = 10$ ns, $T = t_0/3$, $f_c = 500$ MHz).

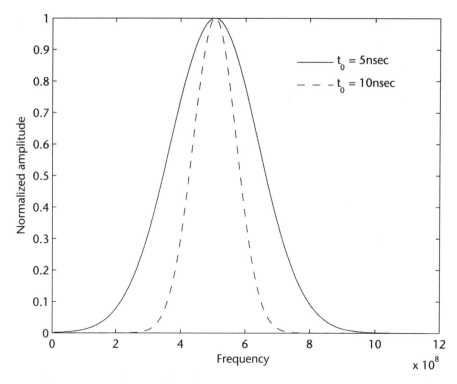

Figure 8.12 Comparison of the frequency spectrum of the modulated Gaussian pulses with different pulse widths, f_c = 500 MHz.

The time waveforms (8.33) can be easily discretized for FDTD simulations. For example, to generate E_y field component at the source node i_s, (8.33b) can be written as

$$E_y^n(i_s) = E_0 \exp\left(-\left[\frac{n - n_0}{n_w}\right]^2\right) \cos\left[2\pi f_c(n - n_0)\right] \qquad (8.34)$$

where n_0 is the center of the pulse and n_w is the width of the pulse, both defined in time step Δt.

Excitation of Modes in Guiding Structures

For many uniform guiding structures, the electromagnetic field distribution in the cross-section is known, and this knowledge may be used to excite fields in the FDTD lattice. Consider a rectangular waveguide and the source plane as shown in Figure 8.14. The TE_{mn}-modes in the waveguide may be excited by using the following electric field distribution at the source plane:

$$\mathbf{E}_s = \hat{y} E_y(x, y) P(t) \qquad (8.35)$$

where $P(t)$ is a time varying source pulse, and $E_y(x, y)$ is the space variation. The space variation for the TE_{10}-mode is shown in Figure 8.14.

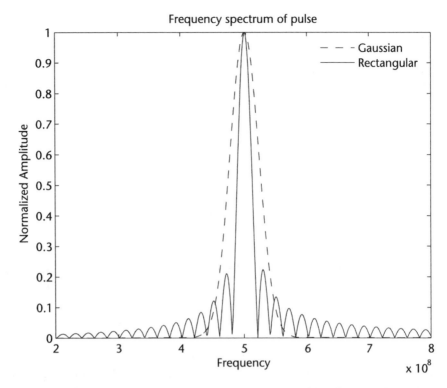

Figure 8.13   Comparison of the frequency spectrum of the modulated rectangular and Gaussian pulses with identical pulse widths, 60 ns.

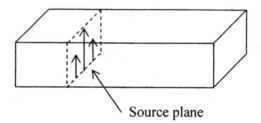

Figure 8.14   Source plane for the excitation of modes in waveguide.

Plane-Wave Excitation

Another important class of excitations is the plane wave excitation. Plane waves are an essential part of electromagnetic wave analysis. We first come across plane waves at the undergraduate level when we study propagation and reflection of waves in unbounded media. The radiation from antennas at a far-off distance can be approximated as a plane wave. The radar cross-section studies require illumination of objects by plane waves. We next discuss excitation of plane waves.

The plane wave may be simulated by means of a one-dimensional array of nodes in the *x-y* plane, and is shown in Figure 8.15. The nodes are located at $(i = i_a, j_a \leq j \leq j_b)$. Each of the nodes may be excited by the field component, say, E_z, as discussed earlier. The time variation may be Gaussian, pulse, or sinusoidal according to the bandwidth requirement. The direction of propagation is determined by the phase difference between the excitation at various nodes. If all the

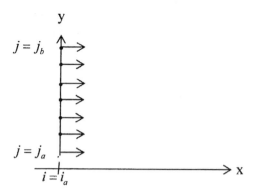

Figure 8.15 Plane wave excitation using a linear array of nodes.

nodes are excited in phase and with the same amplitude, it will simulate a phase front coincident with the y-axis and propagating along the x-direction. The simulated wave is called uniform plane wave. The direction of propagation of the wave may be controlled by introducing constant phase difference between the nodes, $j_a \leq j \leq j_b$.

The most popular method of plane wave generation is based on total-field/scattered-field formulation [3, p. 186]. In this formulation the device to be illuminated by the plane wave is placed at the center of domain and is surrounded by a rectangular shaped (for convenience) enclosure which is followed by another enclosure as shown in Figure 8.16. The scattered field around the device is accompanied by the incident field also, and their sum is called total field. Therefore, total fields are defined in region 1. Only the scattered field is defined in region 2. The incident wave is generated from the connecting condition at the interface between the total-field region and the scattered-field region. This condition needs corrective measures while updating the fields near the interface and the incident wave is introduced as a correction term. This is illustrated by way of one-dimensional example next.

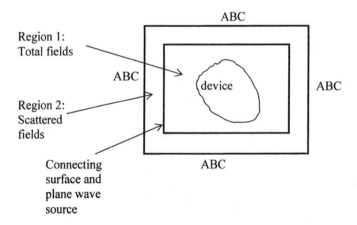

Figure 8.16 Division of two-dimensional domain in total-field and scattered-field regions for the application of total-field/scattered-field formulation to generate plane waves.

Figure 8.17 shows the discretized one-dimensional space along the x-axis. It is terminated by absorbing boundary conditions (ABC) at the ends. According to the total-field/scattered-field formulation, the total-field region (region 2) is placed in the middle where the scatterer is also located and is surrounded by scattered-field region 1. The nodes i_a and i_b are assumed to be part of total-field region 2. The normal FDTD field update at node i_a gives the following:

$$E_z^{total}(i_a, n+1) = E_z^{total}(i_a, n) + \frac{\Delta t}{\epsilon_0 \Delta x}\left(H_y^{total}\left(i_a + \frac{1}{2}, n+\frac{1}{2}\right) - H_y^{total}\left(i_a - \frac{1}{2}, n+\frac{1}{2}\right)\right)$$

(8.36)

The terms on the right are stored in the memory. However, field update based on Figure 8.17 requires that

$$E_z^{total}(i_a, n+1) = E_z^{total}(i_a, n) + \frac{\Delta t}{\epsilon_0 \Delta x}\left(H_y^{total}\left(i_a + \frac{1}{2}, n+\frac{1}{2}\right) - H_y^{scat}\left(i_a - \frac{1}{2}, n+\frac{1}{2}\right)\right)$$

(8.37)

since the node $i_a - 1/2$ is located in the scattered-field region. The difference between (8.36) and (8.37) lies in the last term. In order that the normal FDTD update can still be applied and the last term is the scattered field, we modify the implementation of FDTD update at i_a as follows:

$$E_z^{total}(i_a, n+1) = E_z^{total}(i_a, n) + \frac{\Delta t}{\epsilon_0 \Delta x}\left(H_y^{total}\left(i_a + \frac{1}{2}, n+\frac{1}{2}\right) - H_y^{scat}\left(i_a - \frac{1}{2}, n+\frac{1}{2}\right)\right)$$
$$- \frac{\Delta t}{\epsilon_0 \Delta x}H_y^{inc}\left(i_a - \frac{1}{2}, n+\frac{1}{2}\right)$$

(8.38)

since

$$H_y^{total} = H_y^{scat} + H_y^{inc}$$

(8.39)

Equation (8.38) implies that the field at i_a should be corrected by the term $- \frac{\Delta t}{\epsilon_0 \Delta x}H_y^{inc}\left(i_a - \frac{1}{2}, n+\frac{1}{2}\right)$ after the normal FDTD field update.

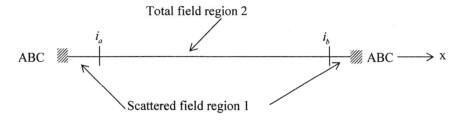

Figure 8.17 Division of one-dimensional domain in total-field and scattered-field regions for the application of total-field/scattered-field formulation.

Similarly, the H_y field update at $i_a - 1/2$ should be corrected as follows since it involves a node in the total field region also,

$$H_y^{scat}\left(i_a - \frac{1}{2}, n + \frac{1}{2}\right) = H_y^{scat}\left(i_a - \frac{1}{2}, n - \frac{1}{2}\right) + \frac{\Delta t}{\mu_0 \Delta x}\left(E_z^{total}(i_a, n) - E_z^{scat}(i_a - 1, n)\right)$$

$$- \frac{\Delta t}{\mu_0 \Delta x} E_z^{inc}(i_a, n) \quad \text{(correction term)} \tag{8.40}$$

A similar correction is applied for updating the field components about the right side of junction between regions 1 and 2, and yields the following:

$$E_z(i_b, n + 1) = E_z(i_b, n) + \frac{\Delta t}{\epsilon_0 \Delta x}\left(H_y\left(i_b + \frac{1}{2}, n + \frac{1}{2}\right) - H_y\left(i_b - \frac{1}{2}, n + \frac{1}{2}\right)\right)$$

$$+ \frac{\Delta t}{\epsilon_0 \Delta x} H_y^{inc}\left(i_b + \frac{1}{2}, n + \frac{1}{2}\right) \quad \text{(correction term)} \tag{8.41}$$

and

$$H_y\left(i_b + \frac{1}{2}, n + \frac{1}{2}\right) = H_y\left(i_b + \frac{1}{2}, n - \frac{1}{2}\right) + \frac{\Delta t}{\mu_0 \Delta x}\left(E_z(i_b + 1, n) - E_z(i_b, n)\right)$$

$$+ \frac{\Delta t}{\mu_0 \Delta x} E_z^{inc}(i_b, n) \quad \text{(correction term)} \tag{8.42}$$

By specifying $E_z^{inc}(i_a, n)$ we can generate a waveform of arbitrary duration and shape. It may be noted that the correction terms in (8.38) through (8.42) amount to inserting excitation at the left junction and deleting it at the right junction. An advantage of this formulation is that the ABC used in this case is not exposed to the incident wave because it lies in the scattered-field zone, thus eliminating the contamination produced by residual reflection of incident wave.

A source code in C for simulating plane TM wave based on total-field/scattered-field formulation is available in [5, p. 72].

Example 8.2. Retro-reflection due to a pulsed hard source in FDTD method can be avoided by removing it from the algorithm after the pulse has decayed essentially to zero. Now consider a Gaussian hard source located at $i_s = 0$ in the region specified by

$$-50 \leq i \leq 50, \ E_y^n(i_s) = E_0 e^{-[(n - n_0)/n_w]^2}; \ n_0 = 3n_w \text{ and } n_w = 30$$

1. Determine if the source would have decayed to zero before the leading edge of the reflected signal reaches the source? Assume zero numerical dispersion, $\Delta x = 2$ mm, $c\Delta t = \Delta x/2$ and $E_y = 0$ if it is $\leq E_0 e^{-9}$.
2. Determine the pulse width in space domain.
3. Calculate the value of n when the source should be removed from the grid and the normal field update can be resumed.

4. It is desired to observe the reflected signal at $i = 40$. Calculate the value of n when the peak of the reflected signal passes this grid point.

5. Write a simple algorithm to remove the hard source after the source has decayed to zero.

Solution.

1. For $n = 0$, $E_y^0(i_s) = E_0 e^{-9} \approx 0$; for $n = 2n_0$, $E_y^{2n_0}(i_s) = E_0 e^{-9} \approx 0$. Therefore, the pulse width in time between zero amplitude points is $2n_0 = 6n_w = 180$. The distance traveled by the leading edge of the reflected signal to reach the source node is $100\Delta x$. Also, the time taken to travel a distance Δx is $2\Delta t$ because $c\Delta t = \Delta x/2$. Therefore, the time taken by the leading edge of the reflected signal to reach the source node is $100 \times 2\Delta t = 200\Delta t$. Since the source pulse requires $180\Delta t$ to decay to zero, which is less than $200\Delta t$, the source would have decayed to zero before the reflected signal reaches the source node.

2. Source width in time, $2n_0 = 180$. The distance traveled in one time step is $\Delta x/2$. Therefore, pulse width in space: $90\Delta x$.

3. Pulse width $2n_0 = 180$. Therefore, source decays to zero for $n > 180$. It can be removed at $n > 180$ and normal field update can be resumed. Also, the leading edge of the source reaches the source node at $n = 200$. Therefore the required value of n for removal of source is $180 < n < 200$.

4. The peak of the source is formed at $i_s = 0$ when $n = n_0 = 90$. The peak takes $100\Delta t$ to arrive at $i = 50$. Therefore, peak of the reflected signal reaches $i = 40$ at $n = 90 + 100 + 20 = 210$.

5. The statement (8.32) may be employed to remove the hard source condition.

Review Question 8.2.   The distance traveled by a wave is given by $x = v_p t$, where v_p is the phase velocity of the wave. In the discretized medium, v_p depends on Δx. For design purposes we may assume $v_p = \alpha c/\sqrt{\epsilon_r}$. Now consider a rectangular pulse with pulse width $20\Delta x$ and centered at $i = 10\Delta x$ at $t = 0$. Let $i_{max} = 200$ and the pulse is propagating in free space from left to the right side. The center of the pulse will hit the right side boundary at

1. $n = 180$, for $\alpha = 1/2$
2. $n = 380$, for $\alpha = 1/2$
3. $n = 180$, for $\alpha = 1$
4. $n = 200$, for $\alpha = 1$

Choose the correct entries and explain.

Answer.   2.

8.2.4   Absorbing Boundary Conditions for One-Dimensional Propagation

A large number of problems are open region problems and require infinite amount of memory for storage of data and equally large amount of processor time for computations unless the domain of the problem is truncated. Absorbing boundary conditions are used for truncation of the domain. These conditions allow the signal to propagate forward without producing any reflection at the boundary.

Mur's First Order Absorbing Boundary Condition
Analysis for this absorbing boundary condition is based on the work of Enquist and Majda [7], and the optimal implementation given by Mur [8]. A reflectionless boundary is created if a wave incident on it continues to propagate in the forward direction only. We know that the Dirichlet and Neumann boundary conditions described by $\phi = 0$ and $\partial\phi/\partial n = 0$, respectively, generate perfect reflections at the boundary. One may wonder if appropriate specification of ϕ at the boundary may produce zero reflection. Let us now discuss how a reflectionless boundary can be described using the appropriate boundary condition there.

Consider a plane wave traveling along the $-x$ direction as shown in Figure 8.18. Heuristically, this wave will not suffer reflection at $x = $ constant plane if

$$\varphi(x, t) = \varphi(x - \Delta x, t + \Delta t) \tag{8.43}$$

and the velocity of wave propagation is $c = \Delta x/\Delta t$. The Taylor's series expansion $\varphi(x - \Delta x, t + \Delta t) = \varphi(x, t) - (\partial\varphi/\partial x)\Delta x + (\partial\varphi/\partial t)\Delta t$ and the use of $\Delta x = c\Delta t$ results in

$$\frac{\partial\phi(x, t)}{\partial x} = \frac{1}{c}\frac{\partial\phi(x, t)}{\partial t} \tag{8.44a}$$

This expression is called the ABC or reflectionless boundary condition for a plane wave incident from the right. Similarly, the ABC for the wave incident from the left on the $x = $ constant plane is obtained as

$$\frac{\partial\phi(x, t)}{\partial x} = -\frac{1}{c}\frac{\partial\phi(x, t)}{\partial t} \tag{8.44b}$$

Similar partial differential equations representing traveling waves along $+y$-axis and $-y$-axis can be written down by inspection of (8.44).

In the finite difference form, (8.44a) can be written as

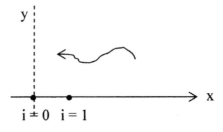

Figure 8.18 Propagation of wave along the $-x$-direction in the two-dimensional space. The node $i = 0$ is at the boundary.

$$\frac{\phi_1^{n+1/2} - \phi_0^{n+1/2}}{\Delta x} = \frac{1}{c} \frac{\phi_{1/2}^{n+1} - \phi_{1/2}^n}{\Delta t} \tag{8.45}$$

for the two-point finite difference approximation centered at $x = \Delta x/2$ and $t = (n + 1/2)\Delta t$. In this form, the finite difference approximation is accurate to the second order in Δx and Δt. If the values at the half grid points and half time steps are not available, for example, for the electric field nodes, they may be obtained by averaging as,

$$\phi_m^{n+1/2} = \frac{\phi_m^{n+1} + \phi_m^n}{2} \quad \text{and} \quad \phi_{m+1/2}^n = \frac{\phi_{m+1}^n + \phi_m^n}{2} \tag{8.46}$$

Use of (8.46) in (8.45) produces the following algorithm:

$$\phi_0^{n+1} = \phi_1^n + \left(\frac{c\Delta t - \Delta x}{c\Delta t + \Delta x}\right)\left(\phi_1^{n+1} - \phi_0^n\right) \tag{8.47a}$$

for updating the tangential fields at the boundary node at $x = 0$ node. Similarly, the expression for updating the tangential fields at the boundary at $x = (N + 1)\Delta x$ is derived as

$$\phi_{N+1}^{n+1} = \phi_N^n + \left(\frac{c\Delta t - \Delta x}{c\Delta t + \Delta x}\right)\left(\phi_N^{n+1} - \phi_{N+1}^n\right) \tag{8.47b}$$

Expressions (8.47) are called the *first order analytical or Mur's boundary condition* and are found to be unconditionally stable [8]. The expressions for the field component E_y are obtained by replacing ϕ by E_y.

The boundary conditions derived above are exact only for a plane wave at normal incidence for the continuous case, and $\alpha = 1$ for the discrete case. For all other values of $\alpha(= c\Delta t/\Delta x)$, the residual reflection arises due to the phase velocity being different from c, and the approximations made in Taylor series expansion at the boundary node.

The residual reflection produced by the Mur's ABC may be determined by assuming the presence of a reflected wave at the boundary node such that

$$\varphi(x, t) = e^{j\omega t}(e^{jkx} + Re^{-jkx}) \tag{8.48}$$

This expression is substituted in (8.47) after discretization, and solved for the reflection coefficient R. A similar procedure was used in Section 8.2.1 for dispersion analysis. The reflection coefficient is found to be

$$R = \left| \frac{e^{j\bar{k}\Delta x}\left(1 + \dfrac{\alpha - 1}{\alpha + 1}e^{j\omega\Delta t}\right) - e^{j\omega\Delta t} - \dfrac{\alpha - 1}{\alpha + 1}}{-e^{-j\bar{k}\Delta x}\left(1 + \dfrac{\alpha - 1}{\alpha + 1}e^{j\omega\Delta t}\right) + e^{j\omega\Delta t} + \dfrac{\alpha - 1}{\alpha + 1}} \right| \tag{8.49}$$

The reflection coefficient (for normal incidence) for the first order ABC and $\alpha = 0.9$ is plotted in Figure 8.19 as a function of $\Delta x/\lambda_0$. Simulations were carried out by terminating free space with Mur ABC and studying the reflected waveform in time domain. Modulated Gaussian pulse with modulation frequency of 1 GHz was used for simulations. The ratio of the peaks of the reflected and the incident waveforms gives the value of reflection coefficient at 1 GHz. The simulations are compared in Figure 8.19 with the values obtained from (8.49). The comparison appears to be very good. The software used for simulations is called *mur_abc.m*. The ABC was located at $i = 400$. A typical waveform sampled at $i = 350$ is plotted in Figure 8.20 for $\Delta x/\lambda_0 = 0.14$. The other parameters for the simulation are: $t_0 = 3$ ns, $T = t_0/3$, $\alpha = 0.9$. The magnitude of reflection coefficient is found to be 0.0106.

Review Question 8.3. The propagation phenomenon is the basis of FDTD method. The following differential equations in one-dimension are suggested to describe the propagation either along the $+x$ or $-x$ directions or both. Choose the correct entries and explain why.

1. $\dfrac{d^2E}{dx^2} + k^2E = 0$

2. $\dfrac{\partial^2 E}{\partial x^2} = \dfrac{1}{c^2}\dfrac{\partial^2 E}{\partial t^2}$

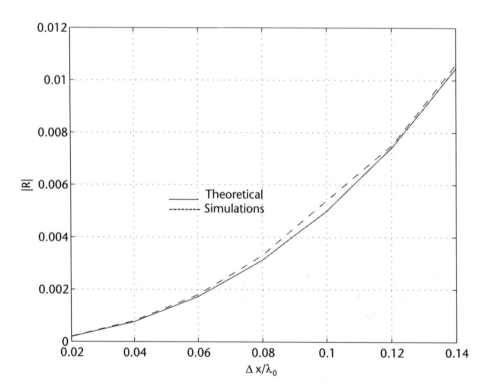

Figure 8.19 Comparison of theoretical, (8.49), and the value obtained from simulations for the reflection coefficient for first order Mur ABC, $\alpha = 0.9$.

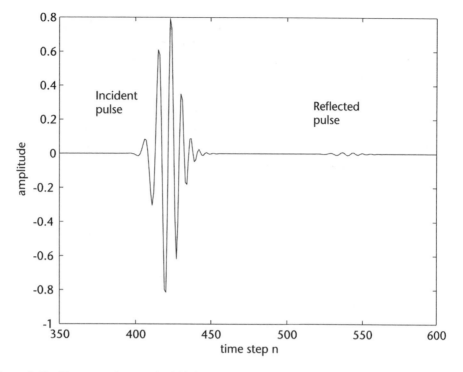

Figure 8.20   Time waveform at $i = 350$ for $\Delta x = 0.042m$, $f_c = 1$ GHz, $\alpha = 0.9$. The magnitude of reflection coefficient is 0.0106.

3. $\dfrac{\partial E}{\partial x} = \dfrac{1}{c} \dfrac{\partial E}{\partial t}$

4. $\dfrac{\partial E}{\partial x} = -\dfrac{1}{c} \dfrac{\partial E}{\partial t}$

Answer.   2, 3, and 4, because propagation involves time variation. 1 produces steady state solution and not time varying.

The waves which are obliquely incident to the boundary suffer a higher reflection coefficient. For this, one needs to use differential equation in two dimensions to arrive at the analytical absorbing boundary condition [8, 9], and this is described in Section 8.4.

8.3   Applications of One-Dimensional FDTD Analysis

We now describe some of the applications of one-dimensional FDTD analysis. These include reflection at an interface, determination of propagation constant in the medium, design of material absorber, and extraction of frequency domain information from the time domain data.

8.3.1   Reflection at an Interface

One of the common phenomena in an electromagnetic device is the reflections suffered by the incident wave at various points of the device. It is important to locate these reflections to improve the device design. FDTD method may be used to locate the centers of reflection and therefore help in diagnostics. We next determine the reflection coefficient at the interface of two media. The procedure may also be used to determine the residual reflection associated with absorbing boundary conditions.

Consider the discretized medium in the form of a number of layers with the first few layers in the free space ($\epsilon_r = 1$) and the rest in a medium characterized by $\epsilon_0 \epsilon_r$ and μ_0. Let a TEM wave be incident normally on the interface with electric field tangential to it. The reflection coefficient due to impedance mismatch is given by

$$\Gamma = \frac{Z^{medium} - Z^{air}}{Z^{medium} + Z^{air}} = \frac{\sqrt{\dfrac{\mu_0}{\epsilon_0 \epsilon_r}} - \sqrt{\dfrac{\mu_0}{\epsilon_0}}}{\sqrt{\dfrac{\mu_0}{\epsilon_0 \epsilon_r}} + \sqrt{\dfrac{\mu_0}{\epsilon_0}}} = \frac{1 - \sqrt{\epsilon_r}}{1 + \sqrt{\epsilon_r}} \qquad (8.50a)$$

The transmission coefficient is obtained as

$$T = 1 + \Gamma = \frac{2}{1 + \sqrt{\epsilon_r}} \qquad (8.50b)$$

Let us choose a Gaussian pulse and a layered medium with free space for $i \leq 249$ and $\epsilon_r = 4$ for $250 \leq i \leq 350$. The other parameters are: $\Delta x = 1.5$ mm, $\Delta t = 2.5$ ps, pulse width at nearly zero amplitudes $= 400$ ps, and $\alpha = 0.5$. The FDTD simulations for this case can be carried out as described by (8.11) and its subsequent modifications for the compact soft source. The other change in the code is to define the medium parameters properly and modify the value of a as

$$a = \frac{\Delta t}{\epsilon_0 \epsilon_r \Delta x} \qquad (8.51)$$

Equations (8.11) are modified to include the effect of dielectric loading, and are given by

$$H_z^{n+1/2}\left(i + \frac{1}{2}\right) = H_z^{n-1/2}\left(i + \frac{1}{2}\right) - \frac{\Delta t}{\mu_0 \Delta x}\left[E_y^n(i+1) - E_y^n(i)\right] \quad (8.52a)$$

$$E_y^{n+1}(i) = E_y^n(i) - \frac{\Delta t}{\Delta x}\frac{1}{\epsilon_0 \epsilon_r}\left[H_z^{n+1/2}\left(i + \frac{1}{2}\right) - H_z^{n+1/2}\left(i - \frac{1}{2}\right)\right] \quad (8.52b)$$

The simulation results for $n = 400$ is shown in Figure 8.21(a). The pulse is centered at about $160\Delta x$ and therefore has not yet reached the interface. The pulse at $n = 700$ is shown in Figure 8.21(b). It comprises a reflected pulse centered near $190\Delta x$ and a transmitted pulse peaking near $275\Delta x$. The peak amplitudes provide the values of transmission and reflection coefficients. Since the reflected pulse is inverted with respect to the incident pulse, we obtain $\Gamma = -0.333$. Similarly,

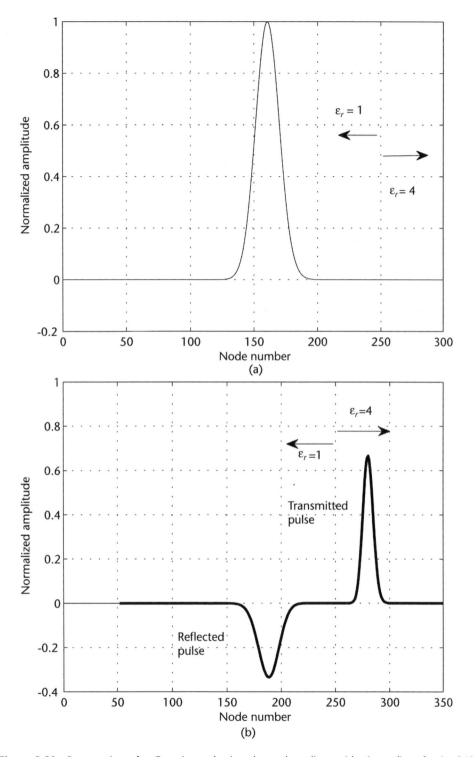

Figure 8.21 Propagation of a Gaussian pulse in a layered medium with air medium for $i \leq 249$ and $\epsilon_r = 4$ for $250 \leq i \leq 350$: (a) the incident pulse for $n = 400$; and (b) the reflected and transmitted pulses for $n = 700$.

$T = 0.667$. These values are consistent with the values obtained from (8.50). The widths of the transmitted and reflected pulses depend on the ϵ_r value of the medium. The pulse width is $49\Delta x$ for the reflected pulse and $27\Delta x$ for the transmitted pulse. The narrow width for the transmitted pulse is due to the higher dielectric loading of the medium which reduces the velocity of propagation. The software used for the simulation is *slab_reflection.m*.

8.3.2 Determination of Propagation Constant

The propagation constant of a wave defined as $\gamma = \alpha + j\beta$ can be determined using FDTD analysis. For this we may employ CW sinusoidal excitation and study the propagation behavior in the medium. The medium is assumed to be the same as used in the last section, and conductor loss in the form of σ is added to the layers for $250 \le i \le 309$ so that attenuation constant α may also be computed along with the phase constant β. The set of parameters selected for the study are: freq = 50 MHz, $\alpha = 0.5$, $\epsilon_r = 1$, $\Delta x = 15$ cm, and $\Delta t = 250$ ps for $i < 250$; and $\epsilon_r = 4$, $\sigma = 0.001$ S/m for $250 \le i \le 309$. The field updating equations (8.52) are modified to include the effect of medium losses, and are given as

$$H_z^{n+1/2}\left(i + \frac{1}{2}\right) = H_z^{n-1/2}\left(i + \frac{1}{2}\right) - \frac{\Delta t}{\mu_0 \Delta x}\left[E_y^n(i + 1) - E_y^n(i)\right] \quad (8.53a)$$

$$E_y^{n+1}(i) = \frac{\epsilon_0 \epsilon_r - \dfrac{\sigma \Delta t}{2}}{\epsilon_0 \epsilon_r + \dfrac{\sigma \Delta t}{2}} E_y^n(i) \quad (8.53b)$$

$$- \frac{\Delta t}{\Delta x}\frac{1}{\epsilon_0 \epsilon_r + \dfrac{\sigma \Delta t}{2}}\left[H_z^{n+1/2}\left(i + \frac{1}{2}\right) - H_z^{n+1/2}\left(i - \frac{1}{2}\right)\right]$$

The simulation results for $n = 800$ is displayed in Figure 8.22. The waveform consists of three distinct portions: the incident wave of amplitude unity for $0 \le i \le 100$, the standing waveform between $100 \le i \le 250$ with amplitude 0.667, and the transmitted waveform for $i > 250$ with decaying amplitude. The amplitude of the standing wave depends on the phase relationship between the two waves and therefore on n. Since the reflection coefficient at the interface is $\Gamma = -1/3$, the amplitude of the wave in this region is $1 - 1/3 = 2/3$. The transmitted wave has amplitude $T = 1 + \Gamma = 2/3$ at $i = 250$; thereafter it decays exponentially as $E_y = E_0 e^{-\alpha_d x}$ due to losses in the medium, where α_d is the attenuation constant.

The phase constant can be obtained from the wavelength in the medium λ as $\beta = 2\pi/\lambda$. The wavelengths in the media can be determined from the separation distance between the consecutive peaks in Figure 8.22. It is found to be 6m in free space, and 3m in the dielectric medium, consistent with the excitation at 50 MHz and dielectric loading of $\epsilon_r = 4$. The attenuation constant α_d in the dielectric is obtained from the amplitude decay of the signal. The amplitude of the wave is 0.5686 at $i = 260$ and 0.3175 at $i = 300$. Modeling the decay in amplitude as

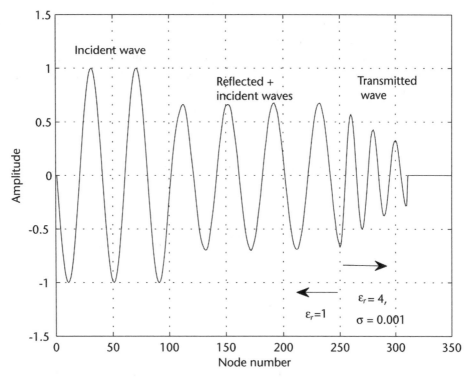

Figure 8.22   Propagation of a CW sine wave through a layered medium. The interface at node 250 separates air region from the lossy dielectric. The parameters are: $\Delta x = 15$ cm, $\Delta t = 250$ ps, $\alpha = 0.5$, $n = 800$, freq = 50 MHz.

$E_y = E_0 e^{-\alpha_d x}$, gives the value $\alpha_d = 0.097$ Np/m. The analytical value of the attenuation constant may be obtained from the following expression if conductivity σ of the medium is known [10, p. 420]:

$$\frac{\alpha_d}{k_0} = \sqrt{\frac{\epsilon_r}{2}} \left[\sqrt{1 + \left(\frac{\sigma}{\omega \epsilon_0 \epsilon_r} \right)^2} - 1 \right]^{1/2} \text{Np/m} \qquad (8.54)$$

and is found to be $\alpha_d = 0.0963$ Np/m at 50 MHz for $\sigma = 0.001$ S/m. This value of α_d is comparable to the value based on FDTD data. The difference may be due to the discretization error. Alternatively, one can obtain the value of σ from the computed value of α_d and (8.54). It may be noted that for the simulations carried out, $\Delta x/\lambda = 1/40$ in free space and $1/20$ in the dielectric. The simulation results are based on the code *propagation_constant.m*.

8.3.3   Design of Material Absorber

The problem discussed in the last section is characterized by a reflected wave at the interface and an exponentially attenuated transmitted wave. If the parameters of the lossy medium are designed properly, it may be possible to develop a reflectionless interface. Let us design an artificial dielectric for the lossy medium with parameters

$(\mu', \epsilon', \sigma^*, \sigma)$ where σ is the electrical conductivity and σ^* is the fictitious magnetic conductivity of the medium. The material parameters are selected such that when a wave is incident normal to the interface separating free space and lossy medium, there should be no reflection. This is possible only if the intrinsic impedance of the two media is same; that is,

$$Z_0 = Z_m \tag{8.55}$$

where Z_0 is the free space impedance $\sqrt{\mu_0/\epsilon_0}$, and Z_m is the impedance of the lossy medium defined as

$$Z_m = \sqrt{\frac{\mu' - j\sigma^*}{\epsilon' - j\sigma}} \tag{8.56}$$

Equation (8.55) therefore becomes

$$\sqrt{\frac{\mu_0}{\epsilon_0}} = \sqrt{\frac{\mu' - j\sigma^*}{\epsilon' - j\sigma}} \tag{8.57a}$$

For convenience, if we choose the parameters of the lossy medium as $\mu' = \mu_0$ and $\epsilon' = \epsilon_0$, the matching condition (8.57a) gives

$$\frac{\sigma^*}{\mu_0} = \frac{\sigma}{\epsilon_0} \tag{8.57b}$$

With this design, the layers $250 \leq i \leq 400$ were designed as lossy air medium with σ increasing linearly from 0.001 to 0.01 S/m as defined below

$$\sigma_i = 0.001 + \frac{(i - 250)(0.01 - 0.001)}{(400 - 250)} \tag{8.58}$$

and $\sigma_i^* = (120\pi)^2 \sigma_i$ to satisfy (8.57b). The value of σ^* was slowly increased from 0 to the maximum value given by (8.57b) in order to minimize reflection due to discretization of σ^*. The updating expressions for the fields (8.53) are modified to include the effect of magnetic conductivity σ^* as well, and are obtained as

$$H_z^{n+1/2}\left(i + \frac{1}{2}\right) = \frac{\mu_0\mu_r - \dfrac{\sigma^*\Delta t}{2}}{\mu_0\mu_r + \dfrac{\sigma^*\Delta t}{2}} H_z^{n-1/2}\left(i + \frac{1}{2}\right) \tag{8.59a}$$

$$- \frac{\Delta t}{\Delta x} \frac{1}{\mu_0\mu_r + \dfrac{\sigma^*\Delta t}{2}}\left[E_y^n(i + 1) - E_y^n(i)\right]$$

$$E_y^{n+1}(i) = \frac{\epsilon_0 \epsilon_r - \dfrac{\sigma \Delta t}{2}}{\epsilon_0 \epsilon_r + \dfrac{\sigma \Delta t}{2}} E_y^n(i) \tag{8.59b}$$

$$- \frac{\Delta t}{\Delta x} \frac{1}{\epsilon_0 \epsilon_r + \dfrac{\sigma \Delta t}{2}} \left[H_z^{n+1/2}\left(i + \frac{1}{2}\right) - H_z^{n+1/2}\left(i - \frac{1}{2}\right) \right]$$

The reflection coefficient as a function of frequency for the Gaussian pulse incident on the layered lossy medium is shown in Figure 8.23. The pulse width is assumed to be 250 ps. The other parameters are: $\Delta x = 1.5$ mm and $\alpha = 0.5$. It may be observed from this figure that the reflection coefficient is of the order of 10^{-4} for frequencies up to 6 GHz. At 3 GHz, $\Delta x / \lambda = 0.015$. The material absorber is employed to truncate the size of the open region problem. The resulting absorbing boundary condition is called the material absorbing boundary condition and provides the basis for the development of perfectly matched layer (PML). The extraction of reflection coefficient as a function of frequency from the time domain data is described in Section 8.3.5. The software used for this simulation is *material_absorber.m*.

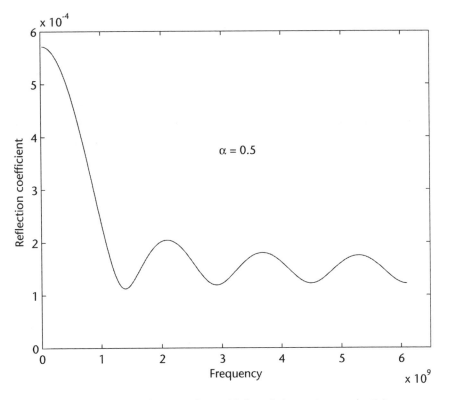

Figure 8.23   Reflection due to a lossy medium with linearly increasing conductivity.

8.3.4 Exponential Time-Stepping Algorithm in the Lossy Region

The material absorber is in general characterized by high values of σ and σ^* in order to reduce the number of cells in the absorber region. For example, only 8 to 12 cells may be sufficient to attenuate the waves to an acceptable value. Therefore, fields decay rapidly (exponentially) and the linear differencing used in the lossless medium is no longer appropriate for the absorber region. Instead, exponential time-differencing is used and the update equations corresponding to (8.59) are written as [3]

$$H_z^{n+1/2}\left(i+\frac{1}{2}\right) = e^{-\sigma^*\Delta t/(\mu_0\mu_r)} H_z^{n-1/2}\left(i+\frac{1}{2}\right) \tag{8.60a}$$

$$-\frac{\Delta t}{\Delta x}\frac{e^{-\sigma^*\Delta t/(2\mu_0\mu_r)}}{\mu_0\mu_r}\left[E_y^n(i+1) - E_y^n(i)\right]$$

$$E_y^{n+1}(i) = e^{-\sigma\Delta t/(\epsilon_0\epsilon_r)} E_y^n(i) - \frac{\Delta t}{\Delta x}\frac{e^{-\sigma\Delta t/(2\epsilon_0\epsilon_r)}}{\epsilon_0\epsilon_r}\left[H_z^{n+1/2}\left(i+\frac{1}{2}\right) - H_z^{n+1/2}\left(i-\frac{1}{2}\right)\right]$$
$$\tag{8.60b}$$

These equations reduce to (8.59) for small losses and were used in computing the data for Figure 8.23.

8.3.5 Extraction of Frequency Domain Information from the Time Domain Data

The FDTD method has been used extensively for calculating the frequency domain characteristics such as propagation constant, S-parameters, driving point impedance, and so on. When the frequency response over a broadband spectrum is of interest, a broadband pulse excitation can provide this frequency response with a single FDTD simulation. The conversion from the time domain to frequency domain is achieved using either discrete Fourier transform (DFT) or fast Fourier transform (FFT). For ease of implementation and the use of an arbitrary number of samples, DFT is preferred generally. This transformation is defined as

$$G(x, f) = \Delta t \sum_{n=1}^{N} g(x, n\Delta t) \exp(-j2\pi f(n-1)\Delta t) \tag{8.61}$$

where $g(x, n\Delta t)$ is the pulse response, Δt is the sampling interval or time step in the FDTD process, and N is the total number of samples. The frequency resolution of the waveform depends on both N and Δt, and is given by

$$\Delta f = \frac{1}{N\Delta t} \tag{8.62}$$

Since commonly used pulse shapes do not have a uniform frequency spectrum, the final values $G(x, f)$ must be normalized by the DFT of the incident pulse to obtain frequency domain data equivalent to data which would have been obtained if the given problem had been excited by a uniform frequency spectrum [11].

As an example, let us determine the reflection coefficient as a function of frequency for the problem discussed in Section 8.3.1. The number of layers is 400 with the first 250 layers described by $\epsilon_r = 1$ and the other 150 layers by $\epsilon_r = 4$. The change in permittivity changes the intrinsic impedance and gives rise to reflection. The voltage reflection coefficient for this configuration is found to be $-1/3$, as shown in the time waveform of Figure 8.24(b). However, the phase velocity variation due to discretization of the medium makes the reflection coefficient vary with frequency about the value $-1/3$. To determine this variation, the time history of the field at a node near the interface is recorded twice; once with free space for all the layers, and second time with the actual medium configuration. The first record gives the incident waveform, while the second record includes the reflected waveform also. The incident pulse (corresponding to free space at all the nodes) at $i = 210$ is shown in Figure 8.24(a) for n varying from 1 to 800. We label this waveform as $E_i^n(i_0)$. The time waveform at the same node with dielectric layers in place is plotted in Figure 8.24(b). The waveform shows both the incident and reflected pulses corresponding to the Gaussian input pulse of width 250 ps. Let us designate this waveform as $E_f^n(i_0)$.

The DFT of these waveforms was taken with $N = 16{,}384$ points. Since $\Delta t = 25$ ps in the FDTD simulations corresponding to $\alpha = 0.5$, $\Delta x = 1.5$ mm, the frequency resolution therefore was $\Delta f = 1/(N\Delta t) = 24.414$ MHz. Let us call the frequency spectrum corresponding to the incident pulse as $E_i(f)$. Since the waveform in Figure 8.24(b) includes both the incident and reflected pulses, the DFT of this waveform gives the spectrum $E_i(f) + E_r(f)$, where $E_r(f)$ is the spectrum for the reflected pulse. Subtracting $E_i(f)$ from this spectrum gives $E_r(f)$. The magnitude of reflection coefficient, defined as

$$\Gamma(f) = \frac{E_r(f)}{E_i(f)} \tag{8.63}$$

is plotted in Figure 8.25. The reflection coefficient increases from 0.333 at zero frequency to 0.34 at 6 GHz. This value compares with the time domain data of Figure 8.24(b) and analytical result $\Gamma = -1/3$ obtained earlier. An alternative method to analyze the time waveforms for reflection coefficient is to use time windowing to separate the incident and reflected pulses in Figure 8.24(b); that is, DFT of the waveform between $400 \leq n \leq 520$ provides the frequency spectrum $E_i(f)$, and that between $580 \leq n \leq 700$ provides $E_r(f)$.

8.3.6   Simulation of Lossy, Dispersive Materials

In developing the FDTD equations, we have assumed that the fields E and H are linearly related to the fluxes through the equations

$$\mathbf{D} = \epsilon \mathbf{E} \qquad \mathbf{B} = \mu \mathbf{H} \tag{8.64}$$

This relationship is exact for free space and can be used approximately for other materials over a narrow frequency range. For problems involving common materials

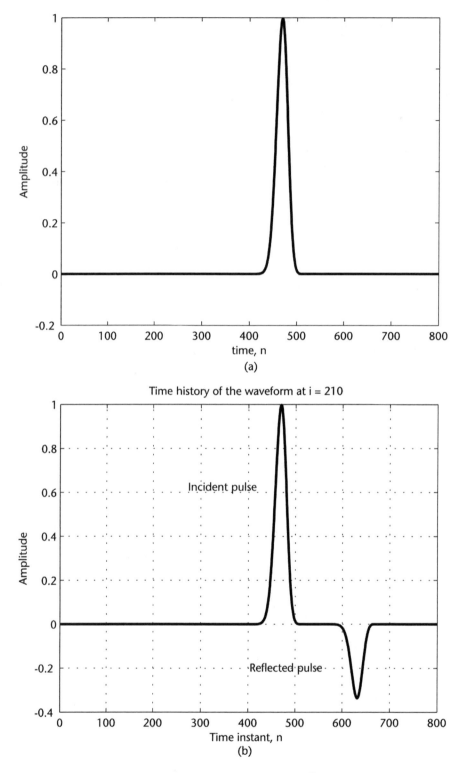

Figure 8.24   Time waveforms at $i_0 = 210$: (a) incident time pulse, $E_i^n(i_0)$; and (b) pulse after reflection from the medium, $E_f^n(i_0)$.

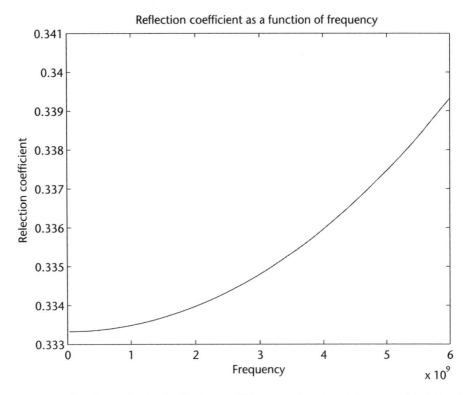

Figure 8.25   Plot of magnitude of reflection coefficient as a function of frequency for a Gaussian pulse incident at an interface.

(e.g., water) and when broadband operation is to be studied, the linear relationship (8.64) does not hold. A complex flux-field relationship needs to be modeled in time domain for use in FDTD equations. The time domain model for the dielectric constant is discussed next.

Material parameters are generally described in the literature as a function of frequency,

$$\mathbf{D}(\omega) = \epsilon_0 \epsilon_r(\omega) \mathbf{E}(\omega) \tag{8.65}$$

where $\epsilon_r(\omega)$ is complex relative dielectric constant. The time domain equivalent of the above expression can be obtained in the form of a convolution integral as

$$\mathbf{D}(t) = \epsilon_0 \int_0^t \mathbf{E}(t - \tau) \epsilon_r(\tau) \, d\tau \tag{8.66}$$

This integral requires the storage of time history of the electric field. The resultant computational costs are too high especially for three-dimensional simulations with dispersive materials. An efficient approach to determine $\mathbf{D}(t)$ is presented next for linearly dispersive materials. The linearly dispersive materials or Debye materials are characterized by the following frequency dependent permittivity with a complex pole:

$$\epsilon_r(\omega) = \epsilon_r + \frac{\sigma}{j\omega\epsilon_0} + \frac{\chi}{1 + j\omega t_0} \tag{8.67}$$

where t_0 is the relaxation time for the material, and χ is its susceptibility. The last term contributes to the frequency dependent permittivity. Let us develop expression for $\mathbf{D}(t)$ for the lossy material first. This will be followed by the expression for the lossy, dispersive material.

Lossy, Dielectric Material Simulation
The flux density for the lossy material characterized by $\epsilon_r(\omega) = \epsilon_r + \sigma/(j\omega\epsilon_0)$ is written as

$$\mathbf{D}(\omega) = \epsilon_0\epsilon_r\mathbf{E}(\omega) + \frac{\sigma}{j\omega}\mathbf{E}(\omega) \tag{8.68}$$

The corresponding time domain expression is obtained by taking the Fourier transform and observing that $f(\omega)/j\omega \leftrightarrow \int f(t)\, dt$,

$$\mathbf{D}(t) = \epsilon_0\epsilon_r\mathbf{E}(t) + \sigma\int_0^t \mathbf{E}(t')\, dt' \tag{8.69}$$

In the sampled time domain, the above expression at the instant $t = n\Delta t$ may be written as

$$\mathbf{D}^n = \epsilon_0\epsilon_r\mathbf{E}^n + \sigma\Delta t\sum_{i=0}^n \mathbf{E}^i \tag{8.70}$$

For the FDTD analysis, however, we need to determine $\mathbf{E}^n$. For this, we rewrite (8.70) as

$$\mathbf{D}^n = \epsilon_0\epsilon_r\mathbf{E}^n + \sigma\Delta t\mathbf{E}^n + \sigma\Delta t\sum_{i=0}^{n-1} \mathbf{E}^i$$

or

$$\mathbf{E}^n = \frac{\mathbf{D}^n - \sigma\Delta t\sum_{i=0}^{n-1} \mathbf{E}^i}{\epsilon_0\epsilon_r + \sigma\Delta t} \tag{8.71}$$

The current value of $\mathbf{E}^n$ can be computed from the current value of $\mathbf{D}^n$ and the previous values of E. For brevity in writing we may denote

$$I^n = \sigma\Delta t\sum_{i=0}^n \mathbf{E}^i \tag{8.72}$$

so that

$$\mathbf{E}^n = \frac{\mathbf{D}^n - I^{n-1}}{\epsilon_0 \epsilon_r + \sigma \Delta t} \tag{8.73a}$$

and

$$I^n = I^{n-1} + \sigma \Delta t E^n \tag{8.73b}$$

In this formulation, we create another array I in addition to that for E and H. The current value I^n is obtained by adding $\sigma \Delta t E^n$ to the previous value, which is already stored. The FDTD formulation for one-dimensional propagation along the x-direction may therefore be written as

$$D_y^{n+1}(i) = D_y^n(i) - \frac{\Delta t}{\Delta x}\left[H_z^{n+1/2}\left(i + \frac{1}{2}\right) - H_z^{n+1/2}\left(i - \frac{1}{2}\right)\right] \tag{8.74a}$$

$$E_y^{n+1}(i) = gay(i)\left[D_y^{n+1}(i) - I_y^n(i)\right] \tag{8.74b}$$

$$I_y^n(i) = I_y^{n-1}(i) + gby(i)E_y^n(i) \tag{8.74c}$$

$$H_z^{n+1/2}\left(i + \frac{1}{2}\right) = H_z^{n-1/2}\left(i + \frac{1}{2}\right) - \frac{\Delta t}{\mu_0 \Delta x}\left[E_y^n(i+1) - E_y^n(i)\right] \tag{8.74d}$$

where

$$gby(i) = \sigma(i)\Delta t \tag{8.75a}$$

$$gay(i) = \frac{1}{\epsilon_0 \epsilon_r(i) + gby(i)} \tag{8.75b}$$

The properties of the dielectric medium are included in (8.75). For free space, $gay = 1/\epsilon_0$ and $gby = 0$. The formulation of Maxwell's equations in terms of D and H has helped us avoid convolution in time.

Lossy, Dispersive Material Simulation
The last term in (8.67) contributes to the dispersive behavior of materials. We shall follow the auxiliary differential equation (ADE) method [12] to convert this term to time domain. Let us define [5, p. 26]

$$S(\omega) = \frac{\chi}{1 + j\omega t_0} E(\omega) \tag{}$$

or

$$S(\omega)(1 + j\omega t_0) = \chi E(\omega) \tag{8.76}$$

Its Fourier transform is obtained as

$$S(t) + t_0 \frac{dS(t)}{dt} = \chi E(t) \tag{8.77}$$

In the sampled time domain about $t = n\Delta t$ one can write

$$\frac{S^n + S^{n-1}}{2} + t_0 \frac{S^n - S^{n-1}}{\Delta t} = \chi E^n \tag{8.78}$$

The first term is averaged over two time steps for consistency in form with the next term. Solving for S^n gives

$$S^n = \frac{1 - \dfrac{\Delta t}{2t_0}}{1 + \dfrac{\Delta t}{2t_0}} S^{n-1} + \frac{\dfrac{\Delta t}{t_0}\chi E^n}{1 + \dfrac{\Delta t}{2t_0}} \tag{8.79}$$

Using $e^{-\delta} \approx 1 - \delta$ and $1/(1 + \delta) \approx e^{-\delta}$ for $\delta \ll 1$, we can approximate

$$S^n \approx e^{-\Delta t/t_0} S^{n-1} + \frac{\Delta t}{t_0}\chi E^n \qquad \text{for } \Delta t \ll t_0 \tag{8.80}$$

Introducing its contribution in (8.70) gives

$$\mathbf{D}^n = \epsilon_0 \epsilon_r \mathbf{E}^n + I^n + \epsilon_0 S^n$$

or

$$\mathbf{D}^n = \epsilon_0 \epsilon_r \mathbf{E}^n + (\sigma \Delta t E^n + I^{n-1}) + \epsilon_0 \left(\chi E^n \frac{\Delta t}{t_0} + e^{-\Delta t/t_0} S^{n-1} \right) \tag{8.81}$$

Solving for E^n gives

$$\mathbf{E}^n = \frac{\mathbf{D}^n - I^{n-1} - \epsilon_0 e^{-\Delta t/t_0} S^{n-1}}{\epsilon_0 \epsilon_r + \sigma \Delta t + \chi \epsilon_0 \dfrac{\Delta t}{t_0}} \tag{8.82}$$

where I^n and S^n have been defined earlier. The convolution is thus avoided, and the frequency-dependent or linear dispersive materials can be handled efficiently.

The FDTD formulation for one-dimensional propagation in a lossy, dispersive material may therefore be written as

$$D_y^{n+1}(i) = D_y^n(i) - \frac{\Delta t}{\Delta x}\left[H_z^{n+1/2}\left(i + \frac{1}{2}\right) - H_z^{n+1/2}\left(i - \frac{1}{2}\right) \right] \tag{8.83a}$$

$$E_y^{n+1}(i) = gay(i) \left[D_y^{n+1}(i) - I_y^n(i) - \epsilon_0 e^{-\Delta t/t_0} S_y^n(i) \right] \tag{8.83b}$$

$$I_y^n(i) = I_y^{n-1}(i) + gby(i) E_y^n(i) \tag{8.83c}$$

$$S_y^{n+1}(i) = gcy(i) E_y^{n+1}(i) + S_y^n(i) e^{-\Delta t/t_0} \tag{8.83d}$$

$$H_z^{n+1/2}\left(i + \frac{1}{2}\right) = H_z^{n-1/2}\left(i + \frac{1}{2}\right) - \frac{\Delta t}{\mu_0 \Delta x}\left[E_y^n(i+1) - E_y^n(i) \right] \tag{8.83e}$$

where

$$gcy(i) = \frac{\chi(i)\Delta t}{t_0} \tag{8.84a}$$

$$gby(i) = \sigma(i)\Delta t \tag{8.84b}$$

$$gay(i) = \frac{1}{\epsilon_0 \epsilon_r(i) + gby(i) + \epsilon_0 gcy(i)} \tag{8.84c}$$

The media is fully characterized by (8.84). A source code in C for the FDTD simulation of frequency dependent materials is available in [5, p. 43].

The FDTD analysis in one dimension is useful to understand the concepts without coding and mathematical complexity. However, some aspects of FDTD can be described properly only when we consider propagation in two or more dimensions. These include anisotropy in dispersion and reflection from the absorbing boundary conditions. Also, most of the electromagnetic devices are two or three dimensional in nature. We discuss two-dimensional FDTD analysis next.

8.4 FDTD Analysis in Two Dimensions

The starting point for two-dimensional analysis is again the Maxwell's equations, (1.1). For simplicity, we describe propagation in two dimensions by means of either TM mode consisting of H_x, H_y, and E_z components or TE mode consisting of E_x, E_y, and H_z. Let us begin with TM mode. The Maxwell's equations now reduce to (with $\partial/\partial z = 0$)

$$\frac{\partial H_x}{\partial t} = -\frac{1}{\mu_0}\frac{\partial E_z}{\partial y} \tag{8.85a}$$

$$\frac{\partial H_y}{\partial t} = \frac{1}{\mu_0}\frac{\partial E_z}{\partial x} \tag{8.85b}$$

$$\frac{\partial E_z}{\partial t} = \frac{1}{\epsilon_0}\left(\frac{\partial H_y}{\partial x} - \frac{\partial H_x}{\partial y}\right) \tag{8.85c}$$

consider the unit cell of Figure 8.28. This is redrawn here as Figure 8.29 in the form of a rectangular E-field loop.

The Faraday's law, (1.3a), when applied to the loop ABCD may be written as (σ^* is assumed zero)

$$\oint_{ABCD} E.dl = -\frac{\partial}{\partial t} \iint_{ABCD} B.ds \tag{8.87}$$

In the FDTD analysis, the cell dimensions are usually selected to be much smaller than the wavelength so that the field distribution may be assumed to be constant over the cell. We can therefore write the left side integral as

$$\oint_{ABCD} E.dl = E_x\left(i, j-\frac{1}{2}\right)\Delta x + E_y\left(i+\frac{1}{2}, j\right)\Delta y - E_x\left(i, j+\frac{1}{2}\right)\Delta x \tag{8.88a}$$

$$- E_y\left(i-\frac{1}{2}, j\right)\Delta y$$

and the right side integral as

$$\iint_{ABCD} B.ds = \mu_0 H_z(i, j)\Delta x \Delta y \tag{8.88b}$$

Substituting (8.88) in (8.87) gives

$$E_x\left(i, j-\frac{1}{2}\right)\Delta x + E_y\left(i+\frac{1}{2}, j\right)\Delta y - E_x\left(i, j+\frac{1}{2}\right)\Delta x \tag{8.89}$$

$$- E_y\left(i-\frac{1}{2}, j\right)\Delta y = -\mu_0 \frac{\partial H_z(i, j)}{\partial t}\Delta x \Delta y$$

Dividing both sides of this expression by $\Delta x \Delta y$ gives

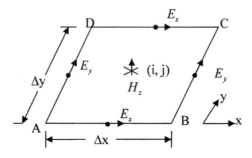

Figure 8.29   Illustration of Faraday's law for the unit cell in two dimensions.

$$\mu_0 \frac{\partial H_z(i, j)}{\partial t} = \frac{E_x\left(i, j + \frac{1}{2}\right) - E_x\left(i, j - \frac{1}{2}\right)}{\Delta y} - \frac{E_y\left(i + \frac{1}{2}, j\right) - E_y\left(i - \frac{1}{2}, j\right)}{\Delta x}$$

or

$$\mu_0 \frac{\partial H_z}{\partial t} = \frac{\partial E_x}{\partial y} - \frac{\partial E_y}{\partial x} \qquad \text{at the node } i, j \qquad (8.90)$$

which is Faraday's law in the differential form. The Ampere's law can be proved similarly by considering the H-field loop.

8.4.2 Numerical Dispersion in Two Dimensions

For the TM mode, the Maxwell's equations in finite-difference form are defined by (8.86). The effect of discretization on the propagation constant can be carried out on the lines similar to that in Section 8.2.1 for one-dimensional equations. We assume plane, monochromatic, traveling wave solutions, substitute them in (8.86), and simplify to obtain the following dispersion relation for the wave in discretized medium:

$$\left(\frac{\sin\left(\frac{\omega \Delta t}{2}\right)}{c \Delta t}\right)^2 = \left(\frac{\sin\left(\frac{\overline{k}_x \Delta x}{2}\right)}{\Delta x}\right)^2 + \left(\frac{\sin\left(\frac{\overline{k}_y \Delta y}{2}\right)}{\Delta y}\right)^2 \qquad (8.91)$$

where $\overline{k}_x$ and $\overline{k}_y$ are the wave numbers in the x and y directions, respectively, in the discretized medium. Assuming a square grid $\Delta x = \Delta y = \Delta$ and wave propagation at an angle θ with respect to the x-axis ($\overline{k}_x = \overline{k} \cos \theta$, $\overline{k}_y = \overline{k} \sin \theta$), (8.91) may be written as

$$\left(\frac{\Delta}{c \Delta t}\right)^2 \sin^2\left(\frac{\omega \Delta t}{2}\right) = \sin^2\left(\frac{\overline{k} \Delta \cos \theta}{2}\right) + \sin^2\left(\frac{\overline{k} \Delta \sin \theta}{2}\right) \qquad (8.92)$$

This expression may be solved for $\overline{k}$ as a function of θ using Newton's method. Figure 8.30 shows the normalized phase velocity as a function of the angle of propagation for $c \Delta t = \Delta/2$. It is observed from here that numerical dispersion not only depends on discretization size Δ but also on the angle of propagation. The dependence on angle is called anisotropy of the numerical dispersion. The phase error is found to be maximum along the x- and y-directions and minimum along the diagonal (45°). This is because the apparent separation between the nodes experienced by the wave is less than Δ along the diagonal. This phenomenon is explained graphically next.

Grid Anisotropy

The anisotropy in Figure 8.30 can be explained by drawing the phase fronts and determining the spacing between them for a given angle of propagation. For

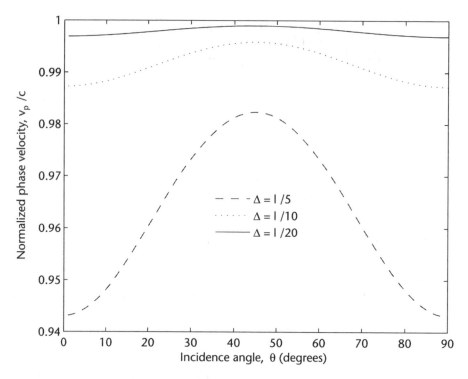

Figure 8.30 Effect of the angle of propagation on the phase velocity with cell size as a parameter. The cells are square in the two-dimensional grid, and $c\Delta t = \Delta x/2$. The propagation along the diagonal corresponds to $\theta = 45°$.

propagation at $\theta = 0°$ the phase fronts are parallel to the y-axis and the spacing between them is equal to Δx as shown in Figure 8.31(a). The phase velocity for this case may also be obtained by employing the one-dimensional dispersion relation (8.23). Next, we consider propagation along the diagonal, $\theta = 45°$. The square grid and the equiphase planes for this case are drawn in Figure 8.31(b). Since the phase front jumps from one row of nodes to the next row as shown, the spacing

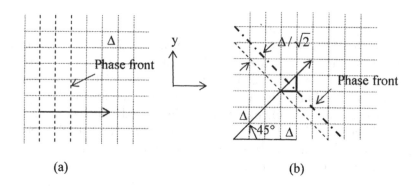

(a)   (b)

Figure 8.31 Phase fronts for two different directions of propagation in two-dimensional propagation. (a) Propagation at $\theta = 0°$. The effective spacing experienced by the wave is Δ. (b) Propagation at $\theta = 45°$. The effective spacing experienced by the wave is $\Delta/\sqrt{2}$.

between the consecutive rows of nodes perpendicular to the direction of propagation is now $\sqrt{2}\Delta/2 = \Delta/\sqrt{2}$, which is less than Δ. The dispersion is therefore less compared to that at $\theta = 0°$ or $90°$. This hypothesis was tested by computing the phase velocity using effective spacing between the nodes and one-dimensional dispersion relation, (8.23). Table 8.1 lists these values for a few directions of propagation. The phase velocity is maximum for $\theta = 45°$ as seen in Figure 8.30.

8.4.3 Time Step Δt for Two-Dimensional Propagation

We again consider square cells of Figure 8.31 to determine the upper limit on the value of Δt for propagation in two dimensions. The value of Δt for propagation along either of the axes is governed by one-dimensional limit (8.27) and is given by $\Delta t < \Delta/v_{max}$. However, this value of Δt is too large for propagation along the diagonal because the effective spacing between the consecutive rows of nodes is $\Delta/\sqrt{2}$. The wave would therefore miss the nodes located on the line marked as dot-dash in Figure 8.31(b) if the time step is greater than $\Delta/(v_{max}\sqrt{2})$. The upper limit on the time step for the stability of solution process is therefore given by

$$v_{max} \cdot \Delta t \leq \frac{1}{\sqrt{\dfrac{1}{(\Delta x)^2} + \dfrac{1}{(\Delta y)^2}}} \tag{8.93a}$$

where v_{max} is the maximum phase velocity of the signal. For square cells, $\Delta x = \Delta y = \Delta$

$$\sqrt{2}v_{max}\Delta t \leq \Delta \tag{8.93b}$$

8.4.4 Absorbing Boundary Conditions for Propagation in Two Dimensions

The general principles for the derivation of ABC for the one-dimensional case presented in Section 8.2.4 may be used to derive ABC for the two-dimensional case also. These ABCs are called Mur's ABC and are discussed next.

Mur's Second Order Absorbing Boundary Condition
In order to develop this boundary condition we consider a second order wave equation in two dimensions,

Table 8.1 Normalized Phase Velocity v_p/c Obtained from (8.23) and Effective Spacing Between the Consecutive Rows of Nodes (Square Grid with $c\Delta t = \Delta/2$, $\Delta/\lambda = 0.2$)

S.N.	Angle of Propagation, θ	Effective Spacing Between the Nodes	v_p/c
1	$0°$, $90°$	Δ	0.9431
2	$\tan^{-1}(1/2) = 26.56°$	$2\Delta/\sqrt{5}$	0.9559
3	$\tan^{-1}(1) = 45°$	$\Delta/\sqrt{2}$	0.9737
4	$\tan^{-1}(2) = 63.43°$	$2\Delta/\sqrt{5}$	0.9559

$$\frac{\partial^2 \phi}{\partial x^2} + \frac{\partial^2 \phi}{\partial y^2} - \frac{1}{c^2}\frac{\partial^2 \phi}{\partial t^2} = 0 \tag{8.94}$$

where ϕ represents any field component. This partial-differential equation supports wave propagation in all the directions in the xy-plane. To develop the one-way wave equations for nonnormal incidence to the planes at $x = 0$ and $x = x_0$, we write the differential operator in (8.94) as [4]

$$\left(\frac{\partial^2}{\partial x^2} - \left(\sqrt{\frac{1}{c^2}\frac{\partial^2}{\partial t^2} - \frac{\partial^2}{\partial y^2}}\right)^2\right)\phi = 0$$

or

$$\left(\frac{\partial}{\partial x} - \sqrt{\frac{1}{c^2}\frac{\partial^2}{\partial t^2} - \frac{\partial^2}{\partial y^2}}\right)\left(\frac{\partial}{\partial x} + \sqrt{\frac{1}{c^2}\frac{\partial^2}{\partial t^2} - \frac{\partial^2}{\partial y^2}}\right)\phi = 0 \tag{8.95}$$

One-way wave equations in two dimensions can be obtained if we apply these factors to ϕ individually [4]. For the boundary at $x = 0$, for instance,

$$\left(\frac{\partial}{\partial x} - \sqrt{\frac{1}{c^2}\frac{\partial^2}{\partial t^2} - \frac{\partial^2}{\partial y^2}}\right)\phi = 0 \tag{8.96}$$

is the proper ABC. The pseudo-differential operator with the square-root sign in (8.96) is not easy to implement. Assuming, $(1/c)\partial/\partial t \gg |\partial/\partial y|$, we may expand the square root in a Taylor series. After some algebra, (8.96) results in

$$\left(\frac{1}{c}\frac{\partial}{\partial t}\frac{\partial}{\partial x} - \frac{1}{c^2}\frac{\partial^2}{\partial t^2} + \frac{1}{2}\frac{\partial^2}{\partial y^2}\right)\phi = 0 \tag{8.97a}$$

It is an approximate one-way wave equation suitable as an ABC at $x = 0$ edge. Expression (8.97a) is called the *second order Mur absorbing boundary condition*. At the $x = x_0$ edge, the ABC may be similarly expressed as

$$\left(\frac{1}{c}\frac{\partial}{\partial t}\frac{\partial}{\partial x} + \frac{1}{c^2}\frac{\partial^2}{\partial t^2} - \frac{1}{2}\frac{\partial^2}{\partial y^2}\right)\phi = 0 \tag{8.97b}$$

The ABC at the edges at $y = 0$ and $y = y_0$ are obtained by the interchange of x and y in (8.97).

For the implementation of the ABC, (8.97) is discretized using central difference approximation. Assuming square grid with $\Delta x = \Delta y = \Delta$ and discretizing (8.97a) about 1/2, j, n yields the following time-stepping algorithm [3, p. 241]

$$\phi_{0,j}^{n+1} = -\phi_{1,j}^{n-1} + \frac{c\Delta t - \Delta}{c\Delta t + \Delta}\left(\phi_{0,j}^{n-1} + \phi_{1,j}^{n+1}\right) + \frac{2\Delta}{c\Delta t + \Delta}\left(\phi_{0,j}^{n} + \phi_{1,j}^{n}\right) \quad (8.98\text{a})$$

$$+ \frac{(c\Delta t)^2}{2\Delta(c\Delta t + \Delta)}\left(\phi_{0,j+1}^{n} - 2\phi_{0,j}^{n} + \phi_{0,j-1}^{n} + \phi_{1,j+1}^{n} - 2\phi_{1,j}^{n} + \phi_{1,j-1}^{n}\right)$$

for updating the tangential fields at the reflectionless boundary at $x = 0$. The corresponding expressions at other boundaries may be obtained by inspection of (8.98a) and are given next.

For the boundary at $x = x_0 = i_0\Delta x$,

$$\phi_{i_0,j}^{n+1} = -\phi_{i_0-1,j}^{n-1} + \frac{c\Delta t - \Delta}{c\Delta t + \Delta}\left(\phi_{i_0,j}^{n-1} + \phi_{i_0-1,j}^{n+1}\right) + \frac{2\Delta}{c\Delta t + \Delta}\left(\phi_{i_0,j}^{n} + \phi_{i_0-1,j}^{n}\right) \quad (8.98\text{b})$$

$$+ \frac{(c\Delta t)^2}{2\Delta(c\Delta t + \Delta)}\left(\phi_{i_0,j+1}^{n} - 2\phi_{i_0,j}^{n} + \phi_{i_0,j-1}^{n} + \phi_{i_0-1,j+1}^{n} - 2\phi_{i_0-1,j}^{n} + \phi_{i_0-1,j-1}^{n}\right)$$

for the boundary at $y = 0$,

$$\phi_{i,0}^{n+1} = -\phi_{i,1}^{n-1} + \frac{c\Delta t - \Delta}{c\Delta t + \Delta}\left(\phi_{i,0}^{n-1} + \phi_{i,1}^{n+1}\right) + \frac{2\Delta}{c\Delta t + \Delta}\left(\phi_{i,0}^{n} + \phi_{i,1}^{n}\right) \quad (8.98\text{c})$$

$$+ \frac{(c\Delta t)^2}{2\Delta(c\Delta t + \Delta)}\left(\phi_{i+1,0}^{n} - 2\phi_{i,0}^{n} + \phi_{i-1,0}^{n} + \phi_{i+1,1}^{n} - 2\phi_{i,1}^{n} + \phi_{i-1,1}^{n}\right)$$

and at $y = y_0 = j_0\Delta y$,

$$\phi_{i,j_0}^{n+1} = -\phi_{i,j_0-1}^{n-1} + \frac{c\Delta t - \Delta}{c\Delta t + \Delta}\left(\phi_{i,j_0}^{n-1} + \phi_{i,j_0-1}^{n+1}\right) + \frac{2\Delta}{c\Delta t + \Delta}\left(\phi_{i,j_0}^{n} + \phi_{i,j_0-1}^{n}\right) \quad (8.98\text{d})$$

$$+ \frac{(c\Delta t)^2}{2\Delta(c\Delta t + \Delta)}\left(\phi_{i+1,j_0}^{n} - 2\phi_{i,j_0}^{n} + \phi_{i-1,j_0}^{n} + \phi_{i+1,j_0-1}^{n} - 2\phi_{i,j_0-1}^{n} + \phi_{i-1,j_0-1}^{n}\right)$$

Residual Reflections from the Mur's ABCs

The Mur's ABCs described above can be evaluated for their residual reflection by using a procedure similar to that used for the dispersion analysis, Section 8.2.1. Since the ABC is expected to be imperfect due to the discretization of medium, the total field at the boundary consists of an incident wave and a reflected wave. For a plane wave incident at an angle θ at the $x = 0$ edge (Figure 8.32), we can write

$$\phi(x, y, t) = e^{j\omega t}(e^{jk_x x} + Re^{-jk_x x})e^{-jk_y y} \quad (8.99)$$

where R is the reflection coefficient. We now substitute (8.99) in the Mur ABC, (8.98a). Use of $k_x = k\cos\theta$, $k_y = k\sin\theta$ gives the following expression for R [3, p. 245]:

$$|R| = \left|\frac{\cos\theta - 1 + 0.5\sin^2\theta}{\cos\theta + 1 - 0.5\sin^2\theta}\right| \quad (8.100)$$

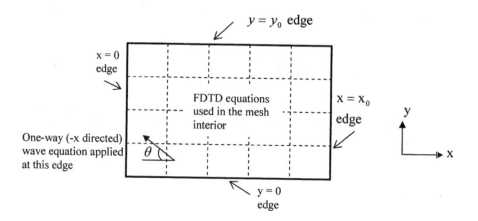

Figure 8.32   Illustration of one-way wave equations used as ABCs.

The actual numerical experiments however show a comparatively poor performance due to the effect of numerical dispersion and is explained next.

In order to derive an expression for R for the discretized version of ABCs, assume a plane wave solution of the form

$$\phi_{lm}^n = e^{j\omega n \Delta t}\left(e^{j\overline{k}_x l \Delta x} + \mathrm{Re}^{-j\overline{k}_x l \Delta x}\right)e^{-j\overline{k}_y m \Delta y} \tag{8.101}$$

where $\overline{k}_x$ and $\overline{k}_y$ are the phase constants for the numerical wave along x- and y-directions, respectively. Using (8.101) in (8.94) leads to [4, p. 514]

$$R = -e^{j\overline{k}_x \Delta x}\,\frac{\Delta x\left[A - \dfrac{1}{2}B\right]\cos\left(\dfrac{\overline{k}_x \Delta x}{2}\right) - C\sin\left(\dfrac{\omega \Delta t}{2}\right)}{\Delta x\left[A - \dfrac{1}{2}B\right]\cos\left(\dfrac{\overline{k}_x \Delta x}{2}\right) + C\sin\left(\dfrac{\omega \Delta t}{2}\right)} \tag{8.102}$$

where

$$A = \sin^2\left(\frac{\omega \Delta t}{2}\right)$$

$$B = \left(\frac{c\Delta t}{\Delta y}\right)^2 \sin^2\left(\frac{\overline{k}_y \Delta y}{2}\right)$$

$$C = c\Delta t \cos\left(\frac{\omega \Delta t}{2}\right)\sin\left(\frac{\overline{k}_x \Delta x}{2}\right)$$

The wavenumbers $\overline{k}$ and $\overline{k}_x, \overline{k}_y$ for the numerical wave are related by $\overline{k}_x = \overline{k}\cos\theta$, $\overline{k}_y = \overline{k}\sin\theta$. Also, $\overline{k}$ is obtained from (8.92) once Δx, Δy, and $\omega \Delta t$ are specified. Figure 8.33 shows the reflection coefficient (8.102) for a square grid,

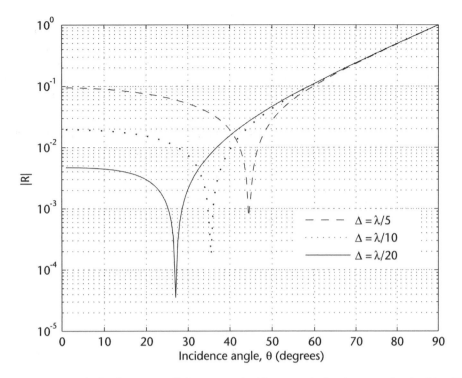

Figure 8.33 Residual reflection coefficient versus incidence angle for the second-order Mur ABC with different (square) grid resolutions and $c\Delta t = \Delta/2$.

$c\Delta t = \Delta/2$, and $\Delta = \lambda/5$, $\lambda/10$, $\lambda/20$. It also demonstrates the effect of angle of incidence on the reflection coefficient. As expected, the quality of ABC improves with the grid resolution. The reflection coefficient is found to depend on the angle of incidence because numerical dispersion is anisotropic (Figure 8.30). Focusing on $\Delta = \lambda_0/10$, we note that $|R| \approx 0.0195$ at normal incidence, becomes minimum at 35°, and increases to unity at 90°.

The high value of reflection coefficient obtained from Mur's second order ABC is generally not acceptable for research applications except for crude experiments. We, discuss the alternatives next.

The design of material absorbers as terminations for one-dimensional propagation was discussed in Section 8.3. We could achieve a reflection coefficient of the order of 10^{-4} by using a lossy medium of sufficient thickness. We may think of a similar approach to reduce the reflection coefficient and grid anisotropy for two-dimensional propagation. This has been devised in the form of what is called perfectly matched layer (PML) ABC by Berenger [13]. This PML formulation involves splitting of field components and increases coding complexity. Nonsplit field formulation for PML is also available and requires use of anisotropic, lossy medium [14, 15].

8.4.5 Perfectly Matched Layer ABC

Let the analytical ABCs in Figure 8.32 be replaced by the absorbing material of sufficient thickness to create an environment of anechoic chamber. The modified configuration with the material absorbers in place is shown in Figure 8.34. The major contribution to the reflection comes from the impedance mismatch between the absorbing medium and the medium in front of it as the angle of incidence of the wave is varied. Berenger could achieve impedance matching by splitting the electric and magnetic field components in the absorbing media, and assigning different conductivities σ and σ^* for different directions. The net effect of this is to create a nonphysical absorbing medium such that the wave impedance in this medium becomes independent of the angle of incidence and frequency of the waves [13, 14, 16–18].

Propagation of TE Plane Wave in the PML Medium
The Maxwell's equations of (1.1) are modified as follows in order to include the effect of losses in the medium characterized by μ, ϵ, σ, and σ^*:

$$\nabla \times \mathbf{E} = -\mu \frac{\partial \mathbf{H}}{\partial t} - \sigma^* \mathbf{H} \qquad \text{Faraday's law} \qquad (8.103a)$$

$$\nabla \times \mathbf{H} = \epsilon \frac{\partial \mathbf{E}}{\partial t} + \sigma \mathbf{E} \qquad \text{Ampere's law} \qquad (8.103b)$$

For the TE-to-z wave in free space with $\partial/\partial z \equiv 0$, the above equations reduce to

$$\mu_0 \frac{\partial H_z}{\partial t} + \sigma^* H_z = \frac{\partial E_x}{\partial y} - \frac{\partial E_y}{\partial x} \qquad (8.104a)$$

$$\epsilon_0 \frac{\partial E_x}{\partial t} + \sigma E_x = \frac{\partial H_z}{\partial y} \qquad (8.104b)$$

$$\epsilon_0 \frac{\partial E_y}{\partial t} + \sigma E_y = -\frac{\partial H_z}{\partial x} \qquad (8.104c)$$

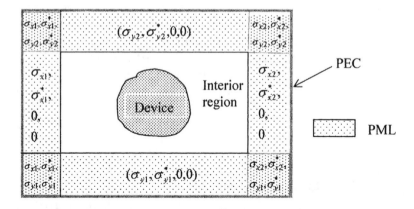

Figure 8.34 Placement of PML for the two-dimensional FDTD domain.

In the formulation proposed by Berenger [13], the incident wave is decomposed into two TEM subwaves, similar to the decomposition of a vector. One of these subwaves propagates along the x-direction, and the other along the y-direction. The respective subwaves travel in a lossy medium characterized by either $\left(\epsilon_0, \mu_0, \sigma_x, \sigma_x^*\right)$ or $\left(\epsilon_0, \mu_0, \sigma_y, \sigma_y^*\right)$. The individual plane waves are normally incident to the lossy medium and are impedance matched if σ_i, σ_i^*, $i = x$, y satisfy (8.57b).

To generate subwaves out of the TE wave, the magnetic field H_z alone is split into H_{zx} and H_{zy} components as

$$H_z = H_{zx} + H_{zy} \tag{8.105}$$

Similarly, the conductivities σ and σ^* are split into x and y components and we replace $\sigma(\sigma^*)$ by the appropriate σ_x or σ_y $\left(\sigma_x^*$ or $\sigma_y^*\right)$ in (8.104). The resulting expressions may be arranged as [13, 14, 16–18]

$$\mu_0 \frac{\partial H_{zy}}{\partial t} + \sigma_y^* H_{zy} = \frac{\partial E_x}{\partial y} \tag{8.106a}$$

$$\epsilon_0 \frac{\partial E_x}{\partial t} + \sigma_y E_x = \frac{\partial (H_{zx} + H_{zy})}{\partial y} \tag{8.106b}$$

$$\mu_0 \frac{\partial H_{zx}}{\partial t} + \sigma_x^* H_{zx} = -\frac{\partial E_y}{\partial x} \tag{8.106c}$$

$$\epsilon_0 \frac{\partial E_y}{\partial t} + \sigma_x E_y = -\frac{\partial (H_{zx} + H_{zy})}{\partial x} \tag{8.106d}$$

The choice of σ_x, σ_y in (8.106) is guided by the consideration that (8.106a) and (8.106b) describe a plane wave (with E_x, H_{zy} components) propagating along the y-direction in a lossy medium characterized by $\left(\sigma_y, \sigma_y^*\right)$. The other plane wave defined by (8.106c) and (8.106d) has E_y, H_{zx} components, and propagates along the x-direction in a lossy medium characterized by $\left(\sigma_x, \sigma_x^*\right)$.

Let the electric vector of the TE wave make an angle φ with respect to the y-axis as shown in Figure 8.35. The plane wave solution of (8.106) for this wave is found to be of the form [3]

$$\psi = \psi_0 e^{j\omega\left(t - \frac{x \cos\phi + y \sin\phi}{cG}\right)} e^{-\frac{\sigma_x \cos\phi}{\epsilon_0 cG}x} e^{-\frac{\sigma_y \sin\phi}{\epsilon_0 cG}y} \tag{8.107}$$

where ψ represents any field component of (8.106) with ψ_0 its magnitude, and

$$G = \sqrt{w_x \cos^2\phi + w_y \sin^2\phi} \tag{8.108a}$$

$$w_x = \frac{1 - \frac{j\sigma_x}{\omega\epsilon_0}}{1 - \frac{j\sigma_x^*}{\omega\mu_0}}, \quad \text{and} \quad w_y = \frac{1 - \frac{j\sigma_y}{\omega\epsilon_0}}{1 - \frac{j\sigma_y^*}{\omega\mu_0}} \tag{8.108b}$$

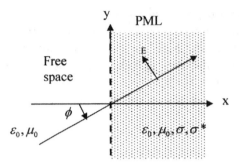

Figure 8.35 Wave incident on PML–free space interface at an arbitrary angle.

The wave impedance of the PML is defined as

$$Z_L = \sqrt{\frac{\mu_0}{\epsilon_0}} \frac{1}{G} \tag{8.109}$$

Reflectionless Matching of the Lossy Medium
The wave impedance of the lossy medium Z_L differs from the free space impedance at the left of interface, by the factor G. If G can be made equal to 1 independent of angle φ and frequency, it will give rise to a perfect match of the two media. The condition $G = 1$ is satisfied for all values of φ if $w_x = w_y$. We conclude from (8.108) that the condition for perfect impedance matching is

$$\frac{\sigma_i^*}{\mu_0} = \frac{\sigma_i}{\epsilon_0}, \qquad i = x, y \tag{8.110}$$

and is the generalization of (8.57b) for the two-dimensional propagation. With this choice of G, (8.107) for the field in the PML reduces to

$$\psi = \psi_0 e^{j\omega\left(t - \frac{x \cos \phi + y \sin \phi}{c}\right)} e^{-\frac{\sigma_x \cos \phi}{\epsilon_0 c}x} e^{-\frac{\sigma_y \sin \phi}{\epsilon_0 c}y} \tag{8.111}$$

The first exponential factor indicates that the wave in the PML medium travels in the direction φ, normal to the electric vector, with the speed of light c. The other two factors account for exponential decay of the field along x- and y-directions, respectively.

Design of PML
Consider the two-dimensional FDTD region shown in Figure 8.34. The interior region where the device under test is to be placed is free space and it is surrounded by PML backed by perfect electric walls (PEC). The different portions of the PML region are described by different values of σ and σ^*. The PML on the sides is characterized by only one pair of conductivities (σ_x, σ_x^*) or (σ_y, σ_y^*); for example, the layers normal to the x-direction are characterized by (σ_x, σ_x^*). The corner

regions are defined by the overlap of respective sides. The values of conductivity pairs (σ_x, σ_x^*) and (σ_y, σ_y^*) are chosen according to (8.110).

The result of field and conductivity splitting is that there is no reflection at the interface between the interior region and the PML irrespective of the angle of incidence and frequency. However, due to the finite thickness of the PML medium, reflection is generated when the outgoing attenuated waves are reflected by the terminating conductors, and these reflections can return to the device. So, an apparent reflection coefficient is defined, which is a function of the PML thickness t and the conductivity profile $\sigma_n(\rho)$, where ρ is the distance from the interface. Henceforth, we shall use σ_n in place of σ_i for the layer normal to the ith direction, $i = x$ or y.

The magnitude of the reflection coefficient for a wave incident at an angle φ with respect to the normal to the layer is [13]

$$R(\phi) = [R(0)]^{\cos(\phi)} \tag{8.112}$$

with $R(0)$ the reflection coefficient for normal incidence at the interior region-PML interface defined as

$$R(0) = \exp\left(\frac{-2}{\epsilon_0 c} \int_0^t \sigma_n(\rho)\, d\rho\right) \tag{8.113}$$

The value of $R(0)$ depends on the PML thickness and the conductivity of the medium. The conductivity profile may be designed as

$$\sigma_n(\rho) = \sigma_{\max}\left(\frac{\rho}{t}\right)^n \tag{8.114}$$

where $n = 1$ gives linear increase, and $n = 2$ the parabolic increase. The reflection coefficient is therefore given by

$$R(0) = \exp\left(\frac{-2\sigma_{\max}t}{(n+1)\epsilon_0 c}\right) \qquad n = 0, 1, 2, \ldots \tag{8.115}$$

where $\sigma_{\max}$ is the maximum conductivity of the absorbing medium.

In the design of PML we have assumed that the medium is continuous. However, in the FDTD analysis the PML medium is discretized into cells and the resulting reflection coefficient may be slightly larger than the designed value. A residual reflection of the order of 60 to 80 dB may be achieved with only 6 to 8 PML cells.

While writing the code, one needs to split some field components and assign different conductivities as described above in the PML region. It is not necessary to split the field components in the interior region. However, this gives rise to different expressions for the Maxwell's curl equations in the two domains. It is possible to unify the formulations in the two domains if we adopt the split field formulation in the interior region also. The integrated approach is computationally more expensive both in computer memory and CPU time requirements, but is simpler to code. It is now possible to design PML without splitting of fields, and the performance of these formulations is shown to be identical with that of Berenger's PML [3, p. 288].

Uniaxial PML

Nonsplit field formulations in PML have been developed to simplify coding. One of these formulations is in the form of complex stretching of the Cartesian coordinates in the frequency domain. The complex stretching variables in three dimensions are defined as [19, p. 285]

$$s_i(i, \omega) = a_i(i) + j \frac{\Omega_i(i)}{\omega}, \qquad i = x, y, z \qquad (8.116)$$

with $a_i \geq 1$ and $\Omega_i \geq 0$ as profile functions for PML. The imaginary part Ω_i is responsible for exponential decay in the PML region. The Maxwell's equations in this coordinate system can be obtained from (1.5) with ∇ replaced by ∇_s with

$$\nabla_s = \hat{x} \frac{1}{s_x} \frac{\partial}{\partial x} + \hat{y} \frac{1}{s_y} \frac{\partial}{\partial y} + \hat{z} \frac{1}{s_z} \frac{\partial}{\partial z} \qquad (8.117)$$

The coordinate stretching of (8.116) may also be viewed as a mapping of the physical space to a complex coordinate space. The stretching variable formulation leads to convolution in time domain, and can be avoided by splitting of fields similar to that in Berenger's PML [19, p. 287].

In a popular unsplit field formulation, PML medium is characterized as aniso-tropic medium with permittivity and permeability tensors, and is called uniaxial PML, UPML for short. For a wave incident on half-space $x > 0$, as shown in Figure 8.35, the uniaxial medium is defined as

$$\bar{\bar{\epsilon}} = \epsilon_0 \bar{\bar{s}}, \qquad \text{and} \qquad \bar{\bar{\mu}} = \mu_0 \bar{\bar{s}} \qquad (8.118)$$

where the media tensor $\bar{\bar{s}}$ for the three-dimensional PML region is defined as [3, p. 287]

$$\bar{\bar{s}} = \begin{bmatrix} s_x^{-1} & 0 & 0 \\ 0 & s_x & 0 \\ 0 & 0 & s_x \end{bmatrix} \qquad (8.119)$$

The medium characterized as above leads to reflectionless propagation from free space to PML for both TE and TM waves, independent of angle of incidence, polarization, and frequency. The medium is called uniaxial because of its anisotropy along one axis only, and its perfect matching property justifies PML label. For exponential decay of the propagating wave in UPML, the media constants should be complex with $s_x = 1 - j\sigma_x/(\omega\epsilon_0)$. The UPML is characterized by $\sigma_x(i)$ alone for layers perpendicular to x-axis, and by $\sigma_y(j)$ alone for layers perpendicular to y-axis, and by $\sigma_x(i)$ and $\sigma_y(j)$ for the corner regions (Figure 8.34). The non-PML region or central region is characterized by $\sigma_x = 0$, thus simplifying coding.

The implementation of UPML in FDTD algorithm is not simple since the medium is dispersive due to frequency dependence of both permittivity and permeability tensors. Direct transformation from frequency domain to time domain involves convolution and is inefficient. Instead, a two-step time marching scheme is suggested in [3, p. 301] for efficient implementation of UPML. In this scheme Maxwell's equations are formulated in terms of $\mathbf{D}$ and $\mathbf{B}$, and $\mathbf{E}$ and $\mathbf{H}$ are derived from these. This process is similar to that described in Section 8.3 for the lossy, dispersive medium. A source code in C for PML in two dimensions is given in [5, p. 67].

8.5   FDTD Analysis in Three Dimensions

The FDTD analysis for one- and two-dimensional problems is useful in developing concepts and analyzing simple problems. The real-world problems are mostly three-dimensional in nature. The concepts and analysis described earlier can be extended to three dimensions easily, except for the coding complexity and increase in processor time.

The Maxwell's equations in three dimensions comprise of all the six field components E_x, E_y, E_z, H_x, H_y, and H_z. These are written as

$$\frac{\partial E_x}{\partial t} = \frac{1}{\epsilon}\left(\frac{\partial H_z}{\partial y} - \frac{\partial H_y}{\partial z}\right) \tag{8.120a}$$

$$\frac{\partial E_y}{\partial t} = \frac{1}{\epsilon}\left(\frac{\partial H_x}{\partial z} - \frac{\partial H_z}{\partial x}\right) \tag{8.120b}$$

$$\frac{\partial E_z}{\partial t} = \frac{1}{\epsilon}\left(\frac{\partial H_y}{\partial x} - \frac{\partial H_x}{\partial y}\right) \tag{8.120c}$$

$$\frac{\partial H_x}{\partial t} = \frac{1}{\mu}\left(\frac{\partial E_y}{\partial z} - \frac{\partial E_z}{\partial y}\right) \tag{8.121a}$$

$$\frac{\partial H_y}{\partial t} = \frac{1}{\mu}\left(\frac{\partial E_z}{\partial x} - \frac{\partial E_x}{\partial z}\right) \tag{8.121b}$$

$$\frac{\partial H_z}{\partial t} = \frac{1}{\mu}\left(\frac{\partial E_x}{\partial y} - \frac{\partial E_y}{\partial x}\right) \tag{8.121c}$$

8.5.1   Yee Cell

The FDTD analysis for three-dimensional devices is carried out by dividing the geometry into a number of cells called Yee cells [1] of dimensions $\Delta x \Delta y \Delta z$. The Yee cell is drawn in Figure 8.36. In this arrangement, E and H nodes are located such that H nodes are displaced in space by half a space step ($\Delta x/2$, $\Delta y/2$, $\Delta z/2$)

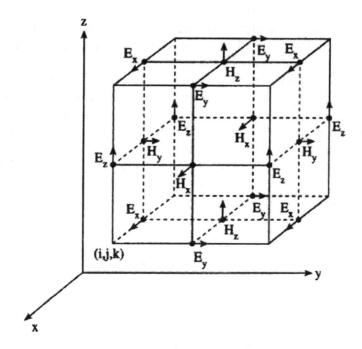

Figure 8.36 Geometry of a Yee cell in FDTD three-dimensional grid.

with respect to the corresponding E nodes (e.g., H_x is displaced with respect to E_x by $(\Delta x/2, \Delta y/2, \Delta z/2)$, and so on for H_y and H_z nodes). The H nodes are surrounded by E nodes and vice versa. The remarkable property of Yee cell is that placement of E and H nodes satisfy Maxwell's equations in integral and differential forms. This property was verified in Section 8.4 for the Yee cell in two-dimensions. The Yee cell saves on computer resources also because E- and H-fields are not to be calculated for all the nodes; that is, either E-field or H-field should be calculated for a node. The arrangement of three-dimensional Yee cells creates an environment in which E-and H-field loops are interlocked with each other in a chain-like fashion. This interlocking is shown in Figure 8.37 for two adjacent Yee cells. The H-field of the gray cell is displaced from the E-field of the white cell by a half space step in each direction. The plane wave propagates over these loops and advances in space. The interlocking simulates the actual propagation phenomenon of electromagnetic waves.

For FDTD analysis, the partial differential equation (8.120a) may be expressed in the finite difference form as

$$
E_x^{n+1}\left(i-\frac{1}{2}, j, k\right) = E_x^n\left(i-\frac{1}{2}, j, k\right)
$$

$$
+ \frac{\Delta t}{\epsilon \Delta y}\left[H_z^{n+1/2}\left(i-\frac{1}{2}, j+\frac{1}{2}, k\right) - H_z^{n+1/2}\left(i-\frac{1}{2}, j-\frac{1}{2}, k\right)\right]
$$

$$
- \frac{\Delta t}{\epsilon \Delta z}\left[H_y^{n+1/2}\left(i-\frac{1}{2}, j, k+\frac{1}{2}\right) - H_y^{n+1/2}\left(i-\frac{1}{2}, j, k-\frac{1}{2}\right)\right]
$$

$$
\text{(8.122a)}
$$

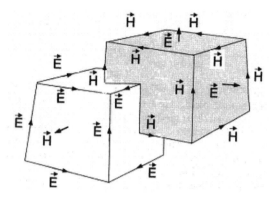

Figure 8.37   Placements of field components on two adjacent Yee cells showing the interlocking of E- and H-field loops. (*From:* [3]. © 2005 Artech House, Inc. Reprinted with permission.)

It may be noted that this expression is consistent with the placement of field components in the Yee cell (Figure 8.36). Similarly,

$$
E_y^{n+1}\left(i, j+\frac{1}{2}, k\right) = E_y^n\left(i, j+\frac{1}{2}, k\right)
$$

$$
+\frac{\Delta t}{\epsilon \Delta z}\left[H_x^{n+1/2}\left(i, j+\frac{1}{2}, k+\frac{1}{2}\right) - H_x^{n+1/2}\left(i, j+\frac{1}{2}, k-\frac{1}{2}\right)\right]
$$

$$
-\frac{\Delta t}{\epsilon \Delta x}\left[H_z^{n+1/2}\left(i+\frac{1}{2}, j+\frac{1}{2}, k\right) - H_z^{n+1/2}\left(i-\frac{1}{2}, j+\frac{1}{2}, k\right)\right]
$$

$$(8.122b)$$

$$
E_z^{n+1}\left(i, j, k+\frac{1}{2}\right) = E_z^n\left(i, j, k+\frac{1}{2}\right)
$$

$$
+\frac{\Delta t}{\epsilon \Delta x}\left[H_y^{n+1/2}\left(i+\frac{1}{2}, j, k+\frac{1}{2}\right) - H_y^{n+1/2}\left(i-\frac{1}{2}, j, k+\frac{1}{2}\right)\right]
$$

$$
-\frac{\Delta t}{\epsilon \Delta y}\left[H_x^{n+1/2}\left(i, j+\frac{1}{2}, k+\frac{1}{2}\right) - H_x^{n+1/2}\left(i, j-\frac{1}{2}, k+\frac{1}{2}\right)\right]
$$

$$(8.122c)$$

Expressions (8.121) may be similarly cast in the following forms

$$
H_x^{n+1/2}\left(i, j+\frac{1}{2}, k+\frac{1}{2}\right) = H_x^{n-1/2}\left(i, j+\frac{1}{2}, k+\frac{1}{2}\right)
$$

$$
-\frac{\Delta t}{\mu \Delta y}\left[E_z^n\left(i, j, k+\frac{1}{2}\right) - E_z^n\left(i, j-1, k+\frac{1}{2}\right)\right]
$$

$$
+\frac{\Delta t}{\mu \Delta z}\left[E_y^n\left(i, j+\frac{1}{2}, k\right) - E_y^n\left(i, j+\frac{1}{2}, k-1\right)\right]
$$

$$(8.123a)$$

$$H_y^{n+1/2}\left(i-\frac{1}{2},j,k+\frac{1}{2}\right) = H_y^{n-1/2}\left(i-\frac{1}{2},j,k+\frac{1}{2}\right)$$

$$-\frac{\Delta t}{\mu\Delta z}\left[E_x^n\left(i-\frac{1}{2},j,k\right) - E_x^n\left(i-\frac{1}{2},j,k-1\right)\right]$$

$$+\frac{\Delta t}{\mu\Delta x}\left[E_z^n\left(i,j,k+\frac{1}{2}\right) - E_z^n\left(i-1,j,k+\frac{1}{2}\right)\right]$$

(8.123b)

$$H_z^{n+1/2}\left(i-\frac{1}{2},j+\frac{1}{2},k\right) = H_z^{n-1/2}\left(i-\frac{1}{2},j+\frac{1}{2},k\right)$$

$$-\frac{\Delta t}{\mu\Delta x}\left[E_y^n\left(i,j+\frac{1}{2},k\right) - E_y^n\left(i-1,j+\frac{1}{2},k\right)\right]$$

$$+\frac{\Delta t}{\mu\Delta y}\left[E_x^n\left(i-\frac{1}{2},j,k\right) - E_z^n\left(i-\frac{1}{2},j-1,k\right)\right]$$

(8.123c)

The difference equations can be coded carefully to develop a software.

Review Question 8.4. In the FDTD method, we compute the electric and magnetic field components at the nodes defined by the Yee cell, and at alternate time steps. However, all the six field components should exist at all the nodes and for all time steps, and are governed by Maxwell's equations. The Yee cell approach thus saves considerable computer resources. Once the FDTD computations have been carried out, the remaining five field components at the nodes and for the remaining type steps may be determined from the FDTD data. With reference to the Yee cell of Figure 8.36, the following options are listed. Score the possibly correct/approximate choices and explain why.

1. $E_x^{n+1/2}\left(i-\frac{1}{2},j,k\right) = \dfrac{E_x^n\left(i-\frac{1}{2},j,k\right) + E_x^{n+1}\left(i-\frac{1}{2},j,k\right)}{2}$

2. $E_x^n(i,j,k) = \dfrac{E_x^n\left(i-\frac{1}{2},j,k\right) + E_x^n\left(i+\frac{1}{2},j,k\right)}{2}$

3. $E_x^{n+1/2}\left(i-\frac{1}{2},j,k\right) = \dfrac{H_y^{n+1/2}\left(i-\frac{1}{2},j,k\right)}{\sqrt{\dfrac{\mu}{\epsilon}}}$

4. Write in for $E_x^{n+1/2}\left(i-\frac{1}{2},j,k\right)$

Answer. 1, 2.

Next we describe some important aspects of FDTD analysis in three dimensions.

8.5.2 Numerical Dispersion in Three Dimensions

The expression for numerical dispersion in three dimensions may be derived from the procedure followed in Section 8.2 for the one-dimensional case. Else, it may be generalized from the corresponding expression for the two-dimensional case, (8.91). One obtains for a plane wave,

$$
\left(\frac{\sin\left(\frac{\omega\Delta t}{2}\right)}{c\,\Delta t}\right)^2 = \left(\frac{\sin\left(\frac{\overline{k}_x\Delta x}{2}\right)}{\Delta x}\right)^2 + \left(\frac{\sin\left(\frac{\overline{k}_y\Delta y}{2}\right)}{\Delta y}\right)^2 + \left(\frac{\sin\left(\frac{\overline{k}_z\Delta z}{2}\right)}{\Delta z}\right)^2
$$

$$(8.124)$$

Qualitatively this expression suggests that numerical dispersion can be reduced to any degree that is desired if one uses a fine enough FDTD mesh. It is also possible to minimize the numerical dispersion by choosing the mesh size in an optimal fashion. It has been found that [20] for optimal dispersion, the mesh dimensions are related as $\Delta x = \Delta y = \Delta z$. The above condition is not always satisfied, because the geometrical dimensions of the device must match the mesh lines and the requirement to use computer resources optimally. However, the influence of unequal mesh size ($\Delta x \neq \Delta y \neq \Delta z$) on dispersion is relatively small, when $\Delta y/\Delta x$, $\Delta z/\Delta x$, and $v_{max}\sqrt{3}\,\Delta t/\Delta x$ ranges from 1.0 to 1.225, where Δx represents the smallest discretization size.

8.5.3 Time Step Δt for Three-Dimensional Propagation

The selection of time step Δt for propagation in three dimensions is based on the same considerations as described earlier for the two-dimensional propagation, Section 8.4. The upper bound on Δt is given by

$$
v_{max}\Delta t \leq \frac{1}{\sqrt{\dfrac{1}{\Delta x^2} + \dfrac{1}{\Delta y^2} + \dfrac{1}{\Delta z^2}}}
$$

$$(8.125)$$

For the special case of cubic cell $\Delta x = \Delta y = \Delta z = \Delta$, the stable FDTD solution requires

$$
v_{max}\Delta t \leq \frac{\Delta}{\sqrt{3}}
$$

$$(8.126)$$

The small value of time step Δt increases the processor time considerably if the structure has fine scale dimensions compared to the wavelength. Alternating-direction implicit (ADI) algorithm has been used to increase Δt by an order of magnitude with a slight increase in storage requirement [21].

8.5.4 Absorbing Boundary Conditions and PML for Three Dimensions

The derivation of ABCs for the three-dimensional case follows the two-dimensional developments closely (Section 8.4). The second-order ABC in three dimensions at $x = 0$ edge, for instance, may be obtained as

$$\left(\frac{1}{c} \frac{\partial}{\partial t} \frac{\partial}{\partial x} - \frac{1}{c^2} \frac{\partial^2}{\partial t^2} + \frac{1}{2} \frac{\partial^2}{\partial y^2} + \frac{1}{2} \frac{\partial^2}{\partial z^2} \right) \phi = 0 \qquad (8.127)$$

This expression can be expanded in central-difference form about the node $(1/2, j, k)$ and used in FDTD code for ABC at $x = 0$.

The generalization of two-dimensional PML medium to three dimensions is shown in Figure 8.38. The various portions of PML surrounding the device may be labeled as face, edge, and corner regions, according to their positions on the surface of interior region [22]. For example, the z-face region is the projection of the z-face of the interior region and is characterized by $\sigma_x = 0$ and $\sigma_y = 0$. Similarly, the xy-edge region surrounds the xy-edge of the interior region, and is characterized by $\sigma_z = 0$. Finally, σ_x, σ_y, σ_z are used to define the corner regions.

Comparison of Mur's ABC with PML
Numerical experiments have been conducted to assess the performance of Mur's ABC and PML. The performance of PML with parabolic profile and second order MUR ABC are compared in Table 8.2 [16]. The test grid consisted of $100 \times 100 \times 50$ cells. The computational resources required are also compared in this table. It may be observed that a PML thickness of 4 to 8 cells is a good trade-off between the ABC effectiveness and computer resources. The broadband nature of PML has been confirmed through numerical experiments [23]. The refection coefficient is found to be almost flat from dc to 10 GHz.

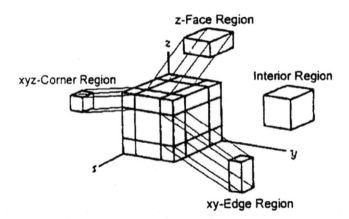

Figure 8.38 Illustration showing the face, edge, and corner regions of the PML medium surrounding the interior region for the three-dimensional propagation. (*From:* [22]. © 1999 IETE. Reprinted with permission.)

Table 8.2   Performance Comparison of Parabolic Profile PML and Second-Order Mur ABC, and Computer Resources Required

ABC	Avg. Local Field Error Reduction Relative to Second-Order Mur	Computer Resources: One CPU, Cray C-90
Mur	1 (0 dB)	10M words, 6.5 sec
4-layer PML	22 (27 dB)	16M words, 12 sec
8-layer PML	580 (55 dB)	23M words, 37 sec
16-layer PML	5,800 (75 dB)	43M words, 87 sec
Source: [16].		

8.6   Implementation of Boundary Conditions in FDTD

The analysis of electromagnetic devices invariably involves satisfying boundary and interface conditions to arrive at a unique solution. The boundary conditions may be in the form of Dirichlet condition, Neumann condition, or mixed boundary condition. The interface condition arises due to the combination of dielectrics in the device. Implementation of some of these conditions in FDTD is discussed next.

8.6.1   Perfect Electric and Magnetic Wall Boundary Conditions

The perfect electric conductor (pec) is characterized by $E_{tan} = 0$ at the surface of the conductor. One simple way to implement this boundary condition is to set $E_{tan} = 0$ for the nodes through which pec passes. This requires that the FDTD grid should be so constructed that pec boundaries coincide with the edges of the unit cells and thus E_{tan}. If it is not possible to arrange the grid structure to meet this requirement and the pec passes through magnetic field nodes as shown in Figure 8.39 for the two-dimensional case, we may use symmetry property of the electric field about pec to realize the boundary condition. The usual FDTD field update for H_x in the two-dimensional case is given by

$$\mu \frac{\partial H_x}{\partial t} \text{ (on grid line)} = -\frac{\partial E_z}{\partial y} = -\frac{E_z \text{ (above)} - E_z \text{ (below)}}{\Delta y} \tag{8.128}$$

The field E_z is expected to be zero at the pec, implying E_z should be anti-symmetric about pec; that is, $E_z(\text{below}) = -E_z(\text{above})$. Hence,

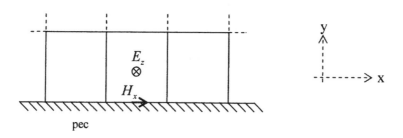

Figure 8.39   Implementation of perfect electric conductor (pec) boundary condition when pec passes through magnetic field nodes.

$$\mu \frac{\partial H_x}{\partial t} \text{ (on grid line)} = -\frac{2E_z \text{ (above)}}{\Delta y} \qquad (8.129)$$

The perfect magnetic conductor (pmc) is described by $H_{\tan} = 0$ at the surface. This boundary condition can also be implemented easily if the pmc boundaries pass through the nodes with $H_{\tan}$. Although pmc may not occur in the electromagnetic device, it is used as a tool to invoke symmetry condition and reduce the size of the problem. We had used symmetry of the expected field distribution and introduced pmc while determining the cutoff wavelength of waveguide modes in Chapter 7.

The implementation of pmc boundary condition when the pmc passes through the electric field nodes may be solved using the duality principle. Consider Figure 8.40 with the pmc passing through the lower E_x node. This figure is the dual of Figure 8.39. The corresponding expression for E_x is obtained as

$$\epsilon \frac{\partial E_x}{\partial t} \text{ (on grid line)} = \frac{2H_z \text{ (above)}}{\Delta y} \qquad (8.130)$$

8.6.2 Interface Conditions

The principal utility of computational methods lies in solving problems in electromagnetics with dielectric inhomogeneity. In these problems, the conditions at the interface of two dissimilar dielectrics imply that the components of E and H tangential to the interface should be continuous. In general, the media within the device is discretized in Yee cells such that the dielectric within each cell is homogeneous and the interface coincides with the edge of the Yee cell. The problem remains now to define the material parameters μ and ϵ for the nodes at the interface; whether it should be that of upper dielectric layer or of lower layer. By the application of Ampere's law to a rectangular loop about the node encompassing both the dielectrics (Figure 8.41), it can be shown that the material parameters used should be the average of the two media; that is, for the node at the interface

$$\mu = \frac{\mu_0}{2} \qquad \epsilon = \frac{\epsilon_1 + \epsilon_2}{2} \qquad (8.131)$$

Some of the useful topics such as processing of FDTD data to determine the S-parameters of circuits, modeling of devices and lumped elements, and near-field

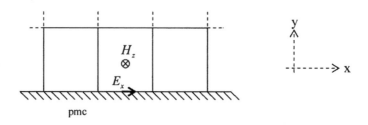

Figure 8.40 Implementation of perfect magnetic conductor (pmc) boundary condition when pmc passes through electric field nodes.

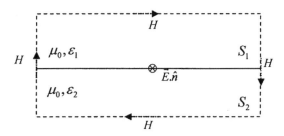

Figure 8.41 Application of Ampere's law to a loop in the inhomogeneous media to determine the effective permittivity at the node.

to far-field transformation could not be included in the text because of its limited scope. We must leave these topics to advanced texts [3].

8.7 Advances in FDTD

The FDTD method has seen robust growth in research activity in the last decade due to its applications in almost all areas, including communications, computing, and bio-medicine. Some of the recent issues in FDTD relate to the modeling of nonlinear dispersive materials used in photonics [24], conformal [25] and unstructured grids [26] to accommodate arbitrary shapes efficiently, pseudo-spectral time domain (PSTD) methods to reduce numerical dispersion [27], unconditionally stable techniques [28], hardware acceleration of computations [29], and hybrid FDTD-FEM techniques [30]. Advances in these techniques are very well described in [3].

8.8 Summary

FDTD method is a versatile method requiring almost no preprocessing of Maxwell's equations to arrive at the governing equations. The derivatives are expressed in finite difference form as in FDM. FDTD analysis simulates propagation behavior in time domain. The FDTD grid consists of electric and magnetic field nodes separated from each other by half space step. The electric field values are computed at integer time steps and magnetic field values at half-integer time steps. This is called staggered grid and it is based on Yee cell. The Yee cell satisfies Maxwell's equations in differential and integral forms. It is efficient also because the field values are computed at alternate nodes and at alternate time steps. The space step Δx is selected to be about $\lambda/10$ or lower to limit the phase error, and the time step Δt is selected such that $\Delta t < \Delta x/(c\sqrt{\epsilon_r})$ to ensure causality and the stability of the solution process. Since the error in phase velocity depends on $\Delta x/\lambda$ and a number of frequencies are contained in a narrow pulse, we choose $\Delta t = 0.5\Delta x/(c\sqrt{\epsilon_r})$ to be on the safer side. The grid is excited by a Gaussian pulse so that useful frequency components are included without wasting computer resources on unwanted frequencies. A point source is a compact source because it requires only a single node for implementing it. Absorbing boundary conditions in the form of differential equations are compact and are used to truncate the infinite space domain to finite size with minimal reflections at the truncation boundary. The one-dimensional FDTD analysis is applied to study basic phenomenon like reflection at a dielectric-air interface, determination of propagation constant in lossy medium, and the design of material absorbers. The pulse undergoes many reflections in to-and-fro manner between the discontinuities and a steady state is reached after a few nanoseconds. The recording of time history of the phenomenon produces animation and is useful as a diagnostic and instructional tool. The steady state behavior is Fourier transformed to obtain the broadband frequency information about the physical process. Similar considerations apply to propagation in two and three dimensions in FDTD grid.

References

[1] Yee, K. S., "Numerical Solution of Initial Boundary Value Problems Involving Maxwell's Equations in Isotropic Media," *IEEE Trans. Antennas Propagat.*, Vol. AP-14, 1966, pp. 302–307.

[2] Bondeson, A., T. Rylander, and P. Ingelstrom, *Computational Electromagnetics*, Berlin: Springer, 2005.

[3] Taflove, A., and S. C. Hagness, *Computational Electrodynamics: The Finite-Difference Time-Domain Method*, 3rd ed., Norwood, MA: Artech House, 2005.

[4] Peterson, A. F., S. L. Ray, and R. Mittra, *Computational Methods for Electromagnetics*, Hyderabad: University Press, 2001, Chapter 12.

[5] Sullivan, D. M., *Electromagnetic Simulation Using the FDTD Method*, New York: IEEE Press, 2000.

[6] Represa, J., C. Pereira, and M. Panizo, "A Simple Demonstration of Numerical Dispersion Under FDTD," *IEEE Trans. Education*, Vol. 40, 1997, pp. 98–102.

[7] Enquist, B., and A. Majda, "Absorbing Boundary Conditions for the Numerical Simulation of Waves," *Mathematics of Computation*, Vol. 31, 1977, pp. 629–651.

[8] Mur, G., "Absorbing Boundary Conditions for the Finite Difference Approximation of the Time Domain Electromagnetic Field Equations," *IEEE Trans.*, EMC-23, 1981, pp. 377–382.

[9] Higdon, R. L., "Numerical Absorbing Boundary Conditions for Wave Equation," *Mathematics of Computation*, Vol. 49, 1987, pp. 65–90.

[10] Cheng, D. K., *Field and Wave Electromagnetics*, Reading, MA: Addison-Wesley, 1992.

[11] Furse, C. M., and O. P. Gandhi, "Why the DFT Is Faster than the FFT for FDTD Time to Frequency Domain Conversions," *IEEE Trans.*, Vol. MGWL-5, 1995, pp. 326–328.

[12] Gandhi, O. P., B. Q. Gao, and Y. Y. Chen, "A Frequency-Dependent Finite-Difference Time-Domain Formulation for General Dispersive Media," *IEEE Trans. Microwave Theory Tech.*, Vol. 41, 1993, pp. 658–665.

[13] Berenger, J. P., "A Perfectly Matched Layer for the Absorption of Electromagnetic Waves," *J. Computational Physics*, Vol. 114, 1994, pp. 185–200.

[14] Gedney, S. D., "An Anisotropic Perfectly Matched Layer Absorbing Medium for the Truncation of FDTD Lattices," *IEEE Trans. Antennas Propagat.*, Vol. AP-44, 1996, pp. 1630–1639.

[15] Zhao, L., and A. C. Cangellaris, "GT-PML: Generalized Theory of Perfectly Matched Layers and Its Application to Reflectionless Truncation of Finite-Difference Time-Domain Grids," *IEEE Trans. Microwave Theory Tech.*, Vol. 44, 1996, pp. 2555–2563.

[16] Katz, D. S., E. T. Thiele, and A. Taflove, "Validation and Extension to Three Dimensions of the Berenger PML Absorbing Boundary Condition for FDTD Meshes," *IEEE Trans.*, Vol. MGWL-4, 1994, pp. 268–270.

[17] Wu, Z., and J. Fang, "Numerical Implementation and Performance of Perfectly Matched Layer Boundary Condition for Waveguide Structures," *IEEE Trans.*, Vol. MTT-43, 1995, pp. 2676–2683.

[18] Berenger, J. P., "Perfectly Matched Layer for the FDTD Solution of Wave Structure Interaction Problems," *IEEE Trans. Antennas and Propagation*, Vol. AP-44, 1996, pp. 110–117.

[19] Chew, W. C., et al., *Fast and Efficient Algorithms in Computational Electromagnetics*, Norwood, MA: Artech House, 2001.

[20] Kodama, M., and M. Kunikaka, "On Precision of Solutions by Finite Difference Time Domain Method of Different Spacing," *IEICE Trans. C*, Vol. E76-B, 1993, pp. 315–317.

[21] Namiki, T., "A New FDTD Algorithm Based on Alternating-Direction Implicit Method," *IEEE Trans. Microwave Theory Tech.*, Vol. 47, 1999, pp. 2003–2007.

[22] Reddy, V. S., and R. Garg, "Finite Difference Time Domain (FDTD) Analysis of Microwave Circuits—A Review with Examples," *IETE J. Research* (India), Vol. 45, 1999, pp. 3–20.

[23] Reuter, C. E., et al., "Ultrawide-Band Absorbing Boundary Condition for Termination of Waveguiding Structures in FDTD Simulations," *IEEE Microwave and Guided Wave Letters*, Vol. 4, 1994, pp. 344–346.

[24] Fujii, M., et al., "High-Order FDTD and Auxiliary Differential Equation Formulation of Optical Pulse Propagation in 2D Kerr and Raman Nonlinear Dispersive Media," *IEEE J. Quantum Electronics*, Vol. 40, 2004, pp. 175–182.

[25] Yu, W., and R. Mittra, "A Conformal FDTD Software Package Modeling Antennas and Microstrip Circuit Components," *IEEE Antennas Propagat. Magazine*, Vol. 42, October 2000, pp. 28–39.

[26] Gedney, S., and D. Rascoe, "A Generalized Yee-Algorithm for the Analysis of MMIC Devices," *IEEE Trans. Microwave Theory Tech.*, Vol. 44, 1996, pp. 1393–1400.

[27] Liu, Q. H., and G. Zhao, "Review of PSTD Methods for Transient Electromagnetics," *Intl. J. Numerical Modeling: Electronic Networks, Devices, Fields*, Vol. 22, 2004, pp. 299–323.

[28] DeRaedt, H., et al., "Solving the Maxwell Equations by the Chebyshev Method: A pne-Step Finite Difference Time-Domain Algorithm," *IEEE Trans. Antennas Propagat.*, Vol. 51, 2003, pp. 3155–3160.

[29] Durbano, J. P., et al., "Hardware Implementation of a Three-Dimensional Finite-Difference Time-Domain Algorithm," *IEEE Trans. Antennas Wireless Propagat. Lett.*, Vol. 2, 2003, pp. 54–57.

[30] Rylander, T., and A. Bondeson, "Stability of Explicit-Implicit Hybrid Time-Stepping Schemes for Maxwell's Equations," *J. Comput. Phys.*, Vol. 179, 2002, pp. 426–438.

Problems

P8.1. Use the software *dispersion.m* for one-dimensional wave propagation. Terminate the grid at its left boundary in ABC and the right boundary in electric field E_y. Excite the grid by a specific Gaussian pulse. Set $E_y = 0$ at the far-right boundary to simulate the presence of a perfect electric conductor. Perform visualization of E_y and H_z distributions within the grid at a number of time snapshots before and after the propagating wave reaches the far-right grid boundary. Show that the perfect electric conductor acts as a mirror that reflects the incident wave. Compare the reflection properties of E_y and H_z components of the wave. (*After*: Problem #3.3 of [3].)

P8.2. Repeat the above problem, but terminate the grid in $H_z = 0$ at the right boundary to simulate the presence of a perfect magnetic conductor. Show that the perfect magnetic conductor acts as a mirror that reflects the incident wave. Compare the reflection properties of E_z and H_y components of the wave. (*After*: Problem #3.4 of [3].)

P8.3. The most common method to introduce excitation at the source plane is to introduce the excitation pulse $P(t)$. An alternative approach is to embed corresponding $P(x)$ into the grid. This can be done by means of the initial condition. Let $P(i, 0) = 1$ for $-10 \le i < 10$, and $P(i, 0) = 0$ elsewhere as the initial condition. Write a code to propagate this pulse in x-direction. You may use $-200 \le i < 200$ for the grid, $N = 70$ and $\alpha = 1$.

P8.4. An expression for the numerical wave number $\overline{k}$ for propagation in one-dimension is obtained as

$$\overline{k} = \frac{2}{\Delta x} \sin^{-1} \left[\frac{\Delta x}{c \Delta t} \sin \frac{\omega \Delta t}{2} \right]$$

1. Plot v_p/c versus $\Delta x/\lambda$ for $c\Delta t/\Delta x = 1, 0.75, 0.5, 0.25$, where $v_p = \omega/\overline{k}$. Comment on the variation of phase velocity with frequency.
2. Use the expression for $\overline{k}$ to compute the group velocity $v_g = d\omega/d\overline{k}$ and plot v_g/c versus $\Delta x/\lambda$ for $c\Delta t/\Delta x = 1, 0.75, 0.5, 0.25$. Comment on the variation of group velocity with frequency.

P8.5. Use Newton's method to solve (8.91) for a square grid $\Delta x = \Delta y = \Delta$. Plot $\Delta/(c\Delta t)$ versus angle of incidence θ for $\Delta x/\lambda = 1/5, 1/10, 1/20$.

Hint [3]: Use $\overline{k}_{i+1} = \overline{k}_i - \dfrac{\sin^2(A\overline{k}_i) + \sin^2(B\overline{k}_i) - C}{A \sin(2A\overline{k}_i) + B \sin(2B\overline{k}_i)}$

with $A = \dfrac{\Delta \cos \theta}{2}$, $B = \dfrac{\Delta \sin \theta}{2}$, $C = \left(\dfrac{\Delta}{c\Delta t} \right)^2 \sin^2 \left(\dfrac{\omega \Delta t}{2} \right)$

P8.6. *Effect of numerical dispersion on pulse broadening* [6]. A sinusoidal packet (with square envelope) is incident on the right-hand boundary of one-dimensional FDTD grid. The packet is described by the starting and end points (expressed in terms of the space step), which can be selected. The general form of the signal is

$$E_y(z, t) = \sin(kz - \omega t)$$

Let us assume the space discretization, $\lambda/\Delta z = m$, where Δz is the space step. The waveform packet is nonzero between $z_{start} \le z \le z_{stop}$ which on discretization may be written as $i_{start}\Delta z \le z \le i_{stop}\Delta z$. Taking into account that $k = 2\pi/\lambda$ and $\omega = ck$, we can write the discretized electric field at $t = 0$ as

$$E_y^0(i) = \begin{cases} \sin\left(\dfrac{2\pi i}{m}\right) & \text{for } i_{start} \le i \le i_{stop} \\ 0 & \text{elsewhere} \end{cases}$$

The magnetic field at the time $n = 1/2$ for all space points may be described as

$$H_x^{1/2}\left(i + \frac{1}{2}\right) = \frac{1}{Z_0}\sin\left(\frac{2\pi\left(i + \dfrac{1}{2} - \dfrac{\alpha}{2}\right)}{m}\right)$$

where Z_0 is the impedance of free space, and $\alpha = c\Delta t/\Delta z$ is the stability factor. Since the spatial domain of simulation must be limited, an absorbing boundary condition is placed at the right-hand boundary. Use the Mur's first order absorbing boundary condition.

Plot the electric field versus the space step number i at $n = 0$ and $n = 640$ for the following cases (assume $m = 8$, $\Delta z = 10^{-3}$ m, $\Delta t = 1.67 \times 10^{-12}$ s, and $s = 0.5$):

1. Narrow square packet defined by $i_{start} = 60$, $i_{stop} = 100$.
2. Broad square packet defined by $i_{start} = 40$, $i_{stop} = 120$.

Interpret the results in terms of numerical dispersion.

P8.7 Calculate and plot the reflection coefficient R as a function of angle of incidence θ for the second-order Mur's ABC, (8.100).

P8.8. Mark the electric field component E_z and magnetic field components H_x and H_y for the grid drawn in the x-y plane in Figure 8.42.

P8.9. Consider the PML medium shown in Figure 8.38. Characterize all the faces, edges, and corners by the respective σ_i, σ_i^* $i = x, y, z$ so that reflection becomes independent of the angle of incidence.

P8.10. Solve Problem P7.7 using FDTD.

P8.11. *Reflection of an electromagnetic wave at an interface.* Propagation of E_y component of a Gaussian pulse through the one-dimensional uniform FDTD grid

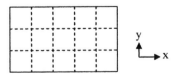

Figure 8.42 Uniform rectangular grid in the x-y plane.

is shown in Figure 8.43. The medium consists of free space ϵ_0 on the left and the dielectric characterized by $\epsilon_0 \epsilon_r$ on the right. The parameters selected for the simulations are: $\alpha = 1/2$ and pulse width $= 400$ ps. Study these figures and determine the following:

1. What do you infer from the plots about the position of interface i_0 and the reflection coefficient Γ?
2. Calculate the approximate value of i_0, and ϵ_r of the medium.
3. What will be the changes in the pulse characteristics such as position, amplitude, and width if the ϵ_r of the medium is increased?

Answer. $\epsilon_r = 9$, $i_0 = 250$, $\Gamma = -0.5$. Reflection coefficient becomes more negative, transmission coefficient decreases, and transmitted pulse becomes narrow and comes nearer to the interface as ϵ_r is increased.

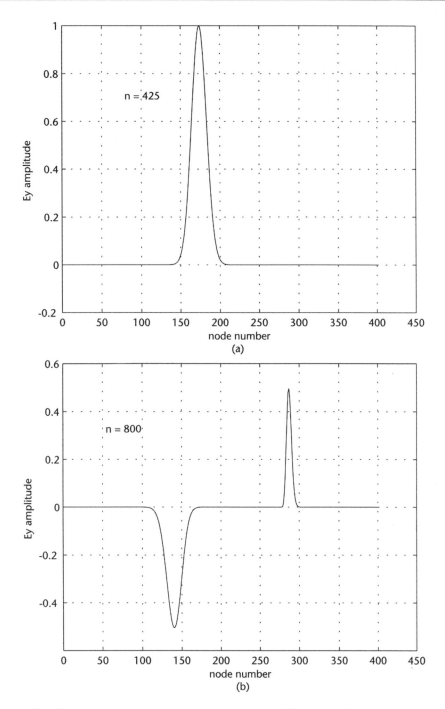

Figure 8.43 Time waveforms at a node near the interface: (a) incident time pulse, and (b) pulse after reflection from the medium.

CHAPTER 9

Variational Methods

One can exactly solve partial differential equations (PDE) for simple or regular shaped geometries with Dirichlet or Neumann boundary conditions. However, when the geometry is not regular, or the dielectric is inhomogeneous, or inhomogeneous boundary conditions are to be satisfied, one cannot solve PDE exactly. In these situations, one may use methods like variational method to solve the problem in an approximate manner. The variational method is discussed in a number of textbooks on electromagnetics and mathematical methods of physics. Some of the notable titles are listed as [1–6].

The solution of a number of problems, such as propagation in transmission lines and waveguides, discontinuities in them, radiation, and scattering of waves, is based on the knowledge of charge distribution, current distribution, or field distribution. In all these cases it is almost impossible to know these distributions exactly. Variational methods may be used to obtain fairly accurate solutions with approximate distributions. *It is the property of variational formulation that a first order error in trial function gives rise to the second order error in the quantity of interest*. Finite element method (FEM) and the method of moments (MoM) employ this property. In variational methods, the demand on computer time and memory are modest. The variational principle is thus a powerful tool in the hands of analytical and computational solution practitioners.

9.1 Calculus of Variations

The variational principle utilizes the ideas of stationarity, extremum, and functional, and are discussed next.

9.1.1 Stationarity

The stationary property of a function is an important concept in the field of calculus of variations. Consider a function $\phi = f(x)$. In the vicinity of point x_0, the function $f(x)$ can be expanded in Taylor series as

$$f(x) = f(x_0) + (x - x_0) \left.\frac{df}{dx}\right|_{x_0} + \frac{1}{2}(x - x_0)^2 \left.\frac{d^2 f}{dx^2}\right|_{x_0} + \ldots \qquad (9.1)$$

$$+ \frac{1}{n!}(x - x_0)^n \left.\frac{d^n f}{\partial x^n}\right|_{x_0} + R_{n+1}$$

where R_{n+1} represents the remainder of the terms in the series expansion, and are expected to contribute insignificantly. The function $f(x)$ is said to be stationary at point x_0 if its first order derivative df/dx vanishes at this point. In the vicinity of stationary point $f(x)$ may therefore be approximated as

$$f(x) - f(x_0) \approx \frac{1}{2}(x - x_0)^2 \frac{d^2 f}{dx^2}\bigg|_{x_0} \tag{9.2}$$

Equation (9.2) implies that a small change in x about the stationary point x_0 results in a still smaller change in the value of $f(x)$. This concept is illustrated in Figure 9.1 for stationary and nonstationary solutions for $\phi = f(x)$. It may be noted that for the same approximate solution x_1 about x_0, the stationary behavior of $f(x)$ gives a smaller error in ϕ then does the nonstationary behavior. It is this property which makes the variational method most useful. The definition of stationarity can be easily generalized to more than one independent variable, say for $f(x, y)$ the condition for stationarity is

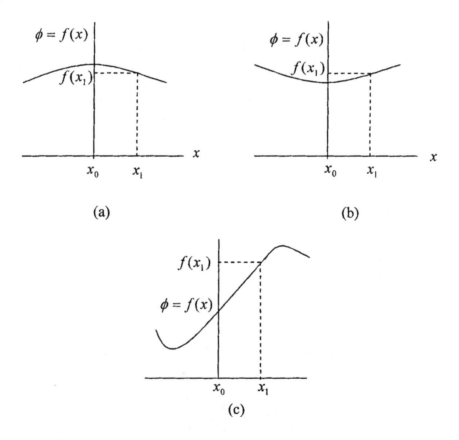

(a)   (b)

(c)

Figure 9.1 Illustration of stationarity and nonstationarity property of a function $\phi = f(x)$: (a) stationarity about the lower-bound solution; (b) stationarity about the upper-bound solution; and (c) nonstationary solution. Note that for the same value of x_1, a stationary solution has a smaller error than does the nonstationary solution.

$$\frac{\partial f}{\partial x} = 0, \quad \text{and} \quad \frac{\partial f}{\partial y} = 0 \qquad (9.3)$$

9.1.2 Extremum

Extremum is a common word used to denote the maximum or minimum of a function. The property of extremum is a consequence of the stationary property of the function. The stationary point (x_0, y_0) of the function $f(x,y)$ corresponds to a local minimum of the function if the changes $(\pm \Delta x, \pm \Delta y)$ in (x, y) about (x_0, y_0) increases the value of the function. Conversely, if the value of the function decreases, then the point (x_0, y_0) corresponds to a local maximum. If these changes are of different sign, the stationary point is a saddle point. An alternative approach to determine the nature of extremum is to calculate $\partial^2 f/\partial x^2$ and $\partial^2 f/\partial y^2$ and carry out the test as follows:

1. $\partial^2 f/\partial x^2 > 0$, and $\partial^2 f/\partial y^2 > 0$, $\Rightarrow$ extremum is a local minimum of $f(x, y)$;
2. $\partial^2 f/\partial x^2 < 0$, and $\partial^2 f/\partial y^2 < 0$, $\Rightarrow$ extremum is a local maximum of $f(x, y)$;
3. Else, saddle point.

Example 9.1.   Let us consider the function $f(x, y) = (x^2 - 6x)(y^2 - 8y)$. The stationary points for this function are obtained as:

$$\frac{\partial f}{\partial x} = (2x - 6)(y^2 - 8y) = 0 \Rightarrow x = 3, y = 8$$

$$\frac{\partial f}{\partial y} = (2y - 8)(x^2 - 6x) = 0 \Rightarrow x = 6, y = 4$$

Both the conditions of (9.3) are satisfied at the stationary points (3, 4), and (6, 8). At point $(3,4)$, $f(x, y) = 144$, and its value is 0 at $(6,8)$. To determine the convexity of the function at the stationary points, let us determine the value of the function at $(x_0 \pm \Delta x, y_0 \pm \Delta y)$. For this let us assume $\Delta x = 0.1x_0$, $\Delta y = 0.1y_0$. We find that $f(x_0 \pm \Delta x, y_0 \pm \Delta y) < f(x_0, y_0)$ about the point $(3,4)$. Therefore, this stationary point is a local maximum of the function. Also, the change in $f(x, y)$ is nearly 2% for a 10% change in (x, y). The point $(6, 8)$ is found to be a local minimum.

Exercise 9.1.   Let $y(x) = 4x^2 - 8x$. Determine if the stationary point $x_0 = 1$ is

1. A local maximum of the function;
2. A local minimum of the function;
3. A saddle point of the function.

Calculate the percentage change in $y(x)$ for 10% change in x_0.

Answer.   2, 1%.

The variational calculus employs the same general idea of stationarity and extremum as applies to a function. However, instead of the particular point x_0 we now have a particular selection of the functions $y(x)$ about which the stationarity is explored, and the function $f(x, y)$ in the preceding analysis is replaced by the functional $F(y)$. It is defined next.

9.1.3 Functional

The functional is another important concept in calculus of variations. *In simple words, a function of other functions is called a functional.* Let us illustrate the concept of functional by means of an example. Consider the following function in the form of an integral

$$F[y(x)] = \int_0^1 y(x) \, dx \tag{9.4}$$

Here, x is the independent variable and $y(x)$ is a function. The function F depends on the function $y(x)$, and is therefore called a functional. Selecting various functions $y(x)$ we shall obtain, in general, different values of the functional F. Thus,

If $y(x) = 1$, then $F[1] = 1$;
If $y(x) = x$, then $F[x] = 1/2$;
If $y(x) = \cos(\pi x)$, then $F[\cos(\pi x)] = 0$.

Example 9.2. Consider the following functional in the differential form:

$$F[y(x)] = y'(x_0) \tag{9.5}$$

If $y(x) = x^2$, and $x_0 = 3$, then $F[x^2] = 6$.
Consider another functional

$$F[y(x)] = \int_{-1}^1 \phi[x, y(x)] \, dx \tag{9.6}$$

If $\phi(x, y) = [\cos^2(x)]/(1 - y^2)$ and $y(x) = \sin x$, then $F[\sin x] = 2$.

Some physically meaningful examples of functional are: the capacitance per unit length, C_0, of transmission lines is dependent on the charge density distribution on the strip, and the functional for this may be denoted as $C_0[\rho(x)]$. The cutoff frequency of waveguide modes, and the resonant frequency of modes in a cavity, is a function of field distribution and may be denoted as $\omega_c[E(x, y)]$, and $\omega_r[E(x, y, z)]$, respectively. Some functionals may not have physical interpretation, but the characteristics derived from it may have.

In order to extend variational concepts from functions to functionals, we need to define the increment of a function (similar to Δx for the function), and the associated variation and stationarity of functionals (similar to the change in the value of functions and their stationarity). These concepts are discussed next.

9.1.4 Variation or Increment of a Function, $\delta\phi(x)$

A small change or increment of a function, denoted by $\delta\phi(x)$ is defined as a change in the functional form of $\phi(x)$. *It is not the change in its variable x.* Let $\phi_0(x)$ be the initial function, and $\phi_\alpha(x)$ be the incremented function; the increment of the function is then defined as

$$\delta\phi(x) = \phi_\alpha(x) - \phi_0(x) \qquad \text{for } x_1 \leq x \leq x_2 \qquad (9.7)$$

Figure 9.2 shows that the curves defined by $\phi = \phi_\alpha(x)$ and $\phi = \phi_0(x)$, specified over the interval $[x_1, x_2]$, differ by a small change. These curves are close in the sense that $|\phi_\alpha(x) - \phi_0(x)|$ is small over $[x_1, x_2]$.

One simple yet systematic method of achieving this small change in the functional form of a function is to add a small multiple of another function $\eta(x)$ to $\phi_0(x)$; that is,

$$\phi_\alpha(x) = \phi_0(x) + \alpha\eta(x) \qquad (9.8)$$

The scale factor or variational parameter α defines the magnitude of variation, and $\eta(x)$ defines the type of variation. The function $\eta(x)$ is almost arbitrary. However, it should be well behaved like $\phi(x)$ and should be zero at the ends of the interval. In this way it is possible to change the functional form of $\phi(x)$ in an arbitrary manner but keeping the new function close to $\phi_0(x)$ by choosing α sufficiently small, as illustrated in Figure 9.3. The Ritz procedure, described in Section 9.3, is a fine example of this implementation. The various terms in the series expansion of a function may also be used to implement variation of a function; for example,

$$\sin(x) = x - \frac{x^3}{3!} + \frac{x^5}{5!} - \frac{x^7}{7!} + \ldots \qquad \text{for } x < 1 \qquad (9.9)$$

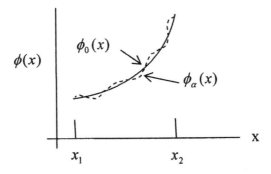

Figure 9.2 Illustration of functions $\phi = \phi_0(x)$ and $\phi = \phi_\alpha(x)$ differing by a small change.

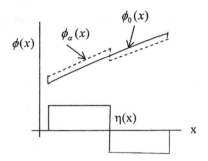

Figure 9.3 Illustration to vary the functional form of function $\phi_0(x)$ in a systematic manner.

The second and subsequent terms may be used to change the functional form. The expansion of an unknown function in terms of entire domain basis functions, discussed in Chapter 6, may be viewed in this manner.

Example 9.3.   Let the function $E_0(x) = x(a - x)$ be defined over $[0, a]$. Then $E_0(x)$ and $E_n(x) = E_0(x) + \{\sin(\pi x/a)\}/n^2$ are close to each other for large values of n only; that is,

$$\delta y(x) = |E_n(x) - E_0(x)| = \left| \frac{\sin\left(\dfrac{\pi x}{a}\right)}{n^2} \right| \le \frac{1}{n^2} \tag{9.10}$$

Here in this example, n is a parameter similar to α.

9.1.5   Variation and Stationarity of Functionals

Consider the functional

$$F[\phi(x)] = \int_{x_1}^{x_2} f[\phi(x), \phi'(x), x] \, dx \tag{9.11}$$

where $f(x, \phi, \phi')$ is an explicitly given function of x, $\phi(x)$, and $\phi'(x)$. An increment or variation in function $\phi(x)$, written as $\delta\phi(x)$, results in a change in the function f. Let us define $\delta\phi(x)$ as

$$\delta\phi(x) = \alpha\eta(x) \tag{9.12}$$

and

$$\delta\phi'(x) = \alpha\eta'(x) \tag{9.13}$$

To the first order in α, the incremented function $f(\phi + \alpha\eta, \phi' + \alpha\eta', x)$ is given by (by Taylor expansion)

$$f(\phi + \alpha\eta, \ \phi' + \alpha\eta', \ x) \cong f(\phi, \ \phi', \ x) + \frac{\partial f}{\partial \phi} \alpha\eta + \frac{\partial f}{\partial \phi'} \alpha\eta' \qquad (9.14)$$

The new functional is denoted by $F[\phi + \delta\phi]$. The corresponding change in the value of the functional is called *variation* of the functional. It is denoted by δF, and is defined as

$$\delta F = F[\phi + \delta\phi] - F[\phi] = \int_{x_1}^{x_2} f(\phi + \alpha\eta, \ \phi' + \alpha\eta', \ x) \ dx - \int_{x_1}^{x_2} f(\phi, \ \phi', \ x) \ dx$$

$$(9.15)$$

Use of (9.14) gives

$$\delta F = \int_{x_1}^{x_2} \left[\frac{\partial f}{\partial \phi} \eta + \frac{\partial f}{\partial \phi'} \eta' \right] \alpha dx \qquad (9.16)$$

Substituting for $\alpha\eta$ and $\alpha\eta'$ gives

$$\delta F = \int_{x_1}^{x_2} \left[\frac{\partial f}{\partial \phi} \delta\phi + \frac{\partial f}{\partial \phi'} \delta\phi' \right] dx \qquad (9.17)$$

Alternatively, the variation of the functional may also be computed as

$$\delta F = \frac{\partial}{\partial k} F[\phi(x) + k\delta\phi(x)]|_{k=0} \qquad (9.18)$$

Stationarity of Functionals
A functional F is said to be stationary if δF vanishes. *In practice, the condition $\delta F = 0$ is enforced by setting*

$$\frac{\partial F}{\partial \phi} = \frac{F(\phi + \delta\phi) - F(\phi)}{\delta\phi}\bigg|_{\delta\phi \to 0} = \frac{\delta F}{\delta\phi}\bigg|_{\delta\phi \to 0} = 0 \qquad (9.19)$$

that is, the derivative of the functional with respect to the function ϕ is set to zero. This stationarity condition is similar to that for the function. Therefore, a functional may be treated as a function for the mathematical purpose, with the difference that its value depends on another function.

To get a feel of the functional, let us compute and plot the following functional as the function is incremented:

$$F[\phi] = \int_0^1 (\phi'^2 - 4\phi^2 + 2x^2\phi) \ dx \qquad (9.20)$$

The function ϕ satisfies the boundary conditions $\phi(0) = \phi(1) = 0$. Let us choose the one-term trial function $\phi = \alpha x(1 - x)$ consistent with the boundary conditions, where α is the variational parameter. Then,

$$F[x(1 - x)] = \int_0^1 [\alpha^2(1 - 2x)^2 - 4\alpha^2(x - x^2)^2 + 2\alpha x^3(1 - x)] \, dx \quad (9.21)$$

$$= \frac{\alpha^2}{5} + \frac{\alpha}{10}$$

Variation of $F[x(1 - x)]$ as a function of variational parameter α is plotted in Figure 9.4. It may be observed from this plot that the functional becomes stationary at $\alpha = -0.25$; that is, the function $\phi = -0.25x(1 - x)$ makes the functional stationary. The optimizing value of α may also be obtained by setting $dF/d\alpha = 0$ in (9.21).

Example 9.4.   Consider the functional

$$F[\phi(x)] = \int_0^1 (\phi'^2 - \pi^2\phi^2) \, dx, \qquad \phi(0) = \phi(1) = 0 \quad (9.22)$$

with $\phi(x) = \sin(\pi x)$, and $\delta\phi(x) = \alpha \sin(\pi x)$ for $\alpha = 0.01$. Then,

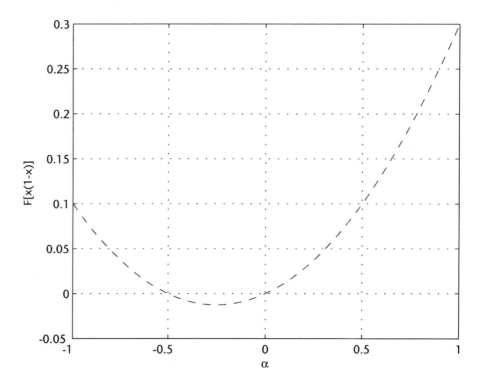

Figure 9.4 Plot of the functional F as a function of the variational parameter α.

$$\delta F = \int\limits_{0}^{1} \left[\frac{\partial f}{\partial \phi} \, \delta\phi + \frac{\partial f}{\partial \phi'} \, \delta\phi' \right] dx \tag{9.23}$$

with $f = \phi'^2 - \pi^2\phi^2$ and $\delta\phi'(x) = \alpha\pi\cos(\pi x)$. Also, $\partial f/\partial \phi' = 2\phi'$, $\phi'(x) = \pi\cos(\pi x)$, and $\partial f/\partial \phi = -2\pi^2\phi$. Therefore,

$$\delta F[\sin(\pi x)] = \int\limits_{0}^{1} [-2\pi^2\phi\alpha\sin(\pi x) + 2\phi'\alpha\pi\cos(\pi x)] \, dx$$

$$= 2\alpha\pi^2 \int\limits_{0}^{a} \cos(2\pi x) \tag{9.24}$$

$$= 0$$

The variation of the functional vanishes independent of the value of α. The functional is therefore stationary with respect to the function $\phi(x) = \sin(\pi x)$, and $\sin(x)$ is called extremetizing function of F.

9.2   Stationary Functionals and Euler Equations

The stationary nature of the functionals gives rise to the most desirable characteristics of the variational formulation of the problem. For a stationary functional F, the error in its value is of second order if the error in the trial function ϕ is of first order. It is possible to determine $\phi(x)$ in a systematic manner. It will now be shown that if the function $f(x, \phi, \phi')$ has continuous partial derivatives up to at least the second order with respect to ϕ and ϕ', the functional $F[\phi] = \int f[\phi, \phi', x] \, dx$ is stationary for a class of functions $\phi(x)$ which satisfy the Euler equation

$$\frac{\partial f}{\partial \phi} - \frac{d}{dx}\left(\frac{\partial f}{\partial \phi'}\right) = 0 \tag{9.25}$$

Proof: Let $\phi(x)$ be a particular class of functions for which the functional F is stationary. The function $\phi(x)$ is as yet unknown. It has continuous derivatives and satisfies the Dirichlet boundary conditions

$$\phi(x_1) = \phi_1 \qquad \phi(x_2) = \phi_2 \tag{9.26}$$

Let us consider a small variation $\delta\phi$ in ϕ such that $\delta\phi(x) = \alpha\eta(x)$ and $\delta\phi'(x) = \alpha\eta'(x)$. The variation of the functional is given by (9.16),

$$\delta F = \alpha \int\limits_{x_1}^{x_2} \left[\frac{\partial f}{\partial \phi} \, \eta + \frac{\partial f}{\partial \phi'} \, \eta' \right] dx \tag{9.27}$$

Integrating the second term on the right side by parts gives

$$\int_{x_1}^{x_2} \frac{\partial f}{\partial \phi'} \eta' \, dx = \left. \frac{\partial f}{\partial \phi'} \eta \right|_{x1}^{x_2} - \int_{x_1}^{x_2} \eta \frac{d}{dx} \left[\frac{\partial f}{\partial \phi'} \right] dx \qquad (9.28)$$

Since the incremented function $\phi + \delta\phi$ must satisfy the same boundary conditions as the original function ϕ, $\eta(x)$ must vanish at the end points; that is, $\eta(x_1) = \eta(x_2) = 0$. Therefore, (9.28) reduces to

$$\int_{x_1}^{x_2} \frac{\partial f}{\partial \phi'} \eta' \, dx = - \int_{x_1}^{x_2} \eta \frac{d}{dx} \left[\frac{\partial f}{\partial \phi'} \right] dx \qquad (9.29)$$

and (9.27) becomes

$$\delta F = \alpha \int_{x_1}^{x_2} \left[\frac{\partial f}{\partial \phi} - \frac{d}{dx} \left(\frac{\partial f}{\partial \phi'} \right) \right] \eta(x) \, dx \qquad (9.30)$$

From the stationarity criterion, we know that the functional $F(\phi)$ is stationary if δF vanishes. Therefore,

$$\int_{x_1}^{x_2} \left[\frac{\partial f}{\partial \phi} - \frac{d}{dx} \left(\frac{\partial f}{\partial \phi'} \right) \right] \eta(x) \, dx = 0 \qquad (9.31)$$

Since $\eta(x)$ is arbitrary, the above equation will be satisfied only if

$$\frac{\partial f}{\partial \phi} - \frac{d}{dx} \left(\frac{\partial f}{\partial \phi'} \right) = 0 \qquad x_1 \le x \le x_2 \qquad (9.32)$$

This equation is known as the Euler equation. It is the governing differential equation for the functional and is an ordinary differential equation of the second order. The functional $F(\phi(x))$ is therefore stationary for the class of functions $\phi(x)$, which satisfy the Euler equation. The functions $\phi(x)$ are called extremetizing functions for the functional F. If the functional F is in the form of an integral, the class of functions may include those functions for which the first derivatives are piecewise continuous. The Euler equation is a necessary *but not sufficient condition* for stationarity of a functional.

Converse: The function $\phi(x)$, for which the functional F is stationary, is also the solution of Euler equations. It means that an alternative way to solve the Euler equation is to express it in the form of a functional. This approach is particularly attractive for geometries which are irregular, inhomogeneously filled, and satisfy inhomogeneous boundary conditions.

Generalization of Euler Equation

We have so far considered functionals of one dependent variable $\phi(x)$, which is a function of one independent variable x. In order to extend the applicability of the variational methods to multidimensional space, the definition of functional and Euler equation are generalized next.

Functionals of several dependent variables: Let us consider the functional F to be a function of several dependent variables $\phi_1(x)$, $\phi_2(x)$, $\phi_3(x)$, ... $\phi_N(x)$, all of which depend on one independent variable x. Then,

$$F = \int_{x_1}^{x_2} f\left[\phi_1(x),\, \phi_2(x),\, \phi_3(x),\, \ldots\, \phi_N(x),\, \phi_1'(x),\, \phi_2'(x),\, \phi_3'(x),\, \ldots\, \phi_N'(x),\, x\right] dx$$

$$(9.33a)$$

The stationarity of (9.33a) leads to a set of Euler equations given as

$$\frac{\phi f}{\partial \phi_i} - \frac{d}{dx}\left(\frac{\partial f}{\partial \phi_i'}\right) = 0 \qquad i = 1, 2, 3, \ldots N \tag{9.33b}$$

This set may be a system of coupled differential equations of second order.

Functionals of several independent variables: Let the functional F be a function of one dependent variable ϕ, which depends on three independent variables x, y, and z; that is, $\phi = \phi(x, y, z)$. Then,

$$F = \iiint f\left[\phi,\, \frac{\partial \phi}{\partial x},\, \frac{\partial \phi}{\partial y},\, \frac{\partial \phi}{\partial z},\, x,\, y,\, z\right] dx\, dy\, dz \tag{9.34a}$$

The corresponding Euler equation is

$$\frac{\partial f}{\partial \phi} - \frac{\partial}{\partial x}\frac{\partial f}{\partial\left(\frac{\partial \phi}{\partial x}\right)} - \frac{\partial}{\partial y}\frac{\partial f}{\partial\left(\frac{\partial \phi}{\partial y}\right)} - \frac{\partial}{\partial z}\frac{\partial f}{\partial\left(\frac{\partial \phi}{\partial z}\right)} = 0 \tag{9.34b}$$

Functionals of more than one dependent and independent variables: The Euler equations for such cases are the combination of (9.33b) and (9.34b).

For the problems in electromagnetics, the function $\phi(x)$ may represent distribution of charge density, electrostatic field, electrostatic potential, vector potential, electromagnetic field, current density, and so on. The function must satisfy, in addition to the Euler equation, the boundary conditions for the given problem. For example, the tangential components of E on a perfect conductor should be zero. In applying the variational method we construct a functional $F[\phi]$ for the problem and choose suitable trial function $\phi(x)$ such that the functional F has a maximum or minimum. The trial function $\phi(x)$ may be selected from the physical considerations (i.e. choose $\phi(x)$ to meet or closely approximate the actual behavior of the distribution it represents). The closer $\phi(x)$ is to the true behavior, less computer time is required to obtain the accurate result for F.

The Euler equation is useful in the systematic search for the extremetizing function $\phi(x)$. The accuracy of the stationary quantity obtainable from a given class of trial functions may be improved by using the Ritz procedure.

Example 9.5. Let us derive the Euler equation corresponding to the functional of Example 9.4; that is,

$$F[\phi(x)] = \int_0^1 (\phi'^2 - \pi^2 \phi^2)\, dx, \qquad \phi(0) = \phi(1) = 0 \tag{9.35}$$

Here, $f(x, \phi, \phi') = \phi'^2 - \pi^2 \phi^2$, $\partial f/\partial\phi = -2\pi^2\phi$, $\partial f/\partial\phi' = 2\phi'$ and the Euler equation is obtained as

$$-2\pi^2\phi - \frac{d}{dx}(2\phi') = 0$$

or

$$\frac{d^2\phi}{dx^2} + \pi^2\phi = 0 \qquad 0 \le x \le 1 \tag{9.36}$$

The solution of (9.36) subject to the boundary conditions $\phi(0) = \phi(1) = 0$ is given by $\phi(x) = \sin(\pi x)$. The same solution was obtained from the stationarity of functional in Example 9.4. Therefore, we can find the solution either by making the functional stationary or by solving the corresponding Euler equation.

9.3   The Ritz Variational Method

This method is sometimes called the Rayleigh-Ritz variational method, and it is an improvement in the implementation of (9.7). *In this method, the function $\phi(x)$ which makes the functional F(ϕ) stationary is expressed in the form of a series with adjustable constants known as variational parameters;* that is,

$$\phi(x) = \sum_{n=1}^N A_n \phi_n(x) \tag{9.37}$$

The functions $\phi_n(x)$ are the known functions and should preferably be orthogonal. The parameters A_n are the variational parameters. It may be noted that our interest is in the value of the functional at the stationary point, and the extremetizing function ϕ is the means to reach it. Therefore, the parameters A_n are chosen to give the minimum or the maximum of the stationary quantity of the functional F. This can be done by requiring

$$\frac{\partial F}{\partial A_k} = 0 \qquad k = 1, 2, 3, \ldots N \tag{9.38}$$

The quality of the Ritz procedure depends crucially on the choice of functions $\phi_n(x)$. It is desirable to keep N, the number of expansion functions, small since the computational labor increases as N^2. Actually, the value of N is decided as a compromise between accuracy and computation time. When a complete set of expansion functions is used in (9.37), the Ritz method may, in principle, lead to an exact solution. We shall discuss applications of Ritz method in Section 9.4.

Example 9.6.   Let us use the Ritz method to determine the first root of the functional

$$F[\phi] = \int\limits_{0}^{1} \left[x\phi'^{2}(x) + \frac{\phi^2}{x} - k^2 x \phi^2(x) \right] dx \tag{9.39}$$

The solution of (9.39) is the Bessel function $J_1(x)$, which has a simple zero at $x = 0$. Using this property we may use a trial function of the form

$$\phi(x) = \alpha x(1 - x)$$

where α is a variational parameter. Also, $\phi'(x) = \alpha(1 - 2x)$. Substituting for ϕ and ϕ' in (9.39) gives

$$F = \alpha^2 \left(\frac{1}{4} - \frac{k^2}{60} \right) \tag{9.40}$$

Setting $\delta F = \partial F/\partial \alpha = 0$, one obtains $k^2 = 15$ so that $k = 3.873$. Let us now compare this value with the first root of $J_1(x)$.

For the correct root of $J_1(x)$, the boundary condition $J_1(kx) = 0$ at $x = 1$ requires that $J_1(k) = 0$ giving the first root as $k = 3.832$. Comparison of this value with that based on Ritz method shows a discrepancy of about 1%. For a better approximation of $F[\phi]$ or $J_1(x)$ we must use a different form for $\phi(x)$.

9.4 Applications of Ritz Approach

In Section 9.2, we derived the Euler's equation or the governing differential equation for the functional. In practice, we know the differential equation for the boundary value problem and we need to develop the functional so that we can apply the variational method.

Physical considerations may be used to derive functionals for some specific problems in electromagnetics. A general expression for the functional is available for the Sturm-Liouville differential equation, and can be used for a large class of problems in electromagnetics.

9.4.1   Variational Solution of Laplace Equation

The energy density U_E of an electrostatic field E is defined as

$$U_E = \frac{1}{2}\,\epsilon E^2 \tag{9.41}$$

Employing $E = -\nabla\phi$, one obtains

$$U_E = \frac{1}{2}\,\epsilon(\nabla\phi)^2 \tag{9.42}$$

Now, let us consider the electrostatic energy stored in a given volume, for example, a rectangular cavity. The energy stored is expected to be minimum for the correct choice of ϕ. Based on this physical consideration, we can use the energy integral as the functional for the problem,

$$\text{Electrostatic energy, } F = \iiint \frac{1}{2}\,\epsilon(\nabla\phi)^2 \, dx\,dy\,dz$$

or

$$F = \frac{1}{2}\,\epsilon \iiint \left[\left(\frac{\partial\phi}{\partial x}\right)^2 + \left(\frac{\partial\phi}{\partial y}\right)^2 + \left(\frac{\partial\phi}{\partial z}\right)^2 \right] dx\,dy\,dz \tag{9.43}$$

for the homogeneous medium. Comparing this functional with (9.33a), we obtain

$$f(\phi_x, \phi_y, \phi_z) = \phi_x^2 + \phi_y^2 + \phi_z^2 \qquad \phi_x = \frac{\partial\phi}{\partial x}, \text{ etc.} \tag{9.44}$$

Use of (9.33b) for minimization of F or the energy yields the following Euler equation:

$$\nabla^2\phi(x, y, z) = 0 \tag{9.45}$$

which is the Laplace equation of electrostatics. It means that the functions ϕ, which are the solutions of the Laplace equation, will also minimize the electrostatic energy functional F. *Therefore, we can either solve the Laplace equation or minimize the electrostatic energy functional F, to determine ϕ.* However, the variational approach is especially useful for those geometries for which the Laplace equation cannot be solved easily (e.g., inhomogeneous dielectric filling and irregular geometries).

The functional corresponding to the Poisson equation $\nabla^2\phi = \rho/\epsilon$ may be written down by inspection of (9.43) and is given by

$$F(\phi) = \iiint\limits_{V} \left(\frac{\epsilon}{2}|\nabla\phi|^2 - \rho\phi \right) dV \tag{9.46}$$

Capacitance of a Transmission Line

The energy functional may be used to determine the capacitance per unit length of a transmission line. The electrostatic energy stored per unit length in the field of a two-conductor transmission line shown in Figure 9.5 can be obtained from (9.43) as

$$W_e = \frac{1}{2} \epsilon \iint \left[\left(\frac{\partial \phi}{\partial x} \right)^2 + \left(\frac{\partial \phi}{\partial y} \right)^2 \right] dx\,dy = \frac{1}{2} \epsilon \iint (\nabla_t \phi)^2 \, dx\,dy \qquad (9.47)$$

Also,

$$W_e = \frac{1}{2} C_0 V_0^2 \qquad (9.48)$$

where C_0 is the line capacitance per unit length, and V_0 is the electrostatic potential between the conductors. Equating (9.47) and (9.48) gives

$$C_0 = \frac{\epsilon \iint (\nabla_t \phi)^2 \, dx\,dy}{V_0^2} \qquad (9.49)$$

or

$$C_0 = \frac{\epsilon \iint (\nabla_t \phi)^2 \, dx\,dy}{\left[\int \nabla_t \phi . dl \right]^2} \qquad V_0 = - \int \nabla_t \phi . dl \qquad (9.50)$$

Similarly, the characteristic impedance Z_0 is given as

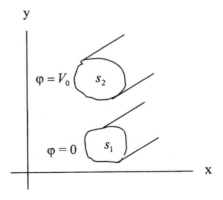

Figure 9.5 Two-conductor transmission line with the cross-sections of conductors, and potentials shown.

$$\frac{1}{Z_0} = \frac{C_0}{\sqrt{\mu\epsilon}} = \frac{\displaystyle\iint (\nabla_t \phi)^2 \, dx \, dy}{\sqrt{\dfrac{\mu}{\epsilon}} \left[\displaystyle\int \nabla_t \phi \cdot \mathbf{dl} \right]^2} \tag{9.51}$$

The above expressions for C_0 and $1/Z_0$ are the functionals corresponding to the Laplace equation $\nabla^2 \phi = 0$. Minimization of the functionals provides upper bound on C_0 and lower bound on Z_0.

The upper bound on Z_0 and the lower bound on C_0 can be obtained from the following functional [4, p. 278]:

$$\frac{1}{C_0} = \frac{\displaystyle\oint_{S_2} \oint_{S_2'} G(x, y; x', y') \rho(x, y) \rho(x', y') \, d\ell_2 \, d\ell_2'}{\left[\displaystyle\oint_{S_2} \rho(x, y) \, d\ell_2 \right]^2} \qquad \text{for } (x, y) \text{ on } S_2 \tag{9.52}$$

The integration is carried out over the surface S_2, and $\rho(x, y)$ is the charge density on this surface. Here $G(.;.)$ is the voltage Green's function for the transmission line.

Example 9.7.   Let us consider the enclosed symmetric strip line of Figure 9.6, in which a strip of width W is embedded in a dielectric enclosed by a rectangular metal box. We shall solve (9.52) for this geometry. Green's function for potential for this geometry is given in Problem 3.15 in Chapter 3,

$$G(x, y; x', y') = \sum_{n=1}^{\infty} \frac{2}{n\pi Y} \sin(\beta_n x) \sin(\beta_n x') \tag{9.53}$$

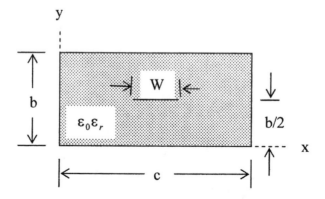

Figure 9.6   Cross-section of a boxed symmetric strip line.

where $\beta_n = n\pi/c$ and the admittance Y at the strip plane, $y = b/2$ is given by

$$Y = 2\epsilon \coth\left(\frac{n\pi b}{2c}\right) \tag{9.54}$$

It is found that the following trial function for the charge density gives accurate results adequate for almost all practical purposes [7, Ch. 3]:

$$\rho(x) = \begin{cases} \dfrac{1}{W}\left[1 + A\left(\dfrac{2}{W}\right)^2 \left|x - \dfrac{c}{2}\right|^3\right] & \text{for } \dfrac{(c-W)}{2} \le x \le \dfrac{(c+W)}{2} \\ 0 & \text{otherwise} \end{cases} \tag{9.55}$$

where A is a variational parameter, and is determined by maximizing the line capacitance C_0 with respect to A.

Substituting for G and $\rho(x)$ in (9.52) gives

$$\frac{1}{C_0} = \frac{1}{\left[\displaystyle\int_{S_2} \rho(x)\,dx\right]^2} \sum_{n=1}^{\infty} \frac{2}{n\pi Y}\left[\int_{(c-w)/2}^{(c+w)/2} \rho(x)\sin\left(\frac{n\pi x}{c}\right)dx\right]^2$$

or

$$C_0 = \frac{\left[\displaystyle\int_{(c-w)/2}^{(c+w)/2} \frac{1}{W}\left\{1 + A\left|\left(\frac{2}{W}\right)\left(x - \frac{c}{2}\right)\right|^3\right\}dx\right]^2}{\displaystyle\sum_{n=1}^{\infty} \frac{2}{n\pi YW^2}\left[\int_{(c-w)/2}^{(c+w)/2}\left\{1 + A\left|\left(\frac{2}{W}\right)\left(x - \frac{c}{2}\right)\right|^3\right\}\sin\left(\frac{n\pi x}{c}\right)dx\right]^2} \tag{9.56}$$

Evaluating the integrals analytically we obtain

$$C_0 = \frac{\left(1 + \dfrac{A}{4}\right)^2}{\displaystyle\sum_{n=1}^{\infty} I_n^2\left[\dfrac{2}{n\pi YW^2}\right]} \tag{9.57}$$

where

$$I_n = \int_{(c-w)/2}^{(c+w)/2} \sin\left(\frac{n\pi x}{c}\right) dx + \left(\frac{8A}{W^3}\right) \int_{(c-w)/2}^{(c+w)/2} \left|x - \frac{c}{2}\right|^3 \sin\left(\frac{n\pi x}{c}\right) dx$$

$$= \left(\frac{2}{\beta_n}\right) \sin\left(\frac{n\pi}{2}\right) \left[\sin p_n + \frac{A}{p_n^3} \left\{ 3\left[p_n^2 - 2\right] \cos p_n + p_n\left[p_n^2 - 6\right] \sin p_n + 6 \right\} \right]$$

(9.58)

Substituting for I_n gives

$$C_0 = \frac{\left(1 + \frac{A}{4}\right)^2}{\displaystyle\sum_{n,\,odd} \frac{(L_n + AM_n)^2 P_n}{Y}}, \qquad \text{since } I_n = 0, \text{ for } n, \text{ even} \qquad (9.59)$$

where

$$L_n = \sin p_n \qquad (9.60a)$$

$$M_n = \frac{1}{p_n^3}\left[3\left(p_n^2 - 2\right)\cos p_n + p_n\left(p_n^2 - 6\right)\sin p_n + 6\right] \qquad (9.60b)$$

$$P_n = \frac{2}{n\pi p_n^2} \qquad (9.60c)$$

$$p_n = \frac{\beta_n W}{2} \qquad (9.60d)$$

The value of C_0 is maximized by setting $dC_0/dA = 0$. This gives [7]

$$A = -\frac{\displaystyle\sum_{n,\,odd} \frac{(L_n - 4M_n) L_n P_n}{Y}}{\displaystyle\sum_{n,\,odd} \frac{(L_n - 4M_n) M_n P_n}{Y}} \qquad (9.61)$$

The above expressions can be numerically solved to obtain C_0 and Z_0.

For the charge distribution specified by (9.55), expression (9.59) combined with (9.60) and (9.61) can be used to evaluate the capacitance of any single conductor multilayered strip line–like transmission line with side walls. The only parameter that needs to be changed is the admittance Y at the strip plane, (9.54).

The characteristics of strip line without side walls can be determined using conformal transformation also. It is described in Section 4.6.1. The expression for the characteristic impedance is given by (4.124)

$$Z_0\sqrt{\epsilon_r} = 29.976\,\pi\frac{K(k)}{K(k')},\ k = \text{sech}\left(\frac{\pi W}{2b}\right),\ k' = \tanh\left(\frac{\pi W}{2b}\right) \qquad (9.62)$$

where the ratio of elliptic functions $K(k)/K(k')$ is given by (4.91). The advantage of the above analysis is that one can introduce offset of the strip in the lateral direction or along the height, and the dielectric filling could be different below the strip and above the strip. If the dielectric above the strip is defined as air, the expression for the admittance at the strip plane becomes

$$Y = \epsilon_0(\epsilon_r + 1)\coth\left(\frac{n\pi b}{2c}\right) \qquad (9.63)$$

An interesting property of this strip line–like microstrip line is that its effective dielectric constant and hence the ratio of guide wavelength to free-space wavelength λ/λ_0 is $(\epsilon_r + 1)/2$, independent of the structural parameters of the line. It behaves like a strip line homogeneously filled with a dielectric of $(\epsilon_r + 1)/2$, and its characteristic impedance is given by

$$Z_0\sqrt{\frac{\epsilon_r + 1}{2}} = 29.976\,\pi\frac{K(k)}{K(k')} \qquad (9.64)$$

Table 9.1 gives the impedance of strip line–like microstrip line for various values of W/b ratio. The table may be used to compare the values obtained from (9.59) for a large value of c/W, say, $c/W = 40$.

It may be pointed out that the procedure adopted in the last example can also be called the integral equation solution using variational method, and (9.52) represents the integral equation.

9.4.2   Cutoff Frequency for Waveguide Modes

For a waveguide with electric walls, the variational expression for the cutoff frequency for the TE and TM modes is given by [2, 8]

$$k_c^2 = \frac{\displaystyle\iint_S (\nabla\phi)^2\,ds}{\displaystyle\iint_S \phi^2\,ds} \qquad (9.65)$$

where $\phi = H_z$ for the TE mode and $\phi = E_z$ field for the TM mode.

Let us verify (9.65) for a rectangular waveguide of width a along the x-axis and height b along the y-axis. Let the trial magnetic field for the TE_{10} mode be $\phi(= H_z) = \cos(\pi x/a)$.

Table 9.1 Characteristic Impedance $Z_0\sqrt{(\epsilon_r + 1)/2}$ of Strip Line–Like Microstrip Line as a Function of W/b

W/b	Impedance	W/b	Impedance	W/b	Impedance
0.1	194.20	0.6	90.60	1.4	51.14
0.2	153.01	0.7	82.58	1.8	42.01
0.3	129.29	0.8	75.89	2.0	38.57
0.4	112.83	0.9	70.22	2.2	35.65
0.5	100.42	0.99	65.80	2.4	33.14

Use of this trial field in (9.65) gives

$$k_c^2 = \left(\frac{\pi}{a}\right)^2 \frac{\displaystyle\int_0^b \int_0^a \sin^2\left(\frac{\pi x}{a}\right) dx\,dy}{\displaystyle\int_0^b \int_0^a \cos^2\left(\frac{\pi x}{a}\right) dx\,dy} = \left(\frac{\pi}{a}\right)^2$$

or

$$k_c = \frac{\pi}{a} \Rightarrow \lambda_c = 2a \tag{9.66}$$

Knowledge of the cutoff frequency of a waveguide *filled homogeneously* in ϵ and μ is sufficient to determine the propagation constant at any other frequency according to

$$\gamma = j\beta = jk_0\sqrt{1 - \left(\frac{k_c}{k_0}\right)^2} \qquad \text{for } f > f_c \tag{9.67}$$

where $k_0 = 2\pi f/c$.

The effect of partial dielectric loading on the cutoff frequency has been analyzed in [5, p. 266]. However, use of (9.67) to determine the propagation constant in this case is not valid because partial loading produces hybrid modes. Very close to the cutoff frequency one may still use (9.67) to approximate the propagation constant [9, p. 251].

The variational nature of the formulation is well suited for perturbation analysis. The effect of perturbation in waveguide geometry on the cutoff frequency is described in [10, p. 53].

9.4.3 Resonant Frequency for Cavity Modes

The resonant cavities with electric walls and filled with a dielectric medium have been analyzed using variational approach [3, p. 331]. The stationary formula for the resonant frequency of a mode in the cavity is given as [3]

$$\omega_r^2 = \frac{\iiint \frac{1}{\mu} (\boldsymbol{\nabla} \times \mathbf{E})^2 \, dv}{\iiint \epsilon \mathbf{E}^2 \, dv} \tag{9.68a}$$

with the condition that $E_{\text{tan}} = 0$ on the surface of the cavity for the trial E-field used in (9.68a).

The H-field functional for the resonant frequency is given as [3]

$$\omega_r^2 = \frac{\iiint \frac{1}{\epsilon} (\boldsymbol{\nabla} \times \mathbf{H})^2 \, dv}{\iiint \mu \mathbf{H}^2 \, dv} \tag{9.68b}$$

with the tangential component of the trial magnetic field satisfying the condition that $E_{\text{tan}} = 0$ on the surface of the cavity.

When the medium filling the cavity is homogeneous (i.e., permittivity and/or permeability of the medium are uniform inside the volume of the cavity), the functionals (9.68) reduce to

$$\omega_r^2 = \frac{\iiint (\boldsymbol{\nabla} \times \mathbf{E})^2 \, dv}{\mu\epsilon \iiint \mathbf{E}^2 \, dv}, \text{ for the } E\text{-field functional} \tag{9.69a}$$

$$\omega_r^2 = \frac{\iiint (\boldsymbol{\nabla} \times \mathbf{H})^2 \, dv}{\mu\epsilon \iiint \mathbf{H}^2 \, dv}, \text{ for the } H\text{-field functional} \tag{9.69b}$$

Now we apply some of the stationary formulae to the problems for which we have an exact answer so that on comparison we will get an idea about the accuracy obtainable from the variational formulation.

Example 9.8   Resonant frequency of a cylindrical cavity [3, p. 336]. Consider a cylindrical cavity of radius a and length d closed at the two ends by perfect conductors as shown in Figure 9.7. The field components for the lowest TM_{010}-mode in the cavity are given as

$$E_z = \frac{k^2}{j\omega\epsilon} J_0 \left(\frac{x_{01}\rho}{a} \right) \tag{9.70a}$$

$$H_\phi = \frac{x_{01}}{a} J_1 \left(\frac{x_{01}\rho}{a} \right) \tag{9.70b}$$

and the resonant frequency is

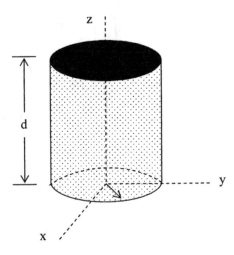

Figure 9.7 A cylindrical cavity of radius a and length d.

$$\omega_r = \frac{1}{a\sqrt{\mu\epsilon}} x_{01} \qquad (9.71)$$

Here, x_{01} is the first zero of the Bessel function $J_0(x)$, and is given as $x_{01} = 2.4048$.

Let us now choose the electric field formula (9.69a) to determine the stationary value of resonant frequency for the TM_{010}-mode. This formula requires that the trial electric field should satisfy the condition $\hat{n} \times \mathbf{E} = 0$ on the cylindrical surface of the cavity; that is, $E_z = 0$ at $\rho = a$. We may therefore choose the trial electric field as

$$\mathbf{E} = \hat{z}\left(1 - \frac{\rho}{a}\right)$$

or

$$\nabla \times \mathbf{E} = \hat{\varphi}\left(\frac{-\partial E_z}{\partial \rho}\right) = \frac{\hat{\varphi}}{a} \qquad (9.72)$$

Substituting for $\mathbf{E}$ and $\nabla \times \mathbf{E}$ in (9.69a), we obtain ($dv = \rho\, d\rho\, d\phi\, dz$, $\mathbf{E}$ and $\nabla \times \mathbf{E}$ are independent of ϕ and z),

$$\omega_r^2 = \frac{\dfrac{1}{a^2}\displaystyle\int_0^a \rho\, d\rho}{\mu\epsilon \displaystyle\int_0^a \left(1 - \frac{\rho}{a}\right)^2 \rho\, d\rho} = \frac{6}{\mu\epsilon a^2}$$

or

$$\omega_r = \frac{2.449}{a\sqrt{\mu\epsilon}} \tag{9.73}$$

This value is about 1.8% higher compared to the correct value $\omega_r = 2.4048/(a\sqrt{\mu\epsilon})$.

In this example we have used a very simple expression for E and still obtained fairly accurate value of resonant frequency. This illustrates the useful feature of the variational method.

Ritz Procedure for Better Accuracy
Let us see if we can improve the accuracy of the resonant frequency of the TM_{010}-mode using the Ritz procedure. The trial electric field satisfying the boundary condition $\hat{n} \times E = 0$ on the surface S of the cavity may be assumed to be of the form

$$E = \hat{z}\left[\left(1 - \frac{\rho}{a}\right) + A\left(1 - \frac{\rho}{a}\right)^2\right] \tag{9.74}$$

Here, A is the variational parameter, and

$$\nabla \times E = \hat{\varphi}\frac{-\partial E_z}{\partial\rho} = \hat{\varphi}\frac{1}{a}\left[1 + 2A\left(1 - \frac{\rho}{a}\right)\right] \tag{9.75}$$

Now, substitute for E and $\nabla \times E$ in the stationary expression (9.69a) to obtain

$$\omega_r^2 = \frac{\dfrac{1}{a^2}\displaystyle\int_0^a \left[1 + 2A\left(1 - \frac{\rho}{a}\right)\right]^2 \rho d\rho}{\mu\epsilon\displaystyle\int_0^a \left[\left(1 - \frac{\rho}{a}\right) + A\left(1 - \frac{\rho}{a}\right)^2\right]^2 \rho d\rho} = \frac{10}{\mu\epsilon a^2}\frac{2A^2 + 4A + 3}{2A^2 + 6A + 5} \tag{9.76}$$

To determine the value of the variational parameter A, we set $\partial\omega_r^2/\partial A = 0$. It yields $2A^2 + 4A + 1 = 0$ or $A = -0.293, -1.707$. Substituting for $A = -0.293$ gives us $\omega_r = 2.4196/(a\sqrt{\mu\epsilon})$, which is about 0.6% higher than the correct value $\omega_r = 2.4048/(a\sqrt{\mu\epsilon})$. This example shows that by adding a variational parameter, the error in the resonant frequency has been reduced from 1.8% to 0.6%. For $A = -1.707$, we obtain $\omega_r = 5.843/(a\sqrt{\mu\epsilon})$. This value of A predicts the resonant frequency for the TM_{020}-mode, which is given exactly as

$$[\omega_r]_{020}^{TM} = x_{02}/\left(a\sqrt{\mu\epsilon}\right) \tag{9.77}$$

Here, $x_{02} = 5.52$, the second root of J_0. Therefore,

$$[\omega_r]_{020}^{TM} = 5.52/\left(a\sqrt{\mu\epsilon}\right) \tag{9.78}$$

The error in this resonant frequency is 5.85%.

Exercise. Consider the TM_{101}-mode in a cylindrical cavity. Find the resonant frequency using variational method and assuming a suitable expression for H_ϕ. Improve the accuracy by using Ritz procedure. Hint: You may use the trial field as $H_\phi = \rho + A\rho^2$ and the Ritz procedure [3].

9.4.4 Variational Formulation in Spectral Domain for the Microstrip Line

Due to the mixed dielectric nature of microstrip line, the analysis in the Fourier domain has several advantages compared to the analysis in the space domain. The formulation in the Fourier domain leads to an algebraic equation for Green's function instead of an integral equation. In addition, the solutions are stationary in character.

Consider the cross-section of a microstrip line of strip width W on a substrate of dielectric constant ϵ_r and dielectric thickness h, as shown in Figure 9.8. The strip and the ground plane are assumed to be perfect conductors of zero thickness, uniform and infinite in the z-direction. The dielectric is assumed to be lossless. The mode of propagation in the microstrip line is quasi-TEM due to the presence of different dielectrics about the strip.

The analysis of microstrip line is very much similar to the one carried out in Section 9.4.1 for the strip line. However, instead of using the Green's function approach, we use here the variational expression for the capacitance per unit length, (9.52) in the following form

$$\frac{1}{C_0} = \frac{1}{Q^2} \int\limits_{strip} \rho_s(x)\,\phi(x, h)\,dx \tag{9.79}$$

where the integration is carried out over the strip. Also, the analysis is performed in the spectral domain due to the mixed dielectric nature of microstrip line.

Parseval's formula is employed to convert the integral from space domain in (9.79) to spectral domain as

$$\frac{1}{C_0} = \frac{1}{Q^2} \int\limits_{-\infty}^{\infty} \tilde{\rho}_s(\alpha)\,\tilde{\phi}(\alpha, h)\,d\alpha \tag{9.80}$$

where the superscript ~ indicates a function in the spectral domain, and is defined as

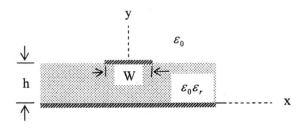

Figure 9.8 Cross-section of a microstrip line.

$$\tilde{f}(\alpha) = \frac{1}{\sqrt{2\pi}} \int\limits_{-\infty}^{\infty} f(x) e^{j\alpha x} \, dx \tag{9.81}$$

It may be pointed out that it is easier to determine $\tilde{\phi}(\alpha, h)$ than $\phi(x, h)$ since

$$\tilde{\phi}(\alpha, h) = \frac{1}{\epsilon} \tilde{\rho}_s(\alpha) \tilde{G}(\alpha, h) \tag{9.82}$$

whereas $\phi(x, h)$ is a convolution integral given by

$$\phi(x, h) = \frac{1}{\epsilon} \int\limits_{-W/2}^{W/2} \rho_s(x') G(x, h; x', h) \, dx' \tag{9.83}$$

The Green's function $\tilde{G}(\alpha, h)$ has been derived in Chapter 5 and is given by (5.160) as

$$\tilde{G}(\alpha, h) = \frac{1}{|\alpha| \left[1 + \epsilon_r \coth\left(|\alpha| h\right)\right]} \tag{9.84}$$

Next, the potential function $\tilde{\varphi}(\alpha, h)$, given by (9.82), is used in (9.80) for calculating the capacitance C_0.

Since $\tilde{\rho}_s(\alpha)$ is unknown, we need to determine it before C_0 can be computed. However, since (9.80) is variational, one may use an approximate trial function for $\tilde{\rho}_s(\alpha)$ and incur only a second order error in the value of capacitance. A trial function that maximizes the value of C_0 gives the closest value to the exact result for the capacitance. For solution in the spectral domain, only those types of trial functions may be used which can be Fourier transformed analytically, otherwise one can use numerical procedure for this. Use of a very simple distribution [11]

$$\rho_s(x) = |x| \qquad \text{for } \frac{-W}{2} \leq x \leq \frac{W}{2} \tag{9.85}$$

gives

$$\frac{\tilde{\rho}_s(\alpha)}{Q} = \frac{\int\limits_{-\infty}^{\infty} \rho_s(x) e^{j\alpha x} \, dx}{\int\limits_{-W/2}^{W/2} \rho_s(x) \, dx} = \frac{2 \sin\left(\dfrac{\alpha W}{2}\right)}{\dfrac{\alpha W}{2}} - \left(\frac{\sin\left(\dfrac{\alpha W}{4}\right)}{\dfrac{\alpha W}{4}}\right)^2 \tag{9.86}$$

The results obtained by this method agree well with those of the other quasi-static methods (e.g., conformal mapping method). The capacitance C_0/ϵ_0 has been plotted in Figure 9.9 as a function of W/h for various values of ϵ_r. The capacitance is seen to increase with the strip width and dielectric constant.

The above method can also be used to take into account the effect of finite strip thickness and metal enclosure. It can be easily extended for microstrip on composite substrates or where dielectric overlay exists over the microstrip [12]. In these cases, one simply has to find an appropriate expression for $\tilde{G}(\alpha)$ and use the above procedure. For a microstrip line with a multilayered substrate and shielded by a top metallic wall as shown in Figure 9.10, Green's function $\tilde{G}(\alpha)$ is given as

$$\tilde{G}(\alpha) = \frac{A + B}{|\alpha|\{A(B + C) + \epsilon_{r2}^2 + BC\}} \qquad (9.87)$$

where $A = \epsilon_{r1} \coth(|\alpha|b)$, $B = \epsilon_{r2} \coth(|\alpha|s)$, and $C = \epsilon_{r3} \coth(|\alpha|d)$.

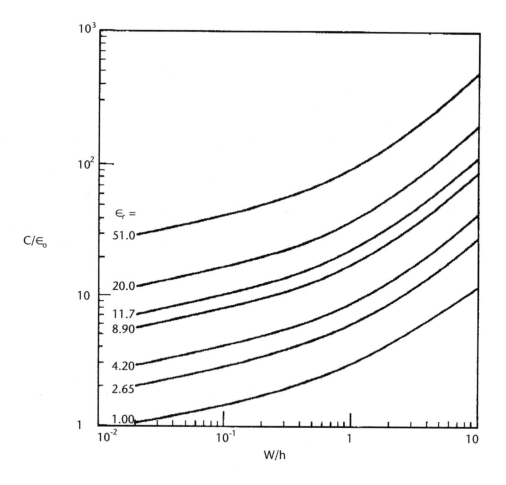

Figure 9.9 Capacitance of microstrip line as a function of W/h for various values of ϵ_r. (*From:* [11]. © 1968 IEEE. Reprinted with permission.)

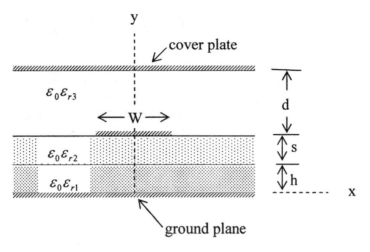

Figure 9.10   A microstrip line with multilayered substrate and a cover plate.

9.5   Construction of Functionals from the PDEs

The functional corresponding to a given differential equation may be constructed by following a general mathematical procedure. It is best illustrated by an example. Let us construct the functional for the Poisson's equation,

$$\nabla^2 \phi = -\rho(x, y) \tag{9.88}$$

As a first step, we multiply this equation by the variation $\delta\phi$ and integrate over the domain of the problem to obtain

$$\delta F = \iint [-\nabla^2 \phi - \rho]\, \delta\phi\, dx\, dy \tag{9.89}$$

$$= -\iint \nabla^2 \phi \delta\phi\, dx\, dy - \iint \rho\delta\phi\, dx\, dy$$

Integrating by parts gives [5, p. 243]

$$\delta F = \frac{\delta}{2} \iint \left[\left(\frac{\partial\phi}{\partial x}\right)^2 + \left(\frac{\partial\phi}{\partial y}\right)^2 - 2\rho\phi \right] dx\, dy - \delta \int \phi \frac{\partial\phi}{\partial x}\, dy - \delta \int \phi \frac{\partial\phi}{\partial y}\, dx \tag{9.90}$$

The last two terms vanish if the boundary conditions are homogeneous, Dirichlet, or Neumann. Hence,

$$\delta F = \delta \iint \frac{1}{2} \left[\left(\frac{\partial\phi}{\partial x}\right)^2 + \left(\frac{\partial\phi}{\partial y}\right)^2 - 2\rho\phi \right] dx\, dy$$

or

$$F(\phi) = \int\int \frac{1}{2}\left[|\nabla\phi|^2 - 2\rho\phi\right] dx\,dy \qquad (9.91)$$

Functional for the wave equation may be derived in an analogous manner. Another approach, called weak-wave equation, is given in the next chapter.

An alternative approach described by Mikhlin shows that the functional corresponding to the Euler equation $L\phi = g$, for a real, self-adjoint, and positive definite operator L is given by [6]

$$F(\phi) = <L\phi,\ \phi> -2 <\phi,\ g> \qquad (9.92)$$

The symbol $< >$ stands for inner product.

It may be verified that (9.92) is equivalent to (9.91) for the Poisson equation. The functionals for the most common partial differential equations in electromagnetics may be obtained from (9.92) and are listed in Table 9.2.

Exercise: Consider the following differential equation:

$$-\frac{d^2f}{dx^2} = 1 + x^2 \qquad 0 \le x \le 1$$

subject to the boundary conditions $f(0) = f(1) = 0$. The stationary functional for the differential equation may be obtained using (9.92). Assume a solution of the form $f(x) = ax(1 - x)^3 + bx^2(1 - x)$, with a and b as variational parameters [13, p. 141]. Use Ritz method and compare the variational solution with the exact solution of (9.93b).

Table 9.2 Partial Differential Equations and Corresponding Functionals in Electromagnetics

PDE	Functional, F
Laplace PDE	
$\epsilon\nabla^2\phi = 0$	$\frac{1}{2}\epsilon\int\int\int_v |\nabla\phi|^2\, dv$
Poisson PDE	
$\nabla^2\phi = -\frac{\rho}{\epsilon}$	$\frac{1}{2}\int\int\int_v \left(|\nabla\phi|^2 - \frac{2\rho\phi}{\epsilon}\right) dv$
Homogeneous wave equation	
$\nabla^2\phi + \epsilon_r k_0^2 \phi = 0$	$\frac{1}{2}\int\int\int_v \left(|\nabla\phi|^2 - \epsilon_r k_0^2 \phi^2\right) dv$
Inhomogeneous wave equation	
$\nabla^2\phi + \epsilon_r k_0^2 \phi = g$	$\frac{1}{2}\int\int\int_v \left(|\nabla\phi|^2 - \epsilon_r k_0^2 \phi^2 + 2g\phi\right) dv$

9.6 Method of Weighted Residuals

The accuracy of the solution based on Ritz variational method depends on the stationary nature of the functional. It is possible to obtain functional for some of the boundary value problems of electromagnetics (Table 9.2). There is an alternative to the Ritz method and this method does not require the knowledge of functional. The method can be applied to any integral or differential equation, and the variational nature of the solution is dependent on the operator. This method is called Galerkin's method, and it is a special case of weighted residual method. The method can be described in terms of operator equation $Lf = g$. We illustrate the method here by seeking the Galerkin's solution of the following differential equation:

$$-\frac{d^2f}{dx^2} = 1 + x^2 \qquad 0 \le x \le 1 \qquad (9.93a)$$

subject to $f(0) = f(1) = 0$. The analytical solution can be worked out by integrating the differential equation twice and using the boundary conditions to determine the constants of integrations; one obtains

$$f(x) = \frac{7x}{12} - \frac{x^2}{2} - \frac{x^4}{12} \qquad (9.93b)$$

When the solution $f(x)$ is known exactly, the left and right sides of the differential equation are equal and the difference between them is zero. In case of approximate solution as in the computational methods, the residual is not zero and is defined as

$$R(x) = \frac{d^2f}{dx^2} + 1 + x^2 \qquad (9.94)$$

The unknown function $f(x)$ is determined by expanding it in a set of basis functions, similar to (9.37), as

$$f = \sum_{n=1}^{N} \alpha_n f_n \qquad (9.95)$$

where the constants α_n are unknown, and need to be determined. *The α_n are found such that the residual $R(x)$ is forced to zero in an average sense*; that is, the weighted integral of the residual is set to zero,

$$\int w(x) R(x) \, dx = 0 \qquad (9.96)$$

where $w(x)$ is called the weight function. In order to determine the constants α_n uniquely, we should have N simultaneous equations. For this, we choose N weight functions $w_m(x)$, $m = 1, 2, 3, \ldots N$, and form the weighted integrals as

$$\int w_m(x) R(x)\, dx = 0 \qquad m = 1, 2, 3, \ldots N \qquad (9.97)$$

$w_m(x)$ is also called test function in connection with MoM. The solution of the resulting set of equations determines α_n and therefore $f(x)$.

Let us try the following basis functions for the solution

$$f_n = x - x^{n+1} \qquad n = 1, 2, 3, \ldots N \qquad (9.98)$$

The use of these basis functions in (9.94) gives the following residual:

$$R(x) = 1 + x^2 + \sum_{n=1}^{N} \alpha_n \left[\frac{d^2}{dx^2}(x - x^{n+1}) \right] \qquad (9.99)$$

$$= 1 + x^2 - \sum_{n=1}^{N} \alpha_n n(n+1) x^{n-1}$$

The weighted residual now becomes

$$\int_0^1 w_m(x) \sum_{n=1}^{N} \alpha_n n(n+1) x^{n-1}\, dx = \int_0^1 w_m(x)(1 + x^2)\, dx \qquad m = 1, 2, 3, \ldots$$

$$(9.100)$$

Remarks. Equation (9.97) may be integrated by parts to reduce the order of derivative in $R(x)$. The resulting expression will require basis functions for which only the first order derivative needs to be continuous. This approach is used in FEM to derive the governing equations [14].

9.6.1 Galerkin's Method

A popular and useful variation of the method of weighted residuals is the *Galerkin's method* [3]. In this method, the test functions are chosen identical to the basis functions. This not only saves us the labor of searching for the test functions but also makes the solution variational in nature. Using the Galerkin's method for the given problem, we have

$$w_m(x) = f_m(x) = x - x^{m+1} \qquad m = 1, 2, 3, \ldots N \qquad (9.101)$$

Use of this in (9.100) gives the following matrix equation:

$$[l][\alpha] = [m] \qquad (9.102)$$

where

$$l_{mn} = \langle x - x^{m+1}, n(n+1) x^{n-1} \rangle = \frac{mn}{m+n+1} \qquad (9.103a)$$

$$m_m = <x - x^{m+1}, 1 + x^2> = \frac{m(3m + 10)}{4(m + 2)(m + 4)} \qquad (9.103b)$$

The matrix equation is solved for the vector $[\alpha]$ and the result substituted in (9.95) to determine $f(x)$. The results for $N = 1, 2, 3$ are given next.

$$f^{(1)}(x) = \frac{13}{20}(x - x^2) \qquad (9.104a)$$

$$f^{(2)}(x) = \frac{2}{5}(x - x^2) + \frac{1}{6}(x - x^3) \qquad (9.104b)$$

$$f^{(3)}(x) = \frac{1}{2}(x - x^2) + \frac{1}{12}(x - x^4) \qquad (9.104c)$$

The superscript on $f(x)$ denotes the value of N. It may be noted that the third order solution is the exact solution, (9.93b). Higher order solutions with $N \geq 4$ are again found to be the exact solutions.

9.6.2   Point Matching Method

In order to simplify calculations of matrix elements, let us satisfy (9.88) at a number of discrete points N in the range $0 \leq x \leq 1$. This procedure is called *point matching* or *collocation*. Besides being simpler, this method provides fairly good solution for sufficiently large values of N. The matching of the solution at selected discrete points is equivalent to choosing Dirac delta functions as test functions.

For the point matching solution, let us choose N equidistant points defined as

$$x_m = \frac{m}{N + 1} \qquad m = 1, 2, 3, \ldots N \qquad (9.105)$$

This division amounts to $N + 2$ points (including 0 and 1) and $N + 1$ intervals over the range $(0, 1)$.

The point matching solution process can be implemented by satisfying (9.93a) at the points x_m; that is,

$$\sum_{n=1}^{N} \alpha_n \left[-\frac{d^2}{dx^2}(x - x^{n+1}) \right]_{x=x_m} = 1 + x_m^2, \qquad m = 1, 2, 3, \ldots N$$

$$(9.106)$$

An alternative is to use Dirac delta functions as test functions in (9.100); that is, $w_m = \delta(x - x_m)$. Either approach results in the following matrix equation:

$$[l][\alpha] = [m] \qquad (9.107)$$

where

$$l_{mn} = n(n + 1)x_m^{n-1} = n(n + 1)\left(\frac{m}{N + 1}\right)^{n-1} \tag{9.108a}$$

$$m_m = 1 + \left(\frac{m}{N + 1}\right)^2 \tag{9.108b}$$

The matrix equation is solved for the vector $[\alpha]$, and the coefficients substituted in (9.95) to determine $f(x)$. The solutions are found to be

$$f^{(1)}(x) = \frac{5}{8}(x - x^2) \tag{9.109a}$$

$$f^{(2)}(x) = \frac{7}{18}(x - x^2) + \frac{1}{6}(x - x^3) \tag{9.109b}$$

$$f^{(3)}(x) = \frac{1}{2}(x - x^2) + \frac{1}{12}(x - x^4) \tag{9.109c}$$

It is worthwhile comparing the errors or residuals incurred in the point matching method (Rp) and the Galerkin's method (Rg) for identical basis functions. The residual is defined by (9.94). Figure 9.11 shows the residual $R(x)$ for $N = 2$

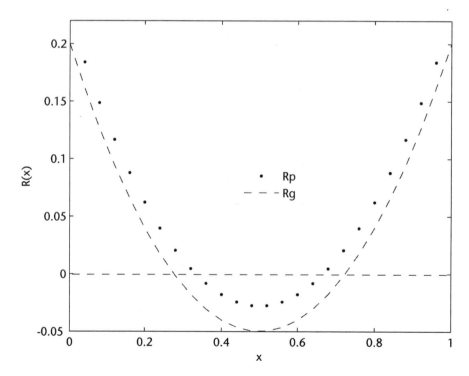

Figure 9.11 Comparison of residuals for point matching (Rp) and the Galerkin's method (Rg) for the same basis functions for $N = 2$.

calculated using (9.104b) for Rg, and (9.109b) for Rp. The residual Rp is found to be zero at the match points $x = 1/3$ and $2/3$. The differential equation is not satisfied for other values of x. The difference between $Rp(x)$ and $Rg(x)$ does not appear to be significant. The error over the domain defined as $\epsilon = \int_0^1 R(x)\, dx$ may be used as the basis for comparison. It is found that $\epsilon_p = 1/18$ and $\epsilon_g = 1/30$, suggesting that the Galerkin's method yields lower error over the domain.

The weighted residual method has been illustrated for the solution of differential equations. It has been applied to the solution of integral equations also, as illustrated in Chapter 11.

Comparison of Galerkin's and the Ritz methods shows the following [8, p. 144; 14, p. 32]:

1. These methods are equivalent for positive-definite differential operators.
2. The Galerkin's method can also be used for nonpositive-definite operators where the solution is not variational.

9.7 Summary

The variational method is an approximation technique and forms the foundation of the computational methods such as FEM and MoM. The calculus of variations is reviewed first to describe stationarity and extremum. Functionals are defined as functions that depend on functions as variables. The variational analysis may be applied to the solution of differential equations and integral equations. The only requirement is that the governing equations should be stationary in nature. The differential equations can be formulated systematically as stationary functionals. The stationary functionals are listed for the important differential equations of electromagnetics. The accuracy of the solution depends on the choice of expansion functions for the unknown function. The use of Ritz method helps us determine the stationary value of functionals. A number of examples illustrate the variational procedure. The variety of examples includes: capacitance of strip line, cutoff frequency of waveguides, resonant frequencies of cavities, and variational solution in spectral domain for microstrip line.

The method of weighted residuals does not require that the governing equations should be stationary in nature. It can be employed to solve differential or integral equations straight-away. These equations are satisfied in an average sense over the domain instead of satisfying them at each and every point. The averaging process selected is the weighted average. Setting the weighted residual to zero defines the governing equation for the problem. The unknown function is then expanded in terms of known functions. The number of weight functions required is equal to the number of expansion functions. The resulting set of simultaneous equations is solved to determine the unknown expansion coefficients. Two popular variants of method of weighted residuals are point matching and Galerkin's method. The point matching method has the advantage of analytical simplicity, whereas Galerkin's method is found to be numerically efficient because of its variational character. Method of moments and finite elements methods are based on the method of weighted residuals. The method of weighted residuals is illustrated here by solving a simple differential equation and comparing the residues for point matching and Galerkin's cases.

References

[1]   Butkov, E., *Mathematical Physics*, Reading, MA: Addison-Wesley, 1968, Chapter 13.

[2]   Van Bladel, J., *Electromagnetic Fields*, Washington, D.C.: Hemisphere Publishing, 1985.

[3]   Harrington, R. F., *Time Harmonic Fields*, New York: McGraw-Hill, 1962.

[4]   Collin, R. E., *Field Theory of Guided Waves*, 2nd ed., New York: IEEE Press, 1991.

[5]   Sadiku, M. N. O., *Numerical Techniques in Electromagnetics*, 2nd ed., Boca Raton, FL: CRC Press, 2001.

[6]   Mikhlin, S. G., *Variational Methods in Mathematical Physics*, New York: Macmillan, 1964.

[7]   Bhat, B., and S. Koul, *Stripline-Like Transmission Lines for Microwave Integrated Circuits*, New Delhi: Wiley Eastern, 1989.

[8]   Bondeson, A., T. Rylander, and P. Ingelstrom, *Computational Electromagnetics*, Berlin: Springer, 2005.

[9]   Jin, J., *The Finite Element Method in Electromgnetics*, 2nd ed., New York: John Wiley, 2002.

[10] Davies, J. B., "The Finite Element Method," in *Numerical Techniques for Microwave and Millimeter-Wave Passive Structures*, T. Itoh, (ed.), New York: John Wiley, 1989.

[11] Yamashita, E., and R. Mittra, "Variational Method for the Analysis of Microstrip Lines," *IEEE Trans. Microwave Theory Tech.*, Vol. MTT-16, 1968, pp. 251–256.

[12] Yamashita, E., "Variational Method for the Analysis of Microstrip-Like Transmission Lines," *IEEE Trans. Microwave Theory Tech.*, Vol. MTT-16, 1968, pp. 529–535.

[13] Volakis, J. L., A. Chatterjee, and L. C. Kempel, *Finite Element Method for Electromagnetics: Antennas, Microwave Circuits, and Scattering Applications*, Hyderabad: University Press (India), 2001.

[14] Chari, M. V. K., and S. J. Salon, *Numerical Methods in Electromagnetism*, New York: Academic Press, 2000.

Problems

P9.1. Consider the following Poisson equation:

$$\nabla^2 \phi(x, y) = -1$$

in the range $0 \leq x \leq a$, $0 \leq y \leq b$, with the condition $\phi = 0$ on the boundary. In order to solve the Poisson equation using the variational method, first determine the functional $F[\phi]$. Now use the trial function $\phi(x, y) = y(b - y)\alpha(x)$ to obtain the approximate solution based on the variational method; that is, determine $\alpha(x)$.

P9.2. (a) Find the variation of the functional (*After:* [14], p. 148)

$$F(V) = \int \left[\left(\frac{dV}{dx} \right)^2 - \frac{2\rho V}{\epsilon_0} \right] dx$$

subject to the boundary condition $V = V_0$ or $dV/dx = 0$. Show that the stationarity of the functional yields the Poisson equation for the electrostatic problem.
(b) Apply the Euler equation

$$\frac{d}{dx}\frac{\partial f}{\partial V'} - \frac{\partial f}{\partial V} = 0 \qquad V' = \frac{dV}{dx}$$

to the integrand $f = \left(\dfrac{dV}{dx}\right)^2 - \dfrac{2\rho V}{\epsilon_0}$ directly to verify the answer to part (a).

P9.3. (a) Use the variational principle and Ritz method to solve the following one-dimensional Poisson equation:

$$\frac{d^2 A}{dx^2} = -\mu J \qquad -d \le x \le d$$

subject to the boundary condition $A(d) = A(-d) = 0$. Hint: You may use the trial function of the form $A = (x - d)(c_1 + c_2 x)$. (*After:* [14], p. 172.)
(b) Show that the exact solution of the above differential equation is

$$A = \frac{\mu J}{2}(d^2 - x^2)$$

(c) Compare the variational solution of (a) with the exact solution (b).

P9.4. Show that the Euler equation for the functional [14, p. 156]

$$F[A_z] = \frac{1}{\mu} \iint \left[\left(\frac{\partial A_z}{\partial x}\right)^2 + \left(\frac{\partial A_z}{\partial y}\right)^2 - 2 A_z J_z \right] dx\, dy$$

is the Poisson equation $\nabla_t^2 A_z(x, y) = -\mu J_z$ for the two-dimensional magnetostatic problem.

P9.5. The cutoff frequency for the dominant mode in a symmetric ridge waveguide of Figure 9.12 may be determined using (9.65). Justify your solution.

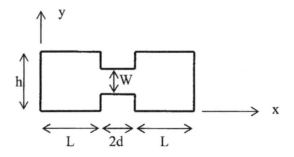

Figure 9.12 Cross-section of a double ridge waveguide.

P9.6. Consider a rectangular waveguide cavity of size $a \times b \times c$ with $0 \le x \le a$, $0 \le y \le b$, $0 \le z \le c$.
(a) Show that the use of the trial electric field

$$E_x = yz(y - b)(z - c)$$

in the variational expression (9.69a) gives

$$\omega_r = \frac{\sqrt{10}}{bc} \sqrt{\frac{b^2 + c^2}{\mu\epsilon}}$$

(b) What are the cavity modes which can be represented by the trial field given in (a) above?

P9.7. For the cylindrical cavity shown in Figure 9.7 the electric field stationary formula for the resonant frequencies ω_r is given by (9.69a).
(a) Determine the resonant frequency for this cavity using the following trial field:

$$\mathbf{E} = \hat{z} \cos\left(\frac{\pi\rho}{2a}\right)$$

Note that this trial field satisfies the boundary condition at $\rho = a$ and is a solution of the wave equation.
(b) Use Ritz procedure to determine the resonant frequency more accurately. (Modify the trial field by adding a term to it.)

P9.8. For the cylindrical cavity shown in Figure 9.7 the magnetic field stationary formula for the resonant frequencies ω_r is given by (9.69b).
(a) Determine the resonant frequency for the dominant mode of the cavity using the following trial field:

$$\mathbf{H} = \hat{\varphi}\left(\rho - \frac{2}{3}\frac{\rho^2}{a}\right)$$

(b) Ritz procedure can be used to improve the accuracy of the solution. You may use the following trial field for this purpose:

$$\mathbf{H} = \hat{\varphi}(\rho + A\rho^2)$$

The exact resonant frequency of the cavity is $\omega_r = \dfrac{2.4048}{a\sqrt{\mu\epsilon}}$. Determine the percentage error in the two cases considered above.

P9.9. Consider the following functional where $y(0) = y(1) = 0$ (After: [2], p. 42.)

$$J(y) = \int\limits_0^1 \left[\frac{1}{2}\left(\frac{dy}{dx}\right)^2 - y \right] dx$$

(a) Show that the Euler's equation for the above functional is

$$\frac{d^2y}{dx^2} = -1$$

(b) Use the Ritz procedure and the following trial function based on Fourier series

$$y(x) = c_1 \sin(\pi x) + c_2 \sin(2\pi x) + c_3 \sin(3\pi x)$$

to verify that the variational solution is

$$y(x) = \frac{4}{\pi^3}\sin(\pi x) + \frac{4}{27\pi^3}\sin(3\pi x)$$

(c) Show that the solution of the Euler's equation subject to the given boundary conditions is

$$y(x) = \frac{x(1-x)}{2}$$

(d) Verify that the variational solution (b) represents the first two terms of the Fourier expansion of the exact solution given in (c).

P9.10. A shielded microstrip line is shown in Figure 9.13. The enclosure and the conducting strip are assumed to be perfect conductors. The strip thickness is assumed to be negligible, and the substrate is assumed to be lossless.
(a) Derive expressions for the capacitance per unit length C, characteristic imped-ance Z_0, and relative effective dielectric constant ϵ_{reff}, using variational principle. Assume the following charge distribution:

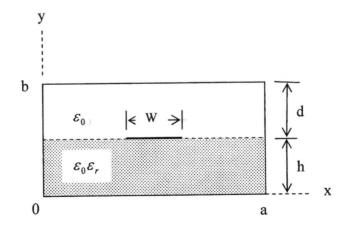

Figure 9.13   Cross-section of a shielded microstrip line.

$$\rho_s(x) = \begin{cases} |x| & \text{for } \dfrac{a - W}{2} \le x \le \dfrac{a + W}{2} \\ 0 & \text{otherwise} \end{cases}$$

(b) Calculate and plot Z_0 and ϵ_{reff} versus W/h from 0.05 to 5 for $\epsilon_r = 2.2$ and 10 and $h = 0.635$ and 1.27 mm, $d = h$, respectively. Compare your results with that obtained from microstrip line with a cover shield (Section 4.6.2).

P9.11. Repeat problem P9.10 with

$$\rho_s(x) = \begin{cases} A & \text{for } \dfrac{a - W}{2} \le x \le \dfrac{a + W}{2} \\ 0 & \text{otherwise} \end{cases}$$

where A is a constant.

CHAPTER 10

Finite Element Method

Finite element method (FEM) was pioneered by structural analysts. Its mathematical base was provided by Courant in 1943 [1], but its use in electromagnetics was not reported until 1968 [2]. Research contributions by Silvester and his group contributed significantly to the use of FEM in electromagnetics [3, 4]. Since then FEM has become popular as a computational method in electromagnetics because of its desirable features like geometrical adaptability and low memory requirements. FEM has been combined with most of the other computational methods to develop more efficient hybrid computational methods (Section 11.5). Some of the recent books provide a clear, conceptual picture of FEM in the general framework of computational techniques [5–7]. A number of computer codes in Fortran, C, and MATLAB are available in [5–7] and are meant primarily for education. Commercial software packages provide user-friendly interface to FEM.

FEM may be described as an adaptation of the variational method (Chapter 9) to devices with irregular shapes and dielectric inhomogeneity. Its variational property implies that the solution is accurate to, for example, second order even if the modeling is accurate to the first order. This translates to coarse discretization for the same accuracy and reduces the computational load. Other than the functional approach, one may use the weighted residual method to develop governing equations for FEM. In this form, FEM is very similar to the method of moments (MoM). This approach is also applicable to problems where the functional for the problem does not exist or cannot be identified. We shall employ the functional approach because of its simplicity.

10.1 Basic Steps in Finite Element Analysis

The FEM analysis of any problem involves four major steps: (1) segmentation or meshing of the device geometry into a finite number of elements such that the dielectric is homogeneous in each element; (2) deriving the governing equations for a typical element; (3) assembly of all elements of the device to generate the system equations; and (4) the solution of the system of equations obtained to determine the unknowns. To these, one may add another step of postprocessing, which involves processing the data generated to obtain characteristics of interest like: S-parameters of circuits, characteristics of transmission lines, resonant frequency, scattering cross-section, radiation characteristics, and so on. The major steps in FEM are described next.

10.1.1 Segmentation or Meshing of the Geometry

The given device geometry when not amenable to solution in the present form is divided into a number of subproblems that are easier to solve. For this, the device geometry is divided into a *finite* number of *elements*. The various elements employed for the analysis in FEM are shown in Figure 10.1. The line segment, straight or curved, is used for one-dimensional problems. For two-dimensional analysis, triangles and/or quadrilaterals are used to model the geometry, and tetrahedron and bricks may be employed to discretize three-dimensional geometry. The discretization for a two-dimensional geometry is illustrated in Figure 10.2. In this figure the circular region is approximated by an inscribed regular octagon, which for the

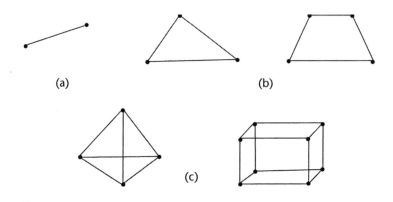

Figure 10.1 Various element types used for FEM analysis: (a) line segments for one dimension; (b) triangles and quadrilaterals for two dimensions; and (c) tetrahedrons and bricks for three dimensions.

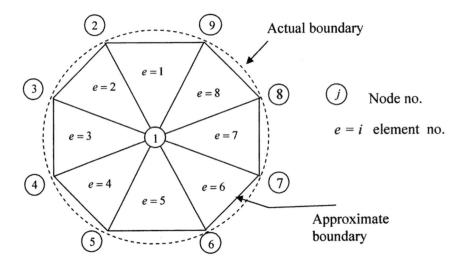

Figure 10.2 Discretization of a circular geometry into a number of triangular elements.

purpose of FEM analysis is modeled in the form of eight nonoverlapping triangular elements and nine nodes. It may be noted that the octagon is used here as an approximation to the circle in order to reduce the computational load. Increasing the number of sides of the polygon improves the approximation so much so that a polygon with an infinite number of sides (each side of the size of a point) will fit the circle exactly. One can use a combination of triangular and quadrilateral elements also for two-dimensional problems, but the triangular elements with variable size are popular because they can model arbitrary geometrical shape better. In commercially available software, the segmentation is carried out by an auto-mesh generator with an option for manual intervention.

If the approximate field (potential) distribution in the element e is φ^e, the distribution in the device is a linear combination of these distributions,

$$\varphi(x, y) = \sum_{e=1}^{N_e} \varphi^e(x, y) \tag{10.1}$$

where N_e is the number of elements of the solution region. The number N_e is chosen such that the longest side length is typically less than $\lambda/10$. Continuity of the solution across the element boundaries incorporates the effect of neighboring elements on the solution.

Next, we determine the distribution φ^e in a typical element.

10.1.2   Derivation of the Element Matrix

We notice from (10.1) that the distribution φ within the device is determined by the distribution φ^e in each of the elements of the device. Next, we expand unknown φ^e in suitable basis functions ψ_i^e. Historically, the basis functions are called *shape functions* in FEM. Two types of basis functions are used in FEM: node-based, and edge-based functions. The edge-based functions are found to be superior and free from spurious solutions [5]. We shall use node-based functions for simplicity of analysis.

Basis Functions
The basis functions are defined over the element only and are zero outside it. They can influence the value of potential in the immediate neighboring elements only through continuity conditions. This property leads to the sparse matrix for the FEM solution.

The basis functions used to expand the unknown function φ^e should form a complete set in order to improve the accuracy of solution. However, due to the limited computer resources the number of terms in the expansion is restricted to a few only. Therefore, there is a trade-off between the desired accuracy and the number of basis functions used. One may use linear, second-order, or third-order polynomial, for the basis functions. In general, the order of the differential equation to be solved determines the maximum order of the polynomial to be employed. In most of the electromagnetic problems, only the continuity of the function (and not the continuity of the derivative) across the elements is imposed. Once the basis functions have been selected, the approximate distribution within a two-dimensional element e with p nodes may be written as

$$\varphi^e(x, y) = \sum_{i=1}^{p} c_i^e \psi_i^e(x, y) \tag{10.2}$$

where $\psi_i^e(x, y)$ are the two-dimensional basis functions, and c_i^e are the unknown complex coefficients to be determined. The equations for c_i^e are obtained by enforcing the governing differential equation for the element, say,

$$\frac{\partial^2 \varphi^e}{\partial x^2} + \frac{\partial^2 \varphi^e}{\partial y^2} + k^2 \varphi^e = 0 \tag{10.3}$$

It is impossible to satisfy the differential equation at each and every point of the element because of the finite memory size. Alternatively, one may satisfy the differential equation in an average sense. For this one may employ the weighted residual method of Section 9.6 and obtain the governing equations for the element [5]. An alternative to this approach is to use the corresponding functional. We shall use the functional for its simplicity.

The functionals for various types of differential equations used in electromagnetics are given in Section 9.5. The functional for the wave equation (10.3) is given by

$$F^e[\varphi^e] = \frac{1}{2} \iint_S \left[|\nabla \varphi^e|^2 - k^2 (\varphi^e)^2 \right] dx\, dy \tag{10.4}$$

The expansion (10.2) is now substituted into (10.4), resulting in a matrix of the form

$$[F^e] = [A^e][\varphi^e] \tag{10.5}$$

where $[A^e]$ is the element matrix and $[\varphi^e]$ is the nodal vector. The form of the element matrix is given below,

$$[A^e] = \begin{bmatrix} A_{11}^e & A_{12}^e \\ A_{21}^e & A_{22}^e \end{bmatrix} \qquad \text{for one-dimensional problems with 2-nodes}$$

$$\tag{10.6}$$

$$[A^e] = \begin{bmatrix} A_{11}^e & A_{12}^e & A_{13}^e \\ A_{21}^e & A_{22}^e & A_{23}^e \\ A_{31}^e & A_{32}^e & A_{33}^e \end{bmatrix} \qquad \text{for two-dimensional problems with 3-nodes}$$

$$\tag{10.7}$$

The matrix element A_{ij}^e describes the coupling between the nodes i and j. The next major step in FEM solution is the assembly of element matrices.

10.1.3 Assembly of Element Matrices

The matrix equation (10.5) represents a set of equations, which are now assembled for all the elements of the device,

$$\sum_{e=1}^{N_e} [A^e][\varphi^e] = \sum_{e=1}^{N_e} [F^e] \tag{10.8}$$

to yield the system matrix

$$[A][\varphi] = [F] \tag{10.9}$$

where $[\varphi]$ is the global nodal vector whose elements are the expansion coefficients c_i^e.

Let us illustrate the effect of assembling various elements by considering Figure 10.2. The octagonal geometry is segmented into eight triangular elements with nine global nodes, and $3 \times 8 = 24$ local nodes. The eight elements give rise to 24 equations. The number of unknowns are nine values of φ, corresponding to the global nodes. Therefore, the number of equations is more than the number of unknowns. This is due to the fact that several elements share the same node (e.g., nodes 1 to 9). The assembly procedure takes care of this multiplicity to yield as many equations as there are unknowns. It also ensures continuity of φ from element to element.

10.1.4 Solution of System Matrix

The matrix $[A][\varphi] = [F]$ should now be solved to determine the unknown vector φ of potentials at the nodes. The correct nodal values are obtained when the functional is made stationary with respect to the expansion coefficients c_i, $i = 1, 2, 3, \ldots N$, where N is the number of global nodes. Using Ritz method (Section 9.3), we set

$$\frac{\partial F}{\partial c_i} = [0], \qquad i = 1, 2, 3, \ldots N \tag{10.10}$$

An equivalent approach is to first take the derivative of F^e with respect to each of c_i^e and then carry out the assembly of elements [8]. This is permissible because of the linearity of assembly process. Equation (10.10) now gets modified as

$$\frac{\partial F}{\partial c_i} = \sum_{e=1}^{N_e} \frac{\partial F^e}{\partial c^e} = [0] \tag{10.11}$$

We shall follow the second approach in which $[\partial F^e / \partial c^e]$ is computed and (10.11) is implemented. The resulting set of simultaneous equations is solved to determine the unknown vector φ. The matrix $[A]$ is a square matrix of size $N \times N$, very sparse and symmetrical also if the medium (material) is reciprocal. The matrix equation obtained from (10.10) or (10.11) may be solved either by using direct matrix solvers such as L-U decomposition or iterative solvers. The matrix solution techniques are described in Appendix A.

10.1.5 Postprocessing

The last major step in the FEM solution process is to process the data in the form of φ vector. The value of φ at the nodes of the element can be used to describe the behavior at any point inside the element by using (10.2). The plot of φ in the device may be used to develop concepts or to obtain the desired characteristics of the device. The electrostatic field distribution can be utilized to determine the stored electric energy from which the capacitance can be obtained. The current distribution on the metallic portion of the antenna may be used to calculate the radiation pattern.

10.2   FEM Analysis in One Dimension

The one-dimensional problems provide the simplest means to learn the mathematical concepts and computational approach, with less programming complexity. However, these problems cannot include arbitrary shape. The potential of FEM is realized by analyzing complex three-dimensional structures with material inhomogeneity and anisotropy. The two-dimensional problems lie in between three-dimensional and one-dimensional problems in terms of complexity and usefulness. This section describes the FEM analysis for one-dimensional problems.

We start with the solution of one-dimensional scalar Helmholtz equation

$$\frac{d^2 E_y}{dx^2} + k^2 E_y(x) = g(x), \qquad 0 \le x \le L \tag{10.12}$$

This differential equation describes steady state field distribution along the x-axis and may be used to describe the wave behavior in free space, transmission lines, parallel plate waveguide, and so on. With appropriate boundary conditions it can describe the field distribution in transmission line circuits including resonators. The Poisson equation follows from (10.12) if we choose $k^2 = 0$.

We now describe the solution of (10.12); the functional for which is listed in Table 9.2 and may be written as

$$F(E_y) = \frac{1}{2} \int \left(\frac{dE_y}{dx} \right)^2 dx - \frac{1}{2} k^2 \int E_y^2 \, dx + \int g E_y \, dx \tag{10.13}$$

The first step in the solution process is to discretize the solution region $(0, L)$ into line segments (elements). We shall use equal length N_e segments/elements of size $\Delta x = L/N_e$. The discretization or segmentation of the domain is shown in Figure

10.3. Nodes are introduced to separate adjacent elements. The coordinate of the ith node denoted as x_i is given by

$$x_i = (i - 1)\Delta x, \qquad i = 1, 2, 3, \ldots N \tag{10.14}$$

where $N = N_e + 1$. This designation of nodes is called global node numbering. Since an element connects two nodes, another number called local node number is assigned to the nodes. The local nodes are denoted as x_1^e and x_2^e with superscript e defining the element number. The local and the global node coordinates are related as

$$x_1^e = (e - 1)\Delta x = x_e \text{ and } x_2^e = e\Delta x = x_{e+1}, \qquad e = 1, 2, 3, \ldots N_e \tag{10.15}$$

The discretization of the domain into elements results in discretization of functional F as

$$F(E_y) = \sum_{e=1}^{N_e} F^e(\varphi^e) \tag{10.16}$$

where F^e follows from (10.13),

$$F^e(\varphi^e) = \frac{1}{2} \int_{x_1^e}^{x_2^e} \left(\frac{d\varphi^e}{dx}\right)^2 dx - \frac{1}{2}k^2 \int_{x_1^e}^{x_2^e} (\varphi^e)^2 \, dx + \int_{x_1^e}^{x_2^e} g\varphi^e \, dx \tag{10.17}$$

For brevity we have changed the notation in (10.16) from E_y to φ.

Next, we describe $\varphi^e(x)$ in terms of known basis/interpolation functions. In order to keep the analysis simple we choose linear basis functions, that is,

$$\varphi^e(x) = a^e + b^e x \tag{10.18}$$

where a^e and b^e are constants, and are determined by expressing the potential at the two nodes as

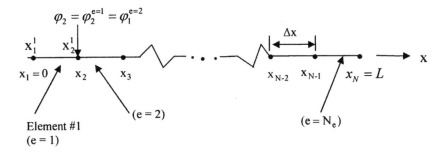

Figure 10.3   Discretization of the range $0 \le x \le L$ into a number of elements of size Δx.

$$\varphi_1^e = a^e + b^e x_1^e, \qquad \varphi_2^e = a^e + b^e x_2^e \tag{10.19}$$

Solving for a^e and b^e gives

$$b^e = \frac{\varphi_2^e - \varphi_1^e}{x_2^e - x_1^e}, \qquad a^e = \frac{\varphi_1^e x_2^e - \varphi_2^e x_1^e}{x_2^e - x_1^e} \tag{10.20}$$

Therefore,

$$\varphi^e(x) = \sum_{j=1}^{2} N_j^e(x) \varphi_j^e \tag{10.21}$$

where N_j^e are the basis/shape functions obtained as

$$N_1^e(x) = \begin{cases} \dfrac{x_2^e - x}{x_2^e - x_1^e}, & x_1^e \leq x \leq x_2^e \\ 0 & \text{otherwise} \end{cases} \tag{10.22a}$$

$$N_2^e(x) = \begin{cases} \dfrac{x - x_1^e}{x_2^e - x_1^e}, & x_1^e \leq x \leq x_2^e \\ 0 & \text{otherwise} \end{cases} \tag{10.22b}$$

The shape functions are plotted in Figure 10.4. It may be noted that there are two overlapping shape functions for each element.

Substituting (10.21) into (10.17) gives

$$F^e(\varphi^e) = \frac{1}{2} \sum_{j=1}^{2} \left(\int_{x_1^e}^{x_2^e} \left(\frac{d(\varphi_j^e N_j^e)}{dx} \right)^2 dx - k^2 \int_{x_1^e}^{x_2^e} (\varphi_j^e N_j^e)^2 dx + 2 \int_{x_1^e}^{x_2^e} g\varphi_j^e N_j^e \, dx \right) \tag{10.23}$$

Taking the derivative with respect to each of the unknowns φ_j^e, we obtain

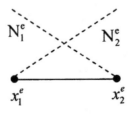

Figure 10.4 One-dimensional linear basis/shape functions.

$$\frac{\partial F^e}{\partial \varphi_i^e} = \sum_{j=1}^{2} \varphi_j^e \left(\int_{x_1^e}^{x_2^e} \frac{dN_i^e}{dx} \frac{dN_j^e}{dx} \, dx - k^2 \int_{x_1^e}^{x_2^e} N_i^e N_j^e \, dx \right) + \int_{x_1^e}^{x_2^e} gN_i^e \, dx \qquad i = 1, 2$$

$$(10.24)$$

This equation can be expressed in the matrix form as

$$\left[\frac{\partial F^e}{\partial \varphi^e} \right] = [A^e][\varphi^e] + [b^e], \qquad e = 1, 2, 3, \ldots N_e \qquad (10.25)$$

where

$$[\varphi^e] = \left[\varphi_1^e, \varphi_2^e \right]^t, \qquad [b^e] = \left[b_1^e, b_2^e \right]^t \qquad (10.26a)$$

$$[A^e] = \begin{bmatrix} A_{11}^e & A_{12}^e \\ A_{21}^e & A_{22}^e \end{bmatrix} \qquad (10.26b)$$

$[A^e]$ is called the *element matrix* and its entries, as obtained from (10.24), are

$$A_{ij}^e = \int_{x_1^e}^{x_2^e} \left[\frac{dN_i^e}{dx} \frac{dN_j^e}{dx} - k^2 N_i^e N_j^e \right] dx \qquad (10.27)$$

The excitation vector entries are obtained as

$$b_i^e = \int_{x_1^e}^{x_2^e} N_i^e(x) g(x) \, dx \qquad i = 1, 2 \qquad (10.28)$$

Since the shape functions are linear, the integration in (10.27) can be carried out analytically while that in (10.28) can also be performed analytically provided $g(x)$ is assumed to be constant over the range of eth element. If we approximate $g(x) \approx g^e$ one obtains

$$A_{11}^e = A_{22}^e = \frac{-1}{\Delta x} + k^2 \frac{\Delta x}{3}, \qquad \Delta x = \frac{L}{N_e} \qquad (10.29a)$$

$$A_{12}^e = A_{21}^e = \frac{1}{\Delta x} + k^2 \frac{\Delta x}{6} \qquad (10.29b)$$

$$b_1^e = b_2^e = g^e \frac{\Delta x}{2} \qquad (10.29c)$$

Assembly of Element Matrices

Since each element generates two equations, the number of equations generated by N_e elements is $2N_e$ whereas the number of unknowns is equal to the number of global nodes $N = N_e + 1$. The redundancy (because $N < 2N_e$) is principally due to the reason that the elements are considered independent although they are part of the domain $0 \leq x \leq L$. The nodal field is continuous across the adjacent elements. Imposing this constraint on the common nodes between the elements will reduce the number of degrees of freedom and therefore the independent number of equations. A simple implementation of the continuity condition is to express (10.25) for the various elements in terms of global nodes as [8]

$$\left[\frac{\partial F^e}{\partial \varphi^e}\right] = [\overline{A}^e][\varphi] + [\overline{b}^e], \qquad e = 1, 2, 3, \ldots N_e \tag{10.30}$$

All the new vectors $[\varphi]$, $[\overline{b}^e]$, and matrix $[\overline{A}^e]$ are of global size N which is greater than the element size, and are obtained by zero filling of the corresponding vector/matrix for the element. The overall effect of the assembly procedure is to generate as many equations as there are global nodes.

The next step in the solution is to sum (10.30) over all the elements and apply the stationarity constraint (10.11) [8] to obtain

$$\frac{\partial F}{\partial \varphi_i} = \sum_{e=1}^{N_e} \frac{\partial F}{\partial \varphi^e} = \sum_{e=1}^{N_e} \left([\overline{A}^e][\varphi] + [\overline{b}^e]\right) = [0] \tag{10.31}$$

The details of the assembly procedure are given in Example 10.1. The system matrix (10.31) can be written as

$$[A][\varphi] = -[b] \tag{10.32}$$

The matrix equation can be solved using either direct-matrix inversion or iterative matrix solution to determine the $[\varphi]$ vector. The distribution $E_y(x)$ over the problem domain is then given by

$$E_y(x) = \sum_{e=1}^{N_e} \sum_{i=1}^{2} \varphi_i^e N_i^e(x) \tag{10.33}$$

10.2.1 Treatment of Boundary and Interface Conditions

The boundary conditions guarantee the uniqueness of the solution, and are applied before solving the system matrix (10.32). The Dirichlet boundary conditions $\varphi = C$ are *essential boundary* conditions and are implemented by specifying the value of global nodes at the boundary. If, for example, E_y is specified to be zero at the end points, φ_1 and φ_N are set to zero a priori. The Neumann boundary conditions $\partial \varphi / \partial n = 0$ in FEM are called *natural boundary conditions*; they are automatically satisfied by the function about which the functional is stationary [6]. Therefore, Neumann boundary conditions are *not* to be imposed separately.

The conditions at the interface of two different media require that the tangential components of electric and magnetic fields should be continuous across the interface. This can be implemented by assigning common nodes at the interface. However, it also enforces continuity of the normal component of the field, and is not correct. Finer discretization near the interface can alleviate this problem [8, p. 195].

Example 10.1.   Use FEM to solve the following differential equation:

$$\frac{d^2 E_y(x)}{dx^2} - \pi^2 E_y(x) = -2\pi^2 \sin(\pi x), \qquad 0 \le x \le 1 \qquad (10.34)$$

subject to the boundary conditions

$$E_y(0) = E_y(1) = 0 \qquad (10.35)$$

Compare the FEM results with the exact solution $E_y(x) = \sin(\pi x)$ [5, p. 83].

Solution.   Based on (10.12) and (10.13), the functional for (10.34) can be written as

$$F(E_y) = \frac{1}{2} \int \left(\frac{dE_y}{dx}\right)^2 dx + \frac{1}{2}\pi^2 \int E_y^2 \, dx - 2\pi^2 \int \sin(\pi x) E_y \, dx \qquad (10.36)$$

Let us divide the domain $0 \le x \le 1$ into four equal segments of size $\Delta x = 0.25$ as in Figure 10.5. The element number, global and local node numbers are also indicated there. The matrix and vector elements are given by

$$[A^e] = \begin{bmatrix} \dfrac{-1}{\Delta x} - \pi^2 \dfrac{\Delta x}{3} & \dfrac{1}{\Delta x} - \pi^2 \dfrac{\Delta x}{6} \\[3mm] \dfrac{1}{\Delta x} - \pi^2 \dfrac{\Delta x}{6} & \dfrac{-1}{\Delta x} - \pi^2 \dfrac{\Delta x}{3} \end{bmatrix}, \qquad e = 1, 2, 3, 4 \qquad (10.37a)$$

and

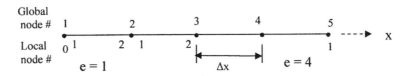

Figure 10.5   Segmentation of the domain $0 \le x \le 1$ into four equal segments of length $\Delta x = 0.25$. The global and local nodes are also shown and numbered.

$$b_i^e = -2\pi^2 \int_{x_1^e}^{x_2^e} N_i^e(x) \sin(\pi x)\, dx \qquad (10.37b)$$

or

$$b_1^e = -2\pi \left(\cos(\pi(e-1)\Delta x) - \frac{\alpha_e}{\Delta x \pi} \right)$$

$$b_2^e = -2\pi \left(-\cos(\pi e \Delta x) + \frac{\alpha_e}{\Delta x \pi} \right)$$

where $\alpha_e = \sin(\pi e \Delta x) - \sin(\pi(e-1)\Delta x)$.

Assembly of Elements
For assembly, the vector $[\varphi^e]$ is replaced by global vector $[\varphi]$ defined as

$$[\varphi] = [\varphi_1 \quad \varphi_2 \quad \varphi_3 \quad \varphi_4 \quad \varphi_5]^t \qquad (10.38)$$

Matrices $[A^e]$ and vectors $[b^e]$ are replaced by the corresponding global *sized* matrices and vectors by appending zeros as shown below for $e = 1, 2$. The global sized matrices and vectors are distinguished by the superscript (-).

$$[A^1] = \begin{bmatrix} A_{11}^1 & A_{12}^1 \\ A_{21}^1 & A_{22}^1 \end{bmatrix} \Rightarrow [\overline{A}^1] = \begin{bmatrix} A_{11}^1 & A_{12}^1 & 0 & 0 & 0 \\ A_{21}^1 & A_{22}^1 & 0 & 0 & 0 \\ 0 & 0 & 0 & 0 & 0 \\ 0 & 0 & 0 & 0 & 0 \\ 0 & 0 & 0 & 0 & 0 \end{bmatrix} \qquad (10.39)$$

$$[b^1] = \begin{bmatrix} b_1^1 \\ b_2^1 \end{bmatrix} \Rightarrow [\overline{b}^1] = \begin{bmatrix} b_1^1 \\ b_2^1 \\ 0 \\ 0 \\ 0 \end{bmatrix} \qquad (10.40)$$

$$[\overline{A}^2] = \begin{bmatrix} 0 & 0 & 0 & 0 & 0 \\ 0 & A_{11}^2 & A_{12}^2 & 0 & 0 \\ 0 & A_{21}^2 & A_{22}^2 & 0 & 0 \\ 0 & 0 & 0 & 0 & 0 \\ 0 & 0 & 0 & 0 & 0 \end{bmatrix}, \quad [\overline{b}^2] = \begin{bmatrix} 0 \\ b_1^2 \\ b_2^2 \\ 0 \\ 0 \end{bmatrix} \qquad (10.41)$$

Similar matrices/vectors are written for the rest of the elements. Next, the matrices/vectors for the various elements are added row by row; that is, the first row of all the elements are added to give the first row of the global matrix/vector, and so on. The resulting matrix equation is

$$
\begin{bmatrix}
A_{11}^1 & A_{12}^1 & 0 & 0 & 0 \\
A_{21}^1 & A_{22}^1 + A_{11}^2 & A_{12}^2 & 0 & 0 \\
0 & A_{21}^2 & A_{22}^2 + A_{11}^3 & A_{12}^3 & 0 \\
0 & 0 & A_{21}^3 & A_{22}^3 + A_{11}^4 & A_{12}^4 \\
0 & 0 & 0 & A_{21}^4 & A_{22}^4
\end{bmatrix}
\begin{bmatrix}
\varphi_1 \\ \varphi_2 \\ \varphi_3 \\ \varphi_4 \\ \varphi_5
\end{bmatrix}
+
\begin{bmatrix}
b_1^1 \\ b_2^1 + b_1^2 \\ b_2^2 + b_1^3 \\ b_2^3 + b_1^4 \\ b_2^4
\end{bmatrix}
=
\begin{bmatrix}
0 \\ 0 \\ 0 \\ 0 \\ 0
\end{bmatrix}
$$

$$(10.42)$$

For coding, one may start with a 5×5 null matrix for $[A]$ and update it element by element.

Boundary Conditions
By replacing the local node numbers by the global node numbers we have ensured continuity of φ from element to element. The boundary conditions $\varphi_1 = 0 = \varphi_5$ are now applied to (10.42) by deleting the first and last equations of (10.42) and set $\varphi_1 = 0 = \varphi_5$ wherever they occur [8]. An alternative is to delete first and last rows in (10.42), and delete first and last columns also from the resulting matrix. One obtains

$$
\begin{bmatrix}
A_{22}^1 + A_{11}^2 & A_{12}^2 & 0 \\
A_{21}^2 & A_{22}^2 + A_{11}^3 & A_{12}^3 \\
0 & A_{21}^3 & A_{22}^3 + A_{11}^4
\end{bmatrix}
\begin{bmatrix}
\varphi_2 \\ \varphi_3 \\ \varphi_4
\end{bmatrix}
= -
\begin{bmatrix}
b_2 \\ b_3 \\ b_4
\end{bmatrix}
\qquad (10.43)
$$

where

$$
b_i = b_2^{i-1} + b_1^i = \frac{2}{\Delta x} \left(\sin\{\pi(i-2)\Delta x\} + \sin(i\pi\Delta x) \right) \qquad (10.44a)
$$

$$
A_{11}^e = A_{22}^e = \frac{-1}{\Delta x} - \frac{1}{3}\pi^2 \Delta x = -4.8225 \qquad (10.44b)
$$

$$
A_{12}^e = A_{21}^e = \frac{1}{\Delta x} - \frac{1}{6}\pi^2 \Delta x = 3.589 \qquad (10.44c)
$$

Solution of (10.43) is obtained as

$$
[\varphi] = [0 \quad 0.7055 \quad 0.9977 \quad 0.7055 \quad 0]^t
$$

and compares favorably with the analytical value $[0 \ 0.7071 \ 1 \ 0.7071 \ 0]^t$ obtained from $\sin(\pi x)$.

The field distribution over x is found from (10.33) and the computed nodal values; it is given by

$$E_y(x) = \sum_{e=1}^{4} \sum_{i=1}^{2} \varphi_i^e N_i^e(x) \tag{10.45}$$

In the given problem, the nodes are located at $x_i = 0.25(i-1)$, and the central three nodes contribute to $E_y(x)$. Therefore,

$$E_y(x) = \varphi_2 N_2^1 + \varphi_2 N_1^2 + \varphi_3 N_2^2 + \varphi_3 N_1^3 + \varphi_4 N_2^3 + \varphi_4 N_1^4$$

or

$$E_y(x) = 0.7055\left(N_2^1(x) + N_1^2(x) + N_2^3(x) + N_1^4(x)\right) + 0.9977\left(N_2^2(x) + N_1^3(x)\right) \tag{10.46}$$

The absence of shape functions $N_1^1(x)$ and $N_2^4(x)$ in (10.46) ensures that the boundary conditions are satisfied. The field distribution, reconstructed from the computed nodal potentials and the basis functions, is plotted in Figure 10.6. It approximates the expected solution fairly well. The solution of the differential equation for large number of segments can be determined using the software *fem_oned.m*. It has been observed that there is hardly any difference between the analytical value $E_y(x) = \sin(\pi x)$ and that computed from the software *fem_oned.m*.

It may be noted from Figure 10.6 that triangular basis functions of size $2\Delta x$ could have been employed for solving this problem with identical results. These basis functions are used in [5] for solving a similar problem. The triangular basis functions are described in Section 6.2.1.

10.2.2 Accuracy and Numerical Dispersion

In this section we discuss the effect of discretization size on the accuracy of results. This particular aspect was discussed earlier in relation to FDM and FDTD. Here, we want to determine numerical dispersion and compare it with that in FDM. The presentation given here follows that in [9].

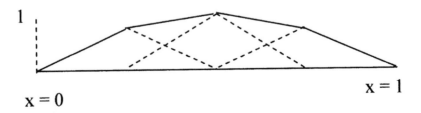

Figure 10.6 FEM solution for the Example 10.1, $N_e = 4$, $\Delta x = 0.25$.

To determine accuracy and resulting dispersion we solve the following one-dimensional scalar Helmholtz equation for the propagation constant:

$$\frac{d^2 E_y}{dx^2} + k^2 E_y(x) = 0 \tag{10.47}$$

where $k = 2\pi/\lambda$ defines the phase constant in the medium.

We assume a large free space with no boundaries and segment it uniformly with element size Δx. Employing the linear basis functions as described in the last section, the equation for the ith node is given by

$$\left[0, 0, 0, \ldots, A^e_{i-1,i}, 2A^e_{i,i}, A^e_{i,i+1}, 0, \ldots 0\right]\{E_y\} = 0 \tag{10.48}$$

where $\{E_y\}^t = \{\ldots, E_{y,i-1}, E_{y,i}, E_{y,i+1}, \ldots\}$ are the coefficients of expansion for the unknown field vector, and the matrix elements are defined by (10.29). Substituting these expressions gives

$$E_{y,i-1}\left[\frac{1}{\Delta x} + \frac{1}{6}k^2\Delta x\right] - E_{y,i}\left[\frac{2}{\Delta x} - \frac{2}{3}k^2\Delta x\right] + E_{y,i+1}\left[\frac{1}{\Delta x} + \frac{1}{6}k^2\Delta x\right] = 0$$

or

$$E_{y,i+1} - 2E_{y,i} + E_{y,i-1} + (k\Delta x)^2\left[\frac{1}{6}E_{y,i+1} + \frac{2}{3}E_{y,i} + \frac{1}{6}E_{y,i-1}\right] = 0 \tag{10.49}$$

We now solve this equation using a discretized version of the traveling wave solution, $E_y(x) = E_0 e^{\pm jkx}$, that is,

$$E_{y,i} = E_0 e^{\pm j\beta(i\Delta x)} \tag{10.50}$$

where β is the wave propagation constant in the discretized medium. Substituting (10.50) in (10.49) results in

$$e^{j\beta\Delta x} - 2 + e^{-j\beta\Delta x} + \frac{1}{6}(k\Delta x)^2(e^{j\beta\Delta x} + 4 + e^{-j\beta\Delta x}) = 0$$

or

$$\beta = \frac{1}{\Delta x}\cos^{-1}\left(\frac{6 - 2(k\Delta x)^2}{6 + (k\Delta x)^2}\right) \tag{10.51}$$

This equation along with (10.50) constitutes the numerical solution for wave propagation on an infinite, uniform mesh. The propagation constant β is found to be different from the propagation constant k and is dependent on cell size Δx

and k. In the limiting case, $\Delta x \rightarrow 0$, $\beta \rightarrow k$ as expected. In the lossless medium, β is real valued for [9, p. 198]

$$k \Delta x \leq \sqrt{12} \qquad (10.52)$$

Therefore, for cell sizes such that $\Delta x < 0.55\lambda$, *the error in the numerical solution is entirely phase error*. This phenomenon is called numerical dispersion. The phase error over a single cell is given by

$$\text{Phase error} = (k - \beta)\Delta x \qquad (10.53)$$

The phase error per wavelength is listed in Table 10.1 as a function of number of cells per wavelength, $\lambda/\Delta x$. For example, the error over a cell of size 0.1λ is about $0.57°$. The phase error increases cumulatively across a mesh, so cells of width 0.1λ spanning a region of 10 cells would produce a total phase error of $5.67°$. Since $\beta/k = v_{cont}/v_{disc} < 1$, the phase velocity is higher in the discretized space compared to that in the continuous medium.

The phase error per wavelength for the FEM and FDM (refer to Table 7.2) for the same discretization size is compared in the last two columns, and shows that the phase error in FEM is slightly lower. The improvement in the accuracy for FEM solution may be due to the use of linear basis functions compared to the pulses employed in FDM. Also, the phase velocity is higher in FEM and lower in FDM compared to that in the continuous medium.

Verification of β/k Through Resonator Simulation. The effect of discretization on phase constant may be determined by carrying out a simple simulation experiment on line resonator. For this, we design a half-wave resonator with shorted ends. Discretize the resonator in segments of length Δx as described earlier, and excite it at the central node. Due to the discretization error in β/k, the resonator will not resonate at the designed wavelength. We need to fine-tune the resonator to determine the actual resonant wavelength. Tuning by changing the line length is difficult because of the coarse discretization employed. We introduce a parameter for changing the dielectric constant of the medium. The wave equation for the resonator, with $L = 1\text{m}$ and therefore resonant wavelength, $\lambda = 2\text{m}$, becomes

$$\frac{d^2 E_y}{dx^2} + C\pi^2 E_y = 0 \qquad 0 < x < 1 \qquad (10.54)$$

with

Table 10.1 Effect of Cell Size Δx on the Phase Velocity and Phase Error

$\lambda/\Delta x$	$\beta\Delta x$ (rad)	$k\Delta x$ (rad)	β/k	Phase Error per λ (deg) FEM	Phase Error per λ (deg) FDM
5	1.186	1.2566	0.9438	20.10°	−29.27°
10	0.6181	0.6283	0.9838	5.67°	−6.19°
20	0.3130	0.3141	0.9965	1.46°	−1.49°

$$E_y(0) = E_y(1) = 0$$

Here C is the parameter introduced to vary the dielectric constant of the medium; $C = 1$ for air medium. FEM analysis of (10.54) shows that the expression for C for which the propagation constant β equals k is given by

$$C = \left(\frac{\sin\left(\frac{k\,\Delta x}{2}\right)}{\frac{k\,\Delta x}{2}} \right)^2 \frac{3}{2 + \cos(k\,\Delta x)} \tag{10.55}$$

For FEM simulation, the resonator was excited at the central segment by setting, for example, $b^t = [0 \ \ 0 \ \ 1 \ \ 0 \ \ 0]$ for five segments. The simulated value of C as a function of $\Delta x/\lambda$ is listed in Table 10.2.

The normalized propagation constant in the discretized medium is then given by $\beta/k = 1/\sqrt{C}$. FEM simulation of the resonator produced the values close to the analytical values. The resonance is marked by the increase in amplitude of the resonator as we approach resonance. The amplitude flips over from high negative to high positive value, or vice versa, at the resonance. The source code *fem_oned.m* was modified for this purpose. The amplitude distribution on the resonator for three values of constant C is shown in Figure 10.7 for $\Delta x = \lambda/20$. The flipping over of the amplitude from high negative to high positive value indicates resonance, and $C = 1.00825$ at resonance. The amplitude is -25 for $C = 1$.

10.3   FEM Analysis in Two Dimensions

The FEM solution of one-dimensional differential equations was discussed in the last section. The emphasis was on developing concepts without running into tedious mathematics and complicated coding, although the practical problems in one dimension are very limited. The two-dimensional problems are more realistic. For example, *TE* and *TM* mode analysis of waveguides of arbitrary cross-section can be carried out using two-dimensional formulation, and capacitance of transmission lines can be found by solving Laplace equation in two dimensions.

The FEM analysis for the two-dimensional problems starts with the use of corresponding two-dimensional functional. The unknown function in the functional is expanded in basis functions, and the expansion coefficients are determined using the Ritz procedure.

Table 10.2 Comparison of the Value of C from (10.55) and Resonator Simulation

$\Delta x/\lambda$	C, Based on Simulation	C, Based on (10.55)	β/k	$(k/\beta)^2$
1/8	1.052–1.053	1.0523	0.9759	1.05
1/12	1.023–1.0235	1.02304	0.989	1.022
1/20	1.0082–1.0083	1.00825	0.9965	1.0070
1/40	1.002–1.0021	1.00206	0.9985	1.003

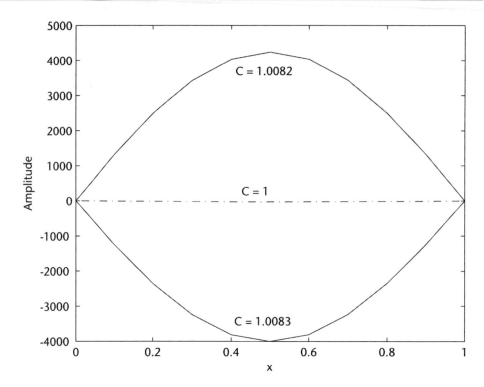

Figure 10.7 Amplitude distribution on half-wave resonator for three values of C for $\Delta x = \lambda/20$. The flipping over of the amplitude indicates resonance.

10.3.1 Solution of Two-Dimensional Wave Equation

It is assumed in the two-dimensional formulation that the device geometry is uniform along the z-direction. This leads to the assumption that the unknown function is independent of the z-coordinate as in a parallel plate capacitor, or the function has known z dependence as in the case of propagation in waveguides. Let us consider the propagation of *TE* modes in a homogeneously filled waveguide. The inhomogeneous wave equation for this case is given by

$$\frac{\partial^2 H_z}{\partial x^2} + \frac{\partial^2 H_z}{\partial y^2} + \gamma^2 H_z = g_1(x, y) \qquad \gamma^2 = k^2 - \beta^2 \qquad (10.56a)$$

Similarly, the wave equation for *TM* modes is given by

$$\frac{\partial^2 E_z}{\partial x^2} + \frac{\partial^2 E_z}{\partial y^2} + \gamma^2 E_z = g_2(x, y) \qquad (10.56b)$$

The functional obtained from Table 9.2 is found to be

$$F(\varphi) = \frac{1}{2} \iint\limits_S \left[|\nabla_t \varphi|^2 - \gamma^2 \varphi^2 + 2g\varphi \right] dx\, dy, \qquad g = g_1 \text{ or } g_2 \quad (10.56c)$$

where φ stands for H_z or E_z as the case may be, and S is the cross-sectional surface of the waveguide. The excitation function of the waveguide is represented here by $g(x, y)$.

10.3.2 Element Matrix for Rectangular Elements

In order to illustrate the analysis procedure we consider FEM analysis of a rectangular waveguide shown in Figure 10.8. First we segment the waveguide into a number of elements. The elements used could be of the same size (uniform discretization) or of different sizes (nonuniform discretization). The shape of the elements could be the same for all or vary from element to element. The preferred element shapes include: triangular, rectangular, and quadrilateral shapes. The triangular elements are the most versatile of these and can be used to model arbitrary shaped geometry. The rectangular elements are efficient for modeling rectangular geometries, and the element matrix can be expressed in a simple closed form. In order to make the analysis amenable to hand calculations, we shall use rectangular elements for segmenting/meshing the waveguide. Also, the rectangular element is natural for this geometry and the number of elements required for a given accuracy is expected to be less.

Rectangular Element
Consider the rectangular element of size $a^e \times b^e$ as shown in Figure 10.9. *The nodes are numbered in the counterclockwise direction.* Unless this convention is followed for numbering the local nodes, the expressions derived next are not valid. However, the local node numbering may start from any node designated as #1. The potential distribution within each element may be approximated using linear variation as

$$\varphi^e(x, y) = a' + b'x + c'y + d'xy \tag{10.57a}$$

The constants a', b', c', and d' can be determined from the potential at the four nodes of the rectangle. If the nodal values are denoted as φ_1^e, φ_2^e, φ_3^e, and φ_4^e, we obtain by enforcing (10.57a) as

$$\varphi_1^e = a' + b'x_1 + c'y_1 + d'x_1y_1 \tag{10.57b}$$

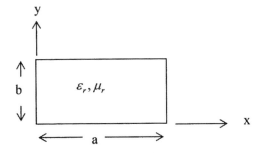

Figure 10.8 Geometry of a rectangular waveguide.

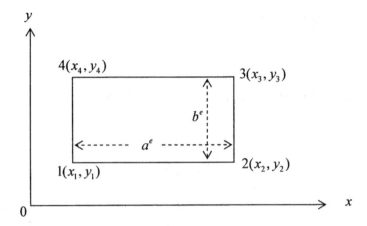

Figure 10.9 Node-based rectangular element. The nodes are numbered in the counterclockwise direction.

$$\varphi_2^e = a' + b'x_2 + c'y_2 + d'x_2y_2 \qquad (10.57c)$$

$$\varphi_3^e = a' + b'x_3 + c'y_3 + d'x_3y_3 \qquad (10.57d)$$

$$\varphi_4^e = a' + b'x_4 + c'y_4 + d'x_4y_4 \qquad (10.57e)$$

Solving for the coefficients and substituting back into (10.57a), the distribution inside a rectangular element is expressed as

$$\varphi^e(x, y) = \sum_{i=1}^{4} \varphi_i^e N_i^e(x, y) \qquad (10.58)$$

where the basis functions $N_i^e(x, y)$ are obtained as

$$N_1^e(x, y) = \frac{(x_2 - x)(y_4 - y)}{\Box} \qquad (10.59a)$$

$$N_2^e(x, y) = \frac{(x - x_1)(y_3 - y)}{\Box} \qquad (10.59b)$$

$$N_3^e(x, y) = \frac{(x - x_4)(y - y_2)}{\Box} \qquad (10.59c)$$

$$N_4^e(x, y) = \frac{(x_3 - x)(y - y_1)}{\Box} \qquad (10.59d)$$

The basis functions are zero outside the element. The symbol $\Box$ denotes the area of the element, and is given by

$$\square = a^e b^e = (x_2 - x_1)(y_3 - y_2) = (x_3 - x_4)(y_4 - y_1), \text{ and so forth}$$

$$(10.60)$$

The basis functions are plotted in Figure 10.10, and are so defined that $N_i^e(x, y)$ are equal to unity at the ith node and taper linearly to zero at the other three nodes; that is, $N_i^e(x_j, y_j) = \delta_{ij}$, for $i, j = 1, 2, 3, 4$. The basis functions are nonorthogonal. Also, $\sum_{i=1}^{4} N_i^e(x, y) = 1$.

For approximate FEM analysis, let us divide the waveguide geometry of Figure 10.8 into four elements of same size $(a^e = a/2, b^e = b/2)$ as shown in Figure 10.11. The discretization of the geometry results in discretization of the functional also. One may therefore write

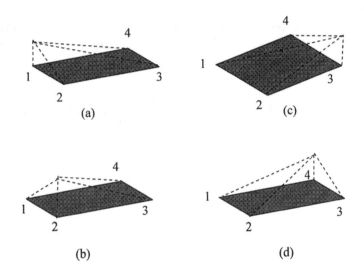

Figure 10.10   Illustration of the node based basis functions $N_i^e(x, y)$ for rectangular element: (a–d) linear basis functions N_1^e, N_2^e, N_3^e, and N_4^e.

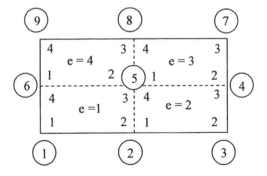

Figure 10.11   Discretization of rectangular waveguide cross-section into four equal rectangular elements of size $a/2 \times b/2$.

$$F(\varphi) = \sum_{e=1}^{N_e} F^e(\varphi^e), \qquad N_e = 4 \tag{10.61}$$

where from (10.56c)

$$F^e(\varphi^e) = \frac{1}{2} \iint \left[\left(\frac{\partial \varphi^e}{\partial x} \right)^2 + \left(\frac{\partial \varphi^e}{\partial y} \right)^2 \right] dx\,dy \tag{10.62}$$

$$- \frac{1}{2} \gamma^2 \iint (\varphi^e)^2 \, dx\,dy + \iint g\varphi^e \, dx\,dy$$

The integration is carried out over the element e.

Element Matrix
The next step in the FEM solution process is to substitute expansion (10.58) into the functional and take its derivative with respect to each of the unknowns φ_j^e yielding

$$\frac{\partial F^e}{\partial \varphi_i^e} = \sum_{j=1}^{4} \varphi_j^e \left(\iint \left\{ \frac{\partial N_i^e}{\partial x} \frac{\partial N_j^e}{\partial x} + \frac{\partial N_i^e}{\partial y} \frac{\partial N_j^e}{\partial y} \right\} dx\,dy - \gamma^2 \iint N_i^e N_j^e \, dx\,dy \right)$$

$$+ \iint g N_i^e \, dx\,dy \tag{10.63}$$

$$e = 1, 2, 3, \ldots N_e, \qquad i = 1, 2, 3, 4$$

Expressing this set of equations in the matrix form one obtains

$$\left[\frac{\partial F^e}{\partial \varphi^e} \right] = [A^e][\varphi^e] + [g^e], \qquad e = 1, 2, 3, \ldots N_e \tag{10.64}$$

where

$$[\varphi^e] = \begin{bmatrix} \varphi_1^e & \varphi_2^e & \varphi_3^e & \varphi_4^e \end{bmatrix}^t, \; [g^e] = \begin{bmatrix} g_1^e & g_2^e & g_3^e & g_4^e \end{bmatrix}^t \tag{10.65}$$

and

$$[A^e] = \begin{bmatrix} A_{11}^e & A_{12}^e & A_{13}^e & A_{14}^e \\ A_{21}^e & A_{22}^e & A_{23}^e & A_{24}^e \\ A_{31}^e & A_{32}^e & A_{33}^e & A_{34}^e \\ A_{41}^e & A_{42}^e & A_{43}^e & A_{44}^e \end{bmatrix} \tag{10.66}$$

$[A^e]$ is called the *element matrix*, and its entries, as obtained from (10.63), are

$$A_{ij}^e = \int\int \left\{ \frac{\partial N_i^e}{\partial x}\frac{\partial N_j^e}{\partial x} + \frac{\partial N_i^e}{\partial y}\frac{\partial N_j^e}{\partial y} \right\} dx\,dy - \gamma^2 \int\int N_i^e N_j^e \, dx\,dy \qquad i, j = 1, 2, 3, 4$$

$$\text{(10.67)}$$

$$g_i^e = \int\int g N_i^e \, dx\,dy \qquad i = 1, 2, 3, 4 \tag{10.68}$$

Since the basis functions are linear, the integrations in (10.67) can be carried out analytically [5, p. 149], while that in (10.68) can also be performed analytically provided $g(x, y)$ is assumed to be constant over the element. If we set $g(x, y) \approx p^e$, where p^e is the value of $g(x, y)$ at the mid-point of element e, one obtains

$$A_{ii}^e = \frac{1}{2}\left(\frac{b^e}{a^e} + \frac{a^e}{b^e} \right) - \gamma^2 \frac{a^e b^e}{9}, \qquad i = 1, 2, 3, 4 \tag{10.69a}$$

$$A_{12}^e = A_{21}^e = A_{34}^e = A_{43}^e = \frac{1}{3}\left(-\frac{b^e}{a^e} + \frac{a^e}{2b^e} \right) - \gamma^2 \frac{a^e b^e}{18} \tag{10.69b}$$

$$A_{23}^e = A_{32}^e = A_{41}^e = A_{14}^e = \frac{1}{3}\left(\frac{b^e}{2a^e} - \frac{a^e}{b^e} \right) - \gamma^2 \frac{a^e b^e}{18} \tag{10.69c}$$

$$A_{24}^e = A_{42}^e = A_{31}^e = A_{13}^e = -\frac{1}{6}\left(\frac{b^e}{a^e} + \frac{a^e}{b^e} \right) - \gamma^2 \frac{a^e b^e}{36} \tag{10.69d}$$

$$g_i^e = p^e \int\int N_i^e \, dx\,dy = \frac{p^e a^e b^e}{4} \qquad i = 1, 2, 3, 4 \tag{10.69e}$$

The dimensions a^e and b^e are defined in Figure 10.9. It may be noted that the grouping of the matrix elements in (10.69) is determined by the distance between the nodes.

10.3.3 Element Matrix for Triangular Elements

The triangular element is a versatile element because it can be used to model an arbitrary shaped geometry especially if elements of different sizes are used. The element matrix for a triangular element may be obtained in a manner similar to that for the rectangular element.

Basis Functions for Triangular Elements
Consider the triangular element shown in Figure 10.12. The nodes are numbered as before in the counterclockwise direction.

The potential distribution inside a triangle may be described by

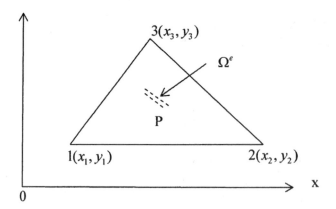

Figure 10.12 Node-based triangular element. The nodes are numbered in the counterclockwise direction.

$$\varphi^e(x, y) = \sum_{i=1}^{3} \varphi_i^e N_i^e(x, y) \tag{10.70}$$

where the linear basis functions $N_i^e(x, y)$ are given as [5]

$$N_i^e(x, y) = \begin{cases} \dfrac{1}{2\Delta^e}(a_i + b_i x + c_i y) & (x, y) \in \Omega^e \\ 0 & \text{otherwise} \end{cases} \tag{10.71}$$

Here Ω^e denotes the surface of the element e. The coefficients a_i, b_i, c_i are introduced for conciseness in writing, and are described in terms of nodal coordinates as

$$a_i = x_j y_k - x_k y_j, \ b_i = y_j - y_k, \ c_i = x_k - x_j \qquad i, j, k = 1, 2, 3 \tag{10.72}$$

where (x_j, y_j) refers to the coordinates of node j as shown in Figure 10.12. The indices (i, j, k) in (10.72) follow the cyclic rule (e.g., $a_1 = x_2 y_3 - x_3 y_2$, $a_2 = x_3 y_1 - x_1 y_3$, $b_1 = y_2 - y_3$, $c_1 = x_3 - x_2$, and so forth). The area of the triangle Δ^e is given by

$$\Delta^e = \frac{1}{2} \begin{vmatrix} 1 & x_1 & y_1 \\ 1 & x_2 & y_2 \\ 1 & x_3 & y_3 \end{vmatrix} = \frac{1}{2}[(x_2 - x_1)(y_3 - y_1) - (x_3 - x_1)(y_2 - y_1)] \tag{10.73}$$

The basis functions are illustrated in Figure 10.13. As seen there the basis functions $N_i^e(x, y)$ are equal to unity at the ith node and taper linearly to zero at the other two nodes. Also, $\sum_{i=1}^{3} N_i^e(x, y) = 1$. A linear combination of the basis functions with assumed nodal values is shown in Figure 10.13(d).

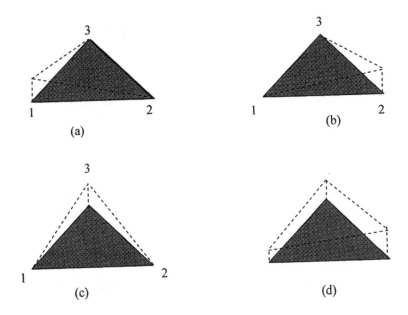

Figure 10.13   Illustration of the node based basis functions $N_i^e(x, y)$ for a triangular element, and the linear approximation: (a–c) linear basis functions N_1^e, N_2^e, and N_3^e; and (d) linear approximation $\Sigma \phi_i^e N_i^e(x, y)$.

Element Matrix
The element matrix may be obtained by following the procedure described earlier for the rectangular element. Specifically, the basis functions (10.71) may be used in (10.63) except that the summation is over three nodes for the triangular element. The elemental matrix $[A^e]$ and the excitation vector $[g^e]$ are given by

$$A_{ij}^e = \int\int \left\{ \frac{\partial N_i^e}{\partial x} \frac{\partial N_j^e}{\partial x} + \frac{\partial N_i^e}{\partial y} \frac{\partial N_j^e}{\partial y} \right\} dx\, dy - \gamma^2 \int\int N_i^e N_j^e \, dx\, dy \qquad i, j = 1, 2, 3$$

$$(10.74)$$

$$g_i^e = \int\int g N_i^e \, dx\, dy \qquad i = 1, 2, 3 \qquad (10.75)$$

Since the basis functions are linear, the integrations can be performed analytically. One obtains

$$\int\int_{\Omega^e} \left\{ \frac{\partial N_i^e}{\partial x} \frac{\partial N_j^e}{\partial x} + \frac{\partial N_i^e}{\partial y} \frac{\partial N_j^e}{\partial y} \right\} dx\, dy = \frac{1}{4\Delta^e} \left[b_i^e b_j^e + c_i^e c_j^e \right] \qquad (10.76a)$$

$$\iint\limits_{\Omega^e} N_i^e N_j^e \, dx \, dy = \frac{\Delta^e}{12} (1 + \delta_{ij}) \qquad (10.76b)$$

or

$$A_{ij}^e = \frac{1}{4\Delta^e} \left[b_i^e b_j^e + c_i^e c_j^e \right] - \gamma^2 \frac{\Delta^e}{12} (1 + \delta_{ij}), \qquad i, j = 1, 2, 3 \qquad (10.77)$$

where the Kronecker delta

$$\delta_{ij} = \begin{cases} 1 & \text{for } i = j \\ 0 & \text{otherwise} \end{cases}$$

It can be shown that for any row or column of the matrix

$$\sum_i \left(b_i^e b_j^e + c_i^e c_j^e \right) = \sum_j \left(b_i^e b_j^e + c_i^e c_j^e \right) = 0 \qquad (10.78a)$$

and therefore

$$\sum_i A_{ij}^e = \sum_j A_{ij}^e = \gamma^2 \frac{\Delta^e}{3} \qquad (10.78b)$$

Equation (10.78) may be used to verify the matrix element calculations.

The matrix $[A^e]$ can therefore be evaluated in closed form. Depending on the form of function $g(x, y)$, numerical integration may be required to evaluate the coefficients g_i^e in (10.75). The matrix for the triangular element may therefore be written as

$$[A^e] = \begin{bmatrix} A_{11}^e & A_{12}^e & A_{13}^e \\ A_{21}^e & A_{22}^e & A_{23}^e \\ A_{31}^e & A_{32}^e & A_{33}^e \end{bmatrix} \qquad (10.79)$$

10.3.4   Assembly of Elements and System Equations

The next step in FEM analysis is to assemble equations (10.64) for all the elements in the domain. Assembly of the element matrices to start with simply implies summation,

$$\frac{\partial F}{\partial \varphi_i} = \sum_{e=1}^{N_e} \frac{\partial F^e}{\partial \varphi^e} = \sum_{e=1}^{N_e} ([A^e][\varphi^e] + [g^e]) \qquad (10.80)$$

In the process of assembly we apply the continuity of $\varphi(x, y)$ at the nodes which are shared by a number of elements. The second aspect of this procedure is to change from local node numbering to the global node numbering. The assembly of element matrices results in a matrix whose size is different from that of the elemental matrix.

The assembly procedure can be simply implemented if the element vector $[\varphi^e]$ is replaced by the global vector $[\varphi]$, and $[A^e]$ and $[g^e]$ are augmented to the global size by adding zeros. If the augmented matrices denoted as $[\overline{A}^e]$ and $[\overline{g}^e]$ are employed for assembly, (10.80) is replaced by

$$\frac{\partial F}{\partial \varphi_i} = \sum_{e=1}^{N_e} ([\overline{A}^e][\varphi] + [\overline{g}^e]) \triangleq [A][\varphi] + [g] \tag{10.81}$$

Now, we apply the constraints arising out of sharing of global nodes by the surrounding elements. It is called continuity condition, and means that the value of φ at the common node should be the same irrespective of the element to which it belongs.

The assembly procedure for the waveguide geometry of Figure 10.11 is illustrated next. The global and local node numbers are also shown in the figure. The global nodes are circled to distinguish them from local nodes. The elements are joined together at node 5.

The matrix equation for the elements is described by

$$\begin{bmatrix} A_{11}^e & A_{12}^e & A_{13}^e & A_{14}^e \\ A_{21}^e & A_{22}^e & A_{23}^e & A_{24}^e \\ A_{31}^e & A_{32}^e & A_{33}^e & A_{34}^e \\ A_{41}^e & A_{42}^e & A_{43}^e & A_{44}^e \end{bmatrix} \begin{bmatrix} \varphi_1^e \\ \varphi_2^e \\ \varphi_3^e \\ \varphi_4^e \end{bmatrix} + \begin{bmatrix} g_1^e \\ g_2^e \\ g_3^e \\ g_4^e \end{bmatrix}, e = 1, 2, 3, 4 \tag{10.82}$$

By the inspection of Figure 10.11, the continuity conditions are as follows:

$$\varphi_2^1 = \varphi_1^2 = \varphi_2; \ \varphi_3^2 = \varphi_2^3 = \varphi_4; \ \varphi_4^3 = \varphi_3^4 = \varphi_8; \ \varphi_1^4 = \varphi_1^1 = \varphi_6; \ \varphi_3^1 = \varphi_4^2 = \varphi_1^3 = \varphi_2^4 = \varphi_5 \tag{10.83}$$

These conditions can be implemented by replacing the local node values φ_i^e in (10.82) by the corresponding global notation (without superscript). One obtains for $e = 1, 2$

$$\begin{bmatrix} A_{11}^1 & A_{12}^2 & A_{13}^3 & A_{14}^4 \\ A_{21}^1 & A_{22}^1 & A_{23}^1 & A_{24}^1 \\ A_{31}^1 & A_{32}^1 & A_{33}^1 & A_{34}^1 \\ A_{41}^1 & A_{42}^1 & A_{43}^1 & A_{44}^1 \end{bmatrix} \begin{bmatrix} \varphi_1 \\ \varphi_2 \\ \varphi_5 \\ \varphi_6 \end{bmatrix} + \begin{bmatrix} g_1^1 \\ g_2^1 \\ g_3^1 \\ g_4^1 \end{bmatrix} \tag{10.84a}$$

$$
\begin{bmatrix} A^2_{11} & A^2_{12} & A^2_{13} & A^2_{14} \\ A^2_{21} & A^2_{22} & A^2_{23} & A^2_{24} \\ A^2_{31} & A^2_{32} & A^2_{33} & A^2_{34} \\ A^2_{41} & A^2_{42} & A^2_{43} & A^2_{44} \end{bmatrix} \begin{bmatrix} \varphi_2 \\ \varphi_3 \\ \varphi_4 \\ \varphi_5 \end{bmatrix} + \begin{bmatrix} g^2_1 \\ g^2_2 \\ g^2_3 \\ g^2_4 \end{bmatrix} \tag{10.84b}
$$

The ordering of rows follows the local node number sequence. Similar expressions are written down for elements 3 and 4.

Each of the matrices of (10.84) may be arranged in the form $[\overline{A}^e][\varphi] + [\overline{g}^e]$, where $[\overline{A}^e]$ is of size 9×9 and vectors $[\varphi]$ and $[\overline{g}^e]$ are of size 9×1. Since the augmented matrices are derived from elemental matrices whose dimension is only four, zeros are added to fill up where necessary. The augmented matrices for $e = 1, 2$ are given next.

For $e = 1$,

$$
\begin{bmatrix} A^1_{11} & A^1_{12} & 0 & 0 & A^1_{13} & A^1_{14} & 0 & 0 & 0 \\ A^1_{21} & A^1_{22} & 0 & 0 & A^1_{23} & A^1_{24} & 0 & 0 & 0 \\ 0 & 0 & 0 & 0 & 0 & 0 & 0 & 0 & 0 \\ 0 & 0 & 0 & 0 & 0 & 0 & 0 & 0 & 0 \\ A^1_{31} & A^1_{32} & 0 & 0 & A^1_{33} & A^1_{34} & 0 & 0 & 0 \\ A^1_{41} & A^1_{42} & 0 & 0 & A^1_{43} & A^1_{44} & 0 & 0 & 0 \\ 0 & 0 & 0 & 0 & 0 & 0 & 0 & 0 & 0 \\ 0 & 0 & 0 & 0 & 0 & 0 & 0 & 0 & 0 \\ 0 & 0 & 0 & 0 & 0 & 0 & 0 & 0 & 0 \end{bmatrix} \begin{bmatrix} \varphi_1 \\ \varphi_2 \\ \varphi_3 \\ \varphi_4 \\ \varphi_5 \\ \varphi_6 \\ \varphi_7 \\ \varphi_8 \\ \varphi_9 \end{bmatrix} + \begin{bmatrix} g^1_1 \\ g^1_2 \\ 0 \\ 0 \\ g^1_3 \\ g^1_4 \\ 0 \\ 0 \\ 0 \end{bmatrix} \tag{10.85a}
$$

For $e = 2$,

$$
\begin{bmatrix} 0 & 0 & 0 & 0 & 0 & 0 & 0 & 0 & 0 \\ 0 & A^2_{11} & A^2_{12} & A^2_{13} & A^2_{14} & 0 & 0 & 0 & 0 \\ 0 & A^2_{21} & A^2_{22} & A^2_{23} & A^2_{24} & 0 & 0 & 0 & 0 \\ 0 & A^2_{31} & A^2_{32} & A^2_{33} & A^2_{34} & 0 & 0 & 0 & 0 \\ 0 & A^2_{41} & A^2_{42} & A^2_{43} & A^2_{44} & 0 & 0 & 0 & 0 \\ 0 & 0 & 0 & 0 & 0 & 0 & 0 & 0 & 0 \\ 0 & 0 & 0 & 0 & 0 & 0 & 0 & 0 & 0 \\ 0 & 0 & 0 & 0 & 0 & 0 & 0 & 0 & 0 \\ 0 & 0 & 0 & 0 & 0 & 0 & 0 & 0 & 0 \end{bmatrix} \begin{bmatrix} \varphi_1 \\ \varphi_2 \\ \varphi_3 \\ \varphi_4 \\ \varphi_5 \\ \varphi_6 \\ \varphi_7 \\ \varphi_8 \\ \varphi_9 \end{bmatrix} + \begin{bmatrix} 0 \\ g^2_1 \\ g^2_2 \\ g^2_3 \\ g^2_4 \\ 0 \\ 0 \\ 0 \\ 0 \end{bmatrix} \tag{10.85b}
$$

Similar expressions may be written down for elements 3 and 4.

The four sets of equations, represented by (10.85a,b,. . .), correspond to the four elements $e = 1, 2, 3, 4$ of the waveguide geometry. These equations are now

added row-wise; that is, the first rows of (10.85a,b, ...) are added, to produce
the first row of (10.86a), and so on. The system matrix $[A]$ defined as
$[A] = \sum_{e=1}^{N_e} [\overline{A}^e]$ is given by

$$
[A] = \begin{bmatrix}
A_{11}^1 & A_{12}^1 & 0 & 0 & A_{13}^1 & A_{14}^1 & 0 & 0 & 0 \\
A_{21}^1 & A_{22}^1 + A_{11}^2 & A_{12}^2 & A_{13}^2 & A_{23}^1 + A_{14}^2 & A_{24}^1 & 0 & 0 & 0 \\
0 & A_{21}^2 & A_{22}^2 & A_{23}^2 & A_{24}^2 & 0 & 0 & 0 & 0 \\
0 & A_{31}^2 & A_{32}^2 & A_{33}^2 + A_{22}^3 & A_{34}^2 + A_{21}^3 & 0 & A_{23}^3 & A_{24}^3 & 0 \\
A_{31}^1 & A_{32}^1 + A_{41}^2 & A_{42}^2 & A_{43}^2 + A_{12}^3 & A_{33}^1 + A_{44}^2 + A_{11}^3 + A_{22}^4 & A_{34}^1 + A_{21}^4 & A_{13}^3 & A_{14}^3 + A_{23}^4 & A_{24}^4 \\
A_{41}^1 & A_{42}^1 & 0 & 0 & A_{43}^1 + A_{12}^4 & A_{44}^1 + A_{11}^4 & 0 & A_{13}^4 & A_{14}^4 \\
0 & 0 & 0 & A_{32}^3 & A_{31}^3 & 0 & A_{33}^3 & A_{34}^3 & 0 \\
0 & 0 & 0 & A_{42}^3 & A_{41}^3 + A_{32}^4 & A_{31}^4 & A_{43}^3 & A_{44}^3 + A_{33}^4 & A_{34}^4 \\
0 & 0 & 0 & 0 & A_{42}^4 & A_{41}^4 & 0 & A_{43}^4 & A_{44}^4
\end{bmatrix}
$$

$$(10.86a)$$

Element-by-element comparison of left and right sides of (10.86a) describes
how the various elements contribute to the system matrix $[A]$. It shows, for example
that, $A_{25} = A_{23}^1 + A_{14}^2$. This expression can be written down by inspection also.
The matrix element A_{25} denotes coupling between global nodes #2 and #5. The
global node #2 is associated with elements #1 and #2, and global node #5 is
associated with all the elements. *Only those elements contribute to A_{25} which
include both the global nodes #2 and #5.* Therefore, elements #1 and #2 only are
eligible; that is, $A_{25} = A_{ij}^1 + A_{ij}^2$. Now, the subscripts ij are replaced by the element
node numbers corresponding to the global nodes #1 and #5 (Figure 10.11). One
obtains $A_{25} = A_{23}^1 + A_{14}^2$. Similarly, the other entries of matrix $[A]$ can be obtained.
For hand calculations, the A-matrix may be written down by inspection of the
geometry. In order to locate the missing elements, if any, in the assembly process,
one observes that matrices of various elements constituting the system are fully
used in the formation of A-matrix. In the example given above the number of
elements are four, each of size 4×4. Therefore, the matrix in (10.86a) should have
64 nonzero entries. Also, the symmetry property of $[A]$ about the diagonal may
be used to correct the mistakes.

The excitation vector $[g]$ may be similarly expressed, and is obtained as

$$
[g] = \begin{bmatrix} g_1^1 & g_2^1 + g_1^2 & g_2^2 & g_3^2 + g_2^3 & g_3^1 + g_4^2 + g_1^3 + g_2^4 & g_4^1 + g_1^4 & g_3^3 & g_4^3 + g_3^4 & g_4^4 \end{bmatrix}^t
$$

$$(10.86b)$$

The system matrix equation is obtained by setting $[A][\varphi] + [g]$ to zero; that
is,

$$[A][\varphi] + [g] = 0 \tag{10.87}$$

The matrix equation may be studied for the characteristics of the waveguide modes.

Assembly of Triangular Elements

The assembly of triangular elements follows the procedure described earlier for the rectangular elements. Again we consider the waveguide geometry of Figure 10.8 for illustration. The geometry is segmented into four triangular elements as shown in Figure 10.14.

By inspection, the system equation resulting from the assembly of elements of Figure 10.14 is obtained as

$$
\begin{bmatrix}
A_{11}^1 + A_{22}^4 & A_{12}^1 & A_{13}^1 + A_{23}^4 & 0 & A_{21}^4 \\
A_{21}^1 & A_{22}^1 + A_{11}^2 & A_{23}^1 + A_{13}^2 & A_{12}^2 & 0 \\
A_{31}^1 + A_{32}^4 & A_{32}^1 + A_{31}^2 & \sum_{i=1}^{4} A_{33}^i & A_{32}^2 + A_{31}^3 & A_{32}^3 + A_{31}^4 \\
0 & A_{21}^2 & A_{23}^2 + A_{13}^3 & A_{22}^2 + A_{11}^3 & A_{12}^3 \\
A_{12}^4 & 0 & A_{23}^3 + A_{13}^4 & A_{21}^3 & A_{22}^3 + A_{11}^4
\end{bmatrix}
\begin{bmatrix}
\varphi_1 \\ \varphi_2 \\ \varphi_3 \\ \varphi_4 \\ \varphi_5
\end{bmatrix}
+
\begin{bmatrix}
g_1^1 + g_2^4 \\
g_2^1 + g_1^2 \\
\sum_{i=1}^{4} g_3^i \\
g_2^2 + g_1^3 \\
g_2^3 + g_1^4
\end{bmatrix}
=
\begin{bmatrix}
0 \\ 0 \\ 0 \\ 0 \\ 0
\end{bmatrix}
\tag{10.88}
$$

10.3.5 Capacitance of a Parallel Plate Capacitor

Let us analyze parallel plate capacitor geometry using FEM for its capacitance per unit length. The geometry is shown in Figure 10.15 and is partially filled with dielectric. The magnetic walls are introduced at the side walls to convert an open region problem into a closed one and thereby reduce the size of the problem. Through this example we can test our skills for the assembly procedure, inclusion of dielectric inhomogeneity in the analysis, and implementation of boundary conditions.

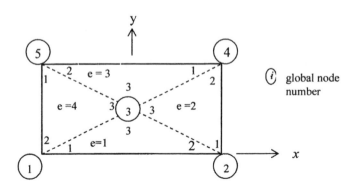

Figure 10.14 Segmentation of the rectangular waveguide geometry of Figure 10.8 into four triangular elements. Local and global node numbers are shown.

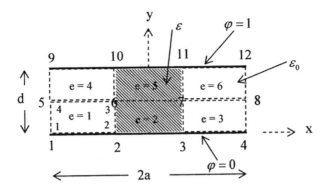

Figure 10.15   An inhomogeneously filled parallel plate capacitor segmented into rectangular elements. Each of the elements has homogeneous dielectric.

Solution.   The potential distribution inside the capacitor is given by the solution of Laplace equation

$$\nabla^2 \varphi = 0 \tag{10.89}$$

subject to the following boundary conditions

$$\varphi = 0 \text{ at } y = 0; \qquad \varphi = 1\text{V at } y = d \tag{10.90a}$$

and magnetic walls at the sides; that is,

$$\frac{\partial \varphi}{\partial x} = 0 \qquad \text{at } x = \pm a \tag{10.90b}$$

The solution for this capacitor problem is well known. With magnetic walls at the sides, there is no fringing field. The electric field is normal to the plates everywhere inside the capacitor region. Therefore, the capacitance per unit length is the sum of the capacitances contributed by each of the three portions and is given by $(2\epsilon_0 + \epsilon)2a/(3d)$; the width of each of the capacitors is $2a/3$. Since the potential varies linearly from 0V at the lower plate to +1V at the upper plate, the potential is expected to be constant at any plane defined by $y = $ constant inside the capacitor. We shall verify these observations through FEM analysis of the capacitor.

The functional for the Laplace equation is the energy functional. The energy per unit length along the z-direction is given by (Table 9.2)

$$F(\varphi) = \frac{1}{2} \iint\limits_{s} \epsilon |\nabla \varphi|^2 \, ds \tag{10.91}$$

where s is the cross-sectional area of the geometry. For the analysis, let us divide the capacitor into six rectangular elements each of size $2a/3 \times d/2$ as shown in

Figure 10.15. The discretization ensures that the dielectric is homogeneous in each of the elements.

The element numbers and the global nodes are shown in the figure. The local node numbering sequence for all the elements is assumed to be the same, and is shown for element #1 only.

The element matrix is given by (10.66) with entries derived from (10.69) by setting $k_0 = 0$ and multiplying each of them by ϵ; that is,

$$A_{ii}^e = \frac{\epsilon}{3}\left(\frac{b^e}{a^e} + \frac{a^e}{b^e}\right), \qquad i = 1, 2, 3, 4 \tag{10.92a}$$

$$A_{12}^e = A_{21}^e = A_{34}^e = A_{43}^e = \frac{\epsilon}{3}\left(-\frac{b^e}{a^e} + \frac{a^e}{2b^e}\right) \tag{10.92b}$$

$$A_{23}^e = A_{32}^e = A_{41}^e = A_{14}^e = \frac{\epsilon}{3}\left(\frac{b^e}{2a^e} - \frac{a^e}{b^e}\right) \tag{10.92c}$$

$$A_{24}^e = A_{42}^e = A_{31}^e = A_{13}^e = -\frac{\epsilon}{6}\left(\frac{b^e}{a^e} + \frac{a^e}{b^e}\right) \tag{10.92d}$$

The matrix equation for the problem is defined as $[A][\varphi] = 0$. Since the number of global nodes is 12, the matrix $[A]$ is 12×12 and the potential vector $[\varphi]$ is 12×1.

It will help reduce the algebra if we start the analysis with the system matrix and make use of symmetry and boundary conditions to reduce the size of the problem.

Boundary Conditions. We first apply the boundary conditions at the lower plate. The condition $\varphi_1 = \varphi_2 = \varphi_3 = \varphi_4 = 0$ reduces the φ vector to $[\varphi] = [\varphi_5, \varphi_6, \ldots \varphi_{11}, \varphi_{12}]^t$. The matrix equation $[A][\varphi] = 0$ therefore becomes

$$
\begin{bmatrix}
A_{5,5} & A_{5,6} & A_{5,7} & A_{5,8} & A_{5,9} & A_{5,10} & A_{5,11} & A_{5,12} \\
A_{6,5} & * & * & * & * & * & * & A_{6,12} \\
A_{7,5} & * & * & * & * & * & * & A_{7,12} \\
A_{8,5} & * & * & * & * & * & * & A_{8,12} \\
A_{9,5} & * & * & * & * & * & * & A_{9,12} \\
A_{10,5} & * & * & * & * & * & * & A_{10,12} \\
A_{11,5} & * & * & * & * & * & * & A_{11,12} \\
A_{12,5} & A_{12,6} & A_{12,7} & A_{12,8} & A_{12,9} & A_{12,10} & A_{12,11} & A_{12,12}
\end{bmatrix}
\begin{bmatrix}
\varphi_5 \\ \varphi_6 \\ \varphi_7 \\ \varphi_8 \\ \varphi_9 \\ \varphi_{10} \\ \varphi_{11} \\ \varphi_{12}
\end{bmatrix}
=
\begin{bmatrix}
0 \\ 0 \\ 0 \\ 0 \\ 0 \\ 0 \\ 0 \\ 0
\end{bmatrix}
\tag{10.93}
$$

Next we apply the boundary conditions at the upper plate, $\varphi_9 = \varphi_{10} = \varphi_{11} = \varphi_{12} = 1$. For this we substitute these values in (10.93) to obtain

$$
\begin{bmatrix}
A_{5,5} & A_{5,6} & A_{5,7} & A_{5,8} \\
A_{6,5} & A_{6,6} & A_{6,7} & A_{6,8} \\
A_{7,5} & A_{7,6} & A_{7,7} & A_{7,8} \\
A_{8,5} & A_{8,6} & A_{8,7} & A_{8,8} \\
A_{9,5} & A_{9,6} & A_{9,7} & A_{9,8} \\
A_{10,5} & A_{10,6} & A_{10,7} & A_{10,8} \\
A_{11,5} & A_{11,6} & A_{11,7} & A_{11,8} \\
A_{12,5} & A_{12,6} & A_{12,7} & A_{12,8}
\end{bmatrix}
\begin{bmatrix}
\varphi_5 \\ \varphi_6 \\ \varphi_7 \\ \varphi_8
\end{bmatrix}
= -
\begin{bmatrix}
A_{5,9} + A_{5,10} + A_{5,11} + A_{5,12} \\
A_{6,9} + A_{6,10} + A_{6,11} + A_{6,12} \\
A_{7,9} + A_{7,10} + A_{7,11} + A_{7,12} \\
A_{8,9} + A_{8,10} + A_{8,11} + A_{8,12} \\
A_{9,9} + A_{9,10} + A_{9,11} + A_{9,12} \\
A_{10,9} + A_{10,10} + A_{10,11} + A_{10,12} \\
A_{11,9} + A_{11,10} + A_{11,11} + A_{11,12} \\
A_{12,9} + A_{12,10} + A_{12,11} + A_{12,12}
\end{bmatrix}
\tag{10.94}
$$

Some of the matrix elements in (10.94) are zero because the associated nodes do not belong to the same element. Setting these matrix elements to zero gives

$$
\begin{bmatrix}
A_{5,5} & A_{5,6} & 0 & 0 \\
A_{6,5} & A_{6,6} & A_{6,7} & 0 \\
0 & A_{7,6} & A_{7,7} & A_{7,8} \\
0 & 0 & A_{8,7} & A_{8,8} \\
A_{9,5} & A_{9,6} & 0 & 0 \\
A_{10,5} & A_{10,6} & A_{10,7} & 0 \\
0 & A_{11,6} & A_{11,7} & A_{11,8} \\
0 & 0 & A_{12,7} & A_{12,8}
\end{bmatrix}
\begin{bmatrix}
\varphi_5 \\ \varphi_6 \\ \varphi_7 \\ \varphi_8
\end{bmatrix}
= -
\begin{bmatrix}
A_{5,9} + A_{5,10} \\
A_{6,9} + A_{6,10} + A_{6,11} \\
A_{7,10} + A_{7,11} + A_{7,12} \\
A_{8,11} + A_{8,12} \\
A_{9,9} + A_{9,10} \\
A_{10,9} + A_{10,10} + A_{10,11} \\
A_{11,10} + A_{11,11} + A_{11,12} \\
A_{12,11} + A_{12,12}
\end{bmatrix}
\tag{10.95}
$$

Symmetry Consideration. The device geometry, including magnetic wall boundary condition, is symmetric about the y-axis. Therefore, the potential distribution is also expected to be symmetric; that is, $\varphi_8 = \varphi_5$ and $\varphi_7 = \varphi_6$. Use of the symmetry conditions results in

$$
\begin{bmatrix}
A_{5,5} & A_{5,6} \\
A_{6,5} & A_{6,6} + A_{6,7} \\
A_{7,8} & A_{7,6} + A_{7,7} \\
A_{8,8} & A_{8,7} \\
A_{9,5} & A_{9,6} \\
A_{10,5} & A_{10,6} + A_{10,7} \\
A_{11,8} & A_{11,6} + A_{11,7} \\
A_{12,8} & A_{12,7}
\end{bmatrix}
\begin{bmatrix}
\varphi_5 \\ \varphi_6
\end{bmatrix}
= -
\begin{bmatrix}
A_{5,9} + A_{5,10} \\
A_{6,9} + A_{6,10} + A_{6,11} \\
A_{7,10} + A_{7,11} + A_{7,12} \\
A_{8,11} + A_{8,12} \\
A_{9,9} + A_{9,10} \\
A_{10,9} + A_{10,10} + A_{10,11} \\
A_{11,10} + A_{11,11} + A_{11,12} \\
A_{12,11} + A_{12,12}
\end{bmatrix}
\tag{10.96}
$$

It can be shown that the following four sets of equations obtained from above give identical values for φ_5 and φ_6:

$$
\begin{bmatrix}
A_{5,5} & A_{5,6} \\
A_{6,5} & A_{6,6} + A_{6,7}
\end{bmatrix}
\begin{bmatrix}
\varphi_5 \\ \varphi_6
\end{bmatrix}
= -
\begin{bmatrix}
A_{5,9} + A_{5,10} \\
A_{6,9} + A_{6,10} + A_{6,11}
\end{bmatrix}
\tag{10.97a}
$$

$$\begin{bmatrix} A_{7,8} & A_{7,7} + A_{7,6} \\ A_{8,8} & A_{8,7} \end{bmatrix} \begin{bmatrix} \varphi_5 \\ \varphi_6 \end{bmatrix} = -\begin{bmatrix} A_{7,10} + A_{7,11} + A_{7,12} \\ A_{8,11} + A_{8,12} \end{bmatrix} \quad (10.97\text{b})$$

$$\begin{bmatrix} A_{9,5} & A_{9,6} \\ A_{10,5} & A_{10,6} + A_{10,7} \end{bmatrix} \begin{bmatrix} \varphi_5 \\ \varphi_6 \end{bmatrix} = -\begin{bmatrix} A_{9,9} + A_{9,10} \\ A_{10,9} + A_{10,10} + A_{10,11} \end{bmatrix}$$

$$(10.97\text{c})$$

$$\begin{bmatrix} A_{11,8} & A_{11,6} + A_{11,7} \\ A_{12,8} & A_{12,7} \end{bmatrix} \begin{bmatrix} \varphi_5 \\ \varphi_6 \end{bmatrix} = -\begin{bmatrix} A_{11,10} + A_{11,11} + A_{11,12} \\ A_{12,11} + A_{12,12} \end{bmatrix}$$

$$(10.97\text{d})$$

We choose (10.97a) to solve for φ_5 and φ_6. The various matrix elements of (10.97a) may be obtained by inspection of Figure 10.15 as

$$\begin{bmatrix} A_{11}^4 + A_{44}^1 & A_{12}^4 + A_{43}^1 \\ A_{21}^4 + A_{34}^1 & A_{33}^1 + A_{44}^2 + A_{11}^5 + A_{22}^4 + A_{12}^5 + A_{43}^2 \end{bmatrix} \begin{bmatrix} \varphi_5 \\ \varphi_6 \end{bmatrix} = -\begin{bmatrix} A_{14}^4 + A_{13}^4 \\ A_{24}^4 + A_{23}^4 + A_{14}^5 + A_{13}^5 \end{bmatrix}$$

$$(10.98)$$

From Figure 10.15 we notice that the elements #1 and #4 are identical in size and dielectric loading. Similarly, elements #2 and #5 are identical. Replacing the superscript 4 by 1, and 5 by 2 gives

$$\begin{bmatrix} A_{11}^1 + A_{44}^1 & A_{12}^1 + A_{43}^1 \\ A_{21}^1 + A_{34}^1 & A_{33}^1 + A_{44}^2 + A_{11}^2 + A_{22}^1 + A_{12}^2 + A_{43}^2 \end{bmatrix} \begin{bmatrix} \varphi_5 \\ \varphi_6 \end{bmatrix} = -\begin{bmatrix} A_{14}^1 + A_{13}^1 \\ A_{24}^1 + A_{23}^1 + A_{14}^2 + A_{13}^2 \end{bmatrix}$$

$$(10.99)$$

Now we use the symmetry and reciprocity properties of A_{ij}^e; that is, $A_{12}^e = A_{21}^e = A_{34}^e = A_{43}^e$, to simplify the above expression as

$$\begin{bmatrix} 2A_{11}^1 & 2A_{12}^1 \\ 2A_{12}^1 & 2A_{11}^1 + 2A_{11}^2 + 2A_{12}^2 \end{bmatrix} \begin{bmatrix} \varphi_5 \\ \varphi_6 \end{bmatrix} = -\begin{bmatrix} A_{14}^1 + A_{13}^1 \\ A_{13}^1 + A_{14}^1 + A_{14}^2 + A_{13}^2 \end{bmatrix}$$

$$(10.100)$$

Also, $A_{ij}^2 = \epsilon_r A_{ij}^1$ (where $\epsilon_r = \epsilon/\epsilon_0$). The matrix equation now reduces to

$$2\begin{bmatrix} A_{11}^1 & A_{12}^1 \\ A_{12}^1 & A_{11}^1(1 + \epsilon_r) + \epsilon_r A_{12}^1 \end{bmatrix} \begin{bmatrix} \varphi_5 \\ \varphi_6 \end{bmatrix} = -\left(A_{13}^1 + A_{14}^1\right)\begin{bmatrix} 1 \\ 1 + \epsilon_r \end{bmatrix}$$

$$(10.101)$$

Solution of the simultaneous equations gives

$$\varphi_5 = \varphi_6 = -\frac{1}{2}\frac{\left(A_{13}^1 + A_{14}^1\right)}{\left(A_{11}^1 + A_{12}^1\right)} \tag{10.102}$$

Substituting for A_{1j}^1, $j = 1, 2, 3, 4$ from (10.69) yields $\varphi_5 = \varphi_6 = 1/2$. This value is true from the expected linear variation of potential between the plates, and the observation that the nodes 5 and 6 are located midway between the plates. The nodal potential vector is therefore given by:

$$[\varphi] = \left[0, 0, 0, 0, \frac{1}{2}, \frac{1}{2}, \frac{1}{2}, \frac{1}{2}, 1, 1, 1, 1\right]^t$$

and the nodal vector for the individual elements is obtained as

$$[\varphi^1] = [\varphi_1 \quad \varphi_2 \quad \varphi_5 \quad \varphi_6]^t = \left[0 \quad 0 \quad \frac{1}{2} \quad \frac{1}{2}\right]^t \tag{10.103a}$$

$$[\varphi^2] = [\varphi^3] = [\varphi^1] \tag{10.103b}$$

$$[\varphi^4] = [\varphi^5] = [\varphi^6] = \left[\frac{1}{2} \quad \frac{1}{2} \quad 1 \quad 1\right]^t \tag{10.103c}$$

Calculation of Capacitance
The nodal values obtained from the FEM analysis is now used to determine the capacitance per unit length of the capacitor. For this we use the energy functional

$$W = F(\varphi) = \frac{1}{2}\iint\limits_S \epsilon|\nabla\varphi|^2 \, ds \tag{10.104}$$

In matrix notation this can be expressed as

$$W = \frac{1}{2}[\varphi]^t[A][\varphi] \tag{10.105}$$

The energy stored in a capacitor is also given by

$$W = \frac{1}{2}CV^2 \tag{10.106}$$

where C is the capacitance per unit length. For $V = 1$ volt, the comparison gives

$$C = [\varphi]^t[A][\varphi] \tag{10.107}$$

This expression is very convenient for implementation on the computer. It requires multiplication of three matrices, which are already stored there.

For hand calculation of capacitance, we may proceed as follows. Since the total energy is additive, we may determine it from the energy stored in the various elements; that is,

$$W = \sum_{e=1}^{6} W^e = \frac{1}{2} \sum_{e=1}^{6} [\varphi^e]^t [A^e][\varphi^e] \tag{10.108}$$

where

$$W^e = \frac{1}{2} \begin{bmatrix} \varphi_1^e & \varphi_2^e & \varphi_3^e & \varphi_4^e \end{bmatrix} \begin{bmatrix} A_{11}^e & A_{12}^e & A_{13}^e & A_{14}^e \\ A_{21}^e & A_{22}^e & A_{23}^e & A_{24}^e \\ A_{31}^e & A_{32}^e & A_{33}^e & A_{34}^e \\ A_{41}^e & A_{42}^e & A_{43}^e & A_{44}^e \end{bmatrix} \begin{bmatrix} \varphi_1^e \\ \varphi_2^e \\ \varphi_3^e \\ \varphi_4^e \end{bmatrix} \tag{10.109}$$

The matrix elements A_{ij}^e are defined in (10.92). All the rectangular elements are identical in size with $a^e = 2a/3$, $b^e = d/2$, leading to

$$A_{ii}^e = \frac{\epsilon}{3} \left(\frac{3d}{4a} + \frac{4a}{3d} \right), \qquad i = 1, 2, 3, 4 \tag{10.110a}$$

$$A_{12}^e = A_{21}^e = A_{34}^e = A_{43}^e = \frac{\epsilon}{3} \left(-\frac{3d}{4a} + \frac{2a}{3d} \right) \tag{10.110b}$$

$$A_{23}^e = A_{32}^e = A_{41}^e = A_{14}^e = \frac{\epsilon}{3} \left(\frac{3d}{8a} - \frac{4a}{3d} \right) \tag{10.110c}$$

$$A_{24}^e = A_{42}^e = A_{31}^e = A_{13}^e = -\frac{\epsilon}{6} \left(\frac{3d}{4a} + \frac{4a}{3d} \right) \tag{10.110d}$$

where $\epsilon = \epsilon_0$ for $e = 1, 3, 4, 6$. The nodal vector $[\varphi^e]$ differs from element to element as described by (10.103).

Use of the elemental matrix and the nodal vector in (10.109) gives rise to

$$W^1 = \frac{1}{2} \begin{bmatrix} 0 & 0 & \frac{1}{2} & \frac{1}{2} \end{bmatrix} [A^1] \begin{bmatrix} 0 \\ 0 \\ \frac{1}{2} \\ \frac{1}{2} \end{bmatrix} = \frac{1}{2} \begin{bmatrix} 0 & 0 & \frac{1}{2} & \frac{1}{2} \end{bmatrix} \frac{1}{2} \begin{bmatrix} A_{13}^1 + A_{14}^1 \\ A_{23}^1 + A_{24}^1 \\ A_{33}^1 + A_{34}^1 \\ A_{43}^1 + A_{44}^1 \end{bmatrix}$$

or

$$W^1 = \left(\frac{1}{2}\right)^2 \left(A_{33}^1 + A_{34}^1 + A_{43}^1 + A_{44}^1\right) = \frac{\epsilon_0 a}{6d} = W^3, \qquad W^2 = \frac{\epsilon a}{6d}$$

$$(10.111)$$

Similarly,

$$W^4 = \frac{1}{2}\begin{bmatrix} \frac{1}{2} & \frac{1}{2} & 1 & 1 \end{bmatrix} [A^1] \begin{bmatrix} \frac{1}{2} \\ \frac{1}{2} \\ \frac{1}{2} \\ 1 \\ 1 \end{bmatrix} \qquad (10.112)$$

$$= \frac{1}{2}\begin{bmatrix} \frac{1}{2} & \frac{1}{2} & 1 & 1 \end{bmatrix} \frac{1}{2} \begin{bmatrix} A_{11}^4 + A_{12}^4 + 2A_{13}^4 + 2A_{14}^4 \\ A_{21}^4 + A_{22}^4 + 2A_{23}^4 + 2A_{24}^4 \\ A_{31}^4 + A_{32}^4 + 2A_{33}^4 + 2A_{34}^4 \\ A_{41}^4 + A_{42}^4 + 2A_{43}^4 + 2A_{44}^4 \end{bmatrix}$$

Using the symmetry and reciprocity properties of the matrix elements we obtain

$$W^4 = \left(\frac{1}{2}\right)^3 \left(10A_{11}^4 + 10A_{12}^4 + 8A_{13}^4 + 8A_{14}^4\right) = \frac{\epsilon_0 a}{6d} = W^6, \qquad W^5 = \frac{\epsilon a}{6d}$$

$$(10.113)$$

Adding the energy stored in all the elements gives

$$W = \sum_{i=1}^{6} W^e = \frac{a}{3d}(2\epsilon_0 + \epsilon) \qquad (10.114)$$

Equating it to $CV^2/2$ (with $V = 1$ volt), the capacitance per unit length is obtained as

$$C = \frac{2a}{3d}(2\epsilon_0 + \epsilon) \qquad (10.115)$$

This value is exactly the same as the analytical value.

Is it surprising that the computed expression for capacitance using FEM is exact; that is, there is no discretization error even with crude discretization used! This accuracy is due to the fact that the expected linear variation of potential between the plates of the capacitor and the modeled variation are matching.

10.3.6   Cutoff Frequency of Waveguide Modes

Consider a rectangular waveguide cross-section with metallic walls as shown in Figure 10.8. Let $a = 1$ cm and $b = 0.5$ cm so that $a/b = 2$. We shall determine the cutoff frequency for the first *TM* mode by segmenting the geometry in four equal rectangular or triangular elements. The wave equation (10.56b) is therefore solved for $\beta = 0$, that is, $\gamma^2 = k^2$.

Solution.   *Case 1, Triangular Elements.* The segmentation of the waveguide geometry into four triangles with 12 local nodes and 5 global nodes is shown in Figure 10.14. The coordinates of global nodes are defined in Table 10.3. Equation (10.72) is now used to determine a_i^e, b_i^e, c_i^e for the elements, and are listed in Table 10.4.

The matrix elements A_{ij}^e, defined by (10.77), are obtained as

$$[A^1] = \begin{bmatrix} -5/8 + k^2/48 & -3/8 + k^2/96 & 1 + k^2/96 \\ -3/8 + k^2/96 & -5/8 + k^2/48 & 1 + k^2/96 \\ 1 + k^2/96 & 1 + k^2/96 & -2 + k^2/48 \end{bmatrix} = [A^3]$$

(10.116a)

$$[A^2] = \begin{bmatrix} -5/8 + k^2/48 & 3/8 + k^2/96 & 1/4 + k^2/96 \\ 3/8 + k^2/96 & -5/8 + k^2/48 & 1/4 + k^2/96 \\ 1/4 + k^2/96 & 1/4 + k^2/96 & -1/2 + k^2/48 \end{bmatrix} = [A^4]$$

(10.116b)

The global matrix (10.88) is now filled up from the element matrices and the matrix equation is obtained as

$$\begin{bmatrix} -5/4 + k^2/24 & -3/8 + k^2/96 & 5/4 + k^2/48 & 0 & 3/8 + k^2/96 \\ -3/8 + k^2/96 & -5/4 + k^2/24 & 5/4 + k^2/48 & 3/8 + k^2/96 & 0 \\ 5/4 + k^2/48 & 5/4 + k^2/48 & -5 + k^2/12 & 5/4 + k^2/48 & 5/4 + k^2/48 \\ 0 & 3/8 + k^2/96 & 5/4 + k^2/48 & -5/4 + k^2/24 & -3/8 + k^2/96 \\ 3/8 + k^2/96 & 0 & 5/4 + k^2/48 & -3/8 + k^2/96 & -5/4 + k^2/24 \end{bmatrix} \begin{bmatrix} \varphi_1 \\ \varphi_2 \\ \varphi_3 \\ \varphi_4 \\ \varphi_5 \end{bmatrix} = \begin{bmatrix} 0 \\ 0 \\ 0 \\ 0 \\ 0 \end{bmatrix}$$

(10.117)

Table 10.3   Node Location Table for Figure 10.14

Global Node #	Coordinates x	y	Global Node #	Coordinates x	y
1	−0.5	0	4	0.5	0.5
2	0.5	0	5	−0.5	0.5
3	0	0.25			

Table 10.4   Coefficients a_i, b_i, c_i for the Elements of Figure 10.14

Element Number	(a_1, b_1, c_1)	(a_2, b_2, c_2)	(a_3, b_3, c_3)
1	1/8, −1/4, −1/2	−1/8,1/4, −1/2	0, 0, 1
2	1/8,1/4, −1/2	−1/8, 1/4, 1/2	1/4, −1/2, 0
3	−1/8, 1/4, 1/2	−1/8, −1/4, 1/2	1/2, 0, −1
4	−1/8, −1/4, 1/2	1/8, −1/4, −1/2	1/4, 1/2, 0

The excitation vector on the right side is a null vector because we are solving an eigenvalue problem. Setting the determinant of the global matrix to zero gives the cutoff wavenumber k_c, from which one obtains $\lambda_c = 2\pi/k_c$.

Eigenvalue Determination
For eigenvalue analysis, the matrix equation may be expressed as

$$[A - \lambda I][x] = [0] \tag{10.118a}$$

or in the generalized form as [10, p. 714]

$$[A - \lambda B][x] = [0] \tag{10.118b}$$

Equation (10.118a) is a special case of (10.118b) in which $[B] = [I]$. We shall prefer (10.118b) because the system matrix can be decomposed as $[A - \lambda B]$. The system matrix (10.117) can be decomposed as

$$
\begin{bmatrix}
-5/4 + k^2/24 & -3/8 + k^2/96 & 5/4 + k^2/48 & 0 & 3/8 + k^2/96 \\
-3/8 + k^2/96 & -5/4 + k^2/24 & 5/4 + k^2/48 & 3/8 + k^2/96 & 0 \\
5/4 + k^2/48 & 5/4 + k^2/48 & -5 + k^2/12 & 5/4 + k^2/48 & 5/4 + k^2/48 \\
0 & 3/8 + k^2/96 & 5/4 + k^2/48 & -5/4 + k^2/24 & -3/8 + k^2/96 \\
3/8 + k^2/96 & 0 & 5/4 + k^2/48 & -3/8 + k^2/96 & -5/4 + k^2/24
\end{bmatrix}
$$

$$
=
\begin{bmatrix}
-5/4 & -3/8 & 5/4 & 0 & 3/8 \\
-3/8 & -5/4 & 5/4 & 3/8 & 0 \\
5/4 & 5/4 & -5 & 5/4 & 5/4 \\
0 & 3/8 & 5/4 & -5/4 & -3/8 \\
3/8 & 0 & 5/4 & -3/8 & -5/4
\end{bmatrix}
+ k^2
\begin{bmatrix}
1/24 & 1/96 & 1/48 & 0 & 1/96 \\
1/96 & 1/24 & 1/48 & 1/96 & 0 \\
1/48 & 1/48 & 1/12 & 1/48 & 1/48 \\
0 & 1/96 & 1/48 & 1/24 & 1/96 \\
1/96 & 0 & 1/48 & 1/96 & 1/24
\end{bmatrix}
\tag{10.119}
$$

Since k is the eigenvalue to be determined, we can write (10.117) in the form

$$[A - \lambda^2 B][\varphi] = [0] \tag{10.120}$$

with $[A]$ and $[B]$ given by (10.119). The above procedure has been adopted in finding the eigenvalues in the software *cutoff_fem_rect.m* and *cutoff_fem_tri.m*.

For the *TM*-modes in the waveguide, the boundary condition is $E_z = 0$ at the boundary walls. Therefore, $\varphi_i = 0$, $i = 1, 2, 4, 5$. Deleting the corresponding rows and columns in the system matrix leaves us with A_{33} only. The expression for the cutoff wavenumber is obtained by setting A_{33} to zero and yields

$$-5 + \frac{k_c^2}{12} = 0 \qquad k_c = \sqrt{60} = 7.746 \tag{10.121}$$

The analytical value of k_c for the TM_{11}-mode in an air-filled waveguide is given as

$$k_c \text{ (analytical)} = \sqrt{\left(\frac{\pi}{a}\right)^2 + \left(\frac{\pi}{b}\right)^2} \qquad (10.122)$$

For $b = a/2$, and $a = 1$ cm, k_c (anal) $= 2.236\,\pi = 7.0246$. The error in k_c is about 10.26%. However, this crude approximation is at a considerable simplification of the problem. The error can be reduced by finer discretization; that is, by increasing the number of triangular elements filling the waveguide geometry.

The cutoff wavenumber for the mode may be used to determine the propagation constant according to (1.35),

$$\beta = \sqrt{k^2 - k_c^2} \qquad k^2 = k_0^2 \epsilon_r \qquad (10.123)$$

The effect of increasing the number of elements on the cutoff wavenumber was studied. The results are given in Table 10.5. The results based on rectangular element division of the waveguide are also given in the table. For eight elements, the waveguide geometry was divided into triangular elements with two divisions along the width and two divisions along the height. For 16 elements, the geometry was divided as shown in Figure 10.16. The software employed for the analysis is *cutoff_fem_tri.m*.

It is observed from Table 10.5 that the error first increases and then decreases with the increase in the number of triangular elements. The computed errors compares with that given in [10, p. 404] for the same discretization size. The error is found to decrease steadily as the number of elements is increased beyond 16. However, the discretization of waveguide geometry into rectangular elements produces better results. It is probably due to the compatibility of geometrical shape of waveguide and the rectangular elements. The number of divisions in the table refers to the divisions of waveguide along the width followed by divisions along the height.

Solution.   *Case 2, Rectangular Elements Discretization.* The division of the waveguide geometry into four rectangular elements of the same size is shown in Figure 10.11. The local and global node numbers are also given there.

Table 10.5   Cutoff Wavenumber $k_c^{11} a$ for the TM_{11}-Mode in a Rectangular Waveguide; $a = 1$ cm, $b = 0.5$ cm

Number of Elements	FEM $\triangle$	FEM $\square$	Exact	% error $\triangle$	% error $\square$
4	7.746	7.746	7.0246	10.266	10.266
No. of divisions	(2×1)	(2×2)			
8	8.944	7.641	7.0246	27.32	8.773
No. of divisions	(2×2)	(4×2)			
16	8.183	7.206	7.0246	16.488	2.586
No. of divisions	(4×2)	(4×4)			

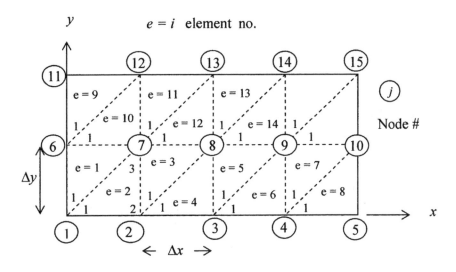

Figure 10.16   Discretization of rectangular waveguide geometry into a uniform mesh and triangular elements.

Following the analysis described above for triangular discretization of rectangular waveguide, the matrix equation reduces to $A_{55}\varphi_5 = 0$. The expression for the cutoff wavenumber is obtained by setting A_{55} of (10.86b) to zero

$$A_{55} = A_{33}^1 + A_{44}^2 + A_{11}^3 + A_{22}^4 = 0 \qquad (10.124)$$

The matrix elements A_{ii}^e are the same for the same sized elements and are given by (10.69a). Setting $A_{ii}^e = 0$ gives

$$\frac{1}{3}\left(\frac{b^e}{a^e} + \frac{a^e}{b^e}\right) - k_c^2\frac{a^e b^e}{9} = 0 \Rightarrow k_c^2 = 3\left(\left(\frac{1}{a^e}\right)^2 + \left(\frac{1}{b^e}\right)^2\right) \qquad (10.125)$$

Also, $a^e = a/2$, $b^e = b/2$; therefore,

$$k_c^2 = 12\frac{a^2 + b^2}{a^2 b^2} \qquad (10.126)$$

For $b = a/2$, and $a = 1$ cm, one obtains $k_c = \sqrt{60}$. When compared with the exact value of k_c (exact) $= 2.236\pi$ for the TM_{11}-mode, the error is about 10.26%. The error reduces to 2.586% for 16 rectangular elements. The software employed is *cutoff_fem_rect.m*.

Comparison with FDM
Let us compare the performance of FDM for the above problem using the rectangular discretization and shown in Figure 10.11 for the FEM solution. The cell size is $\Delta x = a/2$, $\Delta y = b/2$. The finite difference expression for propagation in a waveguide is given by (7.72) for $\Delta x = \Delta y = h$,

$$\varphi_{i+1,j} + \varphi_{i-1,j} + \varphi_{i,j-1} + \varphi_{i,j+1} - (4 - h^2 k^2)\varphi_{i,j} = 0 \qquad (10.127)$$

Since $\Delta x \neq \Delta y$ in this case, the above expression is modified with the help of (7.15) resulting in

$$\varphi_{i,j}\left(\frac{\Delta y}{\Delta x} + \frac{\Delta x}{\Delta y}\right) \approx \frac{1}{2}\left[\frac{\Delta y}{\Delta x}(\varphi_{i+1,j} + \varphi_{i-1,j}) + \frac{\Delta x}{\Delta y}(\varphi_{i,j+1} + \varphi_{i,j-1}) + k_c^2 \varphi_{i,j}\Delta x \Delta y\right]$$

$$(10.128)$$

The boundary condition for the *TM* modes results in $\varphi = 0$ at the boundary nodes. Therefore, $\varphi_i = 0$, $i = 1, 2, 3, 4, 6, 7, 8, 9$. Applying (10.128) to node 5 yields the following expression for the cutoff wavenumber:

$$\frac{ab}{8}k_c^2 = \left(\frac{a}{b} + \frac{b}{a}\right) \qquad (10.129)$$

Using $b = a/2$, and $a = 1$ cm, one obtains $k_c = \sqrt{40}$. When compared with the exact value of k_c (exact) $= 2.236\pi$, the error is about 9.97%. This number compares with the error of about 10.26% computed with the FEM analysis. The relative good performance of FDM is perhaps because of its pulse modeling compared to the linear modeling of FEM.

Results for the cutoff wavenumber for the dominant *TE* mode in rectangular and ridge waveguides are given in Section 10.4.

The manual approach described above for the FEM solution procedure becomes tedious for the devices segmented into a large number of elements. The process should be computerized to avoid human errors and to handle large sized matrices. In this connection look-up tables may be prepared which describe the geometry in the form of global nodes and associated coordinates, discretization of the geometry in elements and material parameters filling the elements, and the boundary conditions. These are discussed in Section 10.4. A simple mesh generator is also described there.

Quiz 10.1. Suggested discretizations of a rectangular waveguide ($a = 2b$) for determining the cutoff wavenumber for the TM_{11}-mode are given in Figure 10.17. Which of these discretizations can be used and why? Which one will you prefer if a higher accuracy of result is desired?

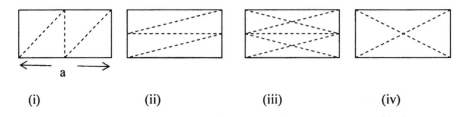

(i)   (ii)   (iii)   (iv)

Figure 10.17 Suggested discretizations of a rectangular waveguide into triangular elements.

Answer.   (iii) and (iv) only because all the nodal values are zero in (i) and (ii). (iii) may yield higher accuracy because of finer discretization along the height of the waveguide.

Quiz 10.2.   Suggested discretizations of a rectangular waveguide ($a = 2b$) in rectangular elements for determining the cutoff wavenumber for TE_{30}-mode are given in Figure 10.18. Which of these discretizations can be used and why? Which one will you prefer if a higher accuracy of result with smallest matrix size is desired?

Answer.   Any of the discretization may be used. (iii) and (iv) should provide the same accuracy and better than provided by (i) and (ii). The matrix size for (iv) is 10×10, whereas it is 15×15 for (iii). Therefore, (iv) is the correct choice.

Quiz 10.3.   A rectangular waveguide is to be analyzed to determine the cutoff wavenumber for the TE_{10}-mode. What is the minimum size of the matrix to be solved for the purpose? You may use considerations like symmetry to reduce the size of the problem. (Hint: Divide the geometry into rectangular elements.)
(i) 8; (ii) 4; (iii) 2; and (iv) 1.

Answer.   (iv). Divide the geometry into two identical rectangular elements about the symmetry plane, and use uniformity of the potential along the height of the waveguide.

Quiz 10.4.   An inhomogeneously filled rectangular waveguide is to be analyzed for the dominant TE/TM mode cutoff. The suggested discretizations in terms of rectangular elements are given in Figure 10.19. The shaded area represents dielectric loading. Select the correct choices with justification.

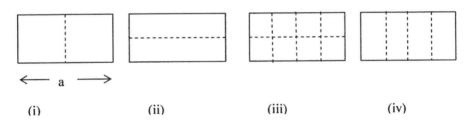

(i)   (ii)   (iii)   (iv)

Figure 10.18 Suggested discretizations of a rectangular waveguide into rectangular elements.

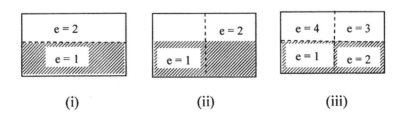

(i)   (ii)   (iii)

Figure 10.19 Suggested discretizations of an inhomogeneously filled rectangular waveguide into rectangular elements.

Answer.   (i) and (iii) for *TE* mode, and (iii) for *TM* mode. The elements should be homogeneously filled and at least one node should have nonzero value.

10.3.7 FEM Analysis of Open Boundary Problems

Open boundary problems are very common in electromagnetics. These include planer transmission lines and circuits based on them, antennas, scatterers, and so on. For analysis of these problems using FEM, boundary walls must be placed to limit the size of the problem domain without affecting its characteristics. One of the commonly used approaches is to apply perfectly matched layer (PML) or analytical absorbing boundary conditions (ABC) at the walls of the device. These techniques are discussed in Chapter 8. PML employs lossy dielectric material to absorb the signal incident on it. ABC simulates reflectionless outgoing wave at the boundary, and is described in the form of a differential equation. For the problems in electrostatics one may use conformal mapping method (Chapter 4) to transform the open region problem into a closed boundary problem, which can be analyzed using FEM. Analysis of planar lines using this hybrid approach has been reported [11].

10.4   Mesh Generation and Node Location Table

The flow chart for the FEM solution process is shown in Figure 10.20. The first block is the mesh generator and is used to generate elements for a given device geometry. The elements are characterized by a set of nodes and edges. This portion of the software assigns the global node numbers, and may also determine their coordinates.

   We shall illustrate the mesh generation procedure by generating mesh for a rectangular geometry of size $a \times b$ and shown in Figure 10.16. The mesh generation procedure is similar to that described in [10, p. 407]. We shall use rectangular grid and triangular elements to discretize the geometry. Let N_x and N_y be the number of divisions in the x and y directions, respectively. Therefore, $\Delta x = a/N_x$, and $\Delta y = b/N_y$. The number of triangular elements and nodes are, therefore, given by

$$N = 2N_x N_y \tag{10.130}$$

$$N_d = (N_x + 1)(N_y + 1) \tag{10.131}$$

We now number the nodes from left to right and bottom to top, as shown in Figure 10.16. The coordinates and the corresponding number for each global node may be assigned in the following manner. We start numbering the nodes with the origin (0, 0) as #1. Next we create arrays: Δx_i, $i = 1, 2, 3, \ldots N_x$, and Δy_j, $j = 1, 2, 3, \ldots N_y$, such that $\Delta x_i = i\Delta x$ and $\Delta y_j = j\Delta y$. The coordinates of the next node is obtained as $x \rightarrow x + \Delta x_i$, while y coordinate remains unchanged at $y = 0$. The process is continued until Δx_i are exhausted. We may start the next row by starting with $x = 0$, $y = y + \Delta y_1$ and increase x until Δx_i are exhausted. This process is repeated until all Δx_i and Δy_j is exhausted (i.e., the last node is reached). The

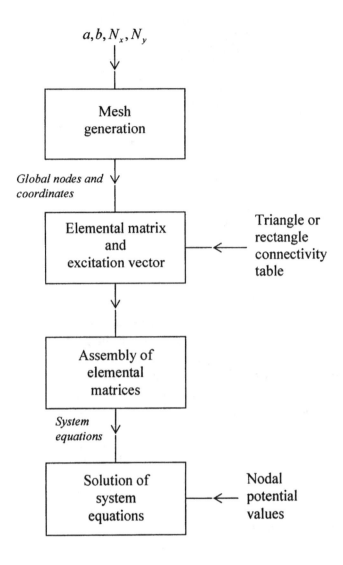

$$a, b, N_x, N_y$$

Mesh
generation

Global nodes and
coordinates

Elemental matrix
and
excitation vector

Triangle or
rectangle
connectivity
table

Assembly of
elemental
matrices

System
equations

Solution of
system
equations

Nodal
potential
values

Figure 10.20   Flow chart for the FEM solution process.

process allows for generating uniform and nonuniform meshes depending on if all Δx_i and Δy_j are equal. The table generated by the mesh generator is called the node location table and specifies the geometry. A code for uniform rectangular mesh generator is included here as *mesh.m*. This software is used to produce the node location Table 10.6 for Figure 10.16. The parameters used are given in the table.

Next, the nodes are joined to create triangular elements taking care that the elements generated do not overlap. A typical combination of nodes is shown in Figure 10.16. The element numbers can be assigned arbitrarily, and a typical assignment is shown in the figure. Each of the triangular elements is assigned local node numbers in counterclockwise fashion, with the choice of the first local node as arbitrary. One choice is shown in the figure. One may prepare this table before-

Table 10.6 Node Location Table for the Geometry of Figure 10.16; $a = 1$, $b = 0.5$, $N_x = 4$, $N_y = 2$, and $\Delta x = 0.25 = \Delta y$

Node Number	x-coordinate	y-coordinate
1.0000	0	0
2.0000	0.2500	0
3.0000	0.5000	0
4.0000	0.7500	0
5.0000	1.0000	0
6.0000	0	0.2500
7.0000	0.2500	0.2500
8.0000	0.5000	0.2500
9.0000	0.7500	0.2500
10.0000	1.0000	0.2500
11.0000	0	0.5000
12.0000	0.2500	0.5000
13.0000	0.5000	0.5000
14.0000	0.7500	0.5000
15.0000	1.0000	0.5000

hand and feed it as an input data. This table is called *Triangle Connectivity Table*, and describes the correspondence between the global node numbers and the local node numbers of the triangular elements. For the geometry of Figure 10.16, the triangle connectivity table is given as Table 10.7. It consists of three arrays corresponding to the local nodes 1, 2, and 3 for the triangles. The element #2 is formed by the global nodes 1, 2, and 7. The corresponding local nodes are 1, 2, and 3. Conversely, the global nodes for element #2 are obtained from the look-up table as 1, 2, 7 in the cyclic order for the local nodes 1, 2, 3. The element label describes the properties of the material filling the element. For lossless dielectric filling, the label denotes the relative dielectric constant, 1 for free space. The material is assumed homogeneous over the element. This table is fed as an input data.

Nodal Potential Table
This table provides the potentials at the fixed nodes as part of the boundary conditions. It may also be used to prescribe initial potential values at the free nodes for speedy convergence. Table 10.8 lists the Dirichlet boundary condition for the waveguide TM modes at the boundary nodes of Figure 10.16. This table may be fed as an input data. A simpler implementation of the Dirichlet condition $\varphi = 0$ is

Table 10.7 Triangle Connectivity Table Between Local and Global Nodes for Figure 10.16

Element Number e	Local Node # 1	Local Node # 2	Local Node # 3	Element Label	e	Local Node # 1	Local Node # 2	Local Node # 3	Element Label
1	1	7	6	1	9	6	12	11	1
2	1	2	7	1	10	6	7	12	1
3	2	8	7	1	11	7	13	12	1
4	2	3	8	1	12	7	8	13	1
5	3	9	8	1	13	8	14	13	1
6	3	4	9	1	14	8	9	14	1
7	4	10	9	1	15	9	15	14	1
8	4	5	10	1	16	9	10	15	1

Table 10.8   Boundary Conditions Table for *TM* mode of Figure 10.16

Global Node#	φ	Node#	φ	Node#	φ	Node#	φ
1	0	4	0	10	0	13	0
2	0	5	0	11	0	14	0
3	0	6	0	12	0	15	0

to prepare the list of free nodes and feed these nodes, implying that the rest of the nodes are fixed nodes with zero value. This approach is followed in the code *cutoff_fem_rect.m.*

An automatic mesh generator for two-dimensional geometries is available in [6, p. 31; 10, p. 414]. The source codes are in FORTRAN, and can divide an arbitrarily shaped geometry in triangular or quadrilateral elements. A source code in FORTRAN for dividing a rectangular geometry into triangular elements and nonuniform mesh is available in [10, p. 409].

Cutoff Frequency for the Dominant TE Mode in Rectangular and Ridge Waveguides
For *TE* modes, the potential function $\varphi = H_z$ is not zero for the boundary nodes, unlike that for *TM* modes. For example, the size of the system matrix is 10×10 for 10 nodes. The software *cutoff_fem_tri.m.* or *cutoff_fem_rect.m.* may be employed to determine the cutoff wavenumber for the dominant *TE* mode in rectangular waveguide. The rectangular mesh is generated automatically as described earlier. The results are summarized in Table 10.9. The number of divisions in the table refers to the divisions of waveguide along the width followed by divisions along the height.

It may be observed from this table that the percentage error between the FEM and analytical values decreases with the increase in the number of elements. Rectangular elements in this case are efficient compared to triangular elements for the number of elements up to eight. However, the accuracy for rectangular elements gets saturated at this number. For 16 elements, the triangular elements result in better accuracy.

The software *cutoff_fem_tri.m* or *cutoff_fem_rect.m.* may be used to determine the cutoff wavenumber for the dominant TE_{10}-mode in a ridge waveguide. The cross-section of ridge waveguide is shown in Figure 10.21. The geometry is discretized in triangular elements as shown there. The number of elements and the triangle

Table 10.9   Cutoff Wavenumber $k_c^{10} a$ for the TM_{10}-Mode in a Rectangular Waveguide; $a = 1m$, $b = 0.5m$

Number of Elements	FEM		Exact	% error	
	Δ	$\square$		Δ	$\square$
4	4.2518	3.7093	3.141	35.34	18.07
No. of division	(2×1)	(2×2)			
8	3.3930	3.2228	3.141	8.004	2.59
No. of divisions	(2×2)	(4×2)			
16	3.2151	3.2228	3.141	2.34	2.59
No. of divisions	(4×2)	(4×4)			

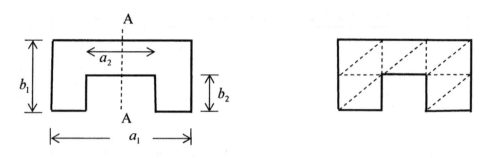

Figure 10.21 Cross-section of a ridge waveguide and its discretization in triangular elements for $a_2 = a_1/3$.

connectivity table differentiate the ridge waveguide geometry from the rectangular waveguide geometry.

The following result is obtained when the ridge waveguide is discretized in 10 equal triangular elements and 12 nodes. The software employed is *cutoff_fem_tri.m*. The values are compared with the results of Hopfer [12].

Waveguide dimensions: $a_1 = 1\text{m}$, $b_1 = 0.5\text{m}$; ridge dimensions: $a_2 = 1/3\text{m}$, $b_2 = 0.25\text{m}$

$$\frac{\lambda_c'}{\lambda_c}(\text{FEM}) = 1.4218 \qquad \frac{\lambda_c'}{\lambda_c}(\text{Hopfer}) = 1.4$$

Here λ_c' is the TE_{10}-mode cutoff wavelength for the ridge waveguide and $\lambda_c = 2a_1$ is the cutoff wavelength for the TE_{10}-mode of the corresponding rectangular waveguide. The ridge waveguide when analyzed using FDM was found to be less efficient. The node density used was 100 (Section 7.3.4).

10.5 Weighted Residual Formulation for FEM

In this section, we describe an alternative formulation for the FEM. It is based on the weighted residual method described in Chapter 9. The details are presented for the one-dimensional scalar wave equation.

The wave equation is given as

$$\frac{d^2 E_y}{dx^2} + k^2 E_y(x) = f(x), \qquad 0 \le x \le x_a \qquad (10.132)$$

The residual R is therefore defined as

$$R(x) = \frac{d^2 E_y}{dx} + k^2 E_y - f(x) \qquad (10.133)$$

Making use of the weighted residual method (Section 9.6), (10.133) is multiplied by the weight function $W(x)$, integrated over the range of x, and set to zero,

$$\int W(x)R(x)dx = \int \frac{d^2E_y}{dx} W(x)dx + k^2 \int E_y(x) W(x)dx - \int f(x) W(x)dx = 0$$

$$(10.134)$$

The integration may be used to reduce the order of derivative in the first term. For this we employ integration by parts, and obtain

$$\int_0^{x_a} W(x) \frac{d^2E_y}{dx} dx = W(x) \frac{dE_y}{dx}\bigg|_{x=0}^{x=x_a} - \int_0^{x_a} \frac{dW}{dx} \frac{dE_y}{dx} dx \qquad (10.135)$$

Substituting (10.135) in (10.134) gives

$$-\int_0^{x_a} \left[\frac{dW}{dx} \frac{dE_y}{dx} - k^2 W(x)E_y(x) + W(x)f(x)\right] dx + W(x) \frac{dE_y}{dx}\bigg|_{x=0}^{x=x_a} = 0$$

$$(10.136)$$

This expression may be called the variational representation of the wave equation. It is also called the weak-form of the differential equation because of the averaging used through the weight function. *It combines in a single mathematical expression the differential equation and the boundary conditions at the end points.* Similar formulations may be arrived at for other types of differential equations.

The subsequent steps for the solution of (10.136) are similar to those used for the method of moments (MoM) and described in the next chapter. The principle advantage of this FEM approach over MoM is that unlike MoM, Green's function is not needed in FEM analysis. This makes FEM a versatile computational tool.

Some of the useful topics such as higher order and edge based elements, three-dimensional analysis, absorbing boundary conditions, and finite element time domain analysis could not be included in the text because of its limited scope. We must leave these topics to advanced texts [5, 8]. Source codes in MATLAB for FEM analysis are available in [5].

10.6 Summary

Finite element method is a versatile, variational method which may be used to analyze complex geometries with arbitrary shape and dielectric inhomogeneity. Two principal formulations for FEM are available; these are based on stationary functional, and weak-wave equation. The emphasis is on functional approach here because of its ready availability for most of the problems. Also, node-based elements are employed for simplicity. The basic steps in the implementation of FEM are discussed in general. Element matrices for triangular and rectangular elements are derived. The assembly of element matrices is an important step in FEM. It is realized in a computer friendly manner by using augmented matrices. The effect of discretization error on the accuracy of the solution is determined. The geometries analyzed are rectangular shaped, for which the analytical solution is available and can be compared with FEM solution. Rectangular elements are used to arrive at the closed-form expressions for the capacitance of an inhomogeneously filled parallel plate capacitor, and cutoff frequency of waveguide modes. The accuracy of FEM is compared with FDM. The software included requires the handy availability of triangle or rectangle connectivity table. The weak-wave equation formulation for FEM is included.

References

[1]   Courant, R., "Variational Methods for a Solution of Problems of Equilibrium and Vibrations," *Bull. Amer. Math. Soc.*, Vol. 49, 1943, pp. 1–23.

[2]   Alett, P. L., A. K. Baharani, and O. C. Zienkiewicz, "Application of Finite Elements to the Solution of Helmholtz's Equation," *Proc. IEE*, Vol. 115, 1968, pp. 1762–1766.

[3]   Silvester, P., "Finite Element Solution of Homogeneous Waveguide Problems," *Alta Frequenza*, Vol. 38, 1969, pp. 313–317.

[4]   Silvester, P., and G. Pelosi, (eds.), *Finite Elements for Wave Electromagnetics: Methods and Techniques*, New York: IEEE Press, 1994.

[5]   Volakis, J. L., A. Chatterjee, and L. C. Kempel, *Finite Element Method for Electromagnetics: Antennas, Microwave Circuits, and Scattering Applications*, Hyderabad: University Press (India) Ltd., 2001.

[6]   Pelosi, G., R. Coccioli, and S. Selleri, *Quick Finite Elements for Electromagnetic Waves*, Norwood, MA: Artech House, 1998.

[7]   Sadiku, M. N. O., "A Simple Introduction to Element Analysis of Electromagnetic Problems," *IEEE Trans. Edu.*, Vol. 32, 1989, pp. 85–93.

[8]   Jin, J., *The Finite Element Method in Electromagnetics*, 2nd ed., New York: Wiley, 2002.

[9]   Peterson, A. F., S. L.Ray, and R. Mittra, *Computational Methods for Electromagnetics*, New York: IEEE Press, 2001.

[10] Sadiku, M. N. O., *Numerical Techniques in Electromagnetics*, 2nd ed., Boca Raton, FL: CRC Press, 2001.

[11] Chang, C. N., Y. C. Wong, and C. H. Chen, "Hybrid Quasistatic Analysis for Multilayer Coplanar Lines," *IEE Proc.-H*, Vol. 138, 1991, pp. 307–312.

[12] Hopfer, S., "The Design of Ridges Waveguides," *IRE Trans.*, Vol. MTT-3, 1955, Figure 5, pp. 20–29.

Problems

P10.1. Write the MATLAB code to solve the following differential equation numerically using FEM:

$$-\frac{d^2f}{dx^2} = 1 + 4x^2 \qquad 0 \leq x \leq 1$$

subject to $f(0) = f(1) = 0$. Compare your solution with the exact value

$$f(x) = \frac{5x}{6} - \frac{x^2}{2} - \frac{x^4}{3}$$

P10.2. Modify the software *fem_oned.m* to solve the resonator problem of Section 10.2. Use $b_m = 1$ at the central segment only and $\Delta x = \lambda_0/10$. The resonance is identified by the flipping over the amplitude of the mode from very high negative value to a very high positive value.

P10.3. Apply FEM to determine the cutoff wavelength for the TE_{10}-mode of an air-filled rectangular waveguide of dimensions $a \times b$, and $b = a/2$. Divide the geometry into eight rectangular elements each of size $a/4 \times b/2$. Half of the wave-guide geometry about the symmetry plane is shown in Figure 10.22. Compare the FEM solution with the analytical expression.

P10.4. Use FEM to determine the capacitance per unit length of the inhomoge-neously filled parallel plate capacitor of Figure 10.23. The capacitor is discretized into four rectangular elements each of size $a \times a/2$. Assume magnetic wall boundary condition at the side walls and $\varphi_1 = \varphi_2 = \varphi_3 = 0V$ and $\varphi_7 = \varphi_8 = \varphi_9 = 1V$. Compare the computed capacitance value with the analytical expression.

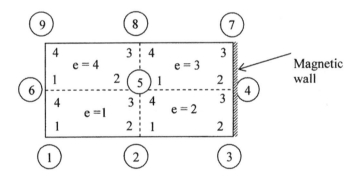

Figure 10.22   Half of the rectangular waveguide geometry about the symmetry plane.

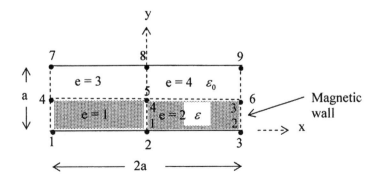

Figure 10.23   A parallel plate capacitor inhomogeneously filled along the height.

P10.5. Analyze the ridge waveguide geometry of Figure 10.21 for TE_{10}-mode cutoff. For this, modify the code *cutoff_fem_tri.m*. Reproduce the results for the normalized cutoff wavelength for the parameters given there.

P10.6. Consider the one-dimensional wave equation (10.47). Discretize the space uniformly and assume a solution of the form $E_y(m) = E_0 e^{-j(\beta m \Delta x)}$.

1. Apply FDM to obtain the expression for β/k.
2. Apply MoM to obtain the expression for β/k.
3. Compare the above solutions with that of FEM, (10.51) and comment.

P10.7. Modify the source code *cutoff_fem_rect.m* for computing the cutoff wavenumber for the dominant TM_{11}-mode for the rectangular waveguide geometry with dimension $a = 1$m and $b = 0.5$m. You may use eight rectangular elements with 15 global nodes. Compare the computed value with the analytical value $k_c = 7.0248$. Plot the electric field in the cross-section.

P10.8. Determine the cutoff wavenumber for TE_{10}-mode in a rectangular waveguide with $a = 1$m, $b = 0.5$m. Use eight square cells and 15 global nodes as shown in Figure 10.20. You may modify the software *cutoff_fem_rect.m* for this purpose. Compare this solution with the solution obtained for 16 triangular elements.

P10.9. The cross-section of a square coaxial line is shown in Figure 10.24. The problem is to be solved for the cutoff wavenumber of the dominant TM mode.

1. Discretize the geometry in triangular elements.
2. Prepare the node location table.
3. Prepare the triangle connectivity table.
4. Prepare the boundary conditions table.

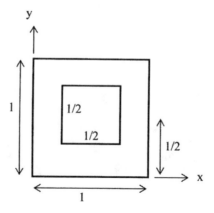

Figure 10.24 Cross-section of a square coaxial line.

CHAPTER 11

Method of Moments

The method of moments (MoM) is a versatile computational method. It can be used to solve differential equations, integral equations, and integro-differential equations. Its use in electromagnetics was pioneered by Harrington [1]. The origin and development of the MoM is very well documented by him [2]. One of the main advantages of this technique lies in its variational nature of the solution, which implies that even if the unknown function (e.g., potential/charge/current) is modeled to first order accuracy, the solution is accurate to the second order [1, p. 18]. The MoM involves a good amount of preprocessing of Maxwell's equations because it makes use of the Green's function. However, the use of Green's function helps solve open region problems such as radiation, scattering, planar circuits, and antennas in an efficient manner. Unlike in other computational methods, the device domain is not discretized, and only the unknown function is discretized in MoM. As a result, this method does not suffer from numerical dispersion and the matrix size is smaller. The limitation of MoM is that this method is generally applied to structures with dimensions of the order of λ. This limitation arises from the computer resources requirements.

11.1 Introduction

The MoM is based on the weighted residual method, which was introduced in Chapter 9, through the solution of differential equation. Here we shall describe the weighted residual method in terms of operator equation, which is symbolic of both the integral and differential equations. We shall first apply the MoM to solve the differential equations with the emphasis on the mathematical concepts and without getting into the physics of the problem. Once these concepts are mastered, the method will be applied to the solution of integral equations.

A general linear, inhomogeneous equation may be described in operator form as

$$L(f) = g \tag{11.1}$$

where L is a linear operator which could be a differential, integral, or integro-differential operator; the function g is known and the corresponding solution f is to be determined. In physical problems, L represents the system, and g the excitation of the system. The unknown function f is expanded in a series of known functions with unknown amplitudes. The amplitudes are determined using a set of *test* functions. The residual corresponding to (11.1) may be defined as

$$R = L(f) - g \tag{11.2}$$

11.1.1 MoM Procedure

The first step in the solution based on MoM is to choose a set of linearly independent basis/expansion functions $f_1, f_2, f_3, \ldots f_m, \ldots$, in the domain of L, and express the unknown function f in the form of a series; that is,

$$f = \sum_n \alpha_n f_n \tag{11.3}$$

where the complex constants α_n are unknown, and are to be determined. This step is called discretization of function f. The basis functions determine the efficiency of MoM. A poor choice may lead to a divergent solution [3]. The basis functions employed may be of entire domain or subdomain type, and these are discussed in Chapter 6.

In the next step we substitute the expansion (11.3) in (11.2). Using the linearity property of L, we can write

$$R = \sum_n \alpha_n L(f_n) - g \tag{11.4}$$

Expression (11.4) is a single equation with a number of unknowns α_n, which can be determined *uniquely* only if we generate sufficient number of simultaneous equations. For this, we choose a set of test functions $w_1, w_2, w_3, \ldots w_m, \ldots$ in the domain of L, and take the inner product of (11.4) with each of w_m. The inner product between two real functions $\Psi(x)$ and $\Phi(x)$ is defined as

$$<\Psi, \Phi> = \int \Psi(x)\Phi(x) \, dx \tag{11.5}$$

The inner product is also called the moment and therefore the name Method of Moments. Physically, (11.5) denotes projecting function Ψ over the function Φ, as explained in Chapter 6. Taking the inner product of residual with weight functions and setting it to zero results in

$$<R, w_m> = 0 \qquad m = 1, 2, 3, \ldots$$

or in expanded form

$$\sum_n \alpha_n <w_m, L(f_n)> = <w_m, g> \qquad m = 1, 2, 3, \ldots \tag{11.6}$$

The set of simultaneous equations represented by (11.6) are independent only if the test functions are linearly independent. The above set of equations can be written in the matrix form as

$$[l][\alpha] = [p] \tag{11.7}$$

where

$$[l] = \begin{bmatrix} <w_1, Lf_1> & <w_1, Lf_2> & \cdots \\ <w_2, Lf_1> & <w_2, Lf_2> & \cdots \\ \cdots & & \cdots & \cdots \end{bmatrix} \tag{11.8a}$$

$$[\alpha] = \begin{bmatrix} \alpha_1 \\ \alpha_2 \\ \cdot \\ \cdot \end{bmatrix} \tag{11.8b}$$

$$[p] = \begin{bmatrix} <w_1, g> \\ <w_2, g> \\ \cdot \\ \cdot \end{bmatrix} \tag{11.8c}$$

Next, the matrix equation (11.7) is solved for the vector $[\alpha]$ by using any of the matrix solution techniques described in Appendix A. Finally, the coefficients α_n are substituted in (11.3) to determine f as

$$f = [f]^t[\alpha] \tag{11.9}$$

where

$$[f]^t = [f_1 \quad f_2 \quad \cdots \quad \cdots] \tag{11.10}$$

is the basis vector.

Remarks. The solution based on the MoM could be an exact solution if the basis set, and the test function set is a complete set. However, it may require an infinite number of functions, and a finite number of functions are used in practice because of the finite memory size of the computer. Therefore, the MoM solution is an approximate solution. *The accuracy and numerical efficiency of the MoM depends on our ingenuity in choosing the appropriate set of basis and test functions.* This point will be illustrated further as we work out some problems.

Considerations for the Choice of Basis Functions

1. Since the solution f satisfies the boundary conditions, the basis functions selected must also satisfy the boundary conditions of the problem if we are solving a differential equation. For integral equation solution, the boundary conditions are built into the integral equation and are therefore not imposed on the basis functions.

2. Since (11.3) is substituted in (11.1) to determine α_n, the basis functions should have continuous derivatives if operator L involves differentiation. If the analysis is carried out in the Fourier domain, the basis functions should be Fourier transformable.

3. The basis functions determine the complexity of the integrals for the matrix elements. These integrals can be determined analytically only if simple basis functions are used.

Various types of basis functions are discussed in Chapter 6.

The considerations for the choice of test functions are discussed next. In this context we shall revisit two popular forms of weighted residual method.

11.1.2 Point Matching and Galerkin's Methods

In general, one can choose any set of test functions in the range of operator L. However, it has been observed that some particular choices of test functions are more popular than others. If Dirac delta functions $\delta(.)$ are used as test functions—that is,

$$w_m = \delta(\mathbf{r} - \mathbf{r_m}) \tag{11.11}$$

the resulting integration for the inner product in (11.6) are avoided and merely becomes a substitution. This choice of test function is called *point matching* or *collocation* because the testing in this sense implies satisfying (11.6) at the test point $\mathbf{r_m}$. Point matching MoM process is a useful simplification. At points other than the match points, (11.6) may not be satisfied. Indeed, it has been found to be so. In between these points one can only hope that the operator equation is not so badly violated that the solution becomes useless and may be called spurious solution. The accuracy of the solution improves as the number of match points is increased.

Remarks.   It may be pointed out that the use of pulse expansion and point matching in MoM solution is equivalent to using the finite difference method [1, p. 153].

In the Galerkin's method, the test function set is taken identical with the basis function set; that is,

$$w_m = f_m \tag{11.12}$$

An advantage of this method is that we do not have to search for test functions once basis functions have been decided. The Galerkin's method is one of the popular choices.

The requirement that the matrix elements should be finite may rule out certain combinations of testing and basis functions. Also, smoothness of the basis/testing functions affects the convergence and accuracy of the numerical solution. The impact of the choice of test functions on the accuracy of numerical solution has been evaluated for specific problems. For *TM* scattering of plane waves by perfectly conducting cylinders, the accuracy is determined by the order of the basis functions and not the test functions [2, Section 5.6].

The inhomogeneous equation of (11.1) contains an excitation term on the right side and the operator part on the left side. The operator describes the system properties in terms of natural modes of the system. The excitation function selects the particular mode (modes) of the system. It is important to study the natural modes for complete information about the system. We next study the natural modes of the operator $-d^2/dx^2$ using MoM with emphasis on the convergence behavior. This study is also called the eigenvalue analysis.

11.1.3 Eigenvalue Analysis Using MoM

Let us consider a simple eigenvalue problem described by

$$-\frac{d^2 f}{dx^2} = \lambda f \qquad 0 \le x \le 1 \tag{11.13}$$

subject to the boundary conditions $f(0) = f(1) = 0$. The above eigenvalue equation may represent the natural modes of vibration of a stretched string fixed at the two ends or a line resonator shorted at the ends. The analytical expressions for the eigenvalues and eigenfunctions are: $\lambda_i = (i\pi)^2$ and $f_i(x) = \sqrt{2}\sin(i\pi x)$, $i = 1, 2, 3, \ldots$

Solution.   Let us attempt a power series solution by choosing the following basis functions:

$$f_n = x - x^{n+1} \qquad n = 1, 2, 3, \ldots N \tag{11.14}$$

Implementing the Galerkin's approach produces the following matrix equation:

$$[l][\alpha] = \lambda[p][\alpha] \tag{11.15}$$

with the matrix elements given by

$$l_{mn} = \int_0^1 (x - x^{m+1})n(n+1)(x^{n-1})\,dx = \frac{mn}{m+n+1} \tag{11.16a}$$

$$p_{mn} = \int_0^1 (x - x^{m+1})(x - x^{n+1})\,dx = \frac{mn(m+n+6)}{3(m+3)(n+3)(m+n+3)} \tag{11.16b}$$

To illustrate the convergence of the numerical solution, let us consider the approximate solutions as the number of basis functions N is increased. For $N = 1$, the matrix element is: $l_{11} = 1/3$ and $p_{11} = 1/30$, and (11.15) gives in this case

$$\frac{1}{3}\alpha_1 = \lambda \frac{1}{30}\alpha_1 \qquad (11.17)$$

Hence our first approximation to λ_1 is $\lambda_1^{(1)} = 10$. The superscript here denotes the value of N. The approximate solution compares well with the exact eigenvalue $\lambda_1 = \pi^2 = 9.8696$. The approximate eigenfunction is given by

$$f_1^{(1)}(x) = \alpha_1(x - x^2) \qquad (11.18)$$

To compare it with the exact normalized eigenfunction $f_1(x) = \sqrt{2}\sin(\pi x)$, we normalize $f_1^{(1)}$ according to

$$\int_0^1 f_1^{(1)}(x)f_1^{(1)}(x)\,dx = 1 \Rightarrow \alpha_1 = \sqrt{30}$$

and

$$f_1^{(1)}(x) = \sqrt{30}(x - x^2) \qquad (11.19)$$

Comparison of this eigenfunction with the exact value $f_1(x) = \sqrt{2}\sin(\pi x)$ shows very little difference over the entire range of x.

Let us now increase the number of basis functions to $N = 2$. Equation (11.15) now becomes

$$\begin{bmatrix} 1/3 & 1/2 \\ 1/2 & 4/5 \end{bmatrix}\begin{bmatrix} \alpha_1 \\ \alpha_2 \end{bmatrix} = \lambda \begin{bmatrix} 1/30 & 1/20 \\ 1/20 & 8/105 \end{bmatrix}\begin{bmatrix} \alpha_1 \\ \alpha_2 \end{bmatrix} \qquad (11.20)$$

The eigenvalues found from the determinant are: $\lambda_1^{(2)} = 10$, $\lambda_2^{(2)} = 42$. To obtain the eigenfunctions corresponding to these values of λ, we first substitute the value of $\lambda_1^{(2)}$ in (11.20) and determine α_1 from normalization; this results in

$$f_1^{(2)}(x) = \sqrt{30}(x - x^2) = f_1^{(1)}(x) \qquad (11.21)$$

Similarly, the eigenfunction corresponding to $\lambda_2^{(2)} = 42$ is obtained as

$$f_2^{(2)}(x) = 3\sqrt{210}(x - x^2) - 2\sqrt{210}(x - x^3) \qquad (11.22)$$

Comparison of this eigenfunction with the exact value $f_2(x) = \sqrt{2}\sin(2\pi x)$ is plotted in Figure 11.1. The agreement for the most part of x appears to be good. Table 11.1 compares the approximate eigenvalues obtained from MoM against the exact analytical solution. It may be noted from this table that the computed eigenvalues $\lambda_i^{(N)}$ are larger than the exact value λ_i for all values of N. This is the property of self-adjoint and positive definite operators like $-d^2/dx^2$ [1]. Also, the

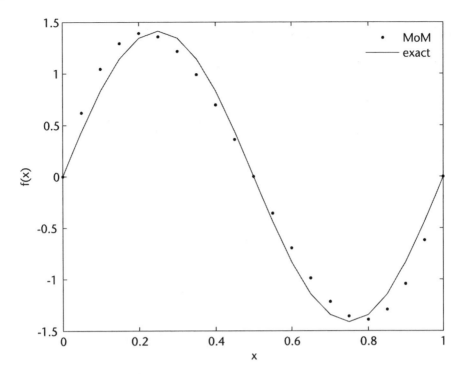

Figure 11.1 Comparison of the MoM solution for $N = 2$ with the exact solution.

Table 11.1 Comparison of Approximate Eigenvalues Based on MoM and Exact Analytical Solution

N	$\lambda_1^{(N)}$	$\lambda_2^{(N)}$	$\lambda_3^{(N)}$	$\lambda_4^{(N)}$
1	10.0000	—	—	—
2	10.0000	42.000	—	—
3	9.8697	42.000	102.133	—
4	9.8697	39.497	102.133	200.583
Exact	9.8696	39.478	88.826	157.914

eigenvalues converge to the exact analytical value as the number of basis functions is increased.

Quiz 11.1.   The following (entire domain or subdomain) basis functions are suggested for expanding the unknown function $f(x)$ for the differential equation

$$-\frac{d^2 f}{dx^2} = 1 + x^2$$ subject to $f(0) = f(1) = 0$. Choose the correct basis functions with justification.

1. $f_n(x) = x - x^n$, $n = 1, 2, 3, \ldots$
2. $f_n(x) = \sin(n\pi x)$, $n = 1, 2, 3, \ldots$
3. Pulse functions $P(x - x_n)$, $n = 1, 2, 3, \ldots$
4. Triangular functions $T(x - x_n)$, $n = 1, 2, 3, \ldots$

Answer.   (1) and (4).

11.2 Solution of Integral Equations Using MoM

The problems in electromagnetics may be formulated as differential equations, or integral equations, or integro-differential equations. The differential equation approach is usually the simpler one and can lead to exact solutions. Integral equation solution usually leads to approximate solutions. The boundary conditions are built into the integral equations rather than imposed through the basis functions. The form of kernel in an integral equation depends on the boundary and interface conditions. We next discuss MoM solution of integral equations.

11.2.1 Integral Equation

An integral equation is one for which the unknown quantity appears under the integral sign. An integral equation of the first kind has the following form:

$$\int I(z')K(z, z')dz' = -E(z) \tag{11.23}$$

where $K(z, z')$ is called the kernel and is known, and $I(z')$ is the unknown function to be determined. The excitation $E(z)$ is known. *The integral equation in electromagnetics represents a boundary condition*. The formulation of integral equation requires the knowledge of Green's function, which is called kernel in (11.23). A well-known example of kernel is the free space Green's function expressed as

$$K = \frac{e^{-jkr}}{4\pi r} \tag{11.24}$$

with r being the distance between the observation point (x, y, z) and the source point (x', y', z'), and is given by

$$r = \sqrt{(x - x')^2 + (y - y')^2 + (z - z')^2} \tag{11.25}$$

As an example of Green's function and the integral equation in electrostatics consider the Poisson equation,

$$\nabla^2 \varphi(\mathbf{r}) = -\frac{\rho(\mathbf{r})}{\epsilon_0 \epsilon_r} \tag{11.26}$$

It has a solution in integral form as

$$\varphi(\mathbf{r}) = \frac{1}{4\pi\epsilon_0 \epsilon_r} \iiint\limits_{V'} \frac{\rho_v(\mathbf{r}')}{|\mathbf{r} - \mathbf{r}'|} dv' \tag{11.27}$$

This expression represents an integral equation if the potential function $\varphi(\mathbf{r})$ is known and we seek the charge density distribution $\rho(\mathbf{r}')$, which produces this potential distribution. The factor $\dfrac{1}{4\pi\epsilon_0\,\epsilon_r}\dfrac{1}{|\mathbf{r}-\mathbf{r}'|}$ is called the Green's function.

The derivation of Green's function is discussed in Chapter 3.

We now illustrate the MoM solution of integral equations.

Example 11.1.   Let us apply the method of moments to obtain the approximate solution for the following integral equation [4]:

$$-\frac{1}{2\pi}\int_{-1}^{1} f(x')\,\ell n|x-x'|\,dx' = \frac{\ell n2}{2} \tag{11.28}$$

and compare the numerical results with the analytical solution

$$f(x) = \frac{1}{\sqrt{1-x^2}} \tag{11.29}$$

Solution.   The given integral equation represents the governing equation for an infinite conducting strip charged to a constant potential if $f(x') = q(x')/\epsilon$. The normalized width of the strip is two units. Let us use pulse expansion functions and point matching in the MoM procedure to determine $f(x)$. The unknown function can be expanded as

$$f(x) = \sum_{n=1}^{N} \alpha_n P_n(x-x_n) \tag{11.30}$$

where α_n are the expansion coefficients and $P_n(.)$ are the pulse functions defined in (6.6). The integral equation (11.28) may therefore be written as

$$\sum_{n=1}^{N} \alpha_n \int_{-1}^{1} P_n(x'-x_n')\,\ell n|x-x'|\,dx' = -\pi\ell n2 \tag{11.31}$$

Consistent with the use of pulse expansion functions we divide the range of x into N equal subintervals of size $h_x = 2/N$. Each of the subintervals is centered at $x_n(=-1+h_x(n-1/2))$. Using (6.6) for the pulse function, we obtain

$$\sum_{n=1}^{N} \alpha_n \int_{x_n-h_x/2}^{x_n+h_x/2} \ell n|x-x'|\,dx' = -\pi\ell n2 \tag{11.32}$$

Point matching at x_m, $m = 1, 2, 3, \ldots N$, is obtained by setting $x = x_m$ and gives

$$\sum_{n=1}^{N} \alpha_n \int_{x_n - h_x/2}^{x_n + h_x/2} \ell n |x_m - x'| \, dx' = -\pi \ell n 2, \qquad m = 1, 2, 3, \ldots N$$

$$(11.33)$$

The above set of equations may be written as

$$\sum_{n=1}^{N} \alpha_n S_{mn} = -\pi \ell n 2 \qquad (11.34)$$

where

$$S_{mn} = \int_{x_n - h_x/2}^{x_n + h_x/2} \ell n |x_m - x'| \, dx' \qquad (11.35)$$

$$= \int_{-h_x/2}^{h_x/2} \ell n |x_m - x_n - x'| \, dx'$$

The integral can be evaluated analytically as

$$-S_{mn} = h_x - \frac{h_x}{2} \ell n \left| (x_m - x_n)^2 - \left(\frac{h_x}{2}\right)^2 \right| - (x_m - x_n) \ell n \frac{\left| x_m - x_n + \frac{h_x}{2} \right|}{\left| x_m - x_n - \frac{h_x}{2} \right|}$$

$$(11.36)$$

Next, we select the center of pulse functions as match points; that is, $x_m = -1 + h_x(m - 1/2)$. Therefore, $x_m - x_n = (m - n) h_x$ and

$$S_{mn} = -h_x \left[1 - \ell n h_x - \frac{1}{2} \ell n \left| (m - n)^2 - \frac{1}{4} \right| \right] + h_x (m - n) \ell n \frac{\left| m - n + \frac{1}{2} \right|}{\left| m - n - \frac{1}{2} \right|}$$

$$(11.37)$$

Also,

$$S_{mm} = h_x \left[-1 + \ell n \left(\frac{h_x}{2}\right) \right] \qquad (11.38)$$

Properties of [S]: It may be observed from (11.37) and (11.38) that:

1. $S_{nm} = S_{mn}$, that is, the matrix is symmetric.
2. The value of S_{mn} depends on $(m-n)$ and note on the individual values of m and n. Therefore, elements along a given diagonal are equal. Such a matrix is called *toeplitz* matrix. From properties (i) and (ii) it can be concluded that there are only N distinct elements in $[S]$, and all elements can be found from the knowledge of first row or first column.
3. The matrix is diagonally dominant because $|S_{mm}| > |S_{mn}|$, $m \neq n$. Usually, computing the inverse of such a matrix or using any other solution method leads to a very stable process.

The set of simultaneous equations (11.34) may be expressed as a matrix equation

$$[S][\alpha] = -\pi \ell n2 [1] \qquad (11.39)$$

where [1] is a column vector with unit entries. The matrix equation (11.39) can be solved using any standard matrix solution technique, and the function $f(x)$ is then given by

$$f(x) = [\alpha]^t [P(x)] \qquad (11.40)$$

The results of the numerical analysis are compared with the analytical solution in Figure 11.2 for $N = 5$, 10, and 15 basis functions. The software used is *integral_eqn.m*. It may be noted from this plot that the computed charge distribution $f(x)$ approaches the analytical solution as the number of basis functions is increased.

The integral equation (11.28) has also been analyzed using entire domain basis functions of the type $(x)^n$. However, it has been observed that the computed solutions are accurate for $N \leq 9$ and the error increases for higher values of N. The condition number $k(N)$ of the matrix $[S]$ is found to increase rapidly and the matrix becomes ill-conditioned.

11.2.2 Static Charge Distribution on a Wire

We consider a cylindrical wire of radius a and length ℓ as shown in Figure 11.3(a). We assume $\ell \gg a$, so that the cylinder can be approximated as a filamentary wire and the effect of the end caps of the cylinder can be neglected. Let the cylinder be charged so that its surface is an equipotential surface with $\phi = V_0$. For the field point on the surface of the wire where $\phi = V_0$, the integral equation satisfying this boundary condition is obtained from (11.27) and is given by

$$V_0 = \frac{1}{4\pi\epsilon_0} \iint_{S'} \frac{\rho_s(\mathbf{r}')}{R} ds' \qquad (11.41)$$

for all points on the surface of the wire described by $\rho = a$, $-\ell/2 \leq z \leq \ell/2$, $0 \leq \varphi \leq 2\pi$ for the field points, and $\rho' = a$, $-\ell/2 \leq z' \leq \ell/2$, $0 \leq \varphi' \leq 2\pi$ for the

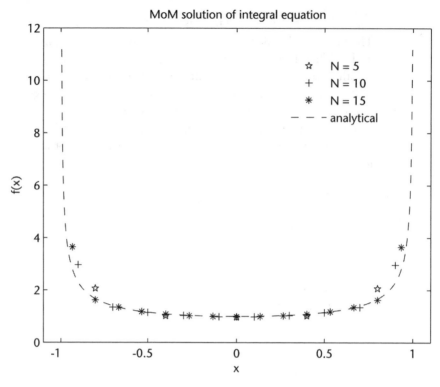

Figure 11.2 Comparison of analytical solution of the integral equation with MoM solution for pulse basis and point matching.

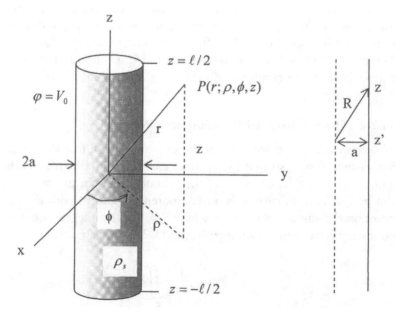

Figure 11.3 (a) A charged conducting wire of radius a and length l with an unknown surface charge density. (b) Simplified model to calculate the distance between the source point and field point. The source point is assumed located at $\rho' = 0$.

source points. Because of the axial symmetry of the wire, we expect the charge density ρ_s to vary with z' and not with φ'. Equation (11.41) can therefore be written as $(ds' = a\,d\varphi'\,dz')$

$$V_0 = \frac{1}{4\pi\epsilon_0} \int\limits_{-\ell/2}^{\ell/2} \rho_s(z') \int\limits_0^{2\pi} \frac{1}{R}\, a\,d\varphi'\,dz' \tag{11.42}$$

where

$$R = |\mathbf{r} - \mathbf{r}'| = \sqrt{2a^2 - 2a^2\cos\varphi' + (z - z')^2} = \sqrt{4a^2\sin^2\left(\frac{\varphi'}{2}\right) + (z - z')^2} \tag{11.43}$$

We have chosen $\varphi = 0$ for convenience, because the potential is independent of φ. Assuming that the wire radius a is small, it can be shown that the integration over φ' can be approximated as [3]

$$\int\limits_0^{2\pi} \left[4a^2\sin^2\left(\frac{\varphi'}{2}\right) + (z - z')^2\right]^{-1/2} d\varphi' \cong 2\pi[a^2 + (z - z')^2]^{-1/2} \tag{11.44}$$

The above approximation amounts to assuming that the potential at $\rho = a$ *arises from the line charge assumed located at* $\rho = 0$ *so that* $R = \sqrt{a^2 + (z - z')^2}$. The simplified model is shown in Figure 11.3(b). This approximation also removes the singularity in the Green's function at $r = r'$. Use of (11.44) in (11.42) gives

$$V_0 = \frac{a}{2\epsilon_0} \int\limits_{-\ell/2}^{\ell/2} \frac{\rho_s(z')\,dz'}{\sqrt{a^2 + (z - z')^2}} \qquad \text{for } -\frac{\ell}{2} \le z \le \frac{\ell}{2} \tag{11.45}$$

The integral equation is to be solved for the unknown charge density $\rho_s(z')$. As a first step, we expand $\rho_s(z')$ into a sum of N linearly independent basis functions,

$$\rho_s(z') = \sum_{n=1}^{N} \alpha_n f_n(z') \tag{11.46}$$

Substituting in (11.45) gives

$$V_0 = \frac{a}{2\epsilon_0} \int\limits_{-\ell/2}^{\ell/2} \sum_{n=1}^{N} \alpha_n \frac{f_n(z')\,dz'}{\sqrt{a^2 + (z - z')^2}} \qquad \text{for } -\frac{\ell}{2} \le z \le \frac{\ell}{2} \tag{11.47}$$

Because the integral operator is linear, we can interchange the order of summation and integration, and obtain

$$\frac{2\epsilon_0 V_0}{a} = \sum_{n=1}^{N} \alpha_n \int_{-\ell/2}^{\ell/2} \frac{f_n(z')\,dz'}{\sqrt{a^2 + (z - z')^2}} \qquad \text{for } -\frac{\ell}{2} \leq z \leq \frac{\ell}{2} \tag{11.48}$$

To simplify numerical calculations we use point matching, and implement it by setting $z = z_m$ at the test points. The test points are taken as equi-spaced points on the wire and are defined as

$$z_m = \frac{\left(m - \frac{1}{2}\right)\ell}{N} - \frac{\ell}{2} \qquad m = 1, 2, \ldots, N \tag{11.49}$$

After point matching, the integral equation takes the following form:

$$\frac{2\epsilon_0 V_0}{a} = \sum_{n=1}^{N} \alpha_n \int_{-\ell/2}^{\ell/2} \frac{f_n(z')\,dz'}{\sqrt{a^2 + (z_m - z')^2}} \qquad m = 1, 2, \ldots N \tag{11.50}$$

Let us denote for conciseness

$$I_{mn} = \frac{a}{2\epsilon_0 V_0} \int_{-\ell/2}^{\ell/2} \frac{f_n(z')\,dz'}{\sqrt{a^2 + (z_m - z')^2}} \tag{11.51}$$

Equation (11.50) thus can be written as

$$1 = \sum_{n=1}^{N} \alpha_n I_{mn} \qquad m = 1, 2, \ldots N \tag{11.52}$$

The set of simultaneous equations may be expressed in matrix form as

$$\begin{bmatrix} 1 \\ 1 \\ 1 \\ \vdots \\ 1 \end{bmatrix} = \begin{bmatrix} I_{11} & I_{12} & I_{13} & \cdots & I_{1N} \\ I_{21} & I_{22} & I_{23} & \cdots & I_{2N} \\ I_{31} & I_{32} & I_{33} & \cdots & I_{3N} \\ \vdots & \vdots & \vdots & \vdots & \vdots \\ I_{N1} & I_{N2} & I_{N3} & \cdots & I_{NN} \end{bmatrix} \begin{bmatrix} \alpha_1 \\ \alpha_2 \\ \alpha_3 \\ \vdots \\ \alpha_N \end{bmatrix}$$

or

$$[1] = [I][\alpha] \tag{11.53}$$

The matrix equation can be solved for the vector $[\alpha]$ using any standard matrix solution technique (see Appendix A).

Evaluation of Matrix Elements.   The evaluation of matrix elements in the MoM solution in general consumes a large amount of processor time. Therefore, it is desirable to reduce this computer time through analytical simplification of the integrals. For this we consider the integral

$$I = \int_{-\ell/2}^{\ell/2} \frac{f_n(z')\,dz'}{\sqrt{a^2 + (z_m - z')^2}} \tag{11.54}$$

It is important to evaluate the integral near $z' = z_m$ more accurately because of near singularity conditions. When $z' = z_m$, the denominator is very small because the radius a is assumed to be small. The integrand, therefore, becomes highly peaked and requires more sampling points (of the order of a few hundred) for accurate evaluation. In order to correct this condition analytically, we use singularity subtraction method of Appendix B and express the integral as

$$I = \int_{-\ell/2}^{\ell/2} \frac{f_n(z') - f_n(z_m) + f_n(z_m)}{\sqrt{a^2 + (z_m - z')^2}}\,dz'$$

or

$$I = f_n(z_m) \int_{-\ell/2}^{\ell/2} \frac{dz'}{\sqrt{a^2 + (z_m - z')^2}} + \int_{-\ell/2}^{\ell/2} \frac{f_n(z') - f_n(z_m)}{\sqrt{a^2 + (z_m - z')^2}}\,dz' \tag{11.55}$$

The first integral can be determined analytically to give

$$I = f_n(z_m) \ln \frac{z_m + \dfrac{\ell}{2} + \sqrt{a^2 + \left(z_m + \dfrac{\ell}{2}\right)^2}}{z_m - \dfrac{\ell}{2} + \sqrt{a^2 + \left(z_m - \dfrac{\ell}{2}\right)^2}} + \int_{-\ell/2}^{\ell/2} \frac{f_n(z') - f_n(z_m)}{\sqrt{a^2 + (z_m - z')^2}}\,dz' \tag{11.56}$$

The integrand is zero at $z' = z_m$ and is smooth elsewhere, and can therefore be evaluated numerically.

Use of Entire Domain Basis Functions

In order to determine the matrix elements of (11.56), we need to decide about the expansion functions $f_s(z)$, the choice of which is critical for the efficiency of MoM solution process. The nature of $\rho_s(z')$ is such that we expect it to be symmetric about $z' = 0$, largest near the wire ends due to the Coulomb forces of repulsion, and smallest in the middle of the wire. The entire domain basis functions in the form of simple power series expansion and satisfying the above conditions may be written as

$$f_n(z') = \left(\frac{z'}{\ell/2}\right)^n \qquad \text{for } -\frac{\ell}{2} \le z' \le \frac{\ell}{2} \tag{11.57}$$

The charge density may therefore be expressed as

$$\rho_s(z') = \sum_{n=0}^{N} \alpha_n \left(\frac{2z'}{\ell}\right)^n \tag{11.58}$$

Computed results for this choice of expansion functions for $\ell = 2$m, $a = 10^{-2}$ m, equally spaced matching points z_m with $N = 6, 12,$ and 20 and $2\epsilon_0 V_0/a = 1$ are shown in Figure 11.4. The software used is *charge_density_wire1.m*. It may be noted from this figure that the use of higher order expansion functions corresponding to $N = 20$ produces the edge singularity better and should therefore lead to a more accurate solution. Also, the edge singularity in charge is described analytically by $1/\sqrt{(\ell/2)^2 - z^2}$.

Use of Subdomain Basis Functions
The use of subdomain basis functions like pulse or linear type makes the evaluation of matrix elements simpler. Let us use pulse expansion functions to model the charge density as,

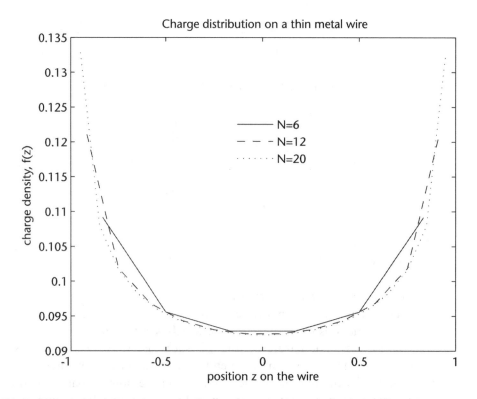

Figure 11.4 Approximate charge density on a conducting wire with length $\ell = 2$m, radius $a = 1$ cm, MoM solution using entire domain expansion functions and point matching.

$$\rho_s(z') = \sum_{n=1}^{N} \alpha_n P_n(z' - z_n) \tag{11.59}$$

where the pulse function has been defined in (6.6). Since $P_n(z' - z_n)$ is constant over each pulse, the integral in (11.54) reduces to [5]

$$I = \int_{z_n - h_z/2}^{z_n + h_z/2} \frac{dz'}{\sqrt{a^2 + (z_m - z')^2}}, \qquad h_z = \frac{\ell}{2N} \tag{11.60}$$

or

$$I = \ln \frac{z_m - z_n + \dfrac{h_z}{2} + \sqrt{a^2 + \left(z_m - z_n + \dfrac{h_z}{2}\right)^2}}{z_m - z_n - \dfrac{h_z}{2} + \sqrt{a^2 + \left(z_m - z_n - \dfrac{h_z}{2}\right)^2}} \tag{11.61}$$

or

$$I = 2 \ln \frac{\dfrac{h_z}{2} + \sqrt{a^2 + \left(\dfrac{h_z}{2}\right)^2}}{a} \qquad \text{for } m = n \tag{11.62a}$$

$$I = \ln \frac{z_m - z_n + \dfrac{h_z}{2} + \sqrt{a^2 + \left(z_m - z_n + \dfrac{h_z}{2}\right)^2}}{z_m - z_n - \dfrac{h_z}{2} + \sqrt{a^2 + \left(z_m - z_n - \dfrac{h_z}{2}\right)^2}} \qquad \text{for } m \neq n, |m - n| \leq 2$$

$$\tag{11.62b}$$

$$I = \ln \frac{z_m - z_n + \dfrac{h_z}{2}}{z_m - z_n - \dfrac{h_z}{2}} \qquad \text{for } |m - n| > 2 \tag{11.62c}$$

The results for pulse expansion functions is shown in Figure 11.5 for $N = 20$ pulses for equi-spaced test points. The software used is *charge_density_wire2.m*.

Comparison of Solutions Based on Entire Domain and Subdomain Basis
The matrix $[l]$ for these basis functions behaves very differently: (1) the matrix for the pulse basis is symmetric, whereas it is asymmetric for the entire basis; (2) the matrix is diagonally dominant for the pulse basis and not for entire domain basis; (3) the condition number of the matrix $k(N)$ increases with N, slowly for pulse basis and very rapidly for entire basis (e.g., $k(5) = 64$, $k(10) = 1.743 \times 10^4$, and $k(15) = 5.427 \times 10^6$ for the entire domain basis, and $k \approx 3$ for pulse basis). However, the accuracy of the solution is similar for the two types of basis functions.

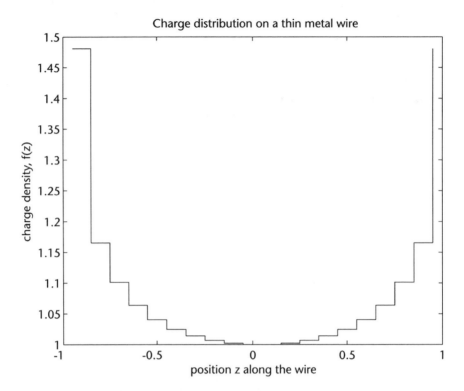

Figure 11.5 Charge density distribution on a wire of length $\ell = 2$m, radius $a = 1$ cm, pulse basis functions and point testing. Number of basis functions, $N = 20$.

Remarks. The MoM solution presented for a charged wire shows the usefulness of this procedure to determine the unknown charge density, which may be used to calculate the capacitance. The calculation of diagonal elements of the matrix is a sticky point in MoM because of the singularity of the integrand. Next we present a two-dimensional case study.

11.2.3 Analysis of Strip Line

Let us determine capacitance per unit length C_0 of a strip line. This example has been chosen because the Green's function for strip line can be expressed in closed form, and the integration for the matrix elements can be carried out analytically. Figure 11.6 shows the cross-section of a strip line with strip width W and ground

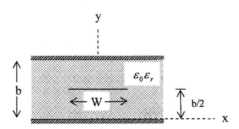

Figure 11.6 Cross-section of a strip line geometry.

planes separation b. For the TEM mode of propagation in strip line, the potential $\varphi(x, y)$ satisfies the Poisson equation

$$\nabla_t^2 \varphi(x, y) = -\frac{\rho(x, y)}{\epsilon} \tag{11.63}$$

and the following boundary conditions:

$$\varphi = 0 \text{ at } y = 0, b \tag{11.64a}$$

$$\varphi = 0 \text{ at } x = \pm\infty \tag{11.64b}$$

$$\varphi = V_0 \text{ at the strip} \tag{11.64c}$$

Here, $\rho(x, y)$ is the charge density on the strip.

The solution of (11.63) may be obtained using the Green's function and the superposition theorem as (see Chapter 3)

$$\varphi(x, y) = \iint G(x, y; x', y') \rho(x', y') \, dx' \, dy' \tag{11.65}$$

where $G(x, y; x', y')$ is the Green's function for the problem and is defined as

$$\nabla_t^2 G(x, y; x', y') = -\frac{\delta(x - x')\delta(y - y')}{\epsilon} \tag{11.66}$$

subject to

$$G = 0 \text{ at } y = 0, b \tag{11.67a}$$

$$G = 0 \text{ at } x = \pm\infty \tag{11.67b}$$

The integral equation for the strip line problem is obtained by employing the boundary condition $\varphi(x, y) = V_0$ on the strip (at $y = b/2$); that is,

$$V_0 = \iint_{strip} G\left(x, \frac{b}{2}; x', y'\right) \rho(x', y') \, dx' \, dy' \text{ for } -\frac{w}{2} \le x \le \frac{w}{2} \tag{11.68}$$

The Green's function for the strip line is found to be (Problem 3.15)

$$G(x, y; x', y') = \frac{1}{\pi\epsilon_0 \epsilon_r} \sum_{n=1}^{\infty} \frac{1}{n} \sin\left(\frac{n\pi y}{b}\right) \sin\left(\frac{n\pi y'}{b}\right) e^{-n\pi|x - x'|/b}$$

$$\tag{11.69}$$

To Determine $\rho(x', y')$ on the Strip. For an infinitely thin strip (strip metal thickness $\to 0$), we can express $\rho(x', y') = \rho(x')\delta(y' - b/2)$. Further assuming $V_0 = 1$ volt, the integral equation (11.68) reduces to

$$1 = \int_{-W/2}^{W/2} G\left(x, \frac{b}{2}; x', \frac{b}{2}\right)\rho(x')\,dx'$$

$$= \frac{1}{\pi\epsilon_0\epsilon_r}\sum_{n=1}^{\infty}\frac{1}{n}\sin\left(\frac{n\pi}{2}\right)\sin\left(\frac{n\pi}{2}\right)\int_{-W/2}^{W/2} e^{-n\pi|x-x'|/b}\rho(x')\,dx'$$

or

$$1 = \frac{1}{\pi\epsilon_0\epsilon_r}\sum_{n,\,odd}^{\infty}\frac{1}{n}\int_{-W/2}^{W/2} e^{-n\pi|x-x'|/b}\rho(x')\,dx' \qquad (11.70)$$

Let us expand the unknown charge distribution $\rho(x')$ as

$$\rho(x') \approx \sum_{m=1}^{M}\alpha_m f_m(x') \qquad (11.71)$$

Substitute this expansion in (11.70) to obtain

$$1 \approx \frac{1}{\pi\epsilon_0\epsilon_r}\sum_{n,\,odd}^{\infty}\frac{1}{n}\int_{-W/2}^{W/2} e^{-n\pi|x-x'|/b}\sum_{m=1}^{M}\alpha_m f_m(x')\,dx' \qquad (11.72)$$

We shall use subdomain expansion functions. Therefore, we divide the strip into M segments each of length $h_x = w/M$. The mth segment or cell is denoted by Δx_m with the center at x_m according to

$$x_m = -\frac{w}{2} + h_x\left(m - \frac{1}{2}\right) \qquad m = 1, 2, 3, \ldots M \qquad (11.73a)$$

and

$$x_m - \frac{h_x}{2} \le \Delta x_m \le x_m + \frac{h_x}{2} \qquad (11.73b)$$

MoM Solution Using Pulse Expansion and Point Matching. Using the property of the pulse functions $P(x' - x_m)$ that they have unit amplitude and are nonzero only over the length of the segment $\Delta x_m = h_x$, (11.72) becomes

$$\frac{1}{\pi\epsilon_0\,\epsilon_r} \sum_{n,\,odd}^{\infty} \frac{1}{n} \sum_{m=1}^{M} \alpha_m \int_{\Delta x_m} e^{-n\pi|x-x'|/b} \, dx' \approx 1 \tag{11.74}$$

For point matching, we enforce the condition at the mid-point of the segment, say, at $x = x_p$. Expression (11.74) therefore reduces to

$$\frac{1}{\pi\epsilon_0\,\epsilon_r} \sum_{n,\,odd}^{\infty} \frac{1}{n} \sum_{m=1}^{M} \alpha_m \int_{\Delta x_m} e^{-n\pi|x_p-x'|/b} \, dx' \approx 1 \tag{11.75}$$

Let us denote for conciseness

$$I(x_p, x_m) \equiv \frac{1}{\pi\epsilon_0\,\epsilon_r} \sum_{n,\,odd}^{\infty} \frac{1}{n} \int_{\Delta x_m} e^{-n\pi|x_p-x'|/b} \, dx' \tag{11.76}$$

The expression (11.75) can then be written as

$$\sum_{m=1}^{M} \alpha_m I(x_p, x_m) \approx 1$$

or

$$\alpha_1 I(x_p, x_1) + \alpha_2 I(x_p, x_2) + \ldots + \alpha_m I(x_p, x_m) + \ldots + \alpha_M I(x_p, x_M) \approx 1 \tag{11.77}$$

This equation implies that at $x = x_p$, the sum of the potentials produced by all the M segments is set equal to unity, the potential at the strip. We enforce (11.77) at all the M points of the strip in order to generate M simultaneous equations as given next:

at $x = x_1$ $\alpha_1 I(x_1, x_1) + \alpha_2 I(x_1, x_2) + \alpha_3 I(x_1, x_3) + \ldots + \alpha_M I(x_1, x_M) \approx 1$

at $x = x_2$ $\alpha_1 I(x_2, x_1) + \alpha_2 I(x_2, x_2) + \alpha_3 I(x_2, x_3) + \ldots + \alpha_M I(x_2, x_M) \approx 1$

at $x = x_3$ $\alpha_1 I(x_3, x_1) + \alpha_2 I(x_3, x_2) + \alpha_3 I(x_3, x_3) + \ldots + \alpha_M I(x_3, x_M) \approx 1$

$\vdots$ $\vdots$

at $x = x_M$ $\alpha_1 I(x_M, x_1) + \alpha_2 I(x_M, x_2) + \alpha_3 I(x_M, x_3) + \ldots + \alpha_M I(x_M, x_M) \approx 1 \tag{11.78}$

The above set of equations may be written in matrix form as

$$\begin{bmatrix} I(x_1, x_1) & I(x_1, x_2) & \cdots & I(x_1, x_M) \\ I(x_2, x_1) & I(x_2, x_2) & \cdots & I(x_2, x_M) \\ \vdots & \vdots & \vdots & \vdots \\ I(x_M, x_1) & I(x_M, x_2) & \cdots & I(x_M, x_M) \end{bmatrix} \begin{bmatrix} \alpha_1 \\ \alpha_2 \\ \vdots \\ \alpha_M \end{bmatrix} = \begin{bmatrix} 1 \\ 1 \\ \vdots \\ 1 \end{bmatrix} \tag{11.79}$$

Let us write

$$I_{pm} = I(x_p, x_m) = \frac{1}{\pi \epsilon_0 \epsilon_r} \sum_{n,\, odd}^{\infty} \frac{1}{n} \int_{\Delta x_m} e^{-n\pi |x_p - x'|/b}\, dx' \qquad (11.80)$$

Each of the equations of (11.78) can be written as

$$\sum_{m=1}^{M} \alpha_m I_m = 1 \qquad p = 1, 2, 3, \ldots M \qquad (11.81)$$

or in matrix form

$$[I][\alpha] = [1] \qquad (11.82)$$

where $[\alpha]$ and $[1]$ are column vectors of size M and $[I]$ is an $M \times M$ square matrix. The vector $[\alpha]$ can be determined using any matrix solution technique. Substituting for $[\alpha]$ in (11.71) gives the approximate solution of the integral equation as

$$\rho(x) \approx [\alpha]^t [P] \qquad (11.83)$$

where $[\alpha]^t$ is the transpose of vector $[\alpha]$.

Determination of Matrix Elements.   The matrix elements I_{pm} may be determined as,

$$I_{pm} = \frac{1}{\pi \epsilon_0 \epsilon_r} \sum_{n,\, odd}^{\infty} \frac{1}{n} \int_{\Delta x_m} e^{-n\pi |x_p - x'|/b}\, dx' \qquad (11.84)$$

$$= \frac{1}{\pi \epsilon_0 \epsilon_r} \sum_{n,\, odd}^{\infty} \frac{1}{n} \int_{x_m - h_x/2}^{x_m + h_x/2} e^{-n\pi |x_p - x'|/b}\, dx'$$

with x_m given by (11.73a). Carrying out the integration gives (independent of $x_p > x'$ or $x_p < x'$)

$$I_{pm} = \frac{1}{\pi \epsilon_0 \epsilon_r} \frac{2b}{\pi} \sum_{n,\, odd}^{\infty} \frac{1}{n^2} e^{-n\pi |x_p - x_m|/b} \sinh\left(\frac{n\pi}{2b} h_x\right) \qquad (11.85)$$

The charge density distribution on the thin strip with $W/b = 1.5$, and $\epsilon_r = 1$ is plotted in Figure 11.7 for $N = 30$. The software employed is *stripline.m*. The number of terms used in the summation for Green's function (11.69) is also 30. The charge distribution shows the expected edge singularity.

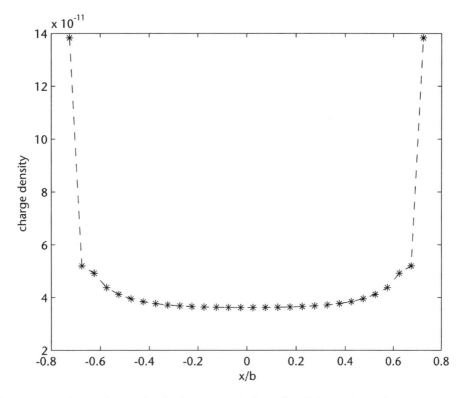

Figure 11.7 Charge density distribution on a strip for $W/b = 1.5$, $\epsilon_r = 1.0$. Pulse expansion and point matching with $N = 30$, the number of terms used in the Green's function is 30.

Determination of C_0.   The charge density computed above may be used to determine the capacitance per unit length of strip line. For the assumed $V_0 = 1$ volt, the capacitance per unit length C_0 is given by

$$C_0 = \frac{Q}{V_0} = Q \tag{11.86}$$

where Q is the total charge on the strip and is given by

$$Q = \iint_{strip} \rho(x', y')\, dx'\, dy' = \int_{-W/2}^{W/2} \rho(x')\, dx' \tag{11.87}$$

for zero thickness metal strip. For pulse modeling of the charge,

$$Q = \sum_{n=1}^{N} \alpha_n \int_{b_x} p_n(x)\, dx = \sum_{n=1}^{N} \alpha_n \tag{11.88}$$

The characteristic impedance is given by

$$Z_0 = \frac{1}{C_0 v} \tag{11.89}$$

where v is the phase velocity of the wave in the transmission line. Since the strip line is homogeneously filled, $v = 1/\sqrt{\mu_0 \epsilon_0 \epsilon_r}$. Therefore,

$$Z_0 = \frac{\sqrt{\mu_0 \epsilon_0 \epsilon_r}}{\sum \alpha_n} \tag{11.90}$$

The computed value of characteristic impedance of the strip line for $N = 20$, $W/b = 1.5$ and $\epsilon_r = 1$ is found to be 47.92 ohms. The conformal mapping solution for the strip line (Chapter 4) gives the following closed-form expression:

$$Z_0 = \frac{29.976\pi}{\sqrt{\epsilon_r}} \frac{K(k)}{K'(k)}, \qquad k = \mathrm{sech}\left(\frac{\pi W}{2b}\right), \qquad k' = \tanh\left(\frac{\pi W}{2b}\right) \tag{11.91}$$

The expression for the ratio of elliptic functions is given in (4.117). The characteristic impedance based on (11.91) is found to be 48.2 ohms, and the comparison with the computed value is very good.

Fringing Field Capacitance

The approximate electric field distribution in strip line is shown in Figure 11.8. The field lines between the strip and the ground planes contribute to the parallel plate capacitance C_p; and the fringing electric field at the edges may be described by C_f. The capacitance per unit length C_0 may therefore be divided as

$$C_0 = C_p + 2C_f \tag{11.92}$$

The capacitance C_p is given by

$$C_p = 2 \frac{\epsilon_0 \epsilon_r W}{\frac{b}{2}} \tag{11.93}$$

The capacitance C_f can be determined from C_0, C_p, and (11.92). For the example considered above, one finds that C_0 is 69.56 pF/m, C_p is 53.125 pF/m, and therefore $C_f = 8.217$ pF/m.

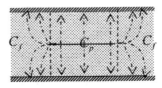

Figure 11.8 Modeling line capacitance C_0 in terms of parallel plate capacitance C_p and fringing field capacitance C_f.

It is seen that C_f, and therefore fringing field, increases with the increase in dielectric thickness b and decrease in the value of dielectric constant ϵ_r of the substrate. This property is useful while selecting the substrate for circuit and antenna applications. Low value of C_f imply lower amount of radiation. Therefore, high value of ϵ_r and thinner substrate is used for circuit applications, and vice versa, for antennas.

Quiz 11.2.   The charge density distribution on a metal strip or a thin wire of finite length is marked by singularity at the ends. While simulating this distribution, if the number of test points per unit length is increased from, say, 12 to 24, the distribution will be marked by:

1. Larger edge singularity;
2. Flatter central region;
3. No change in distribution.

Please choose the correct options and explain why.

Answer.   (1) and (2).

Quiz 11.3.   The integral equation solution based on MoM and pulse basis functions is characterized by:

1. Diagonal matrix elements dominate the matrix
2. Diagonal matrix elements involve pole singularity
3. The matrix is symmetric about the diagonal
4. The matrix is tridiagonal or sparse

Comment on the above and justify the choice.

Answer.   (1), (2), and (3) are correct.

11.2.4 Analysis of Wire Dipole Antenna

When the wire of Section 11.2.1 is excited by a time varying electromagnetic field, a current density $\mathbf{J_s}$ (ampere per meter) is induced on the wire surface. The induced current reradiates and produces an electric field called the radiation or scattered field. If the source of excitation is on the wire itself, the wire dipole is called a transmitting antenna. For a distant source, the dipole may act as a receiving antenna if a receiver is connected to it; otherwise it is called a scatterer. The problem of wire antenna may be analyzed using either Pocklington's integral equation or Hallen's integral equation. The presentation for the Pocklington's integral equation here follows [5].

Consider a very thin wire of radius a and length ℓ as shown in Figure 11.9(a). Let us connect an excitation source across the gap so that the wire behaves as a radiating dipole antenna. The integral equation for the antenna may be obtained from the boundary condition that the sum of the radiated field E_z and the excitation field E_z^i must be zero on the surface of a perfectly conducting wire; that is,

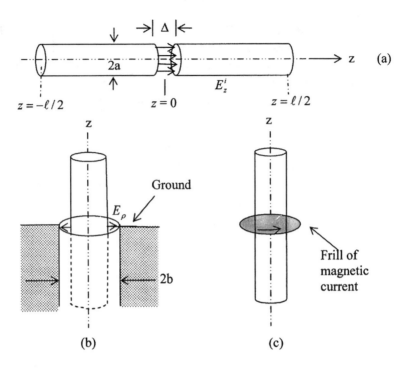

Figure 11.9 Wire antenna and the excitation models: (a) the delta gap model with the impressed electric field; (b) coaxial line feeding a monopole through the ground plane; and (c) wire antenna and magnetic frill generator.

$$E_z = -E_z^i \qquad \text{for } -\frac{\ell}{2} \le z \le \frac{\ell}{2},\, \rho = a,\, 0 \le \varphi \le 2\pi \qquad (11.94)$$

For a very thin wire $(a \ll \lambda)$, one may neglect the circumferential current and approximate the total current by the longitudinal component J_z. Also, due to the circular symmetry of the wire, J_z is not a function of angle ϕ and can be replaced by a filamentary line source of current $I(z')$ given by

$$I(z') = \int_0^{2\pi} J_z(z')\, a d\phi = 2\pi a J_z(z') \qquad (11.95)$$

The electric field produced by this current can be derived from the vector potential **A** using the following expression [see (1.40a)]:

$$j\omega\epsilon_0 \mathbf{E} = \nabla\nabla.\mathbf{A} + k^2\mathbf{A} \qquad (11.96)$$

where the vector potential is related to the current density through (1.39),

$$\nabla^2 A_z + k^2 A_z = -\mu J_z \qquad (11.97)$$

and k is the free space wavenumber. Once the potential A_z is obtained, the electric field produced by it is given by (11.96),

$$E_z = \frac{1}{j\omega\epsilon_0}\left[\frac{d^2 A_z}{dz^2} + k^2 A_z\right]$$

(11.98)

The potential A_z may be determined by solving (11.97) through the Green's function method (Chapter 3), and is given by

$$A_z = \iint_{s'} J_z G(z; z')\, ds' = \int_{-l/2}^{l/2} \int_0^{2\pi} J_z G(z; z')\, a d\phi'\, dz'$$

(11.99)

where $G(z; z') = e^{-jkr}/(4\pi r)$ is the free space Green's function. Using (11.95) for J_z gives the following expression:

$$A_z = \frac{1}{2\pi a} \int_{-l/2}^{l/2} dz' \int_0^{2\pi} I(z')\frac{e^{-jkr}}{4\pi r}\, a d\phi'$$

(11.100)

$$= \frac{1}{2\pi} \int_{-l/2}^{l/2} dz' \int_0^{2\pi} I(z')\frac{e^{-jkr}}{4\pi r}\, d\phi'$$

where r is the distance between the source point and the observation point. Since the current at rf frequencies is concentrated on the metal surface and the observation point should also be on the surface to satisfy the boundary condition, the expression for r is given by

$$r = \sqrt{2a^2 - 2a^2 \cos(\varphi - \phi') + (z - z')^2}$$

(11.101)

Simplification of the Integral for A_z
To simplify the evaluation of the integral in (11.100), we choose $\phi = 0$. As a result,

$$r = \sqrt{2a^2 - 2a^2 \cos\phi' + (z - z')^2} = \sqrt{4a^2 \sin^2\left(\frac{\phi'}{2}\right) + (z - z')^2}$$

(11.102)

We assume further that the current resides along the axis of the wire as discussed in Section 11.2.1. This assumption simplifies the expression for r to [5]

$$r = \sqrt{a^2 + (z - z')^2}$$

(11.103)

and also implies that the contribution of singularity to the integral is not significant for the accuracy considered here.

The expression for r is now independent of ϕ' and (11.100) therefore reduces to

$$A_z = \int\limits_{-l/2}^{l/2} I(z') \frac{e^{-jkr}}{4\pi r} \, dz' = \int\limits_{-l/2}^{l/2} I(z') G(z, z') \, dz' \tag{11.104}$$

Use of this expression in (11.98) gives the following expression for the field produced by the induced current:

$$E_z = \frac{1}{j\omega\epsilon_0} \int\limits_{-\ell/2}^{\ell/2} \left[\frac{\partial^2 G(z; z')}{\partial z^2} + k^2 G(z; z') \right] I(z') \, dz' \tag{11.105}$$

where

$$G(z; z') = \frac{e^{-jkr}}{4\pi r} \tag{11.106}$$

and the distance r is given by (11.103). The integral equation is obtained by equating this electric field to the negative of electric field produced by the excitation current,

$$\frac{1}{j\omega\epsilon_0} \int\limits_{-\ell/2}^{\ell/2} \left[\frac{\partial^2 G(z; z')}{\partial z^2} + k^2 G(z; z') \right] I(z') \, dz' = -E_z^i \qquad \text{at } \rho = a \tag{11.107}$$

This equation is the electric field integral equation (EFIE) for wire antenna and is also called *Pocklington integro-differential equation*. For a very thin wire, the equation may be simplified as [6]

$$\int\limits_{-\ell/2}^{\ell/2} I(z') \frac{e^{-jkr}}{4\pi r^5} [(1 + jkr)(2r^2 - 3a^2) + (kar)^2] \, dz' = -j\omega\epsilon_0 E_z^i \qquad \text{at } \rho = a \tag{11.108}$$

Next we model the excitation field E_z^i on the wire surface.

Gap Generator Model for the Excitation Field, E_z^i
The impressed/excitation field E_z^i at the surface of wire antenna can be simply modeled by a gap voltage generator as shown in Figure 11.9(a). Here, it is assumed that the excitation of the antenna is due to the applied voltage V_i at the feed gap and zero elsewhere. Therefore, the incident electric field is simply given by

$$E_z^i = \begin{cases} V_i/\Delta & \text{over the feed gap } \Delta \\ 0 & \text{elsewhere} \end{cases} \tag{11.109}$$

This model is the simplest but least accurate for input impedance calculations. The accuracy of the model increases with the decrease in gap width Δ.

Frill Magnetic Current Model for Excitation Field.   The feed model based on magnetic frill generator is more accurate [5]. In this model, the excitation is modeled like the coaxial line excitation of a monopole antenna through the ground plane and shown in Figure 11.9(b). The ratio b/a of the coaxial line is designed to produce 50-ohm impedance of coaxial line, and is 2.301 for an air-filled line. The radial electric field in a coaxial line is given by

$$E_f = \hat{\rho}\, \frac{1}{2\rho'}\, \frac{V_i}{\ln\left(\dfrac{b}{a}\right)} \qquad a \leq \rho' \leq b \tag{11.110}$$

For modeling the excitation field, the coaxial aperture is closed off with a perfect conductor using the equivalence principle (Section 1.7), and an equivalent magnetic frill azimuthal current $-\hat{\phi}2E_f$ is placed in the region $a \leq \rho' \leq b$. The frill or annular current at the feed location is shown in Figure 11.9(c). The magnetic current produces the following approximate field at the axis of the wire [5, 7]:

$$E_z^i = -\frac{V_i}{2\ln\left(\dfrac{b}{a}\right)} \left[\frac{e^{-jkR_1}}{R_1} - \frac{e^{-jkR_2}}{R_2} \right] \tag{11.111}$$

where

$$R_1 = \sqrt{z^2 + a^2}, \qquad R_2 = \sqrt{z^2 + b^2} \tag{11.112}$$

MoM Solution.   For conciseness let us write the integral equation of (11.108) in the following form:

$$\int_{-\ell/2}^{\ell/2} I(z')K(z;z')\,dz' = -j\omega\epsilon_0 E_z^i \qquad \text{at } \rho = a \tag{11.113}$$

where

$$K(z;z') = \frac{e^{-jkr}}{4\pi r^5}\left[(1 + jkr)(2r^2 - 3a^2) + (kar)^2\right] \tag{11.114}$$

We shall use pulse expansion for the current $I(z')$ and point matching at the mid-points of segments z_m as described earlier. The integral equation (11.113) therefore reduces to

$$\sum_{n=1}^{N} \alpha_n \int_{\Delta z_n} K(z_m, z')\,dz' \approx -j\omega\epsilon_0 E_z^i(z_m) \qquad m = 1, 2, \ldots N \tag{11.115}$$

where Δz_n denotes the nth segment of the wire. Let us denote

$$I_{mn} = \frac{1}{j\omega\epsilon_0} \int\limits_{\Delta z_n} K(z_m, z')dz' \tag{11.116}$$

$$= \frac{1}{j\omega\epsilon_0} \frac{1}{4\pi} \int\limits_{z_n - \Delta z/2}^{z_n + \Delta z/2} \frac{e^{-jkr}}{r^5}[(1 + jkr)(2r^2 - 3a^2) + (kar)^2]dz'$$

where

$$r = \sqrt{a^2 + (z_m - z')^2} \tag{11.117}$$

Equation (11.115) can therefore be written as

$$\sum_{n=1}^{N} \alpha_n I_{mn} \approx -E_z^i(z_m) \tag{11.118}$$

or

$$\alpha_1 I_{m1} + \alpha_2 I_{m2} + \alpha_3 I_{m3} + \ldots + \alpha_n I_{mn} + \ldots + \alpha_N I_{mN} \approx -E_z^i(z_m) \tag{11.119}$$

The above expression can be interpreted to mean that at $z = z_m$ the sum of the electric field due to all N segments is set equal to the negative of the incident field (to satisfy the boundary condition).

Point matching at N points leads to a set of simultaneous equations in N unknowns, and may be written in the matrix form as

$$[I][\alpha] = \left[-E_z^i\right] \tag{11.120}$$

The matrix equation can be solved for vector $[\alpha]$ by using any standard matrix algorithm. Once we have determined $[\alpha]$, we know the current distribution on the wire, which may be used to calculate the input impedance, radiation pattern, or the radar cross-section of the wire. Next we present results for a half-wave dipole.

Point Matching on a Half-Wave Dipole
A half-wave dipole is expected to be resonant for $\ell = 0.47\lambda$. Let us assume $\lambda = 1$m, and $a = 0.005\lambda$ for the dipole. For convenience we divide the dipole into 11 equal length segments as shown in Figure 11.10. Also, we assume 1 volt excitation at the center of the dipole. The computed electric field and the current obtained using the frill magnetic current model are given in Table 11.2. We note that the current decreases from the center towards the ends as expected.

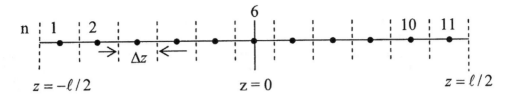

Figure 11.10   A half-wave dipole divided into 11 equal segments of size Δz.

Table 11.2   Computed Electric Field and the
Current Vector α for $N = 11$, $b/a = 2.301$,
$V_i = 1$ volt

Segment Number, n	Excitation Vector $E_z^n, \times 10^{-4}$	Current Vector $\alpha, \times 10^{-3}$
1, 11	$-j1$	$9.7 \angle -40$
2, 10	$-j2$	$17 \angle -39$
3, 9	$-j3$	$22.7 \angle -37$
4, 8	$-j10$	$26.5 \angle -35$
5, 7	$-j67$	$28.3 \angle -32$
6	$-j11333$	$27.7 \angle -27$

Input impedance.   The input impedance of the dipole can be obtained from the above data. For the feed point located at the center of the dipole, the input impedance is given by

$$Z_{in} = \frac{V_i}{\alpha_6} = \frac{1}{27.7 \times 10^{-3} \angle -27°} = 31.97 + j16.7\Omega \qquad (11.121)$$

We observe that the input impedance of a thin dipole of length $0.47\lambda_0$ is inductive. The accurate value of the input impedance for this dipole is $Z_{in} = 80\Omega$.

Effect of Increasing the Number of Segments N.   Table 11.3 compares the input impedance for the dipole as a function of the number of segments for the delta-gap and magnetic frill generators. Software *dipole.m* is used for the purpose. The input impedance converges to the true value for $N > 100$ for both the models, but

Table 11.3   Comparison of the Input Impedance of the
Half-Wave Dipole for Two Different Source Models
($\ell = 0.47\lambda$, $a = 0.005\lambda$)

Number of Segments, N	Delta Gap Source	Magnetic Frill Source
11	$94.17 + j49.0$	$31.96 + j16.7$
21	$77.65 - j0.61$	$47.11 - j0.13$
31	$75.15 - j6.6$	$59.52 - j4.6$
51	$77.18 + j2.6$	$73.11 + j4.3$
71	$79.74 + j8.4$	$77.93 + j11.7$
91	$81.43 + j9.16$	$79.52 + j14.68$
111	$82.62 + j7.1$	$80.08 + j15.76$

the behavior of magnetic frill source is regular. Figure 11.11 shows the effect of increasing N on the amplitude of current distribution for the magnetic frill source. The current converges to the expected half-wave sinusoid for $N > 70$. It is observed that delta gap source also produces convergent current distribution but the dipole current increases with the value of N. The input impedance is plotted in Figure 11.12 for a large number of values of N for the magnetic frill source. The real part of input impedance converges to 80 ohms for $N > 80$.

The MoM analysis of wire antenna in terms of Hallen integral equation is described in [8].

11.2.5 Scattering from a Conducting Cylinder of Infinite Length

The scattering of a propagating wave by an object is one of the fundamental phenomena in electromagnetics. The quantity of interest derived from scattering is the radar cross-section (RCS), which is a measure of the effective area of the scatterer, and is a function of both the angle of incidence and the angle of observation. The larger the RCS, the larger is the power scattered by the object. Stealth technologies are used to reduce the RCS so that the object becomes almost invisible to the incident radiation.

The MoM formulation for scattering from an object is very similar to the formulation for antennas. The difference between the two phenomena lies in the

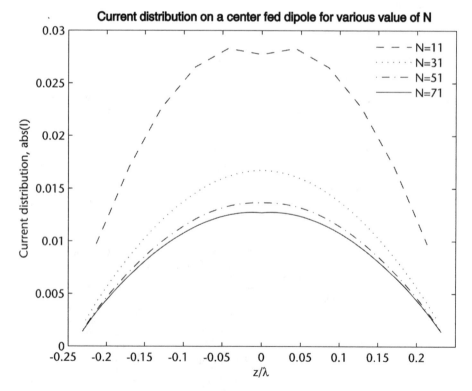

Figure 11.11 Convergence of current distribution on a half-wave wire dipole as a function of N, the number of segments. $\ell = 0.47\lambda$, $a = 0.005\lambda$, $\lambda = 1$m.

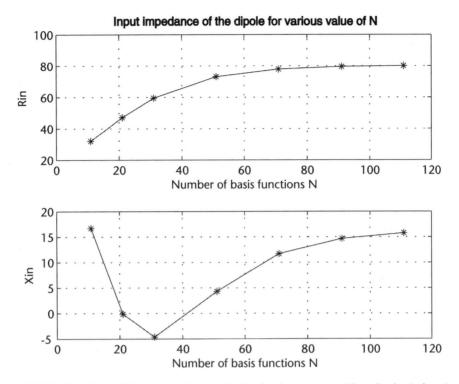

Figure 11.12 Variation of input impedance of wire dipole antenna with pulse basis functions. $\ell = 0.47\lambda$, $a = 0.005\lambda$, $\lambda = 1$m.

source of excitation. In case of antennas, the exciting source is located at the antenna, whereas the exciting source is located at infinity for the purpose of scattering. We shall analyze the problem of scattering by an infinite conducting cylinder. Since there is no reflected wave from the ends of an infinitely long object, the current does not vary along the length, and the problem therefore becomes two dimensional.

An infinitely long, perfectly conducting cylinder is shown in Figure 11.13(a), where the contour C defines the shape of the cylinder. Although the cylinder drawn there is a circular cylinder, the formulation presented next is applicable for cylinders of arbitrary cross-section. The integral equation for the cylinder is obtained by applying the boundary condition that the (total) tangential electric field on its surface is zero; the total electric field consists of the incident field and the scattered field produced by the current induced on the conductor; that is,

$$E_{\tan}^{inc} + E_{\tan}^{scat} = 0 \qquad \text{on the conductor surface} \qquad (11.122)$$

Since the source of incident wave is located at infinity, the incident wave is a plane wave over the cross-section of the cylinder. Let us assume the incident wave to be a plane TM wave with $H_z = 0$ and $\mathbf{E} = \hat{z}E_z(x, y)$. The induced current and the scattered field produced by it are related by the vector wave equation (1.28), which becomes the scalar wave equation in this case,

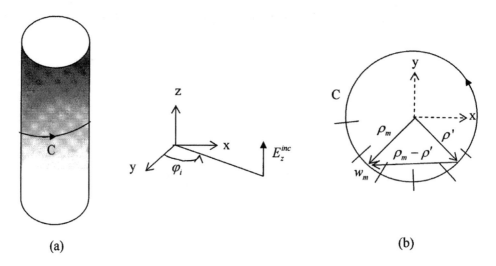

Figure 11.13 (a) TM wave incident on an infinitely long conducting cylinder. (b) Partitioning of contour C and illustration of geometrical parameters.

$$\nabla^2 E_z^{scat} + k^2 E_z^{scat} = j\omega\mu_0 J_z \qquad (11.123)$$

because E_z^{inc} induces J_z only. The Green's function solution for the above wave equation is similar to that given in (3.101), and the scattered field may be expressed as

$$E_z^{scat}(\rho) = -\frac{k\eta_0}{4} \int\limits_C J_z(\rho') H_0^{(2)}(k|\rho - \rho'|) dl' \qquad (11.124)$$

where $\rho = \hat{x}x + \hat{y}y$ is the field point, $\rho' = \hat{x}x' + \hat{y}y'$ is the source point on the contour of cylinder, $H_0^{(2)}(.)$ is the Hankel function of second kind and zero order, $\eta_0 \approx 377\Omega$ is the free space impedance, and $k = 2\pi/\lambda$. The Hankel function of second kind is used here, and not first order, because the scattered wave is traveling outward. The integration is carried out over the contour of the cylinder because the current inside a perfect conductor is zero.

Applying the boundary condition (11.122) on contour C gives the integral equation as

$$E_z^{inc}(\rho) = \frac{k\eta_0}{4} \int\limits_C J_z(\rho') H_0^{(2)}(k|\rho - \rho'|) dl' \qquad (11.125)$$

The unknown surface current can now be determined using MoM. The incident plane wave field is assumed to be propagating normal to the z-axis, the axis of the cylinder. Therefore, $\mathbf{k} = \hat{x}k_x + \hat{y}k_y$ and the incident field may be specified as

$$E_z^{inc} = E_0 e^{-j\mathbf{k}\cdot\mathbf{r}} = E_0 e^{-jk(x\cos\varphi_i + y\sin\varphi_i)} \qquad (11.126)$$

where φ_i is the angle of incidence as shown in Figure 11.13(a). The analysis next follows the discussion given in [2, p. 38].

For the MoM solution we now divide the contour into a number of segments as shown in Figure 11.13(b) and use pulse expansion for the unknown current as,

$$J_z(\rho') \simeq \sum_{m=1}^{M} \alpha_n P(\rho' - \rho_m) \tag{11.127}$$

where $P(\rho - \rho_m)$ is the pulse function centered at ρ_m. If we divide the contour in sufficiently large number of segments, the curved segment may be replaced by a flat segment. Point matching at the mid-point ρ_n of segments reduces the integral equation (11.125) to

$$E_z^{inc}(\rho_n) = \frac{k\eta_0}{4} \sum_{m=1}^{M} \alpha_m \int_{w_m} H_0^{(2)}\big(k|\rho_n - \rho'|\big)dl' \qquad n = 1, 2, 3, \ldots N \tag{11.128}$$

where w_m is the size of the mth segment. The expression (11.128) can be written as a set of simultaneous equations as ($M = N$)

$$\begin{bmatrix} E_z^{inc}(\rho_1) \\ E_z^{inc}(\rho_2) \\ \vdots \\ E_z^{inc}(\rho_N) \end{bmatrix} = \begin{bmatrix} Z_{11} & Z_{12} & \cdots & Z_{1N} \\ Z_{21} & Z_{22} & \cdots & Z_{1N} \\ \vdots & \vdots & \vdots & \vdots \\ Z_{N1} & Z_{N2} & \cdots & Z_{NN} \end{bmatrix} \begin{bmatrix} \alpha_1 \\ \alpha_2 \\ \vdots \\ \alpha_N \end{bmatrix} \tag{11.129}$$

or in matrix form as

$$[E^{inc}] = [Z][\alpha] \tag{11.130}$$

where

$$Z_{mn} \simeq \frac{k\eta_0}{4} \int_{\rho_m - w_m/2}^{\rho_m + w_m/2} H_0^{(2)}\big(k|\rho_n - \rho'|\big)dl' \tag{11.131}$$

Analytical Simplification of Matrix Elements
The matrix elements are now simplified analytically so that they can be expressed in accurate but simpler form. The diagonal elements are relatively large and contribute significantly to the solution. For segment size small enough compared to the wavelength, (11.131) may be approximated as [1, p. 43]

$$Z_{mn} \simeq \frac{k\eta_0}{4} w_n H_0^{(2)}(kR_{mn}) \qquad \text{for } m \neq n \tag{11.132}$$

and

$$R_{mn} = |\rho_n - \rho_m| = \sqrt{(x_m - x_n)^2 + (y_m - y_n)^2}$$

This approximation is found to be accurate to about 2% when the segment size is about $\lambda/10$ [2].

For diagonal and near-diagonal matrix elements, the argument of the Hankel function is small; and the Hankel function may be expressed as [2, p. 39]

$$H_0^{(2)}(t) \approx \left(1 - \frac{t^2}{4}\right) - j\left\{\frac{2}{\pi} \ln\left(\frac{\gamma t}{2}\right) + \left[\frac{1}{2\pi} - \frac{1}{2\pi} \ln\left(\frac{\gamma t}{2}\right)\right]t^2\right\} + O(t^4)$$

(11.133)

where $\gamma = 1.78107. \ldots$. Retaining the dominant terms of (11.133), the integration in (11.131) may be carried out as [2, p. 40]

$$\int_{w_m} H_0^{(2)}(k|\rho_m - \rho'|)\,dl' \simeq 2 \int_0^{w_m/2} \left[1 - j\frac{2}{\pi} \ln\left(\frac{\gamma k u}{2}\right)\right] du$$

(11.134)

$$= w_m - j\frac{2}{\pi}w_m\left[\ln\left(\frac{\gamma k w_m}{4}\right) - 1\right]$$

Using this expression, the diagonal elements are obtained as

$$Z_{mn} \simeq \frac{k\eta_0 w_m}{4}\left\{1 - j\frac{2}{\pi}\left[\ln\left(\frac{\gamma k w_m}{4}\right) - 1\right]\right\}$$

(11.135)

A better accuracy in the evaluation of matrix elements can be achieved by numerical integration. For diagonal elements, singularity subtraction may be employed before integration.

The solution of matrix equation (11.130) yields the induced current distribution J_z on the cylinder and is given by

$$[\alpha] = [Z]^{-1}[E^{inc}]$$

(11.136)

where $E_m^{inc} = e^{jk(x_m \cos \varphi_i + y_m \sin \varphi_i)}$ and (x_m, y_m) denotes the coordinates of the midpoint of the segment.

The induced current density may be used to determine the RCS of the scatterer. For a two-dimensional scatterer, RCS is defined as [2, p. 24]

$$\sigma_{TM}(\varphi, \varphi^{inc}) = \lim_{\rho \to \infty} 2\pi\rho \left|\frac{E_z^{scat}}{E_z^{inc}}\right|^2$$

(11.137)

and can be expressed in terms of induced current on a perfectly conducting cylinder as

$$\sigma_{TM}(\varphi, \varphi^{inc}) = \frac{k\eta_0^2}{4} \left| \int_C J_z(x', y') e^{jk(x'\cos\varphi + y'\sin\varphi)} \, dl' \right|^2 \qquad (11.138)$$

For small sized segments, the integral may be approximated by summation over the segments as,

$$\sigma_{TM}(\varphi, \varphi^{inc}) = \frac{k\eta_0^2}{4} \left| \sum_{n=1}^{N} \alpha_n w_n e^{jk(x_n\cos\varphi + y_n\sin\varphi)} \right|^2 \qquad (11.139)$$

where φ is the angle of observation. In matrix form, (11.139) may be written as [1, p. 45]

$$\sigma_{TM}(\varphi, \varphi^{inc}) = \frac{k\eta_0^2}{4} \left| [V^s][ZS]^{-1}[V^{inc}] \right|^2 \qquad (11.140)$$

where $[V^{inc}]$ is an excitation voltage vector, $[V^s]$ is a voltage matrix at the observation angle, and $[ZS]$ is the scatterer impedance matrix, defined as

$$V_m^{inc} = w_m e^{jk(x_m\cos\varphi_i + y_m\sin\varphi_i)}, \; V_n^s = w_n e^{jk(x_n\cos\varphi + y_n\sin\varphi)}, \; ZS_{mn} = w_m Z_{mn} \qquad (11.141)$$

The above formulation is coded for a circular cylinder of radius a. The software is called *scattering_circular.m*. The plane TM wave is defined by $E_z^{inc} = e^{-jk(x\cos\varphi_i + y\sin\varphi_i)}$ and $\varphi_i = \pi$. The computed induced current on the surface of the cylinder is plotted in Figure 11.14 as a function of φ for two values of ka. The current is maximum on the illuminated side $\varphi = \pi$ of the cylinder and becomes almost negligible in the shadow region. Also, the induced current increases with the increase in the value of ka.

It has been observed that for best accuracy of the above scatterer model, the phase center (x_m, y_m) of each segment should be located on the surface of the original cylinder and the segment width is obtained from the circumference of the original cylinder [2] as $w_m = 2\pi a/N$, where N is the number of segments.

The radar cross-section of the cylinder is computed using (11.140), and the normalized value $\sigma/(2\pi a)$ is plotted in Figure 11.15 as a function of angle of observation φ for $ka = 1$ and 4. The RCS is higher in the direction of shadow region and relatively low on the illuminated side of the cylinder. It is contrary to the distribution of induced current on the cylinder. This paradox is explained by the fact that incident radiation is present for all values of φ, whereas the scattered radiation is more on the illuminated side and cancels the incident radiation there. The scattered field being less on the shadow region, the incident radiation contributes significantly to the RCS in the shadow region [9, p. 129]. Also, the figure

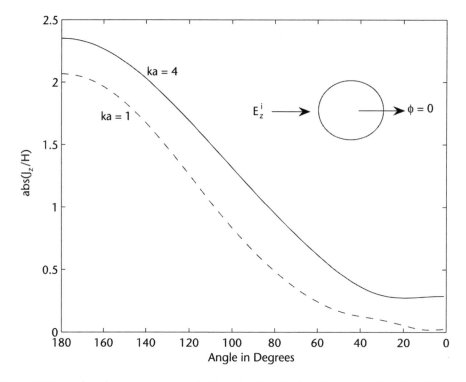

Figure 11.14   Induced current on a conducting circular cylinder of radius $a = 1$m, $N = 160$.

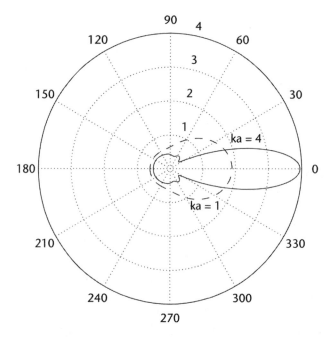

Figure 11.15   RCS ($\sigma/2\pi a$) of a circular cylinder of radius $a = 1$m, $N = 160$.

shows that RCS becomes more directive with the increase in the value of ka. The increase in ka was realized by increasing the frequency of the incident wave and keeping radius a fixed.

Analytical Solution for the Circular Cylinder
The circular cylinder, being a regular shaped geometry, has been analyzed using the method of separation of variables. For this, the incident plane wave is expanded in terms of cylindrical functions as [10, p. 233].

$$E_z^{inc} = E_0 e^{-jk\rho \cos \varphi_i} = E_0 \sum_{n=-\infty}^{\infty} j^{-n} J_n(k\rho) e^{jn\varphi_i} \qquad (11.142a)$$

The scattered field is similarly expressed as

$$E_z^{scat}(\rho) = E_0 \sum_{n=-\infty}^{\infty} j^{-n} a_n H_n^{(2)}(k\rho) e^{jn\varphi_i} \qquad (11.142b)$$

Applying the boundary condition (11.122) at the surface of the cylinder $\rho = a$ gives the following expression for the expansion coefficients a_n:

$$a_n = -\frac{J_n(ka)}{H_n^{(2)}(ka)} \qquad (11.143)$$

The surface current on the cylinder is therefore given by

$$J_z = H_\varphi\big|_{\rho=a} = \frac{1}{j\omega\mu_0} \frac{\partial}{\partial \rho}\left(E_z^{inc} + E_z^{scat}\right)\big|_{\rho=a} \qquad (11.144)$$

Using (11.142) in (11.144) and simplifying the result by means of Wronskian,

$$J_n(x) H_n'^{(2)}(x) - H_n^{(2)}(x) J_n'(x) = \frac{2}{\pi x} \qquad (11.145)$$

yields

$$J_z = -\frac{2E_0}{\omega\mu_0 \pi a} \sum_{n=-\infty}^{\infty} \frac{j^{-n} e^{jn\varphi_i}}{H_n^{(2)}(ka)} \qquad (11.146)$$

The scattered electric field at large distances from the cylinder may be obtained from (11.142b) by using the following asymptotic formula for $H_n^{(2)}(k\rho)$:

$$H_n^{(2)}(x) \xrightarrow[x \to \infty]{} \sqrt{\frac{2j}{\pi x}} j^n e^{-jx} \qquad (11.147)$$

The scattered field is therefore obtained as

$$E_z^{scat} \xrightarrow[k\rho \to \infty]{} E_0 \sqrt{\frac{2j}{\pi k\rho}}\, e^{-jk\rho} \sum_{n=-\infty}^{\infty} a_n e^{jn\varphi_i} \qquad (11.148)$$

The ratio of the scattered field to the incident field is given by

$$\left| \frac{E_z^{scat}}{E_z^{inc}} \right|^2 = \sqrt{\frac{2}{\pi k\rho}} \left| \sum_{n=-\infty}^{\infty} \frac{J_n(ka)}{H_n^{(2)}(ka)} e^{jn\varphi_i} \right| \qquad (11.149)$$

The convergence of this series is slow for large values of ka; for example, six terms give satisfactory result for $ka = 3$, whereas 100 terms are needed for $ka = 100$.

RCS of a Square Cylinder

Next, we study the scattering by a square cylinder to observe the role played by the corners/wedges on the scatterer. The formulation described earlier was coded for a square cylinder of side length $a = 1$m and $ka = 1, 4$. The software is called *scattering_square.m*. The induced current on the cylinder is plotted in Figure 11.16. Due to the symmetry of the cylinder and excitation, the current distribution on only half the cross-section is plotted. The current is found to be singular at the corner facing the incident wave. The current decreases almost linearly towards the shadow region, except at the corner in that region. Also, the induced current increases with ka. The normalized RCS $\sigma/(4a)$ was computed for the square cylinder and is plotted in Figure 11.17. Comparing this plot with Figure 11.15 for the RCS

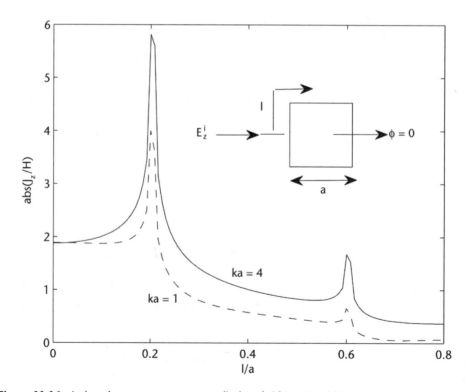

Figure 11.16   Induced current on a square cylinder of side a. $N = 200$.

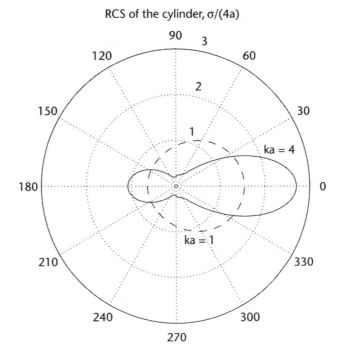

Figure 11.17   RCS ($\sigma/(4a)$) of a square cylinder of side a. $N = 200$.

of a circular cylinder, we find that the RCS of a square cylinder is less directive although the induced current is highly peaked.

11.3   Fast Multipole Solution Methods for MoM

The MoM is an efficient method with less number of unknowns. However, the matrix is dense and the use of direct solution methods, such as LU decomposition, results in large number of floating point operations, which increase with frequency as $O(f^6)$. The processor time can be reduced by using Fourier transformation or CG-FFT techniques for geometries that are thin along one direction [11]. For truly three-dimensional geometries, several methods have been proposed for efficient MoM solution. A very successful scheme involves replacing the matrix-vector multiplication of MoM by the fast multipole method (FMM) and its variant multilevel fast multipole [12]. The first step in this method is to divide the simulation region into a number of subregions, each containing a number of cells. Computations for the cells in the same or adjacent subregions are carried out in the standard way as described earlier. However, the fields produced by the sources far away from the observation cell are computed by multipole expansion of the sources and then projecting this onto a set of plane waves in the observation cell. The efficiency is achieved because only a few terms of the multipole expansion are needed. The number of floating point operations based on FMM increases as $O(f^3)$ and can be reduced to $O(f^2 \log f)$ if the multilevel extension of FMM is implemented. In the multilevel FMM, the FMM algorithm is repeated in a hierarchical fashion similar to that in a telephone network. In a network of N customers, a number of N^2 connections is required if a direct line connects every other customer. The number of connections can be reduced by introducing hubs. To make a telephone call, a customer calls the local hub, which calls another hub, and finally to the recipient of the call.

11.4   Comparison of FDM, FDTD, FEM, and MoM

We have discussed a number of computational methods to solve problems with simple geometries and dielectric inhomogeneities. A common base for FDM, FEM, and MoM is the weighted residual method [13]. The common features of these methods include discretization of the unknown function, approximations, and the solution based on matrix solution techniques. The approximations result in inaccuracies. The convergence and spurious solutions are other problems. We compare the computational methods with these perspectives in Table 11.4. The evaluation is not quantitative but qualitative. With the limited experience we have developed with the computational methods in terms of rectangular grid and one-dimensional and two-dimensional geometries, we can compare the methods in a limited manner only.

Table 11.4   Qualitative Comparison of Computational Methods

Characteristics	FDM	FDTD	FEM	MoM
Preprocessing	Nil	Nil	Moderate	Significant (because of Green's function)
Level of discretization	Surface level	Surface level	Surface level	Contour level
Stability of solution	Good	Poor, if not causal	Good	Very good
Matrix type	Sparse	—	Sparse	Dense
Storage requirement	Large	Very Large	Large	Small to moderate
CPU time	Large	Large	Moderate to large	Small to moderate
Numerical dispersion	Yes	Yes	Yes	No
Analysis of arbitrary shaped geometry	Poor, if rectangular or cylindrical mesh is used	Poor, if rectangular or cylindrical mesh is used	Very good, if triangular elements are employed	Good (only the unknown function is discretized)
Open boundary problems	Convergence is very slow if first order ABC is used	PML or ABC may be used to truncate the domain	PML or ABC may be used to truncate the domain	Efficient (because of Green's function)
Spurious solutions	No	No	Yes, for node based elements	No
Distinctive feature	Easier to understand, but inefficient	Versatile, can produce animation	Analysis of arbitrary shaped geometry	Efficient for open boundary problems.

The various aspects of the computational methods dealt with in this text should be useful in developing the necessary skills. An experienced student can accelerate numerical processing skillfully.

11.5 Hybrid Computational Methods

As discussed in the last section, the computational methods in their classical form suffer from one limitation or the other. MoM results in a dense matrix although the number of unknowns is less, whereas FEM suffers from large number of unknowns. For the unbounded region problems, MoM has the advantage associated with the use of Green's function, whereas FEM is versatile for inhomogeneous medium problems. It is possible to combine the useful features of FEM and MoM so that unbounded region problems with inhomogeneous dielectric could be solved efficiently. The resulting hybrid method is called finite element-boundary integral (FE-BI) method. The formulation of FE-BI by Botha and Jin [14] is based on variational formulation and is very well described in [15, Ch.10]. FEM has been combined with conformal mapping to determine the capacitance of printed lines [16]. Computational efficiency of FDTD may be combined with the body conforming meshing property of FEM [17].

11.6 Summary

The method of moments (MoM) is a powerful computational method which is partly analytical and partly computational. The method is very efficient for the analysis of open region geometries such as antennas, scatterers, and planar circuits. The method is based on weighted residual method in which the residue of the governing equation is set to zero in an average sense over the domain. The unknown function is expanded in a set of expansion functions, and inner product of the residue is set to zero. The number of test functions are taken equal to the number of expansion functions in order to generate as many equations as there are unknowns. The MoM is illustrated by solving a number of examples like, strip line capacitance, charge distribution on a metal wire, current distribution and input impedance of a half-wave dipole, and RCS of a cylindrical scatterer. Various computational methods discussed in the book are compared

References

[1] Harrington, R. F., *Field Computation by Moment Methods*, Malabar, FL: R.E. Kreiger Publishing Co., 1982.

[2] Peterson, A. F., S. L. Ray, and R. Mittra, *Computational Methods for Electromagnetics*, New York: IEEE Press, 2001.

[3] Neff, H. P., Jr., *Basic Electromagnetic Fields*, 2nd ed., New York: Harper & Row, 1987.

[4] Wilton, D. R., and S. Govind, "Incorporation of Edge Condition in the Moment Method Solutions," *IEEE Trans. Antennas Propagat.*, Vol. AP-25, 1977, pp. 845–850.

[5]   Balanis, C. A., *Antenna Theory: Analysis and Design*, 2nd ed., New York: John Wiley, 1982.

[6]   Stutzman, W. L., and G. A. Thiele, *Antenna Theory and Design*, New York: John Wiley, 1981.

[7]   Jordan, E. C., and K. G. Balmain, *Electromagnetic Waves and Radiating Systems*, Englewood Cliffs, NJ: Prentice-Hall, 1968.

[8]   Tsai, L. L., and C. E. Smith, "Moment Methods in Electromagnetics for Undergraduates," *IEEE Trans. Edu.*, Vol. E-31, 1978, pp. 14–22.

[9]   Moore, J., and R. Pizer, (eds.), *Moment Methods in Electromagnetics: Techniques and Applications*, New York: Research Studies Press, 1984.

[10] Harrington, R. F., *Time Harmonic Electromagnetic Fields*, New York: McGraw-Hill, 1961.

[11] Peters, T. J., and J. L. Volakis, "Application of a Conjugate Gradient FFT Method to Scattering from Thin Material Plates," *IEEE Trans. Antennas Propagat.*, Vol. 36, 1988, pp. 518–526.

[12] Chew, W. C., et al., *Fast and Efficient Algorithms in Computational Electromagnetics*, Norwood, MA: Artech House, 2001.

[13] Sadiku, M. N. O., and A. F. Peterson, "A Comparison of Numerical Methods for Computing Electromagnetic Eields," *Proc. 1990 Southeastcon*, pp. 42–47.

[14] Botha, M. M., and J. M. Jin, "On the Variational Formulation of Hybrid Finite Element-Boundary Integral Techniques for Electromagnetic Analysis," *IEEE Trans. Antennas Propagat.*, Vol. 52, 2004, pp. 3037–3047.

[15] Jin, J., *The Finite Element Method in Electromagnetics*, 2nd ed., New York: Wiley, 2002.

[16] Chang, C. N., and J. F. Cheng, "Hybrid Quasistatic Analysis of Multilayer Microstrip Lines," *IEE Proc. H, Microwaves, Antennas and Propagat.*, Vol. 140, April 1993, pp. 79–83.

[17] Rylander, T., and A. Bondeson, "Stability of Explicit-Implicit Hybrid Time-Stepping Schemes for Maxwell's Equations," *J. Comput. Phys.*, Vol. 179, 2002, pp. 426–438.

Problems

P11.1. Solve the following differential equation for the transmission line half-wave resonator using the MoM:

$$\frac{d^2 E_y}{dx^2} + \pi^2 E_y(x) = \delta(x - 0.5) \qquad 0 \leq x \leq 1$$

subject to $E_y(0) = E_y(0) = 0$. Divide the domain into equal segments and use triangular basis and test functions. Obtain the expressions for the matrix element l_{mn} and the excitation vector element p_m.

P11.2. Solve the following ordinary differential equation:

$$-\frac{d^2 f}{dx^2} = 1 + x^2 \qquad 0 \leq x \leq 1$$

subject to $f(0) = f(1) = 0$. Use triangular function expansion-point matching, and Galerkin triangular expansion to determine the function $f(x)$. Comment on the convergence aspect in the two cases, and determine the residual for triangular function expansion point-matching method.

P11.3. The following problem interprets the finite difference method in the language of MoM [2, p. 228].

1. Use MoM to construct discretization of the following scalar wave equation

$$\frac{d^2 E_y}{dx^2} + k^2 E_y(x) = g(x) \qquad 0 \le x \le 1$$

    using triangular basis functions for E_y. Enforce the differential equation with pulse testing functions such that basis functions straddle two cells of dimension Δx, while the test functions are defined between the centers of adjacent cells as shown in Figure 11.18. Determine the mnth entry of the resulting matrix.
2. Construct a second order finite difference discretization of the wave equation of (1) using cells of dimension Δx and central-difference formula

$$\frac{d^2 E_y}{dx^2} = \frac{E_y(i + 1) - 2E_y(i) + E_y(i - 1)}{(\Delta x)^2}$$

    Identify the entries of the tridiagonal finite difference matrix and compare with the entries of the matrix in part (1). Discuss the implications.

P11.4. A parallel strip transmission line consists of two finite width strips running parallel to each other and separated by a distance $2h$, as shown in Figure 11.19. Use the MoM to determine (and plot) the charge density distribution on any of the strips and the characteristic impedance Z_0 of the line for $w = 5$m and $h = 1$m. You may use pulse expansion and point matching with $N = M = 3, 7, 11, 18, 39, 59$, where N and M represent the number of segments on the upper and lower plates, respectively. Also plot Z_0 as a function of w/h for $w/h = 1, 3, 5, 7, 10$ [8].

P11.5. For the center-fed dipole with $\ell = 0.47\lambda$, $a = 0.05\lambda$, and $\lambda = 1$m, determine the input impedance as a function of frequency with frequency ranging from 250 to 350 MHz at an interval of 10 MHz.

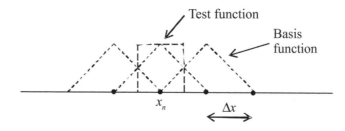

Figure 11.18 Arrangement of triangular and pulse functions to compute matrix element I_{mn}.

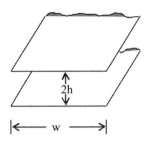

Figure 11.19 Geometry of a parallel strip line.

P11.6. The strip line of Section 11.2 was analyzed using pulse expansion and point matching. Now use triangular expansion and Galerkin's method to determine the expression for matrix elements.

P11.7. Use the expression for matrix elements derived in P11.6 to determine the charge distribution on the strip for the set of parameters considered in Section 11.2. Comment on the convergence behavior.

P11.8. Consider the parallel strip transmission line of Figure 11.19. The MoM analysis of the geometry with pulse expansion and point matching produces the following data for the Z_0 of the line as a function of expansion functions N and w/h:

$N\rightarrow$ $w/h\downarrow$	3	7	11	18	39	59
1	127.72	126.57	126.36	126.27	126.27	126.28
3	69.13	69.30	69.37	69.45	69.57	69.61
5	47.21	48.40	48.69	48.90	49.10	49.17
7	35.19	37.15	37.59	37.89	38.16	38.24
9	27.38	30.03	30.59	30.95	31.27	31.36

1. Determine the most accurate value of Z_0 for each value of w/h. Determine the corresponding capacitance per unit length C.
2. Now model C as a combination of parallel plate capacitance C_p and fringing field capacitance C_f such that $C = C_p + 2C_f$. Determine C_f and plot C_f/C_p as a function of w/h.
3. Explain the behavior of C_f physically.

P11.9. Compare MoM, FDM, and FEM techniques in relation to the solution of

$$-\frac{d^2V}{dx^2} = \pi^2 \sin(\pi x) \qquad 0 \leq x \leq 1$$

subject to $V(0) = V(1) = 0$.

P11.10. Consider the following one-dimensional scalar Helmholtz equation:

$$\frac{d^2 E_y(x)}{dx^2} + k^2 E_y(x) = 0, \qquad 0 \le x \le 1$$

where $k = 2\pi/\lambda$ defines the wavelength in the medium.

1. Apply FDM and the central difference approximation to determine the mnth entry of the resulting matrix.
2. Apply MoM and the triangular basis and testing functions of size $2\Delta x$ to determine the mnth entry of the resulting matrix.
3. Apply FEM and the triangular basis and testing functions of size $2\Delta x$ to determine the mnth entry of the resulting matrix.
4. Compare the entries of (1), (2), and (3) and discuss the implications.

Solution Methods for the Set of Simultaneous Equations

The computational methods discussed in the preceding chapters have one feature in common. These methods, except for FDTD, require the solution of a set of simultaneous equations. Several matrix operations (multiplication, inversion, and so on) are carried out to obtain the potential, field, or current distribution. The analysis may have to be carried out repeatedly to determine the optimum set of parameters. This requires efficient methods to carry out various matrix operations.

The matrix equation may be written as

$$[A][x] = [b] \tag{A.1}$$

where $[A]$ is the $N \times N$ matrix with known elements, $[x]$ is the $N \times 1$ unknown vector, and $[b]$ is the $N \times 1$ known excitation vector.

The matrices we come across during the investigations of electromagnetic problems are of special type. This is due to the physical considerations of the engineering problems. The special matrices may be described variously as: symmetric, hermitian, diagonally dominant, sparse, toeplitz, and triangular, and are defined in Section 2.6.

The computer resources (computer memory and processor time) required to solve (A.1) determine the efficiency of the computational method employed. The efficiency is limited by: (1) the amount of computer storage necessary for the N^2 elements of the matrix, (2) the amount of processor time required to compute those N^2 elements, and (3) the amount of processor time required to solve the matrix equation (A.1). Here we first examine symmetry considerations and algorithms to reduce the execution time and storage requirements.

A.1 Processor Time Considerations [1]

A square matrix of N^2 elements is said to be of order N. The processor time t required to solve the matrix equation is approximately given by [2]

$$t \approx C_1 N^2 + C_2 N^3 \tag{A.2}$$

where the algorithm and computer-dependent factors C_1 and C_2 are defined as follows:

C_1 = processor time required to compute a typical matrix element
C_2 = processor time required to solve $[A][x] = [b]$ by matrix inversion

It is clear from (A.2) that the processor time required for solving the system of equations varies as N^3 and dominates the total processor time. This computer time can be reduced in a number of ways. If instead of using the matrix inversion method, an algorithm such as Gauss-Jordan is used, then the processor time required reduces from $C_2 N^3$ to $C_3 N^2$ and (A.2) becomes

$$t \approx C_1 N^2 + C_3 N^2 \tag{A.3}$$

and represents a significant reduction in the solution time required for a given N. If further, $[A]$ is toeplitz in nature, (A.3) becomes

$$t \approx C_1 N + C_4 N^{5/3} \tag{A.4}$$

Equation (A.4) shows a significant improvement over (A.2) in the first term as well as the second term. The relative CPU time is therefore less if the indirect solution methods are employed for the solution of large sized matrices.

We next examine briefly the matrix properties to reduce the total computer time required.

Toeplitz Matrices

Most of the computational methods produce matrices $[A]$ that are symmetric about the diagonal. Further, if the various segments of the geometry are of the same dimension, all the N^2 matrix elements can be described by any row or column of $[a_{ij}]$, say, the first row. For example, the following matrix is a *toeplitz matrix* and the matrix elements can be obtained by the following algorithm:

$$a_{ij} = a_{1,|i-j|+1} \qquad i \geq 2, j \geq 1 \tag{A.5}$$

$$[A] = \begin{bmatrix} a_{11} & a_{12} & a_{13} & \cdots & a_{1N} \\ a_{21} & a_{22} & a_{23} & \cdots & a_{2N} \\ a_{31} & a_{32} & a_{33} & \cdots & a_{3N} \\ \vdots & \vdots & \vdots & \vdots & \vdots \\ a_{N1} & a_{N2} & a_{N3} & \cdots & a_{NN} \end{bmatrix} = \begin{bmatrix} a_{11} & a_{12} & a_{13} & \cdots & a_{1N} \\ a_{12} & a_{11} & a_{12} & \cdots & a_{1(N-1)} \\ a_{13} & a_{12} & a_{11} & \cdots & a_{1(N-2)} \\ \vdots & \vdots & \vdots & \vdots & \vdots \\ a_{1N} & a_{1(N-1)} & a_{1(N-2)} & \cdots & a_{11} \end{bmatrix} \tag{A.6}$$

Computer algorithms exist for solving toeplitz matrices that are considerably more efficient than the ones for solving a nontoeplitz matrix. An example of toeplitz matrix is given in Section 11.2.

A.2 Matrix Solution Techniques

The most direct solution approach for (A.1) is the matrix inversion technique described as

$$[x] = [A]^{-1}[b] \tag{A.7}$$

where $[A]^{-1}$ can be written in the compact form as

$$[A]^{-1} = adj[A]/|A| \tag{A.8}$$

The ijth element of $adj[A]$ is given by $(-1)^{i+1} \times$ (jth minor of $[A]$). Implementation of this scheme may be very computer intensive. To evaluate the determinant of an $N \times N$ matrix requires $N(N-1)!$ multiplications. For large values of N this is a very computationally intensive formulation and is therefore used for small sized problems only. We next discuss some efficient matrix solution techniques.

The various methods for solving a system of linear equations can be categorized either under direct/elimination methods or iterative methods. The elimination methods constitute the simplest direct approach to the solution of a set of simultaneous equations. Such methods include Gauss's method, Gauss-Jordan, Cholesky's or Crout's method, and the square-root method. The direct methods can be used for $N = 25$ to 60. For very large systems, say $N = 100$ or more, the direct methods become inefficient and suffers from round-off error. Indirect or iterative methods provide an alternative. These include conjugate gradient method, Jacobi, and Gauss-Siedel method.

We discuss two efficient approaches for the solution of matrix equation. These are: (1) Gauss elimination, and (2) L-U decomposition. Some of the sparse matrix techniques used to reduce storage requirements and improve computational efficiency are also discussed. An iterative method based on conjugate gradient is outlined.

A.2.1   Gauss Elimination [3, 4]

The Gauss elimination method is still one of the most popular and efficient methods of solving a system of linear equations if there are no special properties of the matrix to exploit (e.g., sparsity, bandedness, symmetry, and so on). This method involves eliminating one unknown at a time and proceeding with the remaining equations. This process leads to a set of simultaneous equations in triangular form from which the unknowns are determined by back-substitution. We first illustrate the procedure by means of an example.

Example A.1.   Consider the solution of the following set of equations:

$$2x_1 + 2x_2 + 3x_3 = 3 \tag{A.9a}$$

$$4x_1 + 7x_2 + 7x_3 = 1 \tag{A.9b}$$

$$-2x_1 + 4x_2 + 5x_3 = -7 \tag{A.9c}$$

Using (A.9a) to eliminate x_1 from the other two equations, we get

$$2x_1 + 2x_2 + 3x_3 = 3 \tag{A.10a}$$

$$3x_2 + x_3 = -5 \tag{A.10b}$$

$$6x_2 + 8x_3 = -4 \tag{A.10c}$$

We now use (A.10b) to eliminate x_2 from (A.10c), to obtain

$$2x_1 + 2x_2 + 3x_3 = 3 \tag{A.11a}$$

$$3x_2 + x_3 = -5 \tag{A.11b}$$

$$6x_3 = 6 \tag{A.11c}$$

The solutions for $x_1, x_2,$ and x_3 can now be obtained in the reverse order by back-substitution. Equation (A.11c) gives $x_3 = 1$, which on substitution in (A.11b) gives $x_2 = -2$. Finally, from (A.11a), we have $x_1 = 2$. Equations (A.11) are called *triangular equations*, and the process of obtaining the solution in the reverse order is called *back-substitution*.

The basic operations used for triangulation in the above example are:

1. Multiplication of an equation by a constant;
2. Replacement of jth equation by the sum of jth equation and α times the kth equation, where α is any constant and $k \neq j$.

For its implementation on a computer, the operations described above can be performed as follows.

Let us say we wish to place zeros in column 1 below the diagonal a_{11} of (A.6). If we multiply the first equation by $\alpha = a_{21}/a_{11}$ and subtract this equation from the second equation, we shall get a zero in the (2,1) position of $[A]$. We must also perform the same operation on the right side vector $[b]$. Similarly, for the third equation, we multiply the first equation by a_{31}/a_{11} and subtract it from the third equation. The algorithm is then given by

$$a_{jk} = a_{jk} - \frac{a_{ji}}{a_{ii}} a_{ik}, \qquad k = 1, 2, \ldots N; j = i + 1, \ldots N \tag{A.12a}$$

and

$$b_j = b_j - \frac{a_{ji}}{a_{ii}} b_j, \qquad j = i + 1, \ldots N \tag{A.12b}$$

The algorithm for the unknowns can be written as

$$x_N = \frac{b_N}{a_{NN}} \tag{A.13a}$$

$$x_i = \frac{b_i - \displaystyle\sum_{j=i+1}^{N} a_{ij} x_j}{a_{ii}} \qquad i = N - 1, N - 2, \ldots 1 \tag{A.13b}$$

It is assumed in (A.12a) that all the diagonal elements are nonzero. If this is not so we must use *pivot*, as discussed next. If the matrix is nonsingular, we will always find a nonzero pivot.

Pivoting

Here we discuss various criteria which can be used for the selection of pivots at a particular stage. Though, in the previous section, the only restriction mentioned was that the pivots should be nonzero, it has been found that pivots with small magnitudes may lead to numerical errors. This is because of the finite word length used in the computer. The errors caused by pivots of small magnitude can be illustrated by means of the following example.

Example A.2.   Consider the following set of equations:

$$\begin{bmatrix} 0.0001 & 1 \\ 10 & 10 \end{bmatrix} \begin{bmatrix} x_1 \\ x_2 \end{bmatrix} = \begin{bmatrix} 5 \\ 60 \end{bmatrix} \tag{A.14}$$

The solution of (A.14), correct to four decimal places, is $x_1 = 1.0001$, $x_2 = 4.9999$. In Gauss elimination procedure, the first equation of (A.14) is multiplied by 10^5 and subtracted from the second equation, which now becomes

$$-99990x_2 = -499940 \tag{A.15}$$

If the computer uses three-digit arithmetic, (A.15) would become

$$-1.00 \times 10^5 x_2 = -5.00 \times 10^5 \tag{A.16}$$

giving $x_2 = 5.00$. When substituted in the first equation of (A.14), this gives $x_1 = 0.0$, which is grossly incorrect.

Let us also obtain the solution of (A.14) after interchanging rows so that the magnitude of pivot becomes large. Equation (A.14) now becomes

$$\begin{bmatrix} 10 & 10 \\ 0.0001 & 1 \end{bmatrix} \begin{bmatrix} x_1 \\ x_2 \end{bmatrix} = \begin{bmatrix} 60 \\ 5 \end{bmatrix} \tag{A.17}$$

Applying Gauss elimination to (A.17) now gives $x_2 = 5.00$, $x_1 = 1.00$ for a computer with three-digit arithmetic. This is the correct solution for the three-digit arithmetic.

The above example illustrates that, if the absolute value of a pivot is too small, a severe numerical error may result. To avoid these errors, a good strategy is to choose the pivot at the kth stage to be the element with the largest absolute value in column k of all rows from k to N. This consumes some extra computer time but the improvement in the accuracy of the solution is significant. Further improvement in accuracy is possible by interchanging not only the rows but the columns as well. This method is called *complete pivoting* as compared to *partial pivoting* where only rows are interchanged. Additional computer time is required for complete pivoting since the unknowns should also be reordered when the columns are interchanged.

The processor time in Gauss elimination is proportional to N^3. This means that doubling the number of equations will increase the CPU time by up to eight times.

A.2.2 L-U Factorization [3, 4]

In this method, matrix A is decomposed into a lower triangular matrix L ($a_{ij} = 0$ for $i > j$) and an upper triangular matrix U ($a_{ij} = 0$ for $i < j$) such that

$$[A] = [L][U] \tag{A.18}$$

Usually L or U matrix has 1s on the diagonal. These are called unit lower triangular and unit upper triangular matrices, respectively. Equation (A.18) implies

$$
\begin{bmatrix}
a_{11} & a_{12} & \cdots & a_{1N} \\
a_{21} & a_{22} & \cdots & a_{2N} \\
\vdots & \vdots & \vdots & \vdots \\
a_{N1} & a_{N2} & \cdots & a_{NN}
\end{bmatrix}
=
\begin{bmatrix}
1 & 0 & \cdots & 0 \\
\ell_{21} & 1 & \cdots & 0 \\
\vdots & \vdots & \vdots & \vdots \\
\ell_{N1} & \ell_{N2} & \cdots & 1
\end{bmatrix}
\begin{bmatrix}
u_{11} & u_{12} & \cdots & u_{1N} \\
0 & u_{22} & \cdots & u_{2N} \\
\vdots & \vdots & \vdots & \vdots \\
0 & 0 & 0 & u_{NN}
\end{bmatrix}
\tag{A.19}
$$

for the unit lower triangular decomposition. Equation (A.1) can be written as, after decomposition of matrix A,

$$[L][U][x] = [b] \tag{A.20}$$

Solution to (A.20) is obtained in two steps. First, using forward elimination technique, one obtains the solution for $[y]$ satisfying

$$[L][y] = [b]$$

or

$$
\begin{bmatrix}
1 & 0 & \cdots & 0 \\
\ell_{21} & 1 & \cdots & 0 \\
\vdots & \vdots & \vdots & \vdots \\
\ell_{N1} & \ell_{N2} & \cdots & 1
\end{bmatrix}
\begin{bmatrix}
y_1 \\
y_2 \\
\vdots \\
y_N
\end{bmatrix}
=
\begin{bmatrix}
b_1 \\
b_2 \\
\vdots \\
b_N
\end{bmatrix}
\tag{A.21}
$$

The second step requires the solution for $[x]$ from

$$[U][x] = [y]$$

or

$$
\begin{bmatrix} u_{11} & u_{12} & \cdots & u_{1N} \\ 0 & u_{22} & \cdots & u_{2N} \\ \vdots & \vdots & \vdots & \vdots \\ 0 & 0 & 0 & u_{NN} \end{bmatrix} \begin{bmatrix} x_1 \\ x_2 \\ \vdots \\ x_N \end{bmatrix} = \begin{bmatrix} y_1 \\ y_2 \\ \vdots \\ y_N \end{bmatrix} \tag{A.22}
$$

employing the process of back-substitution. The elements of L are unity all along the main diagonal and therefore, for the purpose of storage, L and U can be overlapped and stored as $[T]$ such that

$$
[T] = [L] + [U] - [I] \tag{A.23}
$$

where I is the identity matrix.

The elements of $[L]$ and $[U]$ can be defined in terms of $[A]$ by comparing the two sides of (A.19) term by term [5–8],

$$
u_{1j} = a_{1j}, \qquad j = 1, 2, 3, \ldots, N \tag{A.24a}
$$

$$
\ell_{i1} = \frac{a_{i1}}{u_{11}}, \qquad i = 2, 3, \ldots, N \tag{A.24b}
$$

$$
\ell_{ij} = \frac{1}{u_{jj}} \left[a_{ij} - \sum_{k=1}^{j-1} \ell_{ik} u_{kj} \right], \qquad i \geq j \tag{A.24c}
$$

$$
u_{ij} = a_{ij} - \sum_{k=1}^{i-1} \ell_{ik} u_{kj}, \qquad i \leq j \tag{A.24d}
$$

Example A.3.   Let us consider the solution of the following matrix equation based on LU decomposition:

$$
\begin{bmatrix} 3 & 1 & 1 \\ 1 & 2 & 0 \\ 1 & 0 & 1 \end{bmatrix} \begin{bmatrix} x_1 \\ x_2 \\ x_3 \end{bmatrix} = \begin{bmatrix} 1 \\ 1 \\ 1 \end{bmatrix} \tag{A.25}
$$

Use of (A.24) gives the following matrices for L and U:

$$
[L] = \begin{bmatrix} 1 & 0 & 0 \\ 1/3 & 1 & 0 \\ 1/3 & -1/5 & 1 \end{bmatrix} \tag{A.26}
$$

$$
[U] = \begin{bmatrix} 3 & 1 & 1 \\ 0 & 5/3 & -1/3 \\ 0 & 0 & 3/5 \end{bmatrix} \tag{A.27}
$$

The solution of (A.21) gives

$$
\begin{bmatrix} y_1 \\ y_2 \\ y_3 \end{bmatrix} = \begin{bmatrix} 1 \\ 2/3 \\ 4/5 \end{bmatrix}
\tag{A.28}
$$

Finally, the solution of (A.22) yields

$$
\begin{bmatrix} x_1 \\ x_2 \\ x_3 \end{bmatrix} = \begin{bmatrix} -1/3 \\ 2/3 \\ 4/3 \end{bmatrix}
\tag{A.29}
$$

and is the correct solution of (A.25).

A.3   Sparse Matrix Techniques

A *sparse matrix* is one which has a large number of zero elements. Most matrices resulting from the finite difference or finite element formulations are sparse.

Since the matrices are highly sparse, most of the long operations to be performed for LU factorization will be of the type $0 \times a_{ij}$ and $0/a_{ij}$, which could be avoided. Thus, the computational efficiency of the analysis can be improved if we use techniques that omit operations of these types and perform operations on nonzero elements only. Further, considerable reduction in the storage requirement for the matrix can be achieved if only the nonzero elements are stored. Suitable data structures are designed so that the position of an element in the matrix and its value can be easily retrieved.

A.3.1   Reordering of Equations

After LU decomposition, the L-U factors given by $[T]$ of (A.23) are stored in the same memory location as the original matrix $[A]$. It is possible that a location which is zero in $[A]$ (and so not stored) may become nonzero on decomposition. In LU decomposition, an element a_{ij} is updated as

$$
a_{ij} \rightarrow a_{ij} - a_{ik} a_{kj}
\tag{A.30}
$$

If a_{ij} is zero and a_{ik} and a_{kj} are both nonzero, the element (i, j) which was zero in $[A]$ becomes nonzero in $[T]$. The new nonzero elements thus created are called *fill-ins*. Storage locations should be kept for the fill-ins also.

The number of fill-ins generated is dependent on the ordering of rows and/or columns. If the pivots to be chosen in each row are fixed, then the number of fill-ins generated depends upon the ordering of rows. It is desirable to minimize the fill-ins thus generated. To illustrate the dependence of the number of fill-ins generated on the ordering, we consider the solution of the following equations by LU factorization:

$$\begin{bmatrix} 2 & 4 & -2 & -6 \\ 3 & 9 & 0 & 0 \\ -2 & 0 & 8 & 0 \\ 2 & 0 & 0 & -12 \end{bmatrix} \begin{bmatrix} x_1 \\ x_2 \\ x_3 \\ x_4 \end{bmatrix} = \begin{bmatrix} 6 \\ -6 \\ 14 \\ 26 \end{bmatrix} \qquad (A.31)$$

To determine the number of fill-ins generated, we need not know the actual value of nonzeros, but only their positions. For instance, the matrix of (A.31) for which LU factors are needed is of the following type:

$$\begin{array}{c} \\ 1 \\ 2 \\ 3 \\ 4 \end{array} \begin{array}{cccc} 1 & 2 & 3 & 4 \\ \begin{bmatrix} X & X & X & X \\ X & X & 0 & 0 \\ X & 0 & X & 0 \\ X & 0 & 0 & X \end{bmatrix} \end{array} \qquad (A.32)$$

where X indicates a nonzero element. On LU factorization, the matrix gets completely filled and therefore six fill-ins are generated. The number of long operations needed for this decomposition is 20.

If the equations in (A.31) are ordered as

$$\begin{bmatrix} 9 & 0 & 0 & 3 \\ 0 & -12 & 0 & 2 \\ 0 & 0 & 8 & -2 \\ 4 & -6 & -2 & 2 \end{bmatrix} \begin{bmatrix} x_2 \\ x_4 \\ x_3 \\ x_1 \end{bmatrix} = \begin{bmatrix} 6 \\ 26 \\ 14 \\ 6 \end{bmatrix} \qquad (A.33)$$

with pivot elements being kept the same, the matrix for which LU factors are needed is of the type

$$\begin{array}{c} \\ 1 \\ 2 \\ 3 \\ 4 \end{array} \begin{array}{cccc} 1 & 2 & 3 & 4 \\ \begin{bmatrix} X & 0 & 0 & X \\ 0 & X & 0 & X \\ 0 & 0 & X & X \\ X & X & X & X \end{bmatrix} \end{array} \qquad (A.34)$$

where, as before, X indicates a nonzero element. On LU factorization, the matrix is again of the same type and no fill-ins are generated. The number of nonzero long operations required for factorization is now six. Thus, the rows and columns of $[A]$ should be reordered, not only to minimize the generation of fill-ins, but also to reduce the number of long operations required for LU decomposition. An algorithm for reordering is available [9]. An alternative to this approach is described next.

A.3.2 Preconditioned Conjugate Gradient Method

This method has been extensively used for sparse matrix equations in the last few decades. It combines the useful aspects of iterative methods and direct methods and is a modification of the well-known conjugate gradient method.

The conjugate gradient method is applicable to symmetric, positive definite matrices. The solution is guaranteed to converge in N steps for a matrix equation of order N, provided that round-off errors are not generated in the process. The scheme yields an improved estimate of the solution vector at each step, while permitting restarting at any point during the process. Because the structure of the original matrix is preserved throughout the computation such that the zeros in the matrix are not replaced by fill-ins (as would be the case with Gauss elimination), the conjugate gradient method results in memory savings.

Let us consider solving the matrix equation (A.1) of order N, where $[A]$ is symmetric, positive definite matrix.

The conjugate gradient method essentially consists of finding a set of N vectors w_i, $i = 1, 2, 3, \ldots N$ orthogonal *or conjugate* with respect to $[A]$, such that

$$[w_i]^T [A][w_j] = 0, \qquad \text{for } i \neq j \tag{A.35a}$$

and

$$[w_i]^T [A][w_j] \neq 0, \qquad \text{for } i = j \tag{A.35b}$$

These vectors are independent vectors and span the space in which any solution of the matrix equation must lie. Therefore, any possible solution to matrix equation may be represented as a linear combination of these vectors w_i, such that

$$[x] = \sum_i C_i[w_i] \tag{A.36}$$

where the vectors w_i may be called direction vectors or basis vectors. The details and the algorithm for the conjugate gradient method may be found in [5, 10].

The N-step convergence may be adequate for small values of N. However, for large systems of matrix equations, the number of steps may be too large. The convergence to the actual solution can be accelerated by multiplying the matrix $[A]$ by a preconditioning matrix that will modify the original matrix such that the new matrix will be closer to the identity matrix, or at least the eigenvalues of the new matrix are clustered together [10].

References

[1]   Stutzman, W. L., and G. A. Thiele, *Antenna Theory and Design*, New York: John Wiley, 1981.

[2]   Miller, E. K., and F. J. Deadrick, "Some Computational Aspects of Thin Wire Modeling," Chapter 4 in *Numerical and Asymptotic Techniques in Electromagnetics*, New York: Springer-Verlag, 1975.

[3]   Chua, L. O., and P. M. Lin, *Computer-Aided Analysis of Electronic Circuits: Algorithms and Computational Techniques*, Englewood Cliffs, NJ: Prentice-Hall, 1975, Chapter 16.

[4]   Fox, L., *An Introduction to Numerical Linear Algebra*, New York: Oxford University Press, 1965.

[5]   Chari, M. V. K., and S. J. Salon, *Numerical Methods in Electromagnetism*, San Diego, CA: Academic Press, 2000.

[6]   Al-Khafaji, A. W., and J. R. Tooley, *Numerical Methods in Engineering Practice*, New York: Rinehart and Winston, 1986.

[7]   James, M. L., et al., *Applied Numerical Methods for Digital Computation*, 3rd ed., New York: Harper & Row, 1985.

[8]   Ketter, R. L., and S. P. Prawel, *Modern Methods of Engineering Computation*, New York: McGraw-Hill, 1969.

[9]   Gupta, K. C., R. Garg, and R. Chadha, *Computer-Aided Design of Microwave Circuits*, Dedham, MA: Artech House, 1981.

[10] Sadiku, M. N. O., *Numerical Techniques in Electromagnetics*, 2nd ed., Boca Raton, FL: CRC Press, 2001.

Evaluation of Singular Integrals

We have come across singular integrals in conformal mapping method (Chapter 4), and in the determination of the diagonal matrix elements for MoM (Chapter 11). The square root singularity of mapping functions in Section 4.5 could be evaluated in terms of inverse elliptic function, which are tabulated and can also be determined numerically. However, when the metallization of planar transmission lines is nonplanar or of finite thickness, the integrals are of the form that inverse elliptic functions cannot be used. One of the most common approaches in this connection is the singularity subtraction method, described next.

Singularity Subtraction Method
This method can be used when the integral has a single pole on the path of integration. In this method, the singular part of the integrand is subtracted, thus making the integrand analytic everywhere on the path of integration. The singular part is then computed separately according to the residue calculus and added to the result.

Let

$$I = \int_0^{2c} \frac{f(z)}{z - c}\, dz \tag{B.1}$$

The singular part of the integrand in the above integral is $f(c)/(z - c)$, where $f(c)$ is the residue of the integrand. Therefore, one may write after subtracting the singularity

$$I = \int_0^{2c} \frac{f(z) - f(c)}{z - c}\, dz + \int_0^{2c} \frac{f(z)}{z - c}\, dz \tag{B.2}$$

It may be noted that the first integral is no longer singular, and the value of the second integral is $-j\pi f(c)$ (see Chapter 4). Therefore,

$$I = \int_0^{2c} \frac{f(z) - f(c)}{z - c}\, dz - j\pi f(c) \tag{B.3}$$

The integral in (B.3) will converge rapidly and can be evaluated by a numerical integration routine. However, at $z = c$ the integrand is determined by the Taylor series expansion of $f(z)$ about $z = c$ and is obtained as $f'(c)$. There is no other restriction on the function $f(z)$ except that it should be analytic.

The second integral in (B.2) is singular but simpler than the original integral; it is expected to be evaluated analytically. It is possible that the denominator cannot be expressed simply as $(z\text{-}c)$ (e.g., in the spectral domain analysis of microstrip line). In such cases the second integral cannot be determined analytically, and the numerical evaluation of the integral may impact on accuracy.

Example. Consider the following integral:

$$I = \int_a^b H_0^{(2)}(kx)\,dx \tag{B.4}$$

This integral occurs in the MoM formulation of TM wave scattering by a conducting cylinder [2, p. 39]. For the diagonal elements, the argument kx approaches zero and the Hankel function may be approximated as

$$H_0^{(2)}(t) = J_0(t) - jY_0(t) \simeq 1 - j\frac{2}{\pi}\ln\left(\frac{\gamma t}{2}\right) \tag{B.5}$$

At $t = 0$, the $\ln(.)$ term becomes singular. Use of the singularity subtraction method for I results in [2, p. 42]

$$I = \int_a^b J_0(kx)\,dx - j\int_a^b \left[Y_0(kx) - \frac{2}{\pi}\ln\left(\frac{\gamma kx}{2}\right)\right]dx - j\frac{2}{\pi}\int_a^b \ln\left(\frac{\gamma kx}{2}\right)dx \tag{B.6}$$

The first two integrals can be evaluated numerically. The third integral includes a singularity but can be integrated analytically as

$$\int \ln(t)\,dt = t(\ln t - 1) \tag{B.7}$$

Generalized Gaussian quadrature rules have been developed for integrals of the form [1]

$$I = \int_a^b w(x)f(x)\,dx \tag{B.8}$$

where $w(x)$ is singular. Quadrature rules for handling logarithmic singularity are described in [2, p. 528]. These rules are found to be more efficient than the singularity subtraction method.

The functions with end-point singularity can be integrated numerically using special integration schemes. One such help is available from Mathematica in the form of NIntegrate at [3]. An integral with singularity anywhere in the integration range may be split into integrals with end-point singularity as

$$\int_0^{2c} \frac{f(x)}{x-c}\,dx = \int_0^{c} \frac{f(x)}{x-c}\,dx + \int_c^{2c} \frac{f(x)}{x-c}\,dx \qquad (B.9)$$

References

[1]   Ma, J. H., V. Rokhlin, and S. Wandzura, "Generalized Gaussian Quadrature Rules for Systems of Arbitrary Functions," *SIAM J. Numr. Anal.*, Vol. 33, 1996, pp. 971–996.

[2]   Peterson, A. F., S. L. Ray, and R. Mittra, *Computational Methods for Electromagnetics*, Universities Press (India), 2001.

[3]   http://documents.wolfram.com/mathematica/functions/Nintegrate.

About the Author

Ramesh Garg is a professor in the Department of Electronics and Electrical Communication Engineering at the Indian Institute of Technology, Kharagpur (India). He received a Ph.D. from the Indian Institute of Technology, Kanpur (India) in 1975. He was a visiting assistant professor at the University of Houston, Houston Park, from 1985–1987, and a research associate at the University of Ottawa in 1994. Professor Garg was named a Fellow of the IEEE in 2002 and a Fellow of the IETE (India). He was the chairman of the IEEE Kharagpur Section in 1991. His research interests include printed lines, circuits, and antennas. He is a coauthor of three books and more than 100 publications in these areas.

Index

Recent Titles in the Artech House
Electromagnetic Analysis Series

Tapan K. Sarkar, Series Editor

*Understanding Electromagnetic Scattering Using the Moment Method:
A Practical Approach*, Randy Bancroft

Wavelet Applications in Engineering Electromagnetics, Tapan K. Sarkar,
Magdalena Salazar-Palma, and Michael C. Wicks

For further information on these and other Artech House titles, including previously considered out-of-print books now available through our In-Print-Forever® (IPF®) program, contact:

Artech House
685 Canton Street
Norwood, MA 02062
Phone: 781-769-9750
Fax: 781-769-6334
e-mail: artech@artechhouse.com

Artech House
46 Gillingham Street
London SW1V 1AH UK
Phone: +44 (0)20 7596-8750
Fax: +44 (0)20 7630 0166
e-mail: artech-uk@artechhouse.com

Find us on the World Wide Web at: www.artechhouse.com